Electric Powertrain

Electric Powertrain

Energy Systems, Power Electronics and Drives for Hybrid, Electric and Fuel Cell Vehicles

John G. Hayes
University College Cork, Ireland

G. Abas Goodarzi
US Hybrid, California, USA

This edition first published 2018

Registered Office(s)
John Wiley & Sons, Inc., 111 River Street, Hoboken, NJ 07030, USA
John Wiley & Sons Ltd, The Atrium, Southern Gate, Chichester, West Sussex, PO19 8SQ, UK

Editorial Office
The Atrium, Southern Gate, Chichester, West Sussex, PO19 8SQ, UK

For details of our global editorial offices, customer services, and more information about Wiley products visit us at www.wiley.com.

Wiley also publishes its books in a variety of electronic formats and by print-on-demand. Some content that appears in standard print versions of this book may not be available in other formats.

Library of Congress Cataloging-in-Publication Data

Names: Hayes, John G., 1964– author. | Goodarzi, G. Abas, author.
Title: Electric powertrain : energy systems, power electronics and drives for hybrid, electric and fuel cell vehicles / by John G. Hayes, G. Abas Goodarzi.
Description: Hoboken, NJ : John Wiley & Sons, 2018. | Includes bibliographical references and index. |
Identifiers: LCCN 2017029458 (print) | LCCN 2017043878 (ebook) | ISBN 9781119063667 (pdf) | ISBN 9781119063674 (epub) | ISBN 9781119063643 (cloth)
Subjects: LCSH: Electric vehicles–Power supply. | Hybrid electric vehicles–Power trains. | Power electronics.
Classification: LCC TL220 (ebook) | LCC TL220 .H39 2018 (print) | DDC 629.25/02–dc23
LC record available at https://lccn.loc.gov/2017029458

Cover Design: Wiley
Cover Images: (Bus) Image supplied by G. Abas Goodarzi; (Concept Car) © -M-I-S-H-A-/iStockphoto; (Mars Rover) © NASA

Set in 10/12pt Warnock by SPi Global, Pondicherry, India

10 9 8 7 6 5 4 3 2 1

To all who have contributed to the electrification of the automobile for a cleaner, more sustainable future.

Contents

Preface

"The scientific man does not aim at an immediate result. He does not expect that his advanced ideas will be readily taken up. His work is like that of the planter – for the future. His duty is to lay the foundation for those who are to come, and point the way." Nikola Tesla (1856–1943).

"An inventor is simply a fellow who doesn't take his education too seriously." Charles Kettering (1876–1958)

"A problem well stated is a problem half-solved." Charles Kettering.

This book describes a technological evolution that has major implications around the globe. The objective of this book is to provide the theory behind electric vehicles and insight on the factors motivating the global adoption of these technologies. The story told in the book is largely based on technologies originally developed in Detroit, California, and Japan. However, these technologies are spreading rapidly around the world, having been embraced by German, French, Chinese, Korean and other global manufacturers. While the car is changing, it is worth noting that the foundations of the modern car are anything but new; vehicular technology and electrical machines are products of the nineteenth century, while semiconductors, lithium-ion batteries, and PEM fuel cells are products of the twentieth century. These technologies are significantly impacting transportation in the early twenty-first century and becoming essential components of the modern vehicle.

I had the privilege of working on the General Motors' EV1 electric car program in Southern California for ten years. The EV1 was the first electric car developed for mass production in the modern era. I even met my wife, Mary, a mechanical engineer from Detroit, when we worked together on the EV1 – we were both working on the new wireless charging approach known as inductive coupling. I left the automotive world, returning to Ireland to teach, and yet my teaching and research still revolve around automotive topics. The closest connection to automotive history here on the south coast of Ireland is the ancestral home of the Ford family, from which William Ford fled to the United States during the Great Irish Famine in 1847. His son, Henry Ford, was a semi-literate Michigan farm boy, who grew up to revolutionize an industry and create what we now call mass-market consumer capitalism.

While it can be very useful for writing and teaching to be at a distance from the developing story, it is important not to be detached or isolated from such developments. My co-author, Abas Goodarzi, is a former colleague who is living and working to deliver the new technologies. Abas and I started work together at the General Motors' Hughes Aircraft subsidiary in Culver City, California, in October 1990. After directing the development of the EV1 electric powertrain, Abas pursued an electric vehicle start-up. After working in a few more start-ups, Abas founded US Hybrid, where he remains CEO. US Hybrid is a company specializing in delivering battery, hybrid, and fuel cell solutions for heavy-duty transportation. Between us, we have been part of engineering teams which have developed for mass production all of the technologies discussed in this book.

The modern automobile is a great topic for teaching because it is a consumer product to which all students, family, and friends can relate and discuss. Also, it features engineering marvels such as energy storage, combustion engines, electric drives, power electronics, and more. The structure of this book is set up to explain how these technologies interact in the vehicle as a whole and then becomes more technical as the book or a particular chapter unfolds.

The book features problems and assignments of varying technical difficulty for university students. The reader can attempt them based on his or her level.

The car and electrical technology have a history rich with the contributions of many prominent people. Hence, their quotations are often included at the start of a chapter. They generally tie in with the story or underlying philosophies ... and are often fun and thought-provoking.

Acknowledgments

First, we'd like to thank all our colleagues in industry, government, and academia who have provided us feedback, reviews, comments, suggestions, material, and criticism for the book: Mohamed Alamgir, Peter Bardos, Ted Bohn, Amy Bueno, Tim Burress, Kevin Cadogan, Yue Cao, Paul Carosa, Gilsu Choi, Amgad Elgowainy, Aoife Foley, James Francfort, Mark Gibbons, John Goodenough, Oliver Gross, John Hall, Chris Henze, Silva Hiti, Gerard Hurley for his multiple book launches, Joe Kimmel, John M. Miller for his widely-publicized book review, Tony O'Gorman, Ray Radys, John Renie, Wally Rippel, James Rohan, Brad Rutledge, Steve Schulz, Matthew Shirk, Saeed Siavoshani, Charlie Sullivan, Terry Ward, and George Woody. Thank you to the all the staff at Wiley, especially Michelle Dunckley, Adalfin Jayasingh, Aravind Kannankara and Athira Menon, with a special mention for Peter Mitchell, who answered the question "do you have any textbook which covers all of power electronics and machines and can help me teach an electric vehicle course?" with "No, would you write one?".

A word of thanks to all the supportive staff at University College Cork, especially Michael Egan, and to the former students who have provided help and educated us at times: David Cashman, Kevin Hartnett, Marcin Kacki, Brendan Lyons, Donal Murray, and Marek Rylko; and especially to Brendan Barry, Kevin Davis, Diarmaid Hogan and Robbie Ryan for final proofing and support. A special thank you to the undergraduate and postgraduate students who patiently worked through the various drafts of the book.

We are grateful to the companies and various US agencies for providing us material and would like to acknowledge their great work in the field: AC Propulsion, General Motors, International Council on Clean Transportation (ICCT), Maxon Motors, National Aeronautics and Space Administration (NASA), Jet Propulsion Laboratory (JPL), and the Department of Energy laboratories: Argonne National Laboratory, Oak Ridge National Laboratory, and Idaho National Laboratory.

Abas and I have been lucky to have been supervised in our postgraduate studies by some seminal authors who have led the way in technical education: Ned Mohan of the University of Minnesota, the late John M. D. Murphy of University College Cork, and the late Richard Hoft of the University of Missouri (Columbia).

We wish to acknowledge our former colleagues at Hughes Power Control Systems, and within the General Motors companies and beyond, for their contributions to the EV industry, especially the first commercial battery electric car, featuring the first automotive IGBT traction inverter and an inductive charging infrastructure.

Finally, we thank our extended families and friends for their love, support, and endless patience while we write books or start companies focused on electric vehicles. Mary and Aryan are understanding spouses-Mary is an experienced EV engineer and Aryan is the financial controller at US Hybrid. Thank you to Mary and the girls, Madi, Tasha, and Saoirse, and to Aryan and the boys, Milad and Navid.

Textbook Structure and Suggested Teaching Curriculum

This is primarily an engineering textbook covering the automotive powertrain, energy storage and energy conversion, power electronics, and electrical machines. A significant additional focus is placed on the engineering design, the energy for transportation, and the related environmental impacts. This textbook is an educational tool for practicing engineers and others, such as transportation policy planners and regulators. The modern automobile is used as the vehicle upon which to base the theory and applications, which makes the book a useful educational reference for our industry colleagues, from chemists to engineers. This material is also written to be of interest to the general reader, who may have little or no interest in the power electronics and machines. Introductory science, mathematics, and an inquiring mind suffice for some chapters. The general reader can read the introduction to each of the chapters and move to the next as soon as the material gets too advanced for him or her.

I teach the material across four years here at University College Cork. The material can be taught across various years as outlined in Table I.

The first third of the book (Chapters 1 to 6), plus parts of Chapters 14 and 16, can be taught to the general science or engineering student in the second or third year. It covers the introductory automotive material using basic concepts from mechanical, electrical, environmental, and electrochemical engineering. Chapter 14 on electrical charging and Chapter 16 on electromagnetism can also be used as a general introduction to electrical engineering.

The basics of electromagnetism, ferromagnetism and electromechanical energy conversion (Chapter 16) and dc machines (Chapter 7) are taught to second year (sophomore) engineering students who have completed introductory electrical circuits and physics.

The third year (junior) students typically have covered ac circuit analysis, and so we cover ac machines, such as the induction machine (Chapter 8) and the surface permanent-magnet machine (Chapter 9). As the students typically have studied control theory, we investigate the control of the speed and torque loops of the motor drive (Chapter 15). Power electronics, featuring non-isolated buck and boost converters (Chapter 11), is also introduced in the third year.

The final-year (senior) students then go on to cover the more advanced technologies of the interior-permanent-magnet machine (Chapter 10). Isolated power converters (Chapter 12), such as the full-bridge and resonant converters, inverters (Chapter 13), and power-factor-corrected battery chargers (Chapter 14), are covered

Table I Book content and related teaching.

Chapter	Topic		General	2nd	3rd	4th/PG
1	Vehicles and Energy Sources	Electromobility and the Environment	Y			
2		Vehicle Dynamics	Y			
3		Batteries	Y			
4		Fuel Cells	Y			
5		Conventional and Hybrid Powertrains	Y			
6	Electrical Machines	Introduction to Traction Machines	Y			
7		The Brushed DC Machine		Y		
8		Induction Machines			Y	
9		Surface-Permanent-Magnet AC Machines			Y	
10		Interior-Permanent-Magnet AC Machines				Y
11	Power Electronics	DC-DC Converters			Y	
12		Isolated DC-DC Converters				Y
13		Traction Drives and Three-Phase Inverters				Y
14		Battery Charging				Y
15		Control of the Electric Drive			Y	
16	Basics	Introduction to Electromagnetism, Ferromagnetism, and Electromechanical Energy Conversion		Y		

in the power electronics section. This material can also be covered at the introductory postgraduate level.

Various homework, simulation, and research exercises are presented throughout the textbook. The reader is encouraged to attempt these exercises as part of the learning experience.

About the Companion Web Site

Don't forget to visit the companion web site for this book:

www.wiley.com/go/hayes/electricpowertrain

There you will find valuable material designed to enhance your learning, including:

1) Solutions manual
2) References
3) Slides
4) Simulations

Scan this QR code to visit the companion web site:

Part 1

Vehicles and Energy Sources

1

Electromobility and the Environment

"My first customer was a lunatic. My second had a death wish." Karl Friedrich Benz (1844–1929) is generally credited with pioneering the modern vehicle.

"Practically no one had the remotest notion of the future of the internal-combustion engine, while we were just on the edge of the great electrical development. As with every comparatively new idea, electricity was expected to do much more than we even now have any indication that it can do. I did not see the use of experimenting with electricity for my purposes. A road car could not run on a trolley even if trolley wires had been less expensive; no storage battery was in sight of a weight that was practical ... That is not to say that I held or now hold electricity cheaply; we have not yet begun to use electricity. But it has its place, and the internal-combustion engine has its place. Neither can substitute for the other – which is exceedingly fortunate." Henry Ford in 1923, reflecting on 1899.

"Any customer can have a car painted any color that he wants so long as it is black." Henry Ford (1863–1947) was influenced by slaughterhouse practices when he developed his assembly line for the mass production of the automobile.

"The world hates change, yet it is the only thing that has brought progress." Charles Kettering (1876–1958) invented the electric starter and effectively killed the electric car of that era.

"The spread of civilization may be likened to a fire: first, a feeble spark, next a flickering flame, then a mighty blaze, ever increasing in speed and power." Nikola Tesla (1856–1943).

"Dum spiro, spero." (Latin for "As long as I breathe, I hope.") Marcus Cicero (106–43 BC). A noble aspiration from ancient times ... but what if we can't breathe the air?

Electric Powertrain: Energy Systems, Power Electronics and Drives for Hybrid, Electric and Fuel Cell Vehicles,
First Edition. John G. Hayes and G. Abas Goodarzi.

Companion website: www.wiley.com/go/hayes/electricpowertrain

"It was during that period that I made public my findings on the nature of the eye-irritating, plant-damaging smog. I attributed it to the petrochemical oxidation of organic materials originating with the petroleum industry and automobiles." Aries Jan Haagen-Smit (1900–1977), a pioneer of air-quality control, reflecting in 1970 on his pioneering work from 1952 to explain the Los Angeles smog.

"Tesla's mission is to accelerate the world's transition to sustainable energy." The 2016 mission statement of Tesla, Inc.

In this chapter, the reader is introduced to the factors motivating the development of the electric powertrain. The chapter begins with a brief history of the automobile from an electric vehicle perspective, the various energy sources, and the resulting emissions. Standardized vehicle drive cycles are discussed as drive cycles are used to provide a uniform testing approach to measure the emissions and the fuel economy of a vehicle, both of which are related to the efficiency of the energy conversion from the stored energy to kinetic energy. Government regulations and the marketplace have resulted in strong global trends to reduce these potentially harmful emissions and to increase the fuel economy. These factors of reduced emissions and improved efficiency combine with a greater consumer market appreciation for green technology to motivate the development of the electric powertrain. The competing automotive powertrains are briefly reviewed and discussed in terms of efficiency. The chapter concludes with a brief look at heavy-duty commercial vehicles and other modes of transport.

1.1 A Brief History of the Electric Powertrain

There are three evolutionary eras of electric cars, and we shall now discuss the bigger historical picture.

1.1.1 Part I – The Birth of the Electric Car

The first self-propelled vehicles were powered by steam. Steam vehicles were fueled by coal and wood and took a relatively long time to generate the steam to power the pistons by heating the furnace of an external combustion engine. The modern vehicle, first developed by Karl Benz in the 1880s, is based on the internal-combustion (IC) engine. The early vehicles were unreliable, noisy, polluting, and difficult to start. Meanwhile, modern electrical technologies were being invented as Nikola Tesla, partnering with George Westinghouse, and Thomas Edison battled to invent and establish supremacy for their respective alternating-current (ac) and direct-current (dc) power systems. Battery electric vehicles (BEVs), energized by lead-acid batteries and using a dc power system, competed with IC engine vehicles in the 1890s. Electric vehicles (EVs) did not have the starting problems of the IC engine and had no tailpipe emissions. The low range of the BEVs was not necessarily a problem at the time as the road system was not developed, and so comfortable roads were not available for long driving. In 1900, the sales of gasoline vehicles and EVs in the United States were comparable in quantity, but EV sales were to collapse over the next decade [1–4]. Interestingly, EV sales were poor in the Europe of this period as the French and German auto manufacturers, such as Renault, Peugeot, Daimler, and Benz, were leading the world in the development of the IC engine.

The dominance of the IC engine was to be established with two major developments. First, Henry Ford mass-produced the Model T and drove down the sales price of the gasoline vehicle to significantly below that of both his competitors and of the EVs [5]. However, the gasoline vehicle still needed a manual crank in order to start the engine.

The second major development was the elimination of the manual crank by Charles Kettering's invention of the electric ignition and start. These electric technologies were introduced by Cadillac in 1912 and, ironically, effectively consigned the BEV to history. As the electrically started gasoline cars proliferated, so did road systems. The mobility delivered by the car fostered the development of modern society as it stimulated individualized transportation and suburbanization. California became the poster child for these trends, which have spread globally. Given their low range and high costs, BEVs could no longer compete and the market died, expect for niche applications such as delivery trucks.

1.1.2 Part II – The Resurgent Electric Powertrain

The diesel engine was introduced for vehicles in 1922, 32 years after it was invented by Rudolf Diesel in 1890 as a more efficient compression-ignition (CI) IC engine compared to the spark-ignition (SI) IC engine fueled by gasoline. The first commercial diesel engines were actually developed by a spin-off company of the US brewer Anheuser Busch. The high-torque-at-low-speed characteristic has made the diesel engine the engine of choice for medium and heavy-duty vehicles worldwide. In recent times, the diesel engine became a choice for light vehicles, especially in Europe, due to its reduced carbon emissions compared to gasoline.

Of course, burning fossil fuels in the engine does not come without an environmental cost. A Dutch scientist, Aries Jan Haagen-Smit, had moved to California and was perplexed by the pollution and smog in rapidly urbanizing Southern California. *Smog* is a portmanteau word combining *smoke* and *fog* to describe the hazy air pollution common in urban areas. London-type smog is a term commonly used to describe the smog due to coal, while Los Angeles–type smog is used to describe the smog due to vehicle emissions. Haagen-Smit demonstrated that California smog is the product of a photochemical reaction between IC engine emissions and sunlight to create ozone [6,7]. He is now known as the father of air pollution control and mitigation. The geography of Southern California features valleys, which tend to trap the pollutants for much of the year until the winds from the desert blow through the valleys in the fall. Similar geographic issues worsen the smog situations in other cities, such as Beijing – where the Gobi winds bring dust from the desert to combine with the city's smog.

In the late 1980s, General Motors (GM) decided to develop an all-electric car. The motivations were many. For example, urban pollution in American cities, especially Los Angeles, was severe. An additional significant motivating factor was the success of the solar-powered Sunraycer electric car in the Solar Challenge, a 3000 km race across Australia in 1987. The Sunraycer was engineered by AeroVironment, General Motors, and Hughes Aircraft, who pushed the boundaries to develop the lightweight, low-drag, solar-powered electric car.

The initial GM prototype BEV, known as the Impact, was developed in Southern California, and GM committed to mass-producing the car. The production vehicle, which was to become known as the GM EV1, was developed and produced at GM facilities in Michigan and Southern California, and made its debut in 1996. The vehicle was revolutionary as it featured many of the technologies which we regard as commonplace today. The improved traction motor was a high-power ac induction motor based on the inventions of Nikola Tesla. The car body was built of aluminum in order to reduce vehicle weight. The vehicle aerodynamics were lower than any production vehicle of the day. The vehicle featured advanced silicon technology to control all the electronics in the vehicle and the new IGBT silicon switch to ensure efficient and fast control of the motor. This vehicle introduced electric steering, braking, and cabin heating and cooling. The EV1 featured extensive diagnostics, a feature that is now commonly employed in most vehicles to improve fuel economy and handling. Heavy-duty vehicle prototypes for transit and school buses were also electrified and deployed in public by GM at this time. It is worth noting that electric powertrains have been commonly deployed by the railroad industry for many decades due to the inherent advantages of fuel economy and performance.

However, the GM EV1 went to market powered by lead-acid batteries, a technology which had limited progress over the previous century. The second-generation GM EV1 featured a nickel-metal hydride (NiMH) battery which almost doubled the range of the first-generation vehicle. However, a number of realities were to doom this particular effort – the inadequacy of battery technology, a collapse in the price of gasoline, a lack of consumer demand for energy-efficient and green technologies, a lack of government support, and the advent of the hybrid electric car.

In the early 1990s, Toyota Motor Company was looking ahead at the challenges of the new century and also concluded that transportation had to become more efficient, more electric, and less polluting. Toyota first marketed the Toyota Prius in 1997 in Japan. The vehicle featured an extremely efficient gasoline engine based on the Atkinson cycle. While the efficient Atkinson-cycle engine is not suitable as the engine in a conventional gasoline vehicle, Toyota overcame the engine limitations by hybridizing the powertrain to create a highly fuel-efficient vehicle. The Toyota Prius featured a NiMH battery pack, the efficient IGBT silicon switch, and an ac motor featuring permanent magnets. The interior permanent-magnet (IPM) motor features advanced rare-earth permanent magnets to make the machine more efficient and power dense than its cousin, the induction motor. The Toyota Prius also eliminated the conventional transmission and developed the continuously variable transmission, commonly termed CVT. Toyota realized early on that the hybrid technology enabled a decoupling of the traffic condition from the engine use, such that the overall fuel economy could be maximized. The vehicle introduced technologies such as electric stop-start and idle control, which have become common.

The gasoline-sipping Toyota Prius and its siblings went on to become the mass-market leaders for energy-efficient and green technologies, and effectively ended the brief BEV flurry of the 1990s, while opening the world's eyes to the value of the electric powertrain.

1.1.3 Part III – Success at Last for the Electric Powertrain

The basic limitation for electric cars in the industrial age has been the battery. Significant battery development efforts in the 1970s focused on the lithium battery. John Goodenough was credited with developing the first workable lithium-ion (Li-ion) cell in 1979.

This technology was to be commercialized by Sony Corporation in 1991, and Li-ion technology went on to become the battery of choice for mobile phones and laptop computers due to its high voltage and high energy density. The energy density of the Li-ion cell could be three to five times higher than that for a lead-acid cell, albeit at a higher price.

Pioneers of the GM Impact, Alan Cocconi and Wally Rippel, formed a company known as AC Propulsion and continued to work in the EV field with a focus on drive systems and chargers. AC Propulsion developed a prototype BEV, known as the tzero, shown in Figure 1.1, in the early 2000s. The unique attribute of these prototype vehicles was that they featured a very high number of computer laptop Li-ion cells for the main storage battery. This development produced a very workable EV range whilst demonstrating both high efficiency and high performance.

These prototypes were to be test-driven by Silicon Valley entrepreneurs, who urged commercialization of the technology. Tesla Motors was founded in Silicon Valley, and the first vehicle from Tesla was the Tesla Roadster in 2007. The Tesla Roadster was the first mass-market EV featuring Li-ion cells. It had a very high number of cells, 6,831 Panasonic cells in total.

Tesla built on the Roadster success with the subsequent introduction of the Tesla Model S luxury sedan in 2012, the Tesla Model X in 2015, and the Model 3 in 2017. The company, led by CEO Elon Musk, has very successfully competed in the automotive marketplace, and has attracted buyers globally to EVs. Long-range batteries, high performance, autonomous driving, a more digital driver interface, direct sales, and photovoltaic solar power have all played a part in the Tesla vision for the vehicle [8].

The Nissan Leaf was introduced in 2011 and Nissan became the largest volume seller of EVs. The Nissan Leaf had a much lower price than the Tesla, making the car financially attractive for the mass-market consumer, albeit with a much lower range.

The Tesla and Nissan vehicles were launched in a different era in California from that of the GM EV1. Critical market support was now available from the government. Using a system of credits, the EV manufacturers would effectively receive financial transfers from

Figure 1.1 AC Propulsion tzero. (Courtesy of AC Propulsion.)

the other automotive manufacturers in order to subsidize the business model while the market developed. As the price of batteries continued to drop, more battery EVs came onto the market. In 2016, GM introduced the midrange Chevy Bolt, shown in Figures 1.2 and 1.3, while Tesla introduced the Model 3 in 2017, both vehicles with approximately 200 miles of range.

Hybrid electric vehicles (HEVs) continue to dominate the EV market with multiple products from Toyota, Ford, Honda, Hyundai, BMW, Volkswagen, and others. A number of manufacturers are following General Motors' Chevy Volt, a variant on a plug-in hybrid electric vehicle (PHEV). The PHEV has a large enough Li-ion battery to satisfy most drivers' daily commutes, but it can run an on-board gasoline engine efficiently to recharge the battery and provide propulsion power efficiently when needed.

Figure 1.2 Chevy Bolt. (Courtesy of General Motors.)

Figure 1.3 Battery and propulsion system of a stripped-down Chevy Bolt. (Courtesy of General Motors.)

Fuel cell vehicles with electric powertrains have also been introduced. Fuel cells have been around for a long time. For example, fuel cells have been used on spacecraft for decades. The automotive fuel cell converts stored hydrogen and oxygen from the air into electricity. The hydrogen must be highly compressed to obtain adequate storage on the vehicle. Advances in technology have reduced the size and cost of the fuel cell and the required balance of plant (a term used to describe the additional equipment required to generate the power). The attraction of the fuel cell electric vehicle (FCEV) is that it combines the electric powertrain with energy-dense hydrogen and only emits water at the point of use. Thus, like the battery EV, the FCEV is zero emissions at the point of use. The fuel cell is an attractive option for vehicles as it can increase energy storage compared to the battery. Key challenges for the technology are the generation of hydrogen using low-carbon methods and the development of a distribution and refueling system. Hyundai introduced the Tucson FCEV for lease in California in 2014. Toyota brought the Toyota Mirai FCEV to the market in 2015. Honda introduced the Clarity for limited lease in 2017 (following an earlier version released in 2008).

A major scandal erupted in the car industry in 2015 when it was established that Volkswagen had in effect been cheating in the Environmental Protection Agency (EPA) emissions testing [9]. During an emissions test, the Volkswagen car software would detect that the car was being tested and cause the vehicle to reduce various emissions in order to meet the EPA limits. Once the vehicle software decided that the vehicle was no longer being tested, the emission levels would increase.

Thus, one of the best diesel engine manufacturers in the world had struggled to meet the emission standards, and had done so by manipulating and circumventing the engine controller so as to meet the standards at the specified points, while exceeding the standards during ordinary driving. This case resulted in a multibillion dollar settlement with the US federal government and the state of California to fund the commercialization of zero-emission vehicles, with significant investment in batteries and fuel cells.

In conclusion, the electric car has been well and truly revived in the twenty-first century. There are many variations available ranging from battery electric to hybrid electric to fuel cell electric – hence the content of this book.

Finally, it must be noted that the world changed in the decades following the introduction of the GM EV1 in 1996. Significant consumer interest and a resulting market developed for green technologies. Local, state, and federal governments, the public, and industry awoke to the need to foster, and in some cases to subsidize, greener technologies. The motivations were many: minimize pollution to have tolerable air quality; reduce global warming to minimize climate change; develop local or greener energy sources to reduce energy dependence on more volatile parts of the world; and develop related businesses and industries.

History has burdened many of the oil-producing countries with despotic regimes, wars, volatility, and instability. Many of the energy diversification and efficiency initiatives have been launched as a result of national security considerations due to supply-chain volatility caused by war: the First Gulf War in 1991, the Second Gulf War in 2003, and so on. Together with the changing perceptions on pollution and climate change, these are all significant motivating factors to improve energy efficiency.

Wind and photovoltaic energy sources are commonly used in many countries around the globe. Efficiencies have improved due to regulations and consumer expectations.

Compressed natural gas has become the fossil fuel of choice for electricity generation. The fracking of shale gas and oil has caused an energy revolution in the United States, and a shift away from "dirtier" fuels, such as coal. China has heavily industrialized using coal and now has very severe pollution problems. Nuclear power is once more viewed as problematic after a tsunami hit the Fukushima power plant in 2011 – with Germany deciding to abandon nuclear power as a result. There is increased usage of biofuels in transportation – based on sugarcane in Brazil and corn in the United States. Diesel fuel is viewed as problematic due to the significant NO_x emissions and the high cost per vehicle of treating the NO_x and particulate matter emissions.

Of course, there are many contradictions as countries adopt energy policies. Nuclear power can be perceived negatively due to associated risk factors but does result in low carbon emissions. The use of land for biofuels impacts food output and can raise food prices. Renewable sources can be intermittent and require problematic energy storage. There are many difficult choices, and we are limited by the laws of physics. The **first law of thermodynamics** is the law of conservation of energy. This law states that *energy cannot be created or destroyed, but only changed from one form into another or transferred from one object to another.*

Thus, much has changed since 1996 and will continue to change, as countries around the world adjust their energy sources and usage based on economic, environmental, and security factors ... and the associated politics.

We next investigate some of the characteristics of these energy fuels.

1.2 Energy Sources for Propulsion and Emissions

In this section, we briefly consider the energy sources mentioned above. Characteristics of various fuels are shown in Table 1.1. The first four fuels are all fossil fuels. Gasoline is the most common ground-transportation fuel, followed by diesel. Some related characteristics of gasoline, diesel, and compressed natural gas (CNG) are shown in Table 1.1. The main reference for the specific energy and density is the *Bosch Automotive Handbook* [10]. Coal is also included for reference and has many varieties, of which anthracite is one. There can be minor variations in the actual energy content of a fuel as the fuel is often a blend of slightly different varieties of fuel – and likely also including some biofuel, such as ethanol, in the mix. A formula is presented for a representative compound in the mixture.

Gasoline and diesel have similar energy content per unit weight. Since diesel is a denser fluid than gasoline, it has a higher energy content by volume compared to gasoline. This higher energy content per unit volume, for example, per gallon, accounts for a significant portion of the fuel economy advantage that diesel has over gasoline. The combustion process is additionally more efficient for diesel than for gasoline.

In general, CNG has higher energy densities by mass compared to the liquid fuels, but has significantly lower energy density when measured by volume.

Hydrogen is the fuel of choice for fuel cell vehicles, and its characteristics are included here for comparison. Hydrogen has to be highly compressed. The 2016 Toyota Mirai uses 700 bar or 70.7 MPa pressure for the 5 kg of hydrogen on board. Significant additional vehicle volume is required for the storage components and required ancillaries.

Table 1.1 Energy and carbon content of various fuels.

Fuel	Representative formula	Specific energy (kWh/kg)	Specific energy (kJ/g)	Density (kg/L)	Energy density (kWh/L)	CO_2 emissions ($kgCO_2$/kg fuel)	CO_2 emissions (gCO_2/kWh)
Gasoline	C_8H_{18} (iso-octane)	11.1–11.6	40.1–41.9 [10]	0.72–0.775 [10]	8.0–9.0	3.09	266
Diesel	$C_{12}H_{23}$	11.9–12.0	42.9–43.1 [10]	0.82–0.845 [10]	9.8–10.1	3.16	268
Gas	$C H_4$ (methane)	13.9	50 [10]	0.2	2.8	2.75	198
	Natural (mostly CH_4)	11.2–13.0	40.2–46.7 [10]				
Coal	$C_{240}H_{90}O_4NS$ (anthracite)	8	28.8	0.85	6.8	2.8	350
Hydrogen	H_2	33.3	120 [10]	0.42 (at 700 bar)	14	0	0
Li-ion		0.15	0.54	2.5	0.375	0	0

Some approximate numbers are included for the Li-ion battery pack. The specific energy and the energy density of the Li-ion battery pack are approximately 76 times and 23 times lower than gasoline, respectively. Thus, battery packs have to be relatively large and heavy in order to store sufficient energy for EVs to compete with IC engine vehicles.

Note that the carbon content is calculated based on the formulae presented in column 2 of the table and by using the analysis presented in the next section. The energy content is based on approximate figures for the lower heating value (LHV) which is commonly referenced for the energy content of a fuel.

1.2.1 Carbon Emissions from Fuels

The IC engine works on the principle that the fuel injected into the cylinder can be combined with air and ignited by a spark or pressure. The resulting expansion of the gases due to the heat of the combustion within the cylinder results in a movement of the pistons which is converted into a motive force for the vehicle. The spontaneous thermochemical reaction within the IC engine cylinder has the following formula:

$$C_xH_y + \left(x + \frac{y}{4}\right)O_2 \rightarrow xCO_2 + \frac{y}{2}H_2O \tag{1.1}$$

where the inputs to the reaction are C_xH_y, the generic chemical formula for the fossil fuel, as presented in Table 1.1, and O_2, the oxygen in the injected air [11]. The outputs from the reaction are heat, carbon dioxide (CO_2), and water (H_2O). The indices x and y are governed by the chemical composition of the fuel. Of course, it is air rather than pure oxygen which is input to the engine. As discussed in the next section, the 78% nitrogen content of the air can be a major culprit for emissions.

The engine can operate with various fuel-to-air ratios, and the particular mix can affect the emissions and fuel economy. A low fuel-to-air ratio is termed *lean*, and a high fuel-to-air is termed *rich*. For example, a rich mix can have less molecular oxygen in the combustion reaction, resulting in increased output of carbon monoxide and soot.

The chemical formula for the combustion of iso-octane, part of the gasoline mixture, is

$$C_8H_{18} + \left(8 + \frac{18}{4}\right)O_2 \rightarrow 8CO_2 + 9H_2O \tag{1.2}$$

and for diesel combustion is

$$C_{12}H_{23} + \left(12 + \frac{23}{4}\right)O_2 + \rightarrow 12CO_2 + \frac{23}{2}H_2O \tag{1.3}$$

We can relate the carbon dioxide output from the combustion directly to the carbon content of the fuel and run some simple calculations to determine emissions.

1.2.1.1 Example: Carbon Dioxide Emissions from the Combustion of Gasoline

We can see that the x index of C_8H_{18} results in eight CO_2 molecules for every C_8H_{18} molecule. We next note that the molecular weights of the elements are different.

The atomic mass unit of carbon, hydrogen, and oxygen are 12, 1, and 16, respectively. Thus, the atomic weight of C_8H_{18} is 114 atomic mass units (amu), calculated as follows:

$$\begin{array}{ll} 8\,\text{C atoms} = 8 \times 12\,\text{amu} = & 96\,\text{amu} \\ \underline{18\,\text{H atoms} = 18 \times 1\,\text{amu} =} & \underline{18\,\text{amu}} \\ C_8H_{18} & = 114\,\text{amu} \end{array} \tag{1.4}$$

The atomic weight of the resulting CO_2 is

$$\begin{array}{ll} 8\,\text{C atoms} = 8 \times 12\,\text{amu} = & 96\,\text{amu} \\ \underline{16\,\text{O atoms} = 16 \times 16\,\text{amu} =} & \underline{256\,\text{amu}} \\ 8\,CO_2 & = 352\,\text{amu} \end{array} \tag{1.5}$$

The relative difference of the molecular weights means that for every 1 kg of C_8H_{18} fuel consumed in a thermochemical reaction, $\frac{352}{114} \times 1\,\text{kg} = 3.09\,\text{kg}$ of carbon dioxide is produced. In other units, for every liter of gasoline consumed, about 2.39 kg of CO_2 is emitted. For every US gallon of gasoline consumed, about 9 kg (20 lb) of CO_2 are emitted.

1.2.2 Greenhouse Gases and Pollutants

There are a number of additional emissions from the combustion process – some causing ground-level pollution and others contributing to the greenhouse effect. **After-treatment** is a generic term used to describe the processing of the IC engine emissions on the vehicle, for example, by using a catalytic converter or particulate filter, in order to meet the vehicle emission requirements.

Particulate matter (PM) is a complex mix of extremely small particles that are a product of the combustion cycle. The particles are too small to be filtered by the human throat and nose and can adversely affect the heart, lungs, and brain. They are also regarded as carcinogenic (cancer causing) in humans. A diesel engine can emit significantly more PM than the gasoline engine. The PM emissions can be mitigated by the after-treatment, but at a significant financial cost. These particles are extremely small. In general, particles less than 10 μm in diameter (PM_{10}) are dangerous to inhale. $PM_{2.5}$ particles, which are less than 2.5 μm in diameter, can result from the combustion process and are a significant component of air pollution and a major contributor to cancers.

Carbon monoxide (CO) is a colorless odorless gas that is a product of the combustion cycle. The gas can cause poisoning and even death in humans. Diesel engines produce lower levels of CO than spark-ignition gasoline engines.

A **greenhouse gas (GHG)** is any gas in the earth's atmosphere which increases the trapping of infrared radiation, contributing to a greenhouse effect.

Carbon dioxide (CO_2) is a greenhouse gas as it adds to the concentration of naturally occurring CO_2 in the atmosphere and contributes to the greenhouse effect. It is estimated that approximately 37 billion metric tons of CO_2 are released into the atmosphere every year due to the burning of fossil fuels by human activities [12].

Nitrous oxide (N_2O) and **methane (CH_4)** are additional products of the combustion process which also contribute to the greenhouse effect. Methane, as a GHG, is often a product of the fossil fuel industry but can also result from livestock flatulence and other natural sources.

Nitrogen oxide (NO), **nitrogen dioxide** (NO_2), and **volatile organic compounds** (**VOCs**) are emissions from the combustion process which result in ground-level ozone and other pollutants. They are discussed in the next section.

Total hydrocarbons (THCs) are hydrocarbon-based emissions which contain unburnt hydrocarbons and VOCs. VOCs include alcohols, ketones, aldehydes, and more. THCs also contribute to greenhouse gases.

1.2.2.1 The Impact of NO_x

The air is composed of approximately 78% nitrogen, almost 21% oxygen, about 0.9% argon, about 0.04% CO_2, and minor quantities of other noble gases and water molecules. Although the nitrogen atoms are more tightly bonded together than the oxygen atoms, the elements can react under heat within the cylinders of the IC engine to make nitrogen oxide and nitrogen dioxide:

$$N_2 + O_2 + \text{heat} \rightarrow NO, NO_2 \tag{1.6}$$

Compounds NO and NO_2 are commonly described as NO_x.

The nitrogen dioxide can then react in the sunlight to create atomic oxygen O and nitrogen oxide NO:

$$NO_2 + \text{sunlight} \rightarrow NO + O \tag{1.7}$$

The atomic oxygen O can then react with molecular oxygen O_2 in the air to form **ozone** O_3:

$$O_2 + O \rightarrow O_3 \tag{1.8}$$

Hydrocarbons are also produced by the combustion process. Included within the overall hydrocarbons are VOCs, which are essential for the buildup of ozone and smog at ground level. If VOCs were not present, then the ozone would react with the NO to create oxygen and remove the buildup of ozone. However, the VOC reacts with the hydroxide in the air and with the NO of Equation (1.7) to form NO_2.

The net effect of the overall reaction sequence involving the VOCs is that the ozone continues to build up in the atmosphere. The ground-level ozone is inhaled by humans and other animals. The ozone reacts with the lining of the lung to cause respiratory illnesses such as asthma and lung inflammation. Note that atmospheric ozone is necessary in the upper atmosphere, known as the troposphere, in order to filter the sun's harmful ultraviolet rays.

Another reaction of VOCs and NO_x results in peroxyacyl nitrates (also known as PANs), which can irritate the respiratory system and the eyes. PANs can damage vegetation and are a factor in skin cancer.

Diesel combustion engines are the major source of NO_x in urban environments. Cities such as London have experienced severe pollution in recent times due to the proliferation of diesel engines for light and heavy-duty vehicles. It is projected that 23,500 people die each year in the United Kingdom due to the effects of NO_x [13]. Diesel has been seen by various governments as an important solution to carbon emissions, but the associated NO_x emissions have resulted in a significant degradation of the local air quality.

Many diesel vehicles use **urea** for emissions control. Urea $CO(NH_2)_2$ is a natural component of animal urine, and the compound is particularly useful in reducing NO_x. Industrially produced urea is stored in a tank on the vehicle and is injected into the exhaust system, where it is converted to ammonia. The ammonia reacts in the catalytic converter to reduce the NO_x. The overall reaction is as follows:

$$2CO(NH_2)_2 + 4NO + O_2 \rightarrow 4N_2 + 4H_2O + 2CO_2 \tag{1.9}$$

Undersized on-board urea tanks contributed to the 2015 Volkswagen diesel emissions scandal [9].

A **catalytic converter** is an emissions-control device installed on a vehicle to reduce the emissions of CO, THCs, and NO_x. Lead, also a pollutant, was eliminated from gasoline in order to facilitate the use of catalytic converters because lead coats the surface and disables the converter.

The catalytic converter has two main functions. First, it converts carbon monoxide and unburnt hydrocarbons to carbon dioxide and water:

$$CO,\ C_xH_y \rightarrow CO_2,\ H_2O \tag{1.10}$$

Second, it converts the nitrogen oxides to nitrogen and oxygen:

$$NO,\ NO_2 \rightarrow N_2 + O_2 \tag{1.11}$$

Automotive emissions can impact on public health in many ways, and it is an area of active research (see the Further Reading section of this chapter). Hence, there is a serious movement toward the adoption of electric, hybrid, and fuel cell vehicles as the solution set to mitigate local pollution and carbon emissions.

1.3 The Advent of Regulations

The necessity for reduced vehicle emissions and associated emissions regulation by the government is evidenced here by California, which is only one part of a global problem. A brief outline of related events is presented in Table 1.2. This table is a modified version of the one presented on the web site of the Air Resources Board (ARB) of the California Environmental Protection Agency [14].

As the timeline in the table illustrates, expanding populations in developed countries embracing use of the automobile, and other technological advances, can generate significant pollution and greenhouse gases. The 280 billion miles driven in California in 2000 would have generated about 0.5 kg of CO_2 per mile or 140 billion kg of CO_2. However, this amounts to only a small fraction of the close to 37 billion metric tons of annual CO_2 emissions emitted globally by 2015 [12]. While smog was the original motivating factor, the public interest in controlling CO_2 emissions is significant as the emissions are directly related to fuel economy.

In 1975 the United States Congress introduced the Corporate Average Fuel Economy (CAFE) standards. These standards are resulting in a significant improvement in average fuel economy for cars and light trucks, with the projected fuel economy in 2025 being 54.5 mpg. The standard fuel economy was 18 mpg in 1978.

Table 1.2 Timeline of vehicle-related developments in California.

1940	California's population reached **7 million people** with **2.8 million vehicles** and **24 billion miles** driven.
1943	First recognized episodes of smog occur in Los Angeles in the summer of 1943. Visibility is only three blocks, and people suffer from smarting eyes, respiratory discomfort, nausea, and vomiting. The phenomenon is termed a "gas attack" and blamed on a nearby butadiene plant. The situation does not improve when the plant is shut down.
1952	Over 4,000 deaths attributed to "Killer Fog" in London, England. This was London-type smog.
	Jan Aries Haagen-Smit discovered the nature and causes of photochemical smog. He determined that nitrogen oxides and hydrocarbons in the presence of ultraviolet radiation from the sun form smog (a key component of which is ozone). This was Los Angeles–type smog.
1960	California's population reached **16 million people** with **8 million vehicles** and **71 billion miles** driven.
1966	Auto tailpipe emission standards for HC and CO were adopted by the California Motor Vehicle Pollution Control Board, the first of their kind in the United States.
1968	Jan Aries Haagen-Smit was appointed chairman of the recently formed Air Resources Board by Governor Ronald Reagan. (He was later fired by Reagan in 1973 over policy disagreements.)
	Federal Air Quality Act of 1967 was enacted. It allowed the State of California a waiver to set and enforce its own emissions standards for new vehicles.
1970	US EPA was created to protect all aspects of the environment.
1975	The first two-way catalytic converters came into use as part of the ARB's Motor Vehicle Emission Control Program. CAFE standards introduced by US Congress.
1980	California's population reached **24 million people** with **17 million vehicles** and **155 billion miles driven**.
1996	Big seven automakers commit to manufacture and sell zero-emission vehicles. Debut of GM EV1.
1997	Toyota Prius debuts in Japan.
2000	California's population grows to **34 million** with **23.4 million vehicles** and **280 billion miles** driven.
	A long-term children's health study funded by the ARB revealed that exposure to high air pollution levels can slow down the lung function growth rate of children by up to 10%.
2007	Debut of Tesla Roadster.
2008	ARB adopts two critical regulations aimed at cleaning up harmful emissions from the estimated one million heavy-duty diesel trucks. One requires the installation of diesel exhaust filters or engine replacement, and the other requires installation of fuel-efficient tires and aerodynamic devices.
2010	California adopts the Renewable Energy Standard. One-third of the electricity sold in the state in 2020 to come from clean, green sources of energy.
2011	ARB determines that 9,000 people die annually due to the amount of fine particle pollution in California's air. Debut of Nissan Leaf.
2014	Hyundai fuel cell vehicle debuts in California.
2015	Volkswagen diesel emissions scandal becomes a big news story.

1.3.1 Regulatory Considerations and Emissions Trends

As discussed earlier, vehicles can significantly impact the global and local environments due to the emissions of greenhouse gases and associated urban smog and pollution. In addition, the use of gasoline and diesel as fuels raises considerations related to the security of the supply and the economics of such a critical commodity. Thus, there has been greater regulation of vehicle emissions and fuel economy in recent decades. Global historical and projected trends in CO_2 emissions and fuel consumption are shown in Figure 1.4, as compiled by the International Council on Clean Transportation (ICCT). The horizontal axis represents time, whereas the primary vertical axis represents the carbon emissions in gCO_2 per km normalized to the New European Drive Cycle (NEDC), a commonly referenced drive cycle standard. Globally, there has been a significant reduction in new-vehicle emissions, and these trends are projected to continue to 2025 and beyond. For example, passenger car emissions in the United States are projected to decrease from 205 gCO_2 per km in 2002 to 98 gCO_2 per km in 2025. As fuel consumption is closely related to the carbon emissions, the same plots correlate to the fuel consumption in liters of fuel per 100 km plotted against the secondary vertical axis. Fuel consumption in the United States is projected to decrease from 8.7 L/100 km in 2002 to 4.1 L/100 km in 2025. Clearly, similar trends are happening globally.

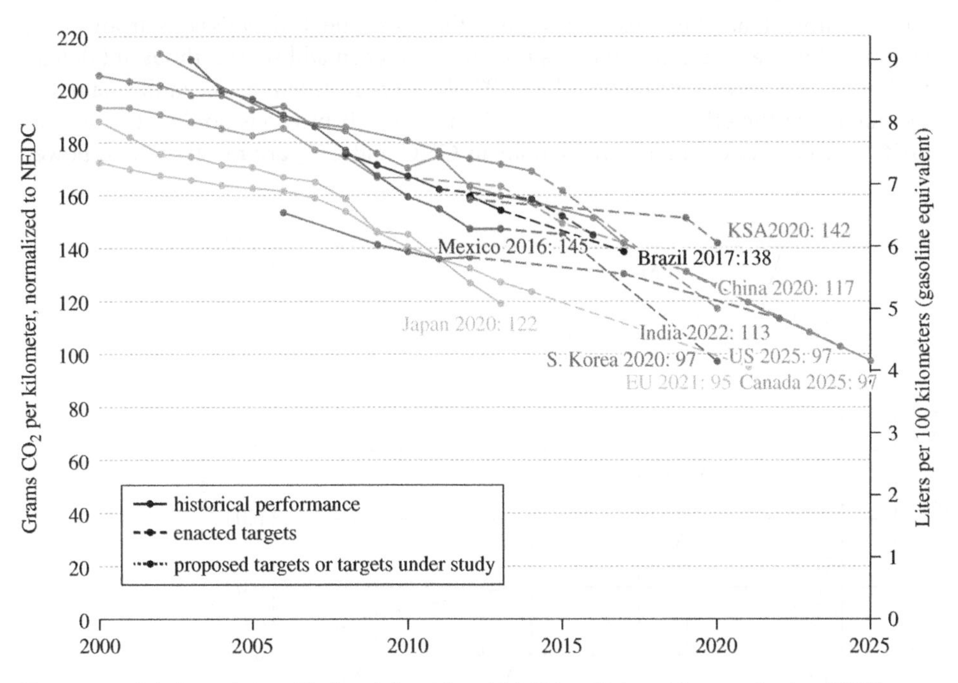

Figure 1.4 Global trends in gCO_2/km (left axis) and L/100 km (right axis) normalized to NEDC. (Courtesy of ICCT.)

1.3.2 Heavy-Duty Vehicle Regulations

As with light-duty vehicles, heavy-duty vehicles are subject to greater emission controls. Light-duty vehicles have a maximum gross vehicle weight of less than 3,855 kg (8,500 lb) in the United States, whereas heavy-duty vehicles typically are up to 36,280 kg (80,000 lb).

European standards are widely referenced. Euro I was introduced for trucks and buses in 1993. The Euro VI standard was introduced for trucks and buses in 2014. As shown in Figure 1.5, the permitted emissions levels for NO_x and PM have been reduced significantly. This pattern of reduction can be credited to costly on-board after-treatment of diesel emissions. The emissions have been reduced from 8 g/kWh and 0.6 g/kWh in Euro I for NO_x and PM, respectively, to 0.4 g/kWh and 0.01 g/kWh in Euro VI.

Commercial vehicles such as freight trucks and passenger buses can be – and are being – hybridized for a number of commercial reasons, which do not necessarily apply to light-duty vehicles:

- These vehicles often have defined drive cycles with defined routes, and locations with available trained user and service personnel.
- There can be significant maintenance and service costs, and so the vehicles can benefit from the lower failure rates and lower service costs associated with electrification.
- Municipal constraints on emissions, especially NO_x and PM, make it more difficult to drive non-zero-emission vehicles in urban areas.
- The heavy-duty vehicles tend to have higher continuous power levels and lower relative peak power levels than the light-duty vehicle.
- After-treatment has been used to reduce the NO_x and PM emissions from diesel engines. However, after-treatment is not cheap and can add several thousand dollars to the cost of a heavy-duty diesel vehicle. While after-treatment has been the main tool for mitigating the effects of NO_x and PM emissions, it provides no direct benefit to GHG or CO_2 reduction. The reductions in GHG and CO_2 are mostly due to power

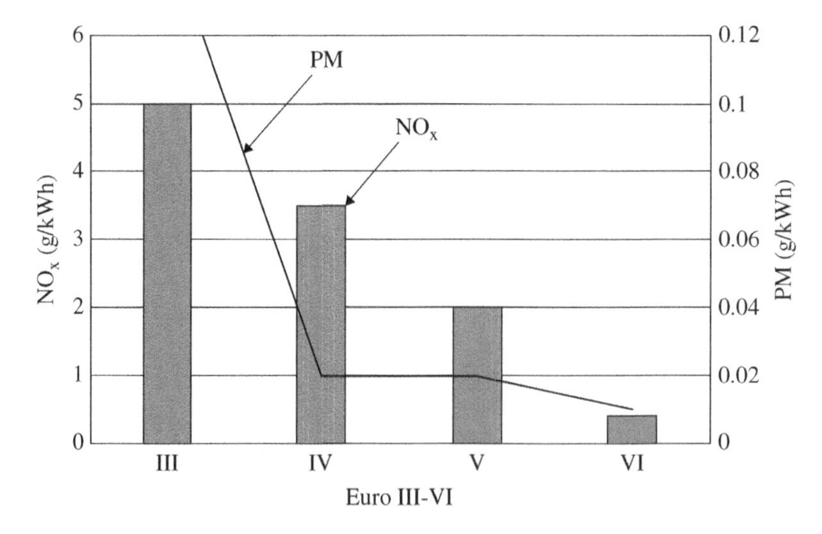

Figure 1.5 Euro III to VI for heavy-duty diesel engines.

plant optimization by utilizing hybrid configurations, or the hydrogen fuel cell, and by downsizing the engines to run more efficiently.

All these factors are resulting in increased electrification of the heavy-duty vehicle.

1.4 Drive Cycles

A drive cycle is a standardized drive profile which can be used to benchmark and compare fuel economy and emissions. Tests are conducted by operating vehicles using a professional driver driving on a flat treadmill-type device called a dynamometer.

There are many types of standardized drive cycles in use around the world. The Worldwide Harmonized Light Vehicles Test Procedures (WLTP) has been developed for adoption in a number of countries. JC08 is a commonly referenced standard in Japan. Drive cycles created and managed by the EPA are commonly referenced in the United States.

This section focuses on the EPA drive cycles as they are widely referenced in the literature and media. The NEDC is commonly referenced in Europe. However, the NEDC is not as representative of real-world driving considerations or as rigorous as the EPA tests, and it can significantly overpredict fuel economy and underestimate fuel consumption and emissions, as will be noted later.

1.4.1 EPA Drive Cycles

The standard US EPA fuel economy estimates are based on testing from five different drive cycles [15–18]. The vehicle exhaust emissions during a test are gathered in bags for analysis. There are four basic cycles used by the EPA, as shown in Table 1.3: FTP, HFET, US06, and SC03. The fifth test is a cold-ambient FTP, termed FTP (cold).

The Federal Test Procedure FTP drive cycle is for city driving and is as shown in Figure 1.6(a). The FTP cycle is based on the earlier LA4 cycle, which is also known as the Urban Dynamometer Drive Schedule or UDDS, and was originally created based on Los Angeles driving. The initial 1400 s of the FTP cycle is the same as that of the full

Table 1.3 Drive cycle parameters.

Drive cycle	Distance (km)	Time (s)	Average speed (km/h)	Maximum speed (km/h)	HVAC
FTP	17.66	1874	33.76	90	
FTP (cold)	17.66	1874	33.76	90	Heat
HFET	16.42	765	77.28	95	
USO6	12.82	596	77.39	127	
SCO3	5.73	596	34.48	88	AC
UDDS (LA4)	11.92	1369	31.34	90	
NEDC	11.02	1180	33.6	120	

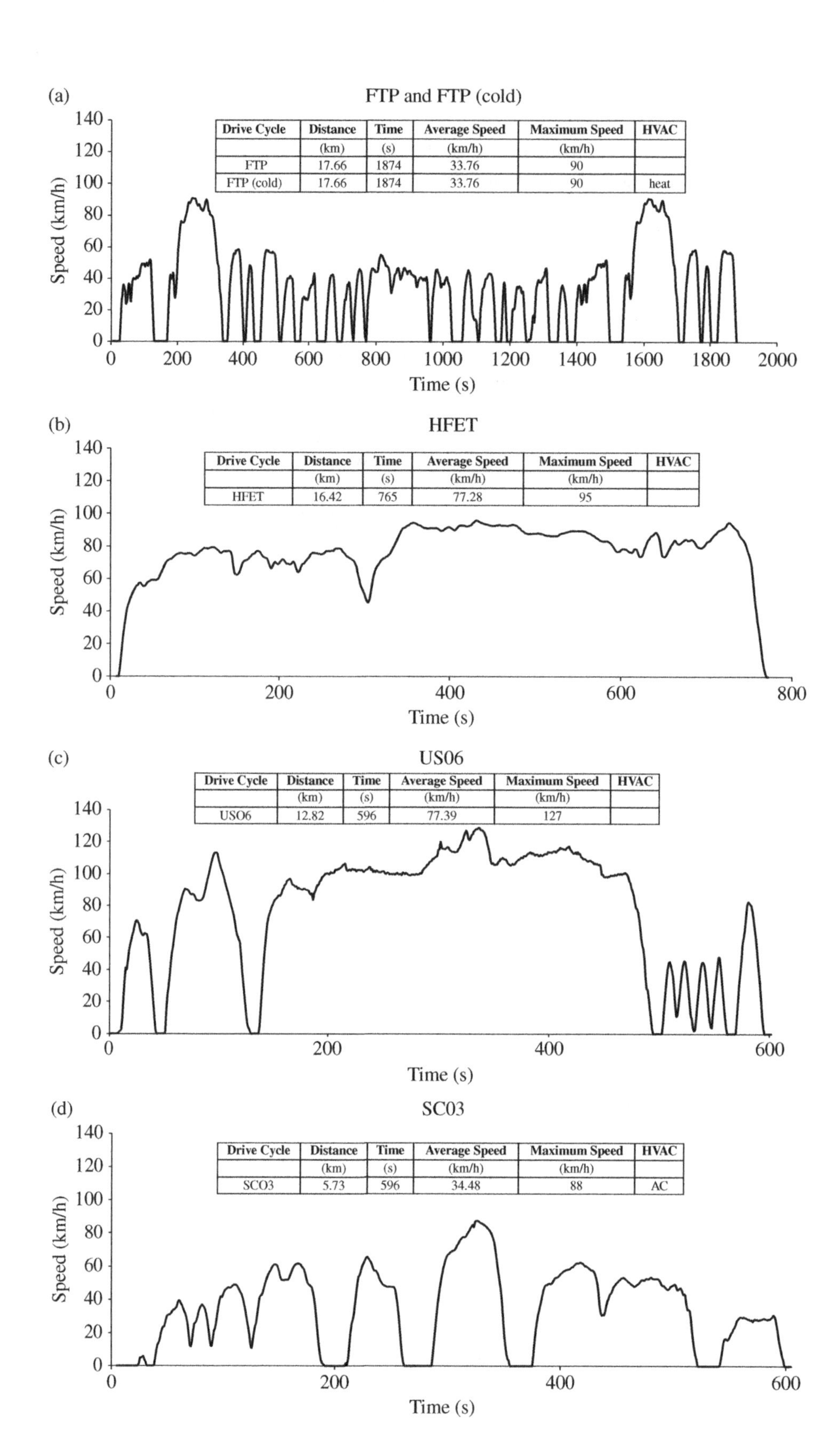

Drive Cycle	Distance	Time	Average Speed	Maximum Speed	HVAC
	(km)	(s)	(km/h)	(km/h)	
FTP	17.66	1874	33.76	90	
FTP (cold)	17.66	1874	33.76	90	heat

Drive Cycle	Distance	Time	Average Speed	Maximum Speed	HVAC
	(km)	(s)	(km/h)	(km/h)	
HFET	16.42	765	77.28	95	

Drive Cycle	Distance	Time	Average Speed	Maximum Speed	HVAC
	(km)	(s)	(km/h)	(km/h)	
US06	12.82	596	77.39	127	

Drive Cycle	Distance	Time	Average Speed	Maximum Speed	HVAC
	(km)	(s)	(km/h)	(km/h)	
SC03	5.73	596	34.48	88	AC

Figure 1.6 EPA drive cycles.

LA4 cycle. The first 450 s of the LA4 cycle is included after the full LA4 cycle to create the full FTP drive cycle. The Highway Fuel Economy Driving Schedule or HFET drive cycle, as shown in Figure 1.6(b), simulates highway driving. A higher-acceleration, more aggressive cycle is based on the USO6 drive cycle, as shown in Figure 1.6(c). The use of air-conditioning in city driving is introduced in the SCO3 drive cycle, as shown in Figure 1.6(d), and the testing is conducted at a hot ambient of 95 °F. The tests factor in varying temperature conditions and the effects of cold and hot starts. The contributions of the various tests are based on driving profiles of the typical American driver, as compiled by the EPA, and the results of the tests are compiled and adjusted in order to generate a realistic estimate of fuel economy, consumption, emissions, and range.

For comparison purposes, all the drive cycles shown in Figure 1.6 are plotted with 140 km/h on the speed axis. Some key parameters for the various tests are noted in Figure 1.6 and Table 1.3. Clearly, HFET and US06 are the highest speed tests. Note that the parameters for the LA4 and NEDC drive cycles are also included in Table 1.3.

The five-cycle tests are based on the IC engine. The tests measure various emissions and fuel economy for given conditions. There can be significant variations in emissions and fuel economy depending on whether a vehicle is starting cold or is already warmed up, or if the vehicle requires heating, ventilation, or air-conditioning (HVAC), as shown in the final column of Table 1.3. The various tests capture the significant conditions.

The testing is conducted by the manufacturer or the EPA, and the data are summarized and published annually by the EPA [16].

Table 1.4 shows EPA test data on the 2017 Toyota Prius Eco. The commonly referenced carbon dioxide emissions (CO_2) are shown in the second column. The emissions in grams per mile (g/mile) related to the THCs, carbon monoxide (CO), the oxides of nitrogen (NO_x), PM, methane (CH_4), and nitrous oxide (N_2O) are shown in the next several columns of the table. The final column presents the fuel economy in miles per US gallon (mpg) for the drive cycle.

From Figure 1.6, it can be seen that the FTP, FTP (cold), and SC03 have similar average and maximum speeds, as they are all for city driving. However, because the vehicle is using additional fuel for heat for FTP (cold) and for air-conditioning for SC03, the emissions and fuel economy can vary relatively significantly. The fuel economy decreases by 33% from FTP at 84.1 mpg to 56.6 mpg for FTP (cold), and by 31% to 58 mpg for SC03. The aggressive driving pattern of US06 results in the lowest fuel economy of 50.1 mpg. The HFET cycle has a relatively high fuel economy of 76.9 mpg.

Table 1.4 EPA data on 2017 Toyota Prius Eco emissions and fuel economy.

Drive cycle	CO_2 (g/mile)	THC (g/mile)	CO (g/mile)	NO_x (g/mile)	PM (g/mile)	CH_4 (g/mile)	N_2O (g/mile)	Unadj. fuel economy (mpg)
FTP	105	0.0117	0.0644	0.0029		0.0019	0.0013	84.1
FTP (cold)	153	0.1906	1.0590	0.0229				56.6
HFET	115	0.0004	0.0156	0.0001		0.0003		76.9
USO6	176	0.0249	0.201	0.0052				50.1
SCO3	152	0.0406	0.2768	0.0036				58

The results of the various tests are combined to generate overall fuel economy numbers for consumer education and vehicle labeling [19]. The fuel economies from the five tests are combined in order to generate fuel economy estimates based on real-world considerations for city, highway, and combined. The fuel economy estimates for various vehicles are as shown later in Table 1.8. The estimates for the 2017 Toyota Prius Eco are 58 mpg for city, 53 mpg for highway, and 56 mpg combined.

While the Toyota Prius Eco is one of the most efficient gasoline-fueled vehicles on the road, diesel engines are generally used for heavy-duty vehicles, and are also very popular in Europe for light vehicles due to increased fuel economy. The data for the 2015 Mercedes-Benz ML250 Bluetec 4MATIC diesel vehicle are shown in Table 1.5. Obviously, given the relative size and enhanced performance of the vehicle compared to the 2017 Toyota Prius Eco, the fuel economy is significantly lower for the ML250. The NO_x and PM columns are significant factors when assessing the diesel engine.

It is worth comparing the emissions for gasoline and diesel options from the same manufacturer, a HEV from Toyota, and the Tesla Model X 90D BEV, as shown in Table 1.6. The gasoline 2015 Mercedes-Benz ML350 4MATIC is compared with its diesel sibling just discussed above for the city (FTP) and highway (HFET) drive cycles. As the diesel engine is larger and heavier than the gasoline engine, the diesel vehicle mass is heavier, as shown in column 4 of the table. The CO_2, THC, and CO emissions are higher for the gasoline engine compared to the diesel engine for city FTP driving, while NO_x and CH_4 drop comparatively for the diesel engine in highway HFET driving. There are significant emissions of PM and NO_x from the diesel vehicle during the city driving.

When the Lexus RX 450 h HEV from Toyota is considered, then the emissions decrease significantly for the HEV compared to the non-hybrid vehicles, and the fuel economy increases significantly.

The advantage of hybridization now becomes clear. The HEV can achieve an efficiency for highway driving that is comparable with diesel, but the HEV can deliver similar fuel economy for stop-and-go city driving, unlike the diesel vehicle. While the carbon emissions are comparable for the HEV and diesel on the highway, the diesel fuel economy of 43.7 mpg is higher than that of the HEV at 38.7 mpg only because diesel fuel contains more energy per unit of volume.

The final option in the table is the Tesla SUV BEV. The advantages of the BEV become clear in this comparison. The equivalent mpg is significantly higher for the BEV than for all the other options, and without tailpipe emissions, albeit at the expense of reduced range.

Table 1.5 EPA data on 2015 Mercedes-Benz ML250 BlueTEC 4MATIC emissions and fuel economy.

Drive cycle	CO_2 (g/mile)	THC (g/mile)	CO (g/mile)	NO_x (g/mile)	PM (g/mile)	CH_4 (g/mile)	N_2O (g/mile)	Unadj. fuel economy (mpg)
FTP	345	0.0185	0.09	0.025	0.0015	0.0130	0.01	29.5
FTP (cold)	471	0.0136	0.2069	0.4839	0.0043			21.6
HFET	233	0.0002	0.0049	0.0001	0.0011	0.0011	0.01	43.7
USO6	388	0	0.0053	0.1276	0.0004			26.2
SCO3	446	0.0017	0.0033	0.0950	0.0026			22.8

Table 1.6 Comparison of conventional gasoline, diesel, hybrid-electric and battery-electric vehicles.

Drive cycle	Model	Fuel	Power (kW)	Mass (kg)	CO_2 (g/mile)	THC (g/mile)	CO (g/mile)	NO_x (g/mile)	PM (g/mile)	CH_4 (g/mile)	N_2O (g/mile)	Unadj. fuel economy (mpge)
FTP	ML250	Diesel	150	2495	345	0.0185	0.09	0.0250	0.0015	0.0130	0.01	29.5
	ML350	Gasoline	225	2381	404	0.0365	0.7445	0.0054		0.0095		21.9
	RX450h	Gas. HEV	183	2268	215	0.0072	0.0756	0.0048				41.2
	Model X 90D	BEV	310	2495								124.2
HFET	ML250	Diesel	150	2495	233	0.0002	0.0049	0.0001	0.0011	0.00112	0.01	43.7
	ML350	Gasoline	225	2381	285	0.019	0.2948	0.0028		0.00728		31.1
	RX450h	Gas. HEV	183	2268	229	0.0003	0.0400	0.0017				38.7
	Model X 90D	BEV	310	2495								135.8

1.5 BEV Fuel Consumption, Range, and mpge

Similarly, BEVs are characterized for fuel economy and consumption – vehicle tailpipe emissions do not need to be considered as there are no emissions from the vehicle.

The fuel economy of a BEV is actually specified using mpg equivalent, or mpge. A US gallon of gasoline nominally contains 33.705 kWh of energy, and so this value is used to determine mpge. Thus, the 90 kWh Tesla Model X 90D contains the equivalent energy of 2.67 US gallons. The BEV powertrain is far more efficient than the IC engine powertrain, and so the mpge of the BEV can be quite high, although the stored energy is low.

There are a number of ways of calculating fuel consumption and emissions. Typically, the BEV fuel economy is calculated using a two-cycle test, as defined by the EPA [18,20]. Basically, the vehicle performance is measured for city and highway driving. The combined fuel economy is based on 55% city driving and 45% highway driving.

As the testing is conducted on a dynamometer, the combined fuel consumption is adjusted by dividing by 0.7 in order to factor in a realistic adjustment for real-world driving and the use of HVAC.

Thus, the adjusted combined fuel consumption (*FC*) is given by

$$\text{Adjusted Combined}FC = (0.55 \times cityFC + 0.45 \times hwyFC) \div 0.7 \tag{1.12}$$

where *cityFC* is the fuel consumption measured for the city drive cycle (FTP) and *hwyFC* is the fuel consumption measured for the highway (HFET) drive cycle.

The manufacturer data on fuel consumption, measured as electrical kWh/100 miles from the charging plug, is published by the EPA and is shown in column 2 of Table 1.7. For example, the 2015 Nissan Leaf consumes 18.65 kWh and 23.28 kWh from the electrical grid in order to complete 100 miles of the city and highway drive cycles, respectively. These numbers are converted to units of Wh/km in column 3. The adjusted combined fuel consumption is calculated as 29.62 kWh/100 miles or 184.1 Wh/km based on Equation (1.12).

We can go further and estimate the vehicle range. If a charging efficiency of 85% is assumed, then the energy consumption from the battery is 85% of 184.1 Wh/km, or

Table 1.7 2015 Nissan Leaf test results.

	(kWh/100 mile)	(Wh/km)	mpge
City FC	18.65	115.8	180.7
Hwy FC	23.28	144.7	144.8
Combined FC	20.73	128.8	162.6
Adj. City FC	26.64	165.6	126.5
Adj. Hwy FC	33.26	206.7	101.3
Adj. Combined FC	29.62	184.1	113.8
Range @ BOL	153 km (95 miles)		
Range @ EOL	123 km (76 miles)		
Range @ Midpoint	138 km (86 miles)		

156.5 Wh/km. Dividing this fuel consumption into the battery capacity of 24 kWh at the battery beginning of life (BOL) provides a range estimate of 153 km (95 miles).

As the battery capacity can decay with a number of factors, we note that the range drops proportionately to 122 km (76 miles) if the battery end-of-life (EOL) capacity is 80% of the BOL capacity. The EPA rating is the midpoint at 90% of BOL for an estimated range of 138 km (86 miles). The actual EPA range for the 24 kWh Nissan Leaf is 137 km, which compares well to the 138 km just estimated. The minor error between the two values can be related to the storage capacity or the charging efficiency assumption.

The NEDC range for the same vehicle is 200 km. The NEDC number is an unadjusted number based on dynamometer testing only. Following on the example above, it is clear that a reduction of at least 30% is needed in order to take into account real-world driving scenarios, and would produce a more useful range estimate for the consumer.

The mpge can be easily calculated using the FC values. The mpge is related to the fuel consumption by the following formula:

$$\text{mpge} = \frac{33.705\dfrac{\text{kWh}}{\text{gal}}}{FC} \tag{1.13}$$

where FC is the fuel consumption in kWh/mile. The published EPA mpge values for adjusted city, highway, and combined driving are 126, 101, and 114 mpge, respectively, and these values are the same as those based on the calculations above, when rounded.

1.6 Carbon Emissions for Conventional and Electric Powertrains

It is useful to compare the overall emissions of a vehicle when considering the overall energy efficiency and environmental impact. In this section, we briefly review the carbon emissions related to the various powertrains. The EPA presents estimated carbon emissions for the various powertrains on its web site [19]. The EPA estimates consider the vehicle tailpipe emissions and the upstream emissions due to the production, transportation, and distribution of the various energy sources.

Note that these numbers are based on EPA data available in early 2017. At the time of publication the upstream emissions figures are not available for fuel cell vehicles, such as the Toyota Mirai. In general, it is expected that vehicle and grid emissions will continue to decrease as both vehicles and the grid become less polluting and more efficient.

EPA data on energy consumption and carbon emissions are presented in Table 1.8 for various midsize and large BEV, HEV, diesel, and gasoline vehicles. The 2017 Nissan Leaf consumes 30 kWh from the electrical grid per 100 miles driven, and this translates in a combined fuel economy of 112 mpge.

The EPA web site provides the option of estimating grid emissions on the basis of the local ZIP or postal code, as well as the national average. Let us compare two contrasting cities. The city of Los Angeles in California, with ZIP code 90013, has a relatively low-emissions electrical grid. The city of Detroit in Michigan, with ZIP code 48201, has relatively high emissions due to the use of coal.

Table 1.8 Fuel economy, upstream carbon emissions, and range for various 2015–2017 vehicles (based on 2017 data).

			Fuel economy			gCO_2e Emissions total (tailpipe)	Range
Vehicle	Type	Size	City (mpge)	Highway (mpge)	Combined (mpge)	(gCO_2e/mile)	miles)
2017 Nissan Leaf	BEV	Midsize	124	101	112	100 (Los Angeles) 180 (United States) 240 (Detroit)	107
2017 Hyundai Ioniq	BEV	Midsize	150	122	136	90 (Los Angeles) 150 (United States) 200 (Detroit)	124
2017 Chevy Volt	PHEV	Compact			106	140 (Los Angeles) 200 (United States) 250 (Detroit)	53 (electricity) 420 (total)
2017 Toyota Prius Eco	HEV	Midsize	58	53	56	190 (158)	633
2017 BMW 328d Auto	Diesel	Compact	31	43	36	346 (285)	540
2017 BMW 320i Auto	Gasoline	Compact	23	35	28	381 (323)	422
2017 Toyota Mirai	Hydrogen	Midsize	67	67	67	Not available	312
2017 Tesla Model X AWD 90D	BEV	SUV	90	94	92	130 (Los Angeles) 220 (United States) 300 (Detroit)	257
2017 Lexus RX 450 h AWD	HEV	SUV	31	28	30	356 (297)	516
2015 Mercedes-Benz ML250 Bluetec 4MATIC	Diesel	SUV	22	29	25	499 (413)	615
2015 Mercedes-Benz ML350 4MATIC	Gasoline	SUV	18	22	19	561 (456)	467
2015 Honda Civic	CNG	Compact	27	38	31	309 (218)	193
2015 Honda Civic	HEV	Compact	43	45	44	242 (202)	581
2015 Honda Civic HF	Gasoline	Compact	31	40	34	314 (259)	449

Thus, while the carbon emissions at the point of use for the 2017 30 kWh Nissan Leaf are zero, the upstream carbon emissions from the electric power stations can result in emissions ranging from 100 gCO_2/mile (62 gCO_2/km) in Los Angeles to 240 gCO_2/mile (149 gCO_2/km) for Detroit, with a national average of 180 gCO_2/mile (112 gCO_2/km).

The 2017 Hyundai Ioniq BEV has a 28 kWh battery pack and a longer range and higher efficiency than the 2017 Nissan Leaf. The EPA range for the vehicle is 124 miles, and the emissions are as low as 150 gCO_2/mile nationally in the United States.

The Chevy Volt PHEV can run 53 miles on battery only and 420 miles in total. The vehicle's carbon emissions are estimated at 140 gCO_2/mile in Los Angeles and up to 250 gCO_2/mile in Detroit. The EPA assumes that the car is driven using the battery for 76.1% of the time and the balance using the gasoline engine.

The Toyota Prius Eco model has a combined fuel economy of 56 mpg with total carbon emissions, including upstream, of 190 gCO_2/mile, or 158 gCO_2/mile from the tailpipe. The BMW 320 series is presented for examples of the midsize diesel and gasoline vehicles. The total emissions are 346 and 381 gCO_2/mile for the diesel 328d and gasoline 320i vehicles, respectively.

EPA data on the 2016 Toyota Mirai fuel cell EV are included. The published fuel economy is 66 miles per kg of hydrogen, or 67 mpge, with a range of 312 miles.

Electric, hybrid, diesel, and gasoline SUV models are next considered. The all-electric high-performance Tesla Model X has a fraction of the upstream emissions compared to its competitors from Lexus and Mercedes-Benz, with the conventional Mercedes vehicles having far higher emissions than the hybrid Lexus.

Finally, CNG is a competitive fuel for use in the spark-ignition IC engine. As discussed earlier, CNG produces less gCO_2 than gasoline per kg of fuel. Even though CNG is less efficient than gasoline when combusted in the engine, a reduced value of the gCO_2/mile is expected overall. Honda Motor Company sold three versions of the Honda Civic in 2015: CNG, gasoline, and gasoline hybrid. A comparison of the vehicles illustrates the advantages and disadvantages of CNG. The tailpipe emissions of the CNG vehicle at 218 gCO_2/mile is an improvement on the conventional gasoline vehicle at 259 gCO_2/mile, but higher than the hybrid gasoline vehicle at 202 gCO_2/mile. The range of the CNG vehicle is less than half of the range of the other vehicles due to the larger storage required for the CNG. The upstream emissions for the CNG vehicle are disproportionately higher than those of the gasoline vehicles. **Equivalent CO_2** or **CO_2e** is the concentration of CO_2 which would cause the equivalent effect with time as the emitted greenhouse gases, especially methane. Leakage of methane during production, transmission, and distribution of CNG has a significantly greater impact as a greenhouse gas over time than CO_2, resulting in increased gCO_2 equivalent emissions for CNG. Thus, the total emissions for the CNG vehicle at 309 gCO_2/mile are only a little lower than those of the gasoline vehicle at 314 gCO_2/mile, and significantly higher than those of the hybrid vehicles at 242 gCO_2/mile.

1.6.1 Well-to-Wheel and Cradle-to-Grave Emissions

Well to wheel is the term used to describe the overall energy flow from the oil or gas well when the fossil fuel comes out of the ground to the ultimate usage in the vehicle to spin the wheels. It is an area of active research and application. For example, Argonne

National Laboratory in the United States has developed the widely referenced GREET (**G**reenhouse gases, **R**egulated **E**missions, and **E**nergy use in **T**ransportation) model for use in this area [21–23]. The GREET model is used to estimate emissions in the EPA models.

There are significant energy and carbon costs incurred during the production of a vehicle. Reference [24] provides a comprehensive study into the equivalent carbon required for the cradle-to-grave emissions assessment of the conventional Ford Focus IC engine vehicles compared to the battery electric version of the Ford Focus. The BEV version incurred about 10.3 metric tons of CO_2e compared to about 7.5 metric tons of CO_2e for the Ford Focus IC engine vehicle. The emissions of CO_2e for the 24 kWh Li-ion battery alone were 3.4 metric tons, or 140 kg per kWh. Reference Problem 1.3.

1.6.2 Emissions due to the Electrical Grid

All EVs have the advantage of zero carbon emissions from the vehicle itself. Clearly, there are emissions at some point in the energy conversion process if fossil fuels are used to generate the electricity for BEVs or compressed hydrogen for FCEVs. Coal is used as the base load in many electricity grids as it is abundant and cheap. The carbon emissions from coal are relatively high compared to the other fossil fuels. It also outputs significant amounts of sulfur dioxide (SO_2) in addition to the nitrogen oxides. Sulfur dioxide is a significant component of urban smog. Atmospheric sulfur dioxide can react with oxygen and then water to create sulfuric acid, the main component of harmful acid rain. Coal fuel has resulted in major problems for populous countries such as China, India, and the United States. Low-sulfur coal can mitigate the problem and is available in abundance in various regions of the world.

Many electrical grids employ coal for the base load and gas as a more flexible and cleaner fuel. The balance of energy is typically provided by nuclear, hydroelectric, and renewable sources. It is useful to estimate the carbon emissions due to the production of electricity. Typical power plant efficiencies range from about 38% for coal to around 50% for gas [25].

1.6.2.1 Example: Determining Electrical Grid Emissions

Determine the carbon emissions from the electrical grid in gCO_2/kWh(electrical) if 35% of the electricity comes from coal and 40% comes from gas, with the balance being carbon-free nuclear, hydroelectric, and renewables. Adjust your answer higher by 20% to allow for fuel production and distribution, and electricity transmission and distribution.

Representative power plant efficiency values are $\eta_{coal} = 38\%$ for coal and $\eta_{gas} = 50\%$ for gas. Carbon emissions for coal and gas are presented in Table 1.1.

Solution:

The carbon emissions are 198 gCO_2/kWh for gas and 350 gCO_2/kWh for coal from Table 1.1. These emission values are based on the primary (or embedded) energy of the fuel. The carbon emissions for the energy output as electricity is lower due to the efficiencies of the power plants.

These emissions numbers must be divided by the power plant efficiencies in order to determine the emissions for each unit of output electrical energy:

$$\text{coal}: \text{gCO}_2/\text{kWh(electrical)} = \frac{350\,\text{gCO}_2/\text{kWh}}{\eta_{coal}} = \frac{350\,\text{gCO}_2/\text{kWh}}{0.38}$$

$$= 920\,\text{gCO}_2/\text{kWh(electrical)}$$

$$\text{gas}: \text{gCO}_2/\text{kWh(electrical)} = \frac{198\,\text{gCO}_2/\text{kWh}}{\eta_{gas}} = \frac{198\,\text{gCO}_2/\text{kWh}}{0.5}$$

$$= 396\,\text{gCO}_2/\text{kWh(electrical)}$$

The emissions are then prorated according to their grid usage:

$$\begin{array}{ll}
35\%\,\text{coal} & = 0.35 \times 920\,\text{gCO}_2/\text{kWh(electrical)} \\
40\%\,\text{gas} & = 0.4\ \times 396\,\text{gCO}_2/\text{kWh(electrical)} \\
25\%\,\text{carbon-free} & = 0.25 \times 0\ \text{gCO}_2/\text{kWh(electrical)} \\
\hline
\text{Emissions} & = 480\,\text{gCO}_2/\text{kWh(electrical) from power plant} \\
 & \times 1.2 \\
 & = 576\,\text{gCO}_2/\text{kWh(electrical)}
\end{array}$$

Thus, the overall emissions are 576 gCO_2/kWh(electrical) for this representative example.

1.7 An Overview of Conventional, Battery, Hybrid, and Fuel Cell Electric Systems

Seven vehicle architectures are considered in this section and over the next four chapters:

1) Conventional vehicles with IC engine:
   i) CI IC engine using diesel
   ii) SI IC engine using gasoline or CNG
2) Battery electric
3) Hybrid electric:
   i) Series
   ii) Parallel
   iii) Series-parallel
4) Fuel cell electric

Simplified vehicle configurations are discussed next. The vehicle and overall well-to-wheel efficiencies are briefly considered. While vehicle efficiencies obviously depend on driving patterns and other factors, representative figures are used in this section to illustrate the differences.

1.7.1 Conventional IC engine Vehicle

The IC engine is coupled via a clutch and gearing through the transmission to the drive axle. The elementary figure of Figure 1.7 simply represents the conventional powertrain. This basic architecture is the basis for the modern vehicle.

The percentage engine efficiency η_{eng} can range from very low values to the mid-to-high 30s for gasoline to the 40s for diesel. An overall tank-to-wheel vehicle efficiency $\eta_{T\text{-}W}$ of 20% is representative for a conventional diesel vehicle, with lower values for gasoline (17%) and CNG (16%). These relatively low engine efficiencies over the speed range are compensated by the high energy density of the fuel, thus enabling long-range and comfortable driving with fast and easy refueling. A representative efficiency for the production, refining, and distribution of the fuel from the well to the tank $\eta_{W\text{-}T}$ is 84% [26]. Thus, the overall well-to-wheel efficiency $\eta_{W\text{-}W}$ is the product of the two efficiencies:

$$\eta_{W-W} = \eta_{W-T} \times \eta_{T-W} \tag{1.14}$$

The overall well-to-wheel efficiencies range from 17% for diesel to 14% for gasoline.

1.7.2 BEVs

The BEV transforms the chemical energy of the battery into mechanical energy using an electric drive, as shown in Figure 1.8. The electric drive features an inverter, electric motor, and controls. The inverter converts the dc of the battery to the ac waveforms

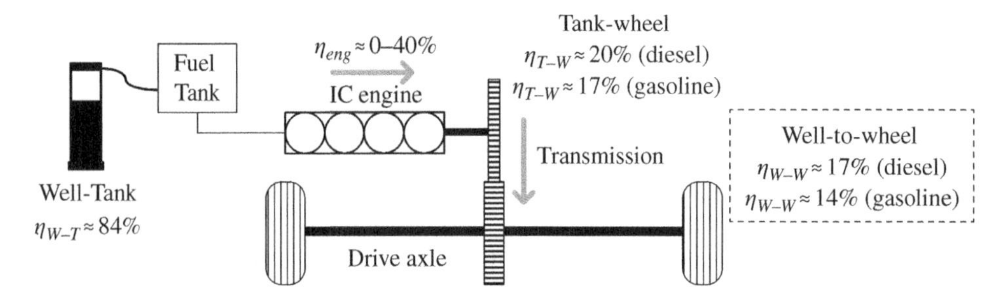

Figure 1.7 Conventional vehicle architecture and energy flow.

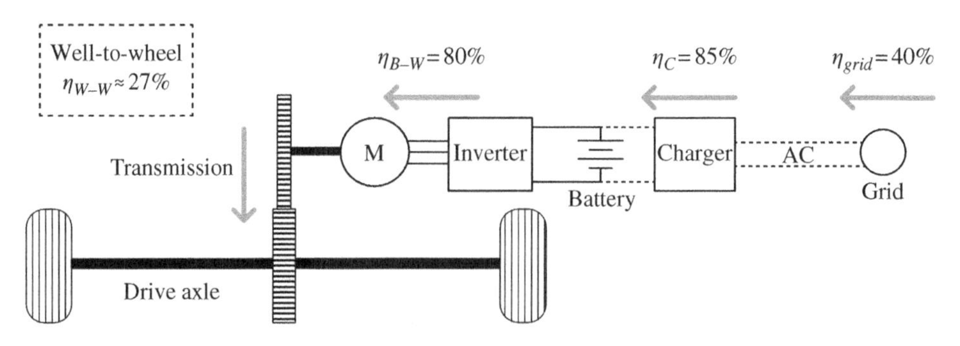

Figure 1.8 BEV architecture and energy flow.

required to optimally power the electric motor. While the BEV is very efficient in on-board energy conversion, the battery range may be limited due to the low battery energy density. A battery-to-wheel powertrain efficiency $\eta_{B\text{-}W}$ of about 80% is a reasonable number for the BEV. The vehicle is refueled by charging the battery with power from the electrical grid. An efficiency for the generation, transmission, and distribution of the electricity from the public grid η_{grid} of about 40% is a reasonable estimate as many grids are primarily dependent on fossil fuels but supplemented by nuclear power and renewables. Notable exceptions exist such as Norway with hydroelectric and France with nuclear as the main electricity sources. A charging efficiency η_C of 85% is a reasonable estimate of efficiency from the plug to the battery.

Thus, the overall well-to-wheel efficiency $\eta_{W\text{-}W}$ is the product of the three efficiencies:

$$\eta_{W-W} = \eta_{grid} \times \eta_C \times \eta_{B-W} \tag{1.15}$$

The overall well-to-wheel efficiency for the BEV is about 27%.

1.7.3 HEVs

HEVs improve the fuel economy of the conventional fossil-fuel-powered vehicles by addressing a number of critical factors which impact fuel economy:

1) Vehicle idling is eliminated as significant fuel can be consumed by the engine as it idles.
2) Regenerative braking energy is recovered and stored in the battery. In a conventional vehicle, the braking energy is dissipated as heat by the braking system and lost to the vehicle.
3) The stop-start, low-speed, low-torque nature of city driving is inefficient for the conventional vehicle, whereas the hybrid vehicle decouples the driving condition from efficient operation of the engine by storing and using battery energy when it is efficient to do so.
4) The engine size can be smaller in a hybrid vehicle compared to the conventional vehicle, and can run more efficiently.
5) Significant low-speed torque can be available in HEVs and BEVs due to the electric traction motor.

The hybrid vehicle has an additional advantage over the BEV:

1) The battery lifetime can be extended and the battery cost reduced as shallower battery discharges can be implemented in a hybrid system compared to a battery electric car.

A **PHEV** is a hybrid electric vehicle with a battery pack which can be recharged from the grid. The PHEV can be designed to run in BEV mode for a significant distance. A PHEV running as a BEV is operating in **charge depleting** (**CD**) mode. A PHEV running as a HEV and maintaining the battery at an average state of charge is operating in **charge-sustaining** (**CS**) mode.

There are a number of different hybrid systems: series, parallel, and series-parallel, and these are discussed next.

1.7.3.1 Series HEV

The series HEV combines the best attributes of the IC engine and the BEV. It combines the powertrain efficiency of the BEV with the high-energy-density fuel of the IC engine, as shown in Figure 1.9. The well-designed series HEV runs the IC engine in a high-efficiency mode, and the engine output is converted via two electric drives in series in order to supply mechanical energy to the drivetrain. However, placing two electric drives in series means that the energy processing can be more inefficient than desired. The engine can be operated with efficiencies in the range of 30% to 40%,while the efficiencies for the electrical generating and motoring stages are estimated at 80% to 90% each.

As with the BEV, a battery-to-wheel powertrain efficiency $\eta_{B\text{-}W}$ of about 80% is a reasonable assumption. The battery is charged using the IC engine as a generator. A charging efficiency η_C of 90% is a reasonable estimate of efficiency from the engine to the battery. The engine efficiency η_{eng} is high as it is run in a high-efficiency mode.

Thus, the overall well-to-wheel efficiency $\eta_{W\text{-}W}$ is the product of four efficiencies:

$$\eta_{W-W} = \eta_{W-T} \times \eta_{eng} \times \eta_{gen} \times \eta_{B-W} \quad (1.16)$$

The overall well-to-wheel efficiency for the series HEV is about 21%.

1.7.3.2 Parallel HEV

The parallel HEV architecture has been implemented using a dual-clutch transmission (DCT) on vehicles such as the Honda Fit and the Hyundai Ioniq. The vehicle runs the engine when it is efficient to do so. The engine or the electric motor can be directly coupled to the drive axle, and the engine can be coupled to the drive motor to recharge the battery. A simple architecture is shown in Figure 1.10. If the vehicle is operating with

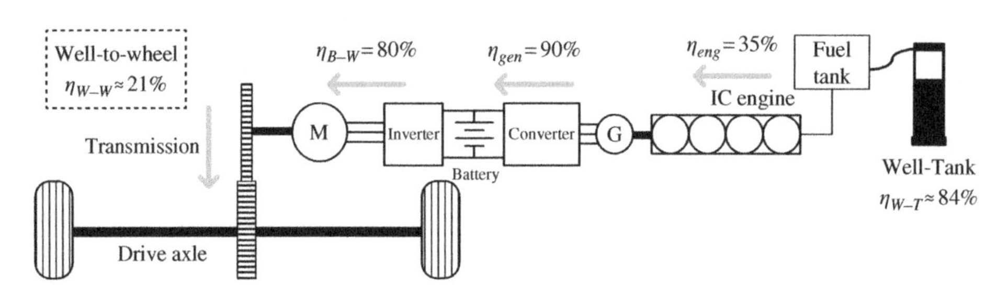

Figure 1.9 Series HEV architecture and energy flow.

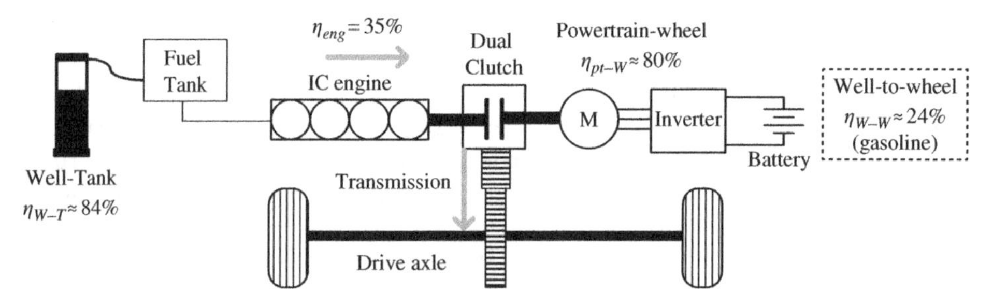

Figure 1.10 Parallel HEV architecture and energy flow.

the engine only, then the motoring efficiency can be high if the engine is operated at peak efficiency. The overall efficiency can drop as the energy is run through the electric system due to the inefficiencies in each direction as the battery is charged and discharged, similar to the series hybrid. The efficiency of the powertrain to wheel η_{pt-W} is assumed to be about 80%.

Thus, the overall well-to-wheel efficiency η_{W-W} is the product of three efficiencies:

$$\eta_{W-W} = \eta_{W-T} \times \eta_{eng} \times \eta_{pt-W} \tag{1.17}$$

The overall well-to-wheel efficiency for the parallel HEV is about 24%:

1.7.3.3 Series-Parallel HEV

The series-parallel HEV typically uses a sun-and-planetary gearing, known as a CVT, to split the engine power such that the vehicle can be optimally controlled to direct the engine output to either the drivetrain for direct propulsion of the vehicle or to the battery for the electric drive, as shown in Figure 1.11. This reduces the inefficiency introduced in the series HEV by having the two electrical stages in series. The series-parallel HEV also has two stages in series but only needs to do so when it is inefficient to drive directly from the IC engine, similar to the parallel hybrid just discussed.

Variants on this architecture are used on the Toyota, Ford, and General Motors hybrids.

The representative well-to-wheel efficiency is the same as in the parallel HEV just discussed, and is about 24%.

1.7.4 FCEV

Similar to the HEV, the fuel cell vehicle, as shown in Figure 1.12, features a battery which is used to absorb the transient power demands and regenerative power. Power cannot be regenerated into the fuel cell, and so the battery system is required for regeneration. A unidirectional boost dc-dc converter interfaces the fuel cell to the high-voltage dc link powering the electric drive. The efficiency of the boost converter is very high, 98% being assumed.

Representational numbers for the production of CNG, and the production, transmission, and supply of hydrogen to the vehicle is about 60%, based on the steam reforming of CNG [21]. A fuel cell system efficiency of about 58% is reasonable for a fuel cell operated in optimum power mode, and buffered by the battery for transients.

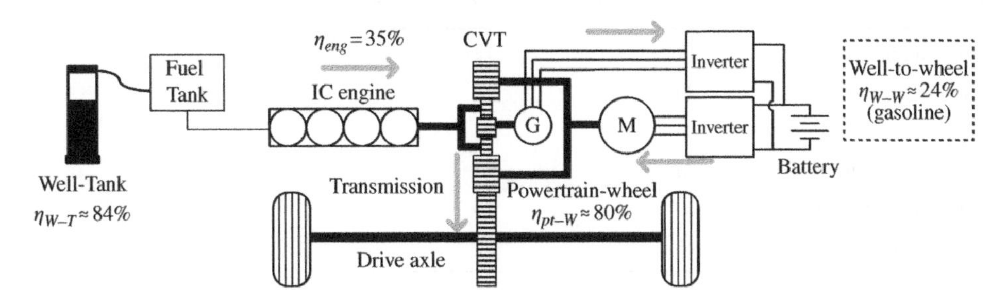

Figure 1.11 Series-parallel HEV architecture and energy flow.

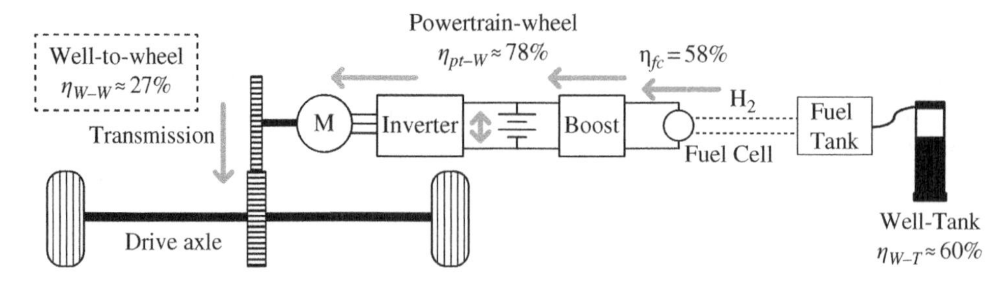

Figure 1.12 FCEV architecture and energy flow.

The efficiency of the powertrain to the wheel is about 78%, a little lower than the value used in the other vehicles due to the boost converter.

Thus, the overall well-to-wheel efficiency $\eta_{W\text{-}W}$ is the product of three efficiencies:

$$\eta_{W-W} = \eta_{W-T} \times \eta_{fc} \times \eta_{pt-W} \tag{1.18}$$

The overall well-to-wheel efficiency for the FCEV is about 27%.

1.7.5 A Comparison by Efficiency of Conventional, Hybrid, Battery, and Fuel Cell Vehicles

The overall on-board powertrain and well-to-wheel efficiencies for the various vehicles are summarized in Table 1.9.

The BEV and FCEV have the highest overall well-to-wheel efficiency at 27%, and are followed by the parallel HEV at 24%. The conventional gasoline vehicle has an efficiency of 14%. Thus, electrification can significantly improve the overall well-to-wheel efficiency. Broader adoption of renewables and nuclear power can reduce and improve the related carbon emissions for the BEV, FCEV, and the PHEV.

Table 1.9 Drivetrain efficiency comparison.

Fuel	Powertrain efficiency (%)	Well-to-wheel efficiency (%)
Gasoline SI	17	14
Diesel CI	20	17
BEV	80	27
Gasoline Series HEV	25	21
Gasoline Parallel HEV	28	24
Hydrogen FCEV	45	27

1.7.6 A Case Study Comparison of Conventional, Hybrid, Battery, and Fuel Cell Vehicles

A case study is conducted over the next four chapters which generates values for the carbon emissions, powertrain efficiency, and fuel economy for a test vehicle on a simple drive cycle. The following configurations are investigated: gasoline, diesel, battery electric, series hybrid electric, series-parallel hybrid electric, and fuel cell electric. The results are summarized in Table 1.10. The vehicles powered by diesel and gasoline have 40 liter fuel tanks. The BEV has a 60 kWh battery, while the FCEV vehicle has 5 kg of hydrogen. The estimated powertrain efficiencies are as shown, with the BEV being most efficient. followed by the FCEV. The conventional gasoline vehicle is the least efficient at 20.4%. Similar results are reflected for the fuel consumption in Wh/km and liters/100 km, and fuel economy in mpge. The carbon emissions are all below 100 gCO_2/km for the BEV, FCEV, and series-parallel HEV, as these are the most efficient architectures. The nominal BEV emissions are 73 gCO_2/km in the United States and as low at 1.5 gCO_2/km in Norway. These two figures illustrate the challenges and opportunities for the BEV. The electrical grid emissions in the United States, and in much of the world, are close to 500 gCO_2/kWh (electrical), and this figure is used to calculate the 73 gCO_2/km value. Nature has endowed Norway with an abundance of renewable hydroelectric power, and the Norwegian emissions are closer to 10 gCO_2/kWh (electrical), resulting in vehicle emissions of 1.5 gCO_2/km. Thus, challenges and opportunities exist for other countries to clean up the power grid and reduce carbon emissions from the grid by increased use of renewables, such as wind, photovoltaics, and hydroelectric. In the final column, we can see the challenges for the BEV. The nominal range of 485 km is significantly lower than those of the other technologies.

Table 1.10 Drivetrain case study comparison.

	Fuel	Storage	η_{pt} (%)	FC_{in} (Wh/km)	FC_{in} (L/100 km)	mpge (US) (mpg)	CO_2 (gCO_2/km)	Range (km)
SI	Gasoline	40 L	20.4	512	5.78	40.7	134	692
CI	Diesel	40 L	24.1	434	4.32	54.4	115	926
BEV	Electricity	60 kWh	84.6	146		144	73/1.5	485
Series HEV	Gasoline	40 L	26.4	396	4.47	52.7	103	895
Series-Parallel HEV	Gasoline	40 L	32.7	320	3.6	65.2	84	1108
FCEV	Hydrogen	5 kg	47.1	222		94	69	750

1.8 A Comparison of Automotive and Other Transportation Technologies

The focus of this chapter, and the book in general, is on light-duty vehicles, although heavy-duty vehicles are discussed with fuel cells in Chapter 4. However, it is worth noting the relative performance of heavier transportation modes. The technologies discussed in this book are being applied to these other transportation options on land, in the sea, and

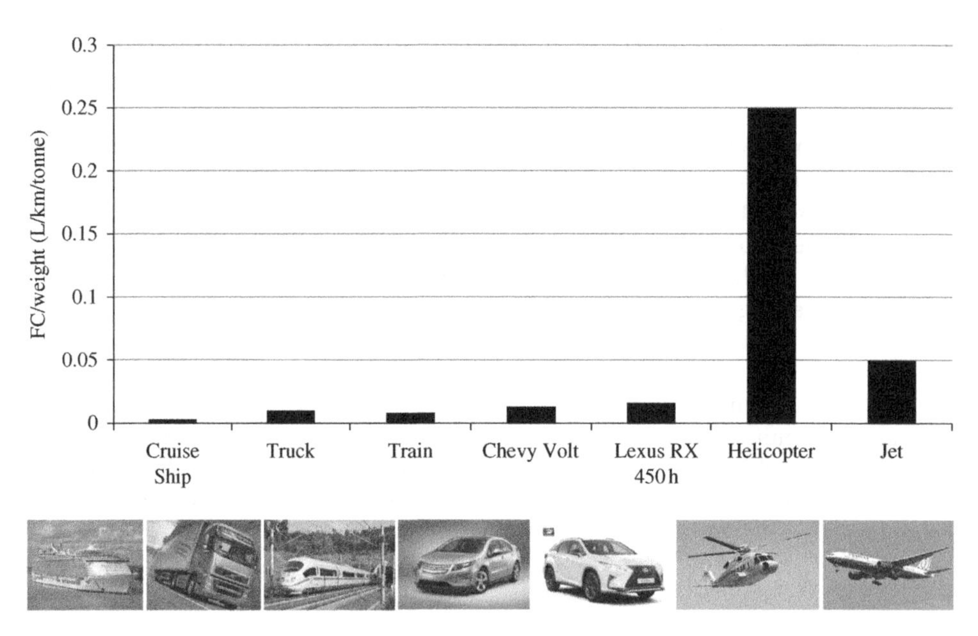

Figure 1.13 Fuel consumption per kilometer per tonne [liter/km/tonne].

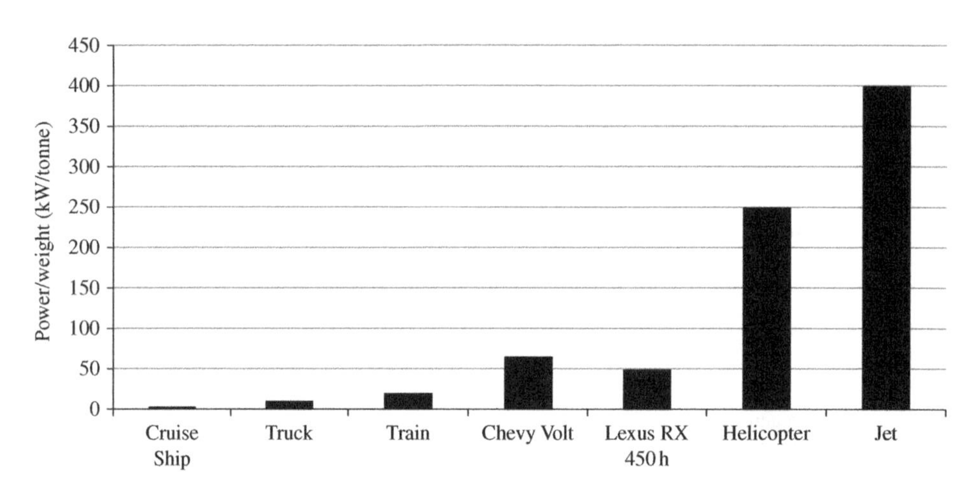

Figure 1.14 Power-to-weight ratio for various transportation options in kW/tonne.

in the air, as powertrains are increasingly electrified and hybridized. The fuel consumption per kilometer per metric ton for various modes of transport is presented in Figure 1.13. The jetliner is relatively competitive with the light-duty vehicles on a relative basis. The shipping, trucking, and train options can be significantly more efficient. Similar plots would result if we were to consider the fuel economy per passenger over distance.

The peak power-to-weight ratios for the various technologies are as shown in Figure 1.14. It can be seen that the commercial truck has a significantly lower power-to-weight ratio compared to the light-duty sedan and SUV. This is also true for commercial buses.

References

1 M. Schnayerson, *The Car That Could*, Random House, 1996, ISBN 0-679-42105-X.

2 R. H. Schallenberg, "Prospects for the electric vehicle: A historical perspective," *IEEE Transactions on Education*, vol. E-23, No. 3, August 1980.

3 D. Yergin, *The Quest: Energy, Security, and the Remaking of the Modern World*, Penguin Books, 2012, ISBN 978-0-241-95777-6.

4 L. Tillemann, *The Great Race: The Global Quest for the Car of the Future*, Simon & Schuster, 2015.

5 H. Ford (with Samuel Crowther), *My Life and Work*, 1923.

6 A. J. Haagen-Smit "A lesson from the Smog Capital of the World," *Proceedings of the National Academy of Sciences*, October 1970.

7 A. J. Haagen-Smit, E. F. Darley, M. Zaitlin, H. Hull, and W. Noble, "Investigation on injury to plants from air pollution in the Los Angeles area," *Plant Physiology*, 27 (18), 1952.

8 A. Vance, *Elon Musk Tesla, SpaceX, and the Quest for a Fantastic Future*, HarperCollins Publishers, 2015.

9 Press release, *EPA's notice of violation of the Clean Air Act to Volkswagen*, International Council on Clean Transportation (ICCT), Sept. 18, 2015.

10 *Bosch Automotive Handbook*, 9th edition, Bentley Publishers, 2014.

11 T. L. Brown, H. E. LeMay, B. E. Bursten, C. J. Murphy, and P. M. Woodward, *Chemistry: The Central Science*, 12th edition, Prentice Hall (Pearson Education).

12 P. Miller, "The pulse of the planet," *National Geographic*, 228 (5), November 2015.

13 Department for Environment, Food and Rural Affairs, *Draft Plan to Improve Air Quality in the UK, Tackling Nitrogen Dioxide in our Towns and Cities*, UK overview document, September 2015.

14 Website of Air Resources Board: http://www.arb.ca.gov

15 Dynamometer Drive Schedules webpage of USA EPA: http://www3.epa.gov/nvfel/testing/dynamometer.htm

16 Test car list data files webpage of USA EPA: https://www.epa.gov/compliance-and-fuel-economy-data/data-cars-used-testing-fuel-economy

17 US EPA, "Fuel Economy of Motor Vehicle Revisions to Improve Calculation of Fuel Economy Estimates," December 2006.

18 US EPA, "EPA Test Procedures for Electric Vehicles and Plug-in Hybrids," January 2015.

19 Fuel economy labeling by USA EPA: http://www.fueleconomy.gov

20 Fuel Economy Guide data at https://www.fueleconomy.gov/feg/download.shtml

21 Natural Gas for Cars, US Department of Energy, December 2015.

22 M. Wang and A. Elgowainy, *Well-to-Wheels GHG Emissions of Natural Gas Use in Transportation: CNGVs, LNGVs, EVs, and FCVs*, Argonne National Laboratory, 2014.
23 GREET at http://greet.es.anl.gov
24 H. C. Kim, T. J. Wallington, R. Arsenault, C. Bae, S. Ahn, and J. Lee, "Cradle-to-Gate Emissions from a Commercial Electric Vehicle Li-Ion Battery: A comparative analysis," *Environmental Science and Technology*, 2016, 50 (14), pp 7715–7722, doi: 10.1021/acs.est.6b00830, June 2016.
25 J. Koornneef, S. Nierop, H. Saehr, and F. Wigand, "International comparison of fossil fuel efficiency and CO_2 intensity – Update 2015 Final Report," Ecofys, 2015.
26 Toyota Motor Company, *Toyota's Fuel Cell Vehicle Achievements and Pathway to Commercialization*, 2010.

Further Reading

1 Sony Pictures, *Who Killed the Electric Car*, 2006.
2 N. Cawthorne, *Tesla vs. Edison: The Life-long Feud that Electrified the World*, Chartwell Books, 2016.
3 T. Barboza and J. Schleuss, "L.A. keeps building near freeways, even though living there makes people sick," *Los Angeles Times*, March 2, 2017.
4 Website of University of Southern California (USC) Environmental Health Centers, www.envhealthcenters.usc.edu.
5 H. Chen, J. C. Kwong, R. Copes, K. Tu, P. J. Villeneuve, A. van Donkelaar, P. Hystad, R. V. Martin, B. J. Murray, B. Jessiman, A. S. Wilton, A. Kopp, and R. T. Burnett, "Living near major roads and the incidence of dementia, Parkinson's disease, and multiple sclerosis: A population-based cohort study," *The Lancet*, Online *January* 4, 2017.

Problems

1.1 Balance the following equation for the combustion of methane:

$$CH_4 + \square\, O_2 \rightarrow \square\, CO_2 + \square\, H_2O$$

1.2 Calculate the CO_2 emissions from the combustion of a kilogram of (a) methane and (b) diesel to verify the numbers presented in Table 1.1.

1.3 The 2014 Ford Focus BEV with a 24 kWh battery pack has upstream emissions of 68 gCO_2/km (110 gCO_2/mile) in Los Angeles, 162 gCO_2/km (260 gCO_2/mile) in Detroit, and 118 gCO_2/km (190 gCO_2/mile) as an average in the United States. The gasoline 2.0 L automatic 2014 Ford Focus has total emissions of 221 gCO_2/km (356 gCO_2/mile).

i) How many kilometers must be driven for an electric Ford Focus to result in lower carbon emissions than the gasoline Ford Focus in (a) Los Angeles,

(b) Detroit, and (c) the United States, if the cradle-to-grave emissions of CO_2 for the electric car and gasoline car are 10,300 kg and 7,500 kg, respectively.

ii) If each car is driven for the US annual average of 21,700 km per year over eight years, determine the combined total for the cradle-to-grave and tailpipe and upstream emissions for the three locations using electric and gasoline.

iii) Norway has electricity emissions of about 50 times less than the United States, or about 2.5 gCO_2/km. What are the combined cradle-grave and upstream emissions if the BEV is driven in Norway for 21,700 km per year over eight years?

iv) A representative HEV has cradle-to-grave emissions of 7,500 kg and total emissions of 118 gCO_2/km. Determine the combined total for the cradle-to-grave and tailpipe and upstream emissions for the various locations using the electric, hybrid and gasoline vehicles when driving 12,000 km per year over eight years.

[Ans. (i) (a) 18,300 km, (b) 47,460 km, (c) 27,180 km, (ii) LA 22,100 kg, Detroit 38,400 kg, United States 30,800 kg, gasoline 45,900 kg, (iii) Norway 10,734 kg, (iv) LA 16,828 kg, Detroit 25,852 kg, USA 21,628 kg, Norway 10,540 kg, gasoline 28716 kg, HEV 18,828 kg]

Assignments

1.1 Research the issues driving transportation emissions and fuel economy in the region where you live. What do you think are the motivating factors: security of supply, environment (GHG or pollution), and or economics?

1.2 Research the well-to-wheel efficiencies in your region. What is the optimum mode of transport if you are to minimize carbon emissions in the overall cycle?

1.3 Research the energy requirements of your home and the options for energy supply and conversion. How do these options differ from each other in terms of performance, efficiency, carbon footprint, and cost?

1.4 Research the provision of electricity in your region. What are the sources? Estimate the carbon emissions in gCO_2/kWh (electrical). Reference Example 1.6.2.1.

1.5 Which vehicle in the EPA database is closest to your vehicle? Which drive cycle most closely represents your driving patterns?

1.6 Research the health issues affecting your community due to automotive pollution.

2

Vehicle Dynamics

"Everybody perseveres in its state of rest, or of uniform motion in a right line, unless it is compelled to change that state by forces impressed thereon."

"The alteration of motion is ever proportional to the motive force impressed; and is made in the direction of the right line in which that force is impressed."

"To every action there is always opposed an equal reaction: or the mutual actions of two bodies upon each other are always equal, and directed to contrary parts."
Eighteenth-century translations of Newton's laws from his book *Principia* (written in Latin and first published in 1687).

"I can calculate the motion of heavenly bodies, but not the madness of people."
Isaac Newton (1643–1727).

In this chapter, we investigate the power and energy requirements of the automotive powertrain by applying the fundamentals of physics to the vehicle's motion. First, the basic vehicle load forces of aerodynamic drag, rolling resistance, and climbing resistance are considered. Vehicle acceleration is then quantified. This initial study enables the reader to quantify the power and energy requirements of a vehicle, and to convert these vehicle requirements of speed and accelerating or braking power into mechanical specifications of torque and speed for the electric motor or internal combustion (IC) engine.

2.1 Vehicle Load Forces

Vehicle drive requirements and performance specifications must be understood in order to develop the electric powertrain. In this section, the main load forces of aerodynamic drag F_D, rolling resistance F_R, and climbing resistance acting on the vehicle F_c, as shown in Figure 2.1, are considered in Sections 2.1.2, 2.1.3, and 2.1.6, respectively.

Electric Powertrain: Energy Systems, Power Electronics and Drives for Hybrid, Electric and Fuel Cell Vehicles,
First Edition. John G. Hayes and G. Abas Goodarzi.

Companion website: www.wiley.com/go/hayes/electricpowertrain

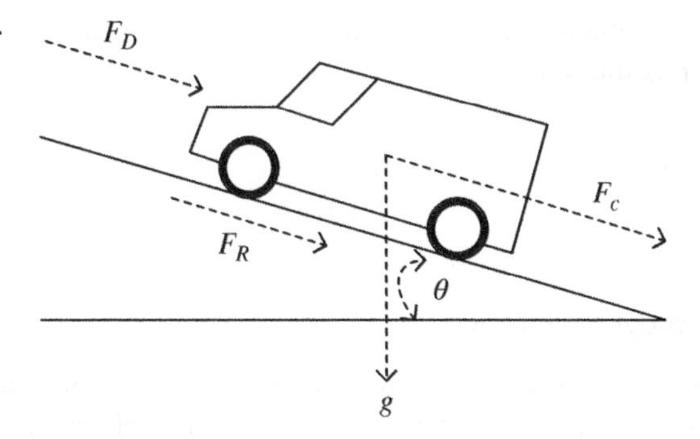

Figure 2.1 Vehicle load forces.

These forces will next be discussed, but first we note some basic relationships.

2.1.1 Basic Power, Energy, and Speed Relationships

Power is defined as work done per second. The unit of power is the watt (symbol W). The unit is named after James Watt (1736–1819), a Scottish inventor who contributed greatly to the development of steam engines and the dawning of the Industrial Revolution.

If a vehicle is traveling at a constant speed v, then the power P required to propel it is equal to the product of the force F and the speed. In equation form:

$$\text{power} = \text{force} \times \text{speed} \tag{2.1}$$

$$P = Fv = F\frac{s}{t} \tag{2.2}$$

where s is the distance, and t is the time required to travel the distance. The metric units for distance, time, and speed are the meter (m), the second (s), and meters/second (m/s), respectively.

The unit of force is the newton (N), named after one of the great minds of science and physics. Isaac Newton (1643–1727) was an English scientist whose monumental findings and writings make him one of the most influential scientists of all time.

An object is said to possess energy if it can do work. The unit of energy and work is the joule (symbol J). The unit is named after the English physicist James Joule (1818–1889), who made important contributions to the science of energy conversion.

The energy E required to propel the vehicle at a constant speed is simply the product of power and time:

$$\text{energy} = \text{power} \times \text{time} \tag{2.3}$$

$$E = Pt = P\frac{s}{v} \tag{2.4}$$

The above equation can be rewritten to express the distance in terms of energy, speed, and power as follows:

$$s = \frac{Ev}{P} \tag{2.5}$$

2.1.2 Aerodynamic Drag

Aerodynamic drag is the resistance of air to the movement of the vehicle. The aerodynamic drag force F_D and power P_D acting on the vehicle are defined as

$$F_D = \frac{1}{2}\rho C_D A (v + v_{air})^2 \tag{2.6}$$

and

$$P_D = F_D v \tag{2.7}$$

where ρ is the density of air, C_D is the aerodynamic drag coefficient, A is the cross-sectional area of the vehicle, v is the vehicle speed in m/s, and v_{air} is the wind speed in m/s.

From Equation (2.7) it can be seen that the aerodynamic drag power is a function of the cube of the speed, and this load generally is the most significant load at high speed. Thus, at high speeds:

$$P_D \propto v^3 \tag{2.8}$$

The energy E_D required to overcome the high-speed drag is directly proportional to the aerodynamic drag force, or the square of the speed. Thus, at high speeds it can be approximated by

$$E_D \propto v^2 \tag{2.9}$$

For example, if the vehicle speed is doubled, the energy required goes up by a factor of four if we consider only the drag force. Similarly, when dominated by drag, the distance which a vehicle can travel for a given energy storage is inversely proportional to the square of the speed:

$$s \propto \frac{1}{v^2} \tag{2.10}$$

The drag force increases with the cross-sectional area of the vehicle and also increases or decreases depending on whether the vehicle is experiencing a headwind or a tailwind. A tailwind results in a negative speed and reduced drag, and the net air velocity relative to the vehicle ($v - v_{air}$) is reduced, whereas a headwind results in a positive speed and increases the drag, as air has a velocity of ($v + v_{air}$) relative to the car.

The variation of power with the cube of the speed is a common relationship in aerodynamics. For example, wind turbines have a similar relationship with wind speed. Vehicle energy usage when cruising at high speeds is dominated by drag loss.

The density of air is a function of temperature, humidity, and altitude, and is specified as 1.204 kg/m^3 at 20°C and a standard atmospheric pressure of 101 kPa at sea level. Air density can have a reasonably large variation over the operating temperature range of a vehicle. The density of air varies from 1.514 kg/m^3 at −40°C to 1.293 kg/m^3 at 0°C to 1.127 kg/m^3 at

+40°C. Thus, additional power and energy are required to propel a vehicle in cold weather. Moreover, vehicle motor oils can become more viscous (i.e., more thick and sticky) at cold temperatures, resulting in additional power and drive requirements at low temperatures.

The aerodynamic drag coefficient C_D is a function of the vehicle design and streamlining. The GM EV1 had an extremely low drag coefficient of 0.19, obtained by using aerodynamic features such as optimized air vents, a tapered "teardrop" body, and a staggered wheel base. Rather than using the standard whip antenna, the GM EV1 had the radio antenna integrated into the roof in order to further reduce the drag and increase the range. Air vents are required for air cooling of the engine of a conventional vehicle, and can increase the drag coefficient, but are less of a problem for an electric vehicle compared to a conventional vehicle as the electric vehicle is significantly more efficient and requires less cooling.

A predecessor of the GM EV1, known as the Impact, set an EV land-speed record of 296 km/h (184 mph) in 1993. This particular vehicle had a number of modifications in order to enable it to run at high speeds with reduced drag. The drag was reduced by removing the side-view mirrors, lowering the ground clearance, covering the wiper-blade cavity, attaching smooth disks to the wheels, and adding an aerodynamic tail cone. Additional batteries were included, and the gearing was changed in order to achieve more speed. Finally, ice was added to the cooling loop to keep the motor and inverter electronics cool. While modifications such as these are impractical for everyday driving, they do lend themselves for consideration in vehicle design.

Vehicle parameters for various battery electric vehicles (BEVs), hybrid electric vehicles (HEVs), plug-in hybrid electric vehicles (PHEVs), and fuel cell electric vehicles (FCEV) vehicles are shown in Table 2.1. The published parameter values are presented where available. The various parameters in Table 2.1 are discussed later in this section. The curb weight is the weight of the vehicle without occupants or baggage. The Environmental Protection Agency (EPA) test weight is the weight of the vehicle, including occupants and baggage, when being tested for the EPA tests discussed in Section 2.1.4.

The lowest drag coefficients for these recent production vehicles are for the Tesla Model S and the Toyota Prius at 0.24 and 0.25, respectively. The Nissan Leaf and Chevy Volt follow closely at 0.28. The Toyota Mirai is specified at 0.29. The Lexus RX 450h is included as it is a sports-utility vehicle (SUV) with a higher drag coefficient and greater weight.

2.1.2.1 Example: Aerodynamic Drag

An electric vehicle has the following attributes: drag coefficient $C_D = 0.25$, vehicle cross section $A = 2\ \text{m}^2$, and available propulsion energy of $E_b = 20$ kWh (1 kWh = 3.6×10^6 J). Let the density of air $\rho_{air} = 1.2\ \text{kg m}^{-3}$.

Instantaneously at a vehicle speed of 120 km/h, calculate the aerodynamic drag force, power, and range, while driving in (a) calm conditions with no wind and (b) windy conditions with a 12 km/h headwind.

Solution:

The vehicle speed v in m/s is obtained by dividing the speed in km/h by 1000/3600 or 3.6:

$$v = 120\frac{\text{km}}{\text{h}} = 120 \times \frac{1000\,\text{m}}{3600\,\text{s}} = \frac{120}{3.6}\,\text{m/s} = 33.33\,\text{m/s}$$

Table 2.1 Vehicle specifications.

Parameters	Symbol	GM EV1	Nissan Leaf	Tesla Model S	Toyota Prius	Lexus RX 450h	Chevy Volt	Toyota Mirai
Model year		1996	2015	2014	2015	2015	2015	2016
Vehicle type		BEV	BEV	BEV	HEV	HEV	PHEV	FCEV
Model			S	85D				
Drag coefficient	C_D	0.19	0.28	0.24	0.25	0.33	0.28	0.29
Rolling resistance coefficient	C_R		0.0083	0.0084	0.0055	0.0064	0.0065	0.0076
Curb weight (kg)		1400	1477	2100	1365	2091	1720	1850
EPA test weight (kg)	m		1645	2155	1531	2268	1814	1928
Rated power (kW)	P_r	100	80	270	73	183	111	113
Rated torque (Nm)	T_r	150	254	440				335
Max. speed (km/h) [mph]		129 [80]	144 [90]	224 [140]	180 [112]	180 [112]	160 [100]	180 [112]
0–60mph (s) *0–100 km/h		8.5	11.5	5.4	*10.4	7.7	9.2 (e-only)	*9.6
A (N)	A		133.3	177.2	82.3	141.8	115.9	143.8
B (N/ms^{-1})	B		0.7094	1.445	0.222	3.273	−0.119	1.990
C (N/m^{2}s^{-2})	C		0.491	0.354	0.403	0.569	0.405	0.407
Gear ratio	n_g	10.9	8.19	9.73	CVT	CVT	CVT	
Wheel radius (m)	r	0.292	0.315	0.352	0.313	0.370	0.334	0.334

a) Under calm conditions, the force, power, and range are given by

$$F_D = \frac{1}{2}\rho C_D A v^2 = \frac{1}{2} \times 1.2 \times 0.25 \times 2 \times 33.33^2\,\text{N} = 333\,\text{N}$$

$$P_D = F_D v = 333 \times 33.33\,\text{W} = 11.1\,\text{kW}$$

$$s = \frac{E_b v}{P_D} = \frac{20 \times 3.6 \times 10^6 \times 33.33}{11.1 \times 10^3}\,\text{m} = 216\,\text{km}$$

b) The wind speed is calculated to be

$$v_{air} = \frac{12}{3.6}\,\text{m/s} = 3.33\,\text{m/s}$$

The headwind reduces the range as the drag force and power increase to

$$F_D = \frac{1}{2}\rho C_D A (v + v_{air})^2 = \frac{1}{2} \times 1.2 \times 0.25 \times 2 \times (33.33 + 3.33)^2\,\text{N} = 403\,\text{N}$$

$$P_D = F_D v = 403 \times 33.33\,\text{W} = 13.4\,\text{kW}$$

while the range drops to

$$s = \frac{E_D v}{P_D} = \frac{20 \times 3.6 \times 10^6 \times 33.33}{13.4 \times 10^3} \text{m} = 179\,\text{km}$$

2.1.2.2 Example: Aerodynamic Drag and Fuel Consumption

Determine the percentage increase in the required power and fuel consumption when a driver who normally drives at 120 km/h increases the speed to 150 km/h. Consider drag forces only.

Solution:

Since the drag power is related to the cube of the speed, the ratio of the drag powers at the two speeds is given by

$$\frac{P_D(150\,\text{km/h})}{P_D(120\,\text{km/h})} \approx \left(\frac{150\,\text{km/h}}{120\,\text{km/h}}\right)^3 = 1.95 \Rightarrow P_D(150\,\text{km/h}) = 1.95 P_D(120\,\text{km/h})$$

Thus, the power required at 150 km/h is 95% greater than the power at 120 km/h.

Similarly, from Equation (2.9), the fuel consumption is related to the square of the speed. Thus,

$$\frac{E_D(150\,\text{km/h})}{E_D(120\,\text{km/h})} \approx \left(\frac{150}{120}\right)^2 = 1.56 \Rightarrow E_D(150\,\text{km/h}) = 1.56 E_D(120\,\text{km/h})$$

Thus, the energy required at 150 km/h is 56% greater than at 120 km/h.

2.1.3 Rolling Resistance

The rolling resistance is the combination of all frictional load forces due to the deformation of the tire on the road surface and the friction within the drivetrain. The rolling resistance F_R is described by the equation

$$F_R = C_R m g \tag{2.11}$$

where m is the mass of the vehicle, g is the acceleration due to gravity, nominally 9.81 m/s^2, and C_R is the coefficient of rolling resistance.

The weight of the vehicle directly affects the rolling resistance. The coefficient of rolling resistance tends to be relatively constant at low speeds. Although it increases at high speeds, the effects are not as significant, as the vehicle losses are dominated by drag.

EVs use high-pressure tires in order to minimize the rolling resistance. A typical value for an EV tire is 0.01 or less. Estimates of C_R are presented in Table 2.1. These estimations are based on test parameters discussed in the following section – coefficient A is simply divided by mg in order to get an estimation of C_R.

Note that the energy dissipated by the rolling resistance increases the tire temperature and pressure.

2.1.3.1 The Ford Explorer Recall

It is interesting to note the role that rolling resistance played in the Ford Explorer vehicle recall in the late 1990s [1]. The Ford Explorer is an SUV and consequently has a relatively high center of gravity. A high center of gravity can make a vehicle more prone to roll over in the case of a sharp or emergency turn. Ford specified a tire pressure lower than that recommended by the tire manufacturer in order to mitigate the effects of rollover and improve road holding. The lower tire pressure would result in greater rolling resistance, but could cause tire overheating and delamination in hot climates, resulting in tire blowouts with fatal consequences.

2.1.3.2 The A-Class Mercedes in the 1990s

A similar problem occurred with the Mercedes A-Class sedan during its early testing in the 1990s. The center of gravity was relatively high, and the vehicle was prone to roll over in an emergency turn. The problem was first publicized after failing the (swerve-to-avoid-a) moose test in Sweden. The problem was ultimately corrected by modifying the car design to have a lower center of gravity.

2.1.3.3 The Tesla Model S in 2013

Similarly, the early Tesla Model S has an active suspension to raise and lower the car. Lowering the car results in a lower drag coefficient and improves road handling. However, the battery pack is arranged along the floor of the vehicle, and becomes vulnerable if the vehicle floor is penetrated, for example, due to debris on the road. There were a small number of related accidents in 2013, and the problem was corrected by a number of fixes, including reinforcing the vehicle floor using a titanium plate [2].

2.1.3.4 Example: Rolling Resistance

An electric vehicle has the following attributes: coefficient of rolling resistance C_R = 0.0085 and vehicle mass of 2000 kg. Instantaneously at a vehicle speed of 10 km/h, calculate the rolling resistance and power. Let $g = 9.81\ \text{m/s}^2$.

Solution:

The rolling resistance is simply calculated from Equation (2.11):

$$F_R = C_R mg = 0.0085 \times 2000 \times 9.81\ \text{N} = 167\ \text{N}$$

The vehicle speed in m/s is

$$v = \frac{10}{3.6}\ \text{m/s} = 2.77\ \text{m/s}$$

The rolling resistance power is

$$P_R = F_R v = 167 \times 2.77\ \text{W} = 463\ \text{W}$$

2.1.4 Vehicle Road-Load Coefficients from EPA Coast-Down Testing

Significant testing is conducted by vehicle manufacturers for the various regulatory agencies. In addition to fuel economy and emissions data, the EPA in the United States requires the manufacturers to supply data on vehicle road-load coefficients based on the "coast-down" test [3]. In this test, the vehicle is allowed to coast down from 120 km/h in neutral, and three coefficients are generated to simulate the rolling, spinning, and aerodynamic resistances.

The vehicle road-load force F_v as a function of speed, is defined as follows:

$$F_v = A + Bv + Cv^2 \tag{2.12}$$

where v is the vehicle speed in m/s, and the A, B, and C coefficients are determined from the coast-down test. Typically, coefficient A correlates to the rolling resistance and coefficient C to the aerodynamic drag. Coefficient B relates to the spinning or rotational losses and tends to be relatively small. Using the coast-down coefficients to predict vehicle road loads is more accurate than simply using values for the rolling resistance and the drag forces. The coefficients are provided in the EPA database in units of pound force (lbf) and mph. These coefficients can be converted using the factors from Table 2.2 to obtain the metric values with units of N and m/s. The values for the 2015 Nissan Leaf are used as an example in Table 2.2.

A plot of the vehicle road-load force as a function of the speed is shown in Figure 2.2 for five electrically propelled vehicles: the 2015 Nissan Leaf S, the 2015 Toyota Prius, the 2015 Lexus RX 450h SUV, the 2015 Chevrolet Volt, and the 2014 Tesla Model S. As shown in the plots, the Lexus SUV has the highest vehicle road-load force above 30 km/h. This is to be expected given the profile of the SUV compared to its smaller sibling, the Toyota Prius, which understandably has a significantly lower road load.

Table 2.2 2015 Nissan Leaf coast-down coefficients in US and metric units.

Coefficient	Value	Unit	Conversion	Value	Unit
A	29.97	lbf	4.448	133.3	N
B	0.0713	lbf/mph	9.950	0.7094	N/ms^{-1}
C	0.02206	lbf/(mph)^2	22.26	0.491	N/(ms^{-1})^2

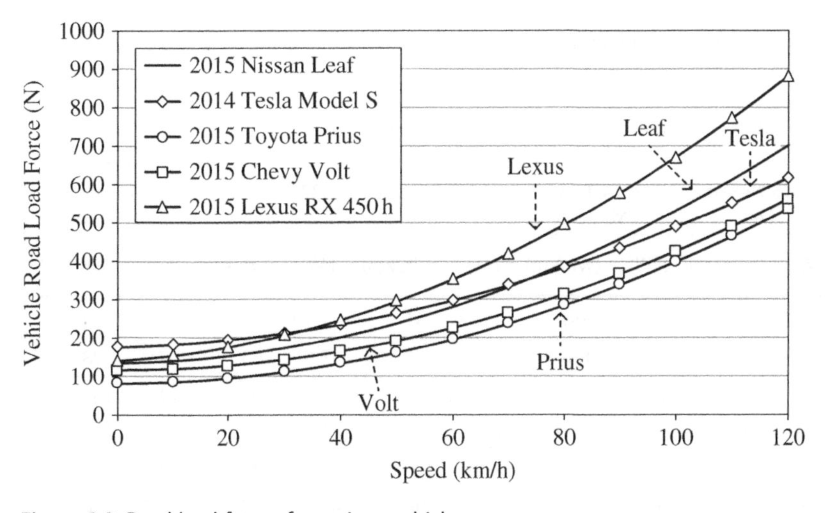

Figure 2.2 Road-load forces for various vehicles.

The Chevy Volt has a slightly higher load force than the Toyota Prius but lower than the Tesla Model S, which has a lower load force than the Nissan Leaf above 80 km/h.

The vehicle road power P_v required can be determined by multiplying the road-load force by the speed:

$$P_v = F_v v = Av + Bv^2 + Cv^3 \tag{2.13}$$

The vehicle road powers for the earlier vehicles are plotted as a function of the speed in Figure 2.3. As can be seen, the power can increase substantially depending on the vehicle. The Tesla Model S requires almost 100 kW in order to overcome road loads at its maximum speed. However, and more importantly, the road power for the Model S is substantially lower for typical road speeds below 100 km/h.

Typical road powers are presented in Table 2.3 for the various vehicles. Clearly, there is a significant variation across the vehicles, especially between the smaller and larger

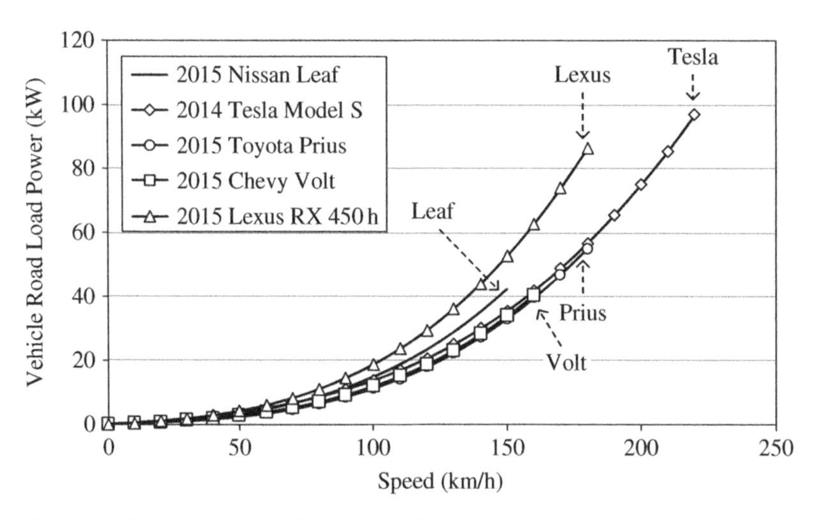

Figure 2.3 Road-load power for various vehicles.

Table 2.3 Vehicle road power and energy consumption examples.

Vehicle (km/h)	2015 Nissan Leaf (kW)	2015 Nissan Leaf (Wh/km)	2014 Tesla Model S (kW)	2014 Tesla Model S (Wh/km)	2015 Toyota Prius (kW)	2015 Toyota Prius (Wh/km)	2015 Chevy Volt (kW)	2015 Chevy Volt (Wh/km)	2015 Lexus RX 450 h (kW)	2015 Lexus RX 450 h (Wh/km)
30	1.44	48	1.78	59	0.93	31	1.19	40	1.74	58
60	4.7	78	5.0	83	3.3	55	3.8	63	5.9	98
90	11.4	127	10.9	121	8.5	94	9.2	102	14.5	161
120	23.4	195	20.6	172	17.9	149	18.7	156	29.4	245
150			35.5	237	33.0	220	33.9	226	52.8	352
Maximum	37.9	263	96.9	433	55.1	306	40.5	253	86.4	480

vehicles at high speeds. Again, the Lexus RX 450h consumes significantly more power than its smaller sibling, the Toyota Prius. The energy consumption in Wh/km is easily calculated and is also worth noting for the various vehicles. The energy consumption is calculated by simply dividing the kilowatt-hours consumed by one hour's driving by the distance driven in that hour. Equation (2.5) can be rearranged to show that the fuel consumption is

$$FC = \frac{E}{s} = \frac{P}{v} \tag{2.14}$$

For example, when driving at 60 km/h the Nissan Leaf consumes 4.7 kWh to drive 60 km for an energy consumption of 4700 Wh/60 km or 78 Wh/km.

2.1.5 Battery Electric Vehicle Range at Constant Speed

The electric vehicle range can be estimated once the load power is known. An estimation of range is quite complex and depends on many factors. However, a simple estimation can be made based on a few basic assumptions, as done in the next example.

2.1.5.1 Example: Plot of BEV Range Versus Speed

A high-performance vehicle has an available battery energy of 90 kWh. Let the efficiency of the powertrain from the battery to the transmission be 85%. An estimate of the vehicle range based on the road-load power and range formulas of Equations (2.13) and (2.14) and the vehicle parameters of the Tesla Model S of Table 2.1 is plotted as a function of speed in Figure 2.4. The approximate range drops from about 800 km at 70 km/h to 360 km at 140 km/h.

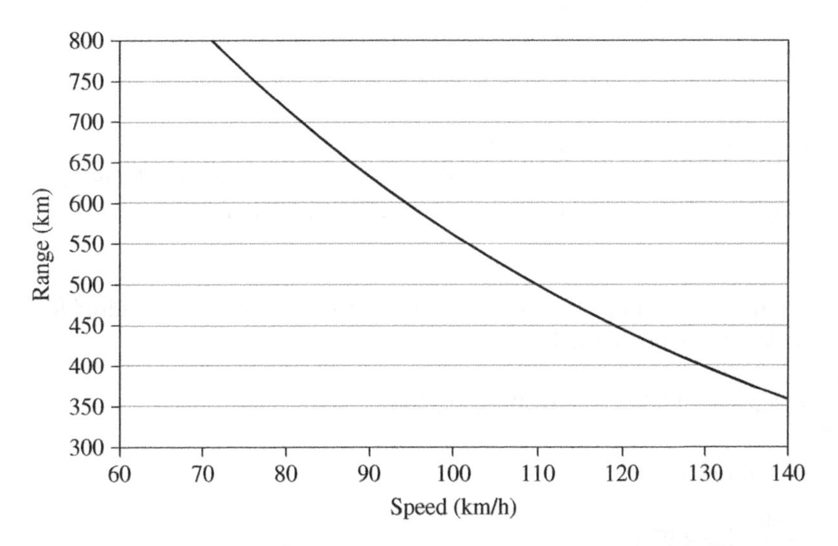

Figure 2.4 Plot of high-performance BEV range at constant cruising speed.

2.1.5.2 Example: Estimate of BEV Range

Estimate the range of the above high-performance vehicle at 120 km/h.

Solution:

The vehicle speed v in m/s is given by

$$v = \frac{120}{3.6}\,\text{m/s} = 33.33\,\text{m/s}$$

The road-load power is given by

$$P_v = Av + Bv^2 + Cv^3$$

$$= 177.2 \times 33.33\,\text{W} + 1.445 \times 33.33^2\,\text{W} + 0.354 \times 33.33^3\,\text{W}$$

$$= 5.906\,\text{kW} + 1.605\,\text{kW} + 13.107\,\text{kW} = 20.618\,\text{kW}$$

The battery power P_b is the road-load power divided by the electric powertrain efficiency η_{pt}:

$$P_b = \frac{P_v}{\eta_{pt}} = \frac{20.618}{0.85}\,\text{kW} = 24.256\,\text{kW}$$

Assuming a constant power draw, the time in hours for which the electric powertrain can source constant power from the battery is given by

$$t = \frac{E_b}{P_b} = \frac{90}{24.256}\,\text{h} = 3.71\,\text{h}$$

The vehicle can travel a distance s in this time at the constant speed v:

$$s = vt = 33.33\,\frac{\text{m}}{\text{s}} \times 3.71\,\text{h} \times 3600\,\frac{\text{s}}{\text{h}}$$

$$= 445\,\text{km}$$

2.1.5.3 Example: Effect of Auxiliary Loads on Range

Determine the reduction in range for the BEV in the previous example if the vehicle has a continuous heating, ventilation and air conditioning (HVAC) load of 6 kW.

Solution:

The HVAC load power P_{HVAC} of 6 kW is simply added to the battery power required for motion of 24.256 kW to estimate the battery power P_b:

$$P_b = \frac{P_v}{\eta_{pt}} + P_{HVAC} = 24.256\,\text{kW} + 6\,\text{kW} = 30.256\,\text{kW}$$

The driving time drops from 3.71 h to

$$t = \frac{E_b}{P_b} = \frac{90}{30.256}\,\text{h} = 2.97\,\text{h}$$

and the range drops from 445 km to

$$s = vt = 33.33\frac{\text{m}}{\text{s}} \times 2.97\,\text{h} \times 3600\frac{\text{s}}{\text{h}}$$
$$= 356\,\text{km}$$

2.1.6 Gradability

The vehicle load power can increase or decrease depending on whether the car is ascending or descending an incline. The climbing resistance or downgrade force is given by

$$F_c = mg\sin\theta \tag{2.15}$$

where θ is the angle of incline and g is the acceleration due to gravity. The climbing force is positive, resulting in motoring operation. The downgrade force is negative and can result in energy regeneration to the battery, a mode commonly used in electrically propelled vehicles rather than friction braking to slow the vehicle.

The gradability is the maximum slope that a vehicle can climb at a certain speed. In simple terms, it is the ratio of the rise to the run, or the tangent of the incline angle. It is often quoted as a percentage, with tan 45° being 100%. Some incline angles and values are shown in Table 2.4.

The vehicle road-load power is plotted as a function of speed for the 2015 Nissan Leaf in Figure 2.5 for incline slopes of 0°, + 6°, and −6°. As shown in the figure, the vehicle requires approximately 21 kW at 40 km/h and 80 kW at 130 km/h for a 6° incline. Significant additional power is available to be regenerated to recharge the battery of an EV when on a negative slope of −6°. A peak regenerative power of −33 kW is available at 120 km/h on the −6° slope. Recovering the regenerated energy rather than dissipating the energy in the brakes is a very significant advantage for EVs over conventional vehicles using standard non-energy-recovery braking systems.

2.1.6.1 Example: Downgrade Force and Regeneration

The 2015 Nissan Leaf is traveling down a −6° slope at 120 km/h (33.33 m/s). Assuming calm conditions, how much regenerative power is available to brake the vehicle while maintaining a constant speed?

Solution:

As the vehicle is required to cruise at a constant speed down the slope, any downgrade power greater than the road-load power can be regenerated back to the battery.

Table 2.4 Incline angles and grade.

Incline (°deg)	Grade (%)
0	0
6	10.5
45	100

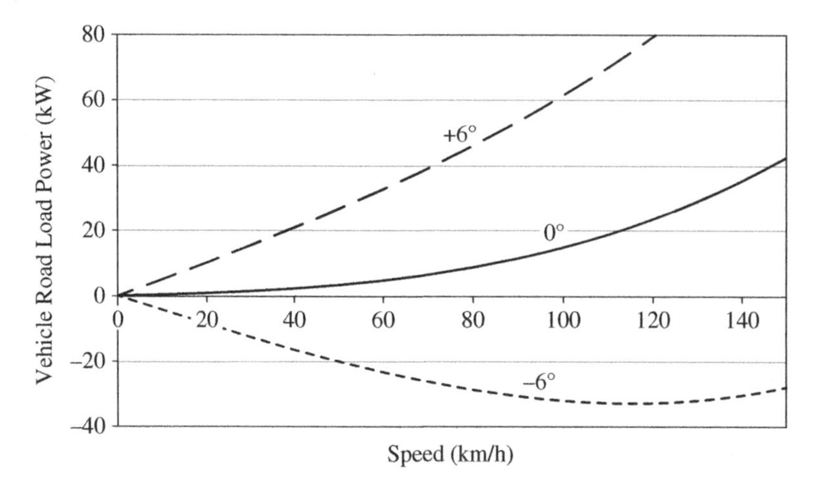

Figure 2.5 Nissan Leaf vehicle road-load power versus speed for various slopes.

The downgrade power is

$$P_c = mg\sin\theta \times v = 1645 \times 9.81 \times \sin(-6^\circ) \times 33.33\,\text{W} = -56.2\,\text{kW}$$

The road load is given by Table 2.3 as

$$P_v = 23.4\,\text{kW}$$

The resultant power P_{regen} can be regenerated back to the battery:

$$P_{regen} = P_c + P_v = -56.2\,\text{kW} + 23.4\,\text{kW} = -32.8\,\text{kW} \tag{2.16}$$

2.2 Vehicle Acceleration

Nominal vehicle power requirements are typically based on vehicle acceleration requirements, usually specified as the time to accelerate from 0 to 100 km/h (62 mph) or from 0 to 60 mph. Under these conditions, the maximum available torque and power of the propulsion system are likely to be required.

Per Newton's second law of motion for a linear system, the force required to accelerate or brake a vehicle F_a is given by

$$F_a = ma = m\frac{dv}{dt} \tag{2.17}$$

where a is the linear acceleration.

The **motive force** F_m required to accelerate the vehicle is the sum of the acceleration, load, and climbing forces and is given by

$$F_m = F_a + F_v + F_c \tag{2.18}$$

Thus, by substituting Equations (2.12), (2.15), and (2.17) into Equation (2.18), we can express the motive force as

$$F_m = m\frac{dv}{dt} + A + Bv + Cv^2 + mg\sin\theta \tag{2.19}$$

The motive force describes the force required for linear motion. The **motive torque** T_m is the torque required at the drive axle and is obtained by multiplying the motive force by the wheel radius r, thus relating the linear motion to rotating motion:

$$T_m = F_m r \tag{2.20}$$

From Newton's second law of motion for a rotating system, we can express torque as

$$T = J\alpha = J\frac{d\omega}{dt} \tag{2.21}$$

where J is the moment of inertia, and ω and α are the angular speed and acceleration.

In addition to the motive torque to move the vehicle, we must also consider the torque to spin the rotating parts within the drivetrain. This is represented by a drive-axle referenced moment of inertia, J_{axle}. Thus, the total torque required at the drive axle T_{axle} is the sum of the motive torque and the torque required to accelerate J_{axle}:

$$T_{axle} = T_m + J_{axle}\alpha_{axle} \tag{2.22}$$

where α_{axle} is the angular acceleration.

By substituting Equations (2.19) and (2.21) into Equation (2.22) and knowing that $v = r\omega$, the axle torque can be rewritten as

$$T_{axle} = r\left\{m\frac{dv}{dt} + (A + mg\sin\theta) + Bv + Cv^2\right\} + \frac{J_{axle}}{r}\frac{dv}{dt} \tag{2.23}$$

or

$$T_{axle} = r\left\{\left(m + \frac{J_{axle}}{r^2}\right)\frac{dv}{dt} + (A + mg\sin\theta) + Bv + Cv^2\right\} \tag{2.24}$$

The traction torque T_t is the torque developed on the output shaft of the IC engine or electric motor. The maximum traction motor or engine torque and power are typically specified by the manufacturer. These values are the maximum traction power and traction torque available on the shaft of the motor or engine. The traction torque is directly geared to the drive-axle torque. The powertrain gearing ratio n_g is specified by the manufacturer in the case of most cars, but is investigated further in Chapter 5 in the case of the hybrids using a continuously variable transmission (CVT). The gearing and transmission efficiency η_g can significantly impact the required motive torque. The axle torque is related to the traction torque T_t, when motoring, as follows:

$$T_{axle} = n_g\eta_g T_t \tag{2.25}$$

as shown in Figure 2.6(a).

The traction torque T_t can be expressed as follows:

$$T_t = \frac{r}{n_g\eta_g}\left\{\left(m + \frac{J_{axle}}{r^2}\right)\frac{dv}{dt} + (A + mg\sin\theta) + Bv + Cv^2\right\} \tag{2.26}$$

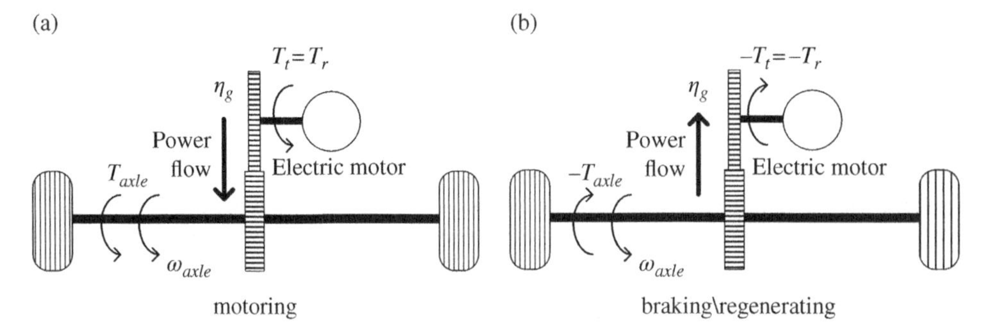

Figure 2.6 Vehicle (a) motoring and (b) regenerating.

The gear losses have a characteristic similar to other spinning components and have two main loss components: (1) the no-load or spinning loss and (2) the load-torque loss. The gearing efficiency in the above equation is an assumed value for the load-torque loss within the gearing and transmission, and ignores the no-load spinning loss. We are assuming that the traction motor and gearing are engaged during the coast-down test and that the *ABC* parameters capture the no-load spinning losses composed of the friction and drag (windage) losses of the motor, gearing, and other transmission components. In this work, the load-torque gear power loss is determined by assuming that it linearly scales with the gearing torque and speed to determine a gearing friction loss for any condition.

2.2.1 Regenerative Braking of the Vehicle

The energy used to brake or slow down a vehicle in a conventional vehicle is dissipated as heat in the braking system and lost to the vehicle. An electric vehicle can capture or regenerate the energy and store it on the vehicle.

The traction motor can develop a negative torque, up to the rated value in the forward direction, which reverses the flow of power such that the kinetic energy of the vehicle is converted to negative mechanical power on the rotor shaft, and subsequently converted to electrical power by the machine, which is used to recharge the battery. This is the principle of regenerative braking, as shown in Figure 2.6(b). The same theory as developed for acceleration applies, except that the efficiency of the gearing must be correctly accounted for as follows:

$$T_{axle} = \frac{n_g T_t}{\eta_g} \text{ (in regen)} \tag{2.27}$$

2.2.2 Traction Motor Characteristics

Given that the maximum traction machine power and torque are published, we can investigate the vehicle acceleration. The electric machines will be analyzed in depth later in the book, but a basic understanding is now required. We should note that the torque versus speed characteristic of an electric machine is a lot simpler and more linear than that of the IC engine, another attribute which makes the electric machine an attractive choice for the automotive powertrain.

As discussed in Chapter 6, the rated conditions of speed, torque, and power are the conditions for which the machine is designed, either continuously or intermittently with time. The conditions also depend on various environmental conditions, such as ambient and coolant temperatures. When discussing the electric powertrain, the rated conditions are typically for the maximum torque or power output.

The electric motor can be characterized by two modes of operation, constant-torque mode and constant-power mode.

In **constant-torque mode**, which is a low-speed mode, the machine can output a constant rated rotor torque $T_{r(rated)}$, and the rotor power P_r increases linearly with speed. Thus, the rotor torque T_r and rotor power P_r are limited as follows:

$$T_r = T_{r(rated)} \text{ and } P_r = T_{r(rated)}\omega_r \tag{2.28}$$

where ω_r is the angular frequency or speed of the rotor. This condition holds until the **rated speed** $\omega_{r(rated)}$ of the machine defined by

$$\omega_{r(rated)} = \frac{P_{r(rated)}}{T_{r(rated)}} \tag{2.29}$$

As presented here, the rated speed is also the base speed of the machine. The **base speed** of the machine is the minimum speed at which the machine can output rated power.

In **constant-power mode**, which is a high-speed mode, when operating above the rated speed, the machine can output a constant rated power $P_{r(rated)}$ and the available rotor torque decreases inversely with rotor speed.

$$P_r = P_{r(rated)} \text{ and } T_r = \frac{P_{r(rated)}}{\omega_r} \tag{2.30}$$

2.2.2.1 Example: 2015 Nissan Leaf Rated Speed

The rated motor torque and power of the electric motor of the 2015 Nissan Leaf are 254 Nm and 80 kW, respectively. Determine the rated speeds of the electric motor and vehicle.

Solution:

The rated speed of the motor is

$$\omega_{r(rated)} = \frac{P_{r(rated)}}{T_{r(rated)}} = \frac{80,000}{254}\text{Nm} = 314.96\,\text{rad/s}$$

Given the gear ratio of 8.19 and a wheel radius of 0.315 m from Table 2.1, the vehicle speed at which the motor reaches the rated speed is given by

$$v = r\frac{\omega_{r(rated)}}{n_g} = 0.315 \times \frac{314.96}{8.19}\text{m/s} = 12.11\,\text{m/s}\,(= 43.61\,\text{km/h or } 27.1\,\text{mph})$$

Table 2.5 is compiled on the basis of the nominal 2015 Nissan Leaf characteristics. Values for the rotor frequency f_r in hertz (Hz) and the rotor speed N_r in revolutions per minute (rpm) are also included and are calculated as follows.

$$f_r = \frac{\omega_r}{2\pi} \text{ and } N_r = 60f_r \tag{2.31}$$

The key characteristic operating points from Table 2.5 are used to generate the torque and power versus speed curve shown in Figure 2.7. The related MATLAB code is provided at the end of this chapter.

The constant-torque and constant-power modes of operation of an electric motor provide it a significant advantage over an IC engine, which has a significantly different torque and power versus speed characteristic, as is discussed in Chapter 5, Section 5.1.

Extremely fast accelerations are often publicized for electric vehicles. This capability is due to the ability of the electrical machine to output maximum torque at standstill or low speeds.

Table 2.5 Simplified characteristics for 2015 Nissan Leaf traction motor.

	Unit	Rated speed	Maximum speed
Vehicle speed, v	km/h	43.61	144
	m/s	12.11	40
Rotor angular speed, ω_r	rad/s	314.96	1040
Rotor frequency, f_r	Hz	50.13	165.52
Rotor rpm, N_r	rpm	3008	9931

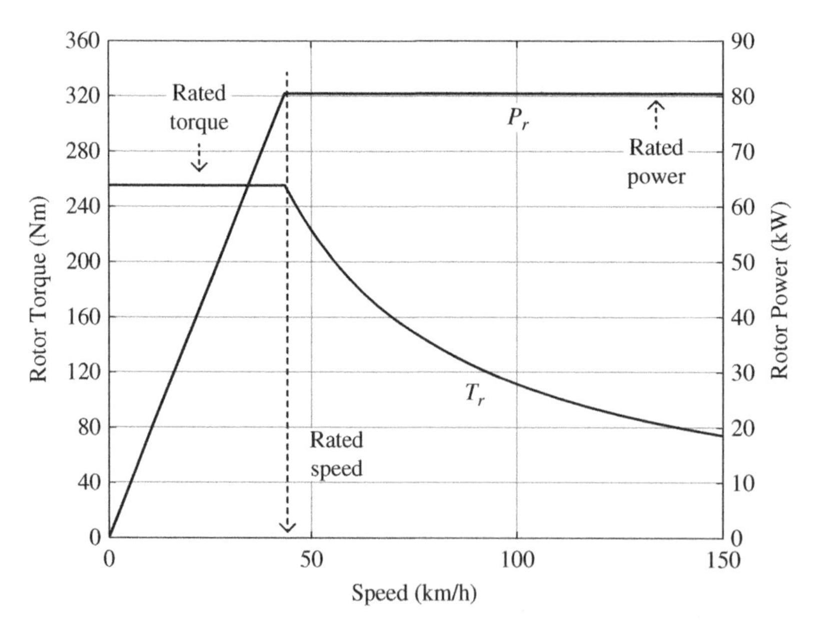

Figure 2.7 Nissan Leaf nominal power and torque characteristics.

2.2.3 Acceleration of the Vehicle

2.2.3.1 Time-Step Estimation of Vehicle Speed

As the key vehicle drive and propulsion parameters are known, the vehicle acceleration can easily be modeled using any software or spreadsheet program. Please refer to Chapter 15 for a discussion of using MATLAB\Simulink to model and control the vehicle.

Equation (2.26) can be rewritten as a differential equation:

$$\frac{dv}{dt} = \left\{n_g\eta_g T_r - r\left[(A + mg\sin\theta) + Bv + Cv^2\right]\right\} \Big/ \left(rm + \frac{J_{axle}}{r}\right) \tag{2.32}$$

where it is assumed that the accelerating torque is from the rotor of an electrical machine ($T_t = T_r$).

This differential equation is easily numerically solved by calculating the differential change in speed for a small differential time step Δt. Let

$$\frac{dv}{dt} = \frac{v(n+1) - v(n)}{\Delta t} \tag{2.33}$$

where $v(n)$ and $v(n+1)$ are the vehicle speeds at the nth and the $(n+1)$th steps, respectively.

Thus,

$$v(n+1) = v(n) + \Delta t\left\{n_g\eta_g T_r - r\left[(A + mg\sin\theta) + Bv + Cv^2\right]\right\} \Big/ \left(rm + \frac{J_{axle}}{r}\right) \tag{2.34}$$

The above equation is solved using the MATLAB code in the appendix and the acceleration profile is generated for the 2015 Nissan Leaf, as shown in Figure 2.8. From the simulation, the code predicts a value of about 10.8 s, which is very close to the published test value of 11.5 s [4].

2.2.3.2 A Simplified Equation Set for Characterizing Acceleration by Ignoring Load Forces

In this section, we develop a simplified closed-form equation for vehicle acceleration by neglecting the load forces. If the vehicle load forces and inertia are ignored, then Equation (2.32) reduces to

$$\frac{dv}{dt} = \frac{n_g\eta_g}{rm} T_r \Rightarrow dv = \frac{n_g\eta_g}{rm} T_r dt \tag{2.35}$$

Let us consider the two main modes of operation for the acceleration characteristic: constant torque from zero to the rated speed, and constant power above the rated speed. By integrating both sides of Equation (2.35), we get a solution for a particular mode:

$$\int_{v_i}^{v_f} dv = \frac{n_g\eta_g}{rm} \int_{t_i}^{t_f} T_r(v)dt \tag{2.36}$$

where v_i and v_f are the initial and final speeds at the initial and final time instances t_i and t_f, respectively.

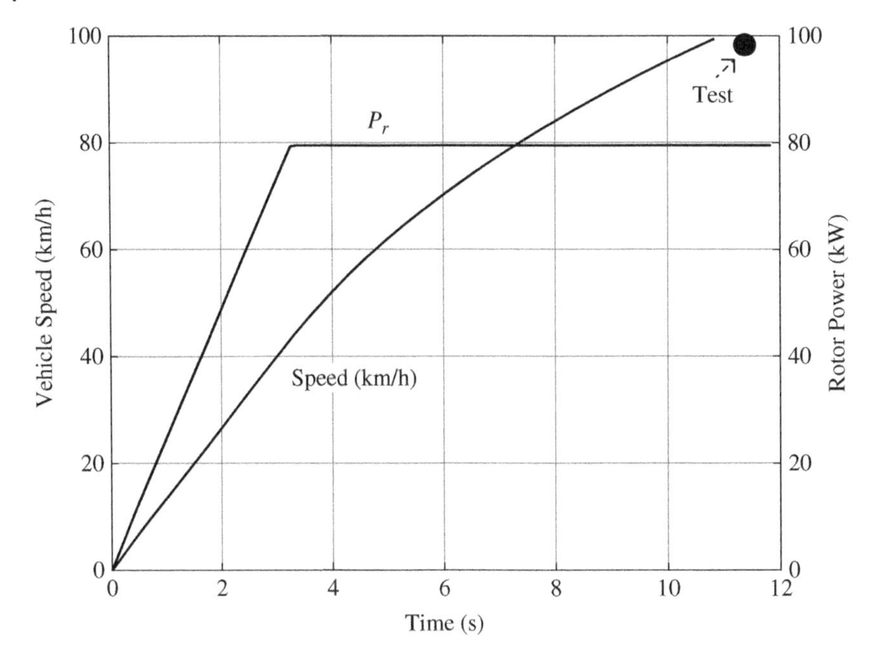

Figure 2.8 Acceleration profile for Nissan Leaf modeled with road load.

Below the rated speed, the maximum torque for acceleration is

$$T_r(v) = T_{r(rated)} \tag{2.37}$$

The integration of Equation (2.36) for constant-torque mode results in

$$\int_{v_i}^{v_f} dv = [v]_{v_i}^{v_f} \quad \text{and} \quad \frac{n_g \eta_g T_{r(rated)}}{rm} \int_{t_i}^{t_f} 1\,dt = \frac{n_g \eta_g T_{r(rated)}}{rm} [t]_{t_i}^{t_f} \tag{2.38}$$

$$\Rightarrow v_f - v_i = \frac{n_g \eta_g T_{r(rated)}}{rm} (t_f - t_i)$$

When starting from zero speed, $v_i = 0$, at an initial time, $t_i = 0$, and accelerating to the rated speed v_{rated} in time t_{rated}, the above integration reduces to

$$v_{rated} = \frac{n_g \eta_g}{rm} T_{r(rated)} t_{rated} \tag{2.39}$$

Thus, the time to accelerate from zero to the rated speed is

$$t_{rated} = \frac{rm v_{rated}}{n_g \eta_g T_{r(rated)}} \tag{2.40}$$

Above the rated speed, when operating in the constant-power mode, the maximum machine torque is

$$T_r(v) = \frac{P_{r(rated)}}{\omega_r} = \frac{T_{r(rated)} \omega_{r(rated)}}{\omega_r} = \frac{T_{r(rated)} v_{rated}}{v} \tag{2.41}$$

since $v = r\omega_r / n_g$.

The integral form for the constant-power mode is

$$\int_{v_i}^{v_f} v dv = \frac{n_g \eta_g}{rm} T_{r(rated)} v_{rated} \int_{t_i}^{t_f} 1 dt \tag{2.42}$$

which has the following solutions:

$$\int_{v_i}^{v_f} v dv = \left[\frac{1}{2} v^2\right]_{v_i}^{v_f} \quad \text{and} \quad \frac{n_g \eta_g T_{r(rated)} v_{rated}}{rm} \int_{t_i}^{t_f} 1 dt = \frac{n_g \eta_g T_{r(rated)} v_{rated}}{rm} [t]_{t_i}^{t_f} \tag{2.43}$$

$$\Rightarrow \frac{1}{2} v_f^2 - \frac{1}{2} v_i^2 = \frac{n_g \eta_g}{rm} T_{r(rated)} v_{rated} \left(t_f - t_i\right) \tag{2.44}$$

The equation can be rewritten as

$$t_f = \frac{rm}{2 n_g \eta_g T_{r(rated)} v_{rated}} \left(v_f^2 - v_{rated}^2\right) + t_{rated} \tag{2.45}$$

where t_i and v_i equal t_{rated} and v_{rated}, respectively.

Thus, the time to go from zero to the final speed is the sum of Equations (2.40) and (2.45), which simplifies to

$$t_f = \frac{rm}{n_g \eta_g T_{r(rated)}} \frac{v_f^2 + v_{rated}^2}{2 v_{rated}} \tag{2.46}$$

Now that we have a simple equation, we can easily calculate the energy E_a required to accelerate the vehicle. The energy is the product of the average power over the time period and the time period itself. During the constant-torque mode, the average power is half the rated power, and during the constant-power period, the average power is the rated power. In equation form:

$$E_a = \frac{P_{r(rated)}}{2} t_{rated} + P_{r(rated)} \left(t_f - t_{rated}\right) = P_{r(rated)} \left(t_f - \frac{t_{rated}}{2}\right) \tag{2.47}$$

2.2.3.2.1 *Example: Simple Calculation for Acceleration Time for the Nissan Leaf*

Estimate the 0 to 60 mph acceleration time and energy for the 2015 Nissan Leaf. Ignore all road loads and the internal moment of inertia. Consult Table 2.1, and assume a gear efficiency of 97%.

Solution:

Per Equation (2.40), the time to get to the rated speed is

$$t_{rated} = \frac{rm v_{rated}}{n_g \eta_g T_{r(rated)}} = \frac{0.315 \times 1645 \times 12.11}{8.19 \times 0.97 \times 254} s = 3.11 s$$

The vehicle speed v in m/s is given by

$$v = 60 \times \frac{1.609}{3.6} \text{m/s} = 26.8 \text{m/s}$$

Applying the simple formula of Equation (2.46) results in a time of

$$t_f = \frac{rm}{n_g \eta_g T_{r(rated)}} \frac{v_f^2 + v_{rated}^2}{2v_{rated}} = \frac{0.315 \times 1645}{8.19 \times 0.97 \times 254} \times \frac{26.8^2 + 12.11^2}{2 \times 12.11} \text{s} = 9.18\,\text{s} \quad (2.48)$$

This simple calculation of 9.18 s for the acceleration time from zero to 60 mph compares reasonably to the simulation and experimental values of approximately 10.8 s and 11.5 s, respectively.

When ignoring the load forces, the energy required at the rotor shaft to accelerate is given by

$$\begin{aligned} E_a &= P_{r(rated)}\left(t_f - \frac{t_{rated}}{2}\right) = 80,000 \times \left(9.18 - \frac{3.11}{2}\right)\text{J} \\ &= 610\,\text{kJ}\,(0.169\,\text{kWh}) \end{aligned} \quad (2.49)$$

2.3 Simple Drive Cycle for Vehicle Comparisons

In the following chapters, the fuel economy, efficiency, carbon emissions, and vehicle range will be determined for the various powertrain architectures and energy sources. The examples use basic engine and powertrain data.

The following assumptions are made in order to simplify the analysis of the drive cycle:

1) The various machines operate in steady state.
2) The engine, generator and the motor response times are negligible.
3) The transient losses of accelerating and braking and the resultant kinetic energy gain and loss are negligible due to the few stops and starts in the drive cycle.
4) The gearing and transmission efficiency, regardless of the vehicle, is 95%.
5) The electric drive efficiency for both motoring and generating is 85%.

The basic vehicle powertrain is as shown in Figure 2.9.

A test vehicle, using the parameters of the 2015 Nissan Leaf as presented in Table 2.1, will be analyzed for the various powertrains based on the elementary drive cycle shown in Figure 2.10. Over the one hour cycle on a flat road, the vehicle is cruising at 50 km/h for a

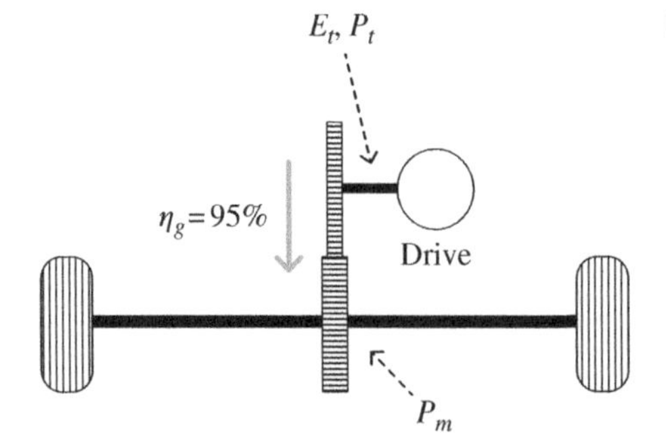

Figure 2.9 Simplified vehicle powertrain.

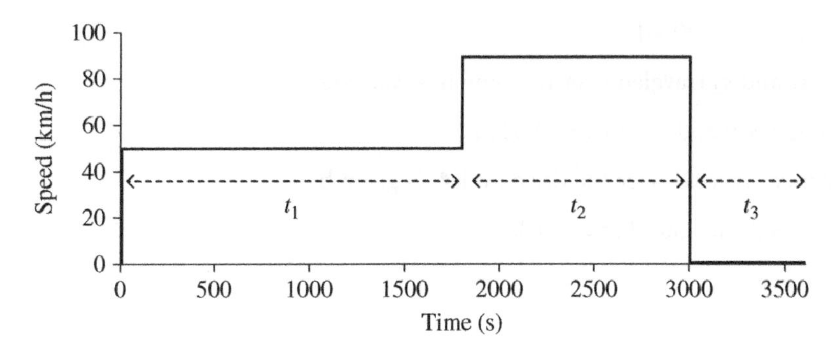

Figure 2.10 Simple drive cycle.

time period t_1= 1800 s, cruising at 90 km/h for t_2 = 1200 s, and in idle mode for t_3 = 600 s. Subscripts 1, 2, and 3 will be used for the various parameters during time intervals t_1, t_2, and t_3, respectively, while subscript C is used for the overall cycle.

First, the vehicle motor power must be determined for the various speeds.

The vehicle speeds v_1 and v_2 in m/s are given by

$$v_1 = \frac{50}{3.6}\,\text{m/s} = 13.89\,\text{m/s} \quad \text{and} \quad v_2 = \frac{90}{3.6}\,\text{m/s} = 25\,\text{m/s}$$

The motive powers P_{m1} at 50 km/h and P_{m2} at 90 km/h are

$$P_{m1} = Av_1 + Bv_1{}^2 + Cv_1{}^3 = 133.3 \times 13.89\,\text{W} + 0.7094 \times 13.89^2\,\text{W} + 0.491 \times 13.89^3\,\text{W} = 3.304\,\text{kW}$$
$$P_{m2} = Av_2 + Bv_2{}^2 + Cv_2{}^3 = 133.3 \times 25\,\text{W} + 0.7094 \times 25^2\,\text{W} + 0.491 \times 25^3\,\text{W} = 11.448\,\text{kW}$$

The traction power P_t is the power at the output of the drive machine, be it an electric motor or an IC engine, and equals the road-load power divided by the gear efficiency:

$$P_t = \frac{P_m}{\eta_g} \tag{2.50}$$

The traction powers P_{t1} and P_{t2} are

$$P_{t1} = \frac{P_{m1}}{\eta_g} = \frac{3.304}{0.95}\,\text{kW} = 3.478\,\text{kW} \text{ and } P_{t2} = \frac{P_{m2}}{\eta_g} = \frac{11.448}{0.95}\,\text{kW} = 12.050\,\text{kW}$$

The traction energy E_t is the energy required at the output of the machine or engine in order to propel the vehicle, and is given by

$$E_t = P_t \times t \tag{2.51}$$

The traction energies required, E_{t1} and E_{t2}, in order to propel the vehicle are given by

$$E_{t1} = P_{t1} \times t_1 = 3.478 \times 1800\,\text{kJ} = 6.260\,\text{MJ}$$
$$E_{t2} = P_{t2} \times t_2 = 12.050 \times 1200\,\text{kJ} = 14.460\,\text{MJ}$$

The total traction energy E_{tC} required for propulsion over the cycle is

$$E_{tC} = E_{t1} + E_{t2} = 20.720\,\text{MJ}$$

The distances s_1 and s_2 traveled over the time intervals are

$$s_1 = v_1 t_1 = 25.0\,\text{km} \text{ and } s_2 = v_2 t_2 = 30.0\,\text{km}$$

The total distance traveled over the drive cycle s is given by

$$s = s_1 + s_2 = 25.0\,\text{km} + 30.0\,\text{km} = 55.0\,\text{km}$$

References

1 R. L. Simison, K. Lundegaard, N. Shirouzu, and J. Heller, "How a tire problem became a crisis for Firestone, Ford," *The Wall Street Journal,* August 10, 2000.
2 E. Musk, "Tesla adds titanium underbody shield and aluminum deflector plates to Model S," Tesla website, www.tesla.com/blog, March 28, 2014.
3 Test car list data files webpage of US EPA: https://www.epa.gov/compliance-and-fuel-economy-data/data-cars-used-testing-fuel-economy
4 T. Nakada, S. Ishikawa, and S. Oki, "Development of an electric motor for a newly developed electric vehicle," SAE Technical Paper 2014-01-1879.
5 C. Morris, "Tesla Semi hits the highway with a bang," *Charged Electric Vehicle Magazine,* November, 2017.

Further Reading

1 E. K. Nam and R. Giannelli, *Fuel Consumption Modelling of Conventional and Advanced Technology Vehicles in the Physical Emission Rate Estimator (PERE),* EPA420-P-05-001, draft February 2005, US EPA.
2 J. G. Hayes, R. P. R. De Oliveira, S. Vaughan, and M. G. Egan, "Simplified electric vehicle powertrain models and range estimation," *IEEE Vehicular Power and Propulsion Conference,* Chicago, 2011.
3 J. G. Hayes and K. Davis, "Simplified electric vehicle powertrain model for range and energy consumption based on EPA coast-down parameters and test validation by Argonne National Lab data on the Nissan Leaf," *IEEE Transportation Electrification Conference,* Dearborn, June 2014.

Problems

2.1 An electric vehicle has the following attributes: mass m = 500 kg, wheel radius r = 0.3 m, gear ratio from rotor to drive axle n_g = 10, and a nominal gear efficiency η_g = 95%. The vehicle is required to accelerate linearly from 0 to 36 km/h in 5 s on a flat road surface under calm wind conditions. Neglecting load forces, instantaneously at 18 km/h calculate the electromagnetic torque from the electric motor to achieve this acceleration torque.

[Ans. 31.6 Nm]

2.2 The vehicle of Problem 2.1 is required to *decelerate* linearly from 36 to 0 km/h in 5 s on a flat road surface under calm wind conditions. Neglecting load forces, instantaneously at 18 km/h calculate the regenerative torque to the electric motor to achieve this braking.

Hint: The acceleration is the same as in Problem 2.1, but the sign is reversed. If you correctly account for the gear efficiency per Figure 2.6(b), you will get the right answer.

[Ans. −28.5 Nm]

2.3 An electric car is climbing at 80 km/h up a 5° incline against a 10 km/h headwind. The vehicle has the following attributes: mass m = 1400 kg, drag coefficient C_D = 0.19, vehicle cross section A = 2.4 m^2, coefficient of rolling resistance C_R = 0.0044, wheel radius r = 0.3 m, gear ratio from rotor to drive axle n_g = 11, and a nominal gear efficiency η_g = 95%. Assume a density of air ρ_{air} = 1.2 kg m^{-3}.
i) Calculate the rotor output torque and speed.
ii) How much greater is the power requirement for climbing the 5° slope compared to a flat road?

[Ans. 41.3 Nm, 7740 rpm, approx. six times greater]

2.4 Determine the approximate time for the GM EV1 to accelerate from 0 to 60 mph, while ignoring vehicle load forces. Use parameters from Table 2.1, and let η_g = 95%.

[Ans. 7.65 s]

2.5 The Nissan Leaf (parameters in Table 2.1) is traveling at 144 km/h. At what slope will the rotor power reach 80 kW?

[Ans. 3.53°]

2.6 An electric drive features a 24 kWh battery pack and has a range of 170 km at a constant speed of 88 km/h. What is the new range when traveling at 88 km/h if the heating, ventilation, and air conditioning draw a constant power of 6 kW?

[Ans. 115 km]

2.7 The drive cycle of Section 2.3 is modified such that the idle period is replaced by a highway cruise at 120 km/h for 600 s.
i) Determine the tractive power required at 120 km/h.
ii) Determine the total motive energy and distance for the revised drive cycle.

[Ans. 24.64 kW, 35.5 MJ, 75 km]

2.8 Elon Musk of Tesla Inc. introduced the new Tesla truck-trailer in late 2017 [5]. He announced the following vehicle parameters: gross vehicle weight m = 36,280 kg (80,000 lb) when fully loaded, drag co-efficient C_D = 0.36, and range when fully loaded of 804.5 km (500 miles) at an average speed of 96.54 km/h (60 mph). Assume the vehicle has the following attributes: vehicle cross-section A = 9 m^2, co-efficient of rolling resistance C_R = 0.006, nominal efficiency of the powertrain and transmission η_{pt} = 85%, auxiliary load P_{aux} = 2 kW, tire radius r = 0.55 m, four Tesla Model S traction motors combined and rated at 1760 Nm and 1080 kW, gearing ratio n_g = 20, and gearing efficiency η_g = 95%. Let the density of air ρ_{air} = 1.2 kg m^{-3}.
i) Determine the battery energy required to meet the range when fully loaded.
ii) Determine the approximate time for the fully-loaded truck-trailer to accelerate from 0 to 60 mph, while ignoring vehicle load forces.

[Ans. 945.5 kWh, 17.85 s]

Sample MATLAB Code

Simplified Traction Machine Torque-Speed Characteristic (John Hayes)

```
%Plotting the torque and speed characteristics vs. vehicle speed
%In the example the Nissan Leaf parameters from Table 2.1 are used

close all; clear all; clc;
Prrated     = 80000;              %rated rotor power in W
Trrated     = 254;                %rated rotor torque in Nm
r           = 0.315;              %wheel radius
ng          = 8.19;               %gear ratio
vmax        = 150;                %vehicle max speed at full power in
                                   km/h
wrrated     = Prrated/Trrated;    %angular speed at rated condition
wrmax       = ng*vmax/(3.6*r);    %rotor speed at maximum
                                   vehicle speed
N           = 1000;               %number of steps
wr          = linspace(1,wrmax,N);      %array of values for wr
speed       = r*3.6/ng*wr;        %vehicle speed array in km/h

T           = zeros(1,N);         %initialize torque array
P           = zeros(1,N);         %initialize power array
v           = zeros(1,N);         %initialize speed array

for n = 1:N                       %Looping N times
    if wr(n) < wrrated            %Less than rated speed
        T(n) = Trrated;           %torque array equation (2.28)
        P(n) = Trrated*wr(n);     %power array equation (2.28)
        v(n) = wr(n)/ng*r;
    elseif (wr(n) >= wrrated)     %More than base speed
        T(n) = Prrated/wr(n);     %torque array equation (2.30)
        P(n) = Prrated;           %power array equation (2.30)

    end;
end;
[hAx,hline1,hline2] = plotyy(speed,T,speed,P/1000);
%title('Full power acceleration of Nissan Leaf and Power
 vs. Speed');
set(hline1,'color','black','linewidth',3)
set(hline2,'color','black','linewidth',3)
set(hAx,{'ycolor'},{'black';'black'})
xlabel('Speed (km/h)');
ylabel(hAx(1),'Rotor Torque (Nm)');
ylabel(hAx(2),'Rotor Power (kW)');
%legend('Torque','Power','Location','northwest');
ylim([0 300]);

grid on
ha = findobj(gcf,'type','axes');
set(ha(1),'ytick',linspace(0,200,10));
```

```
set(hAx(1),'YLim',[0 360])
set(hAx(1),'YTick',[0:40:360])
set(hAx(2),'YLim',[0 90])
set(hAx(2),'YTick',[0:10:90])

%set(hline1,'linestyle','--','color','black','linewidth',3)
```

Simplified BEV Acceleration Profile

```
%Plotting the acceleration vs. time for the Nissan Leaf (John Hayes)
%In the example the Nissan Leaf parameters from Table 2.1 are used

close all;clear all; clc;

Prrated     = 80000;          %max rotor power in W
Trrated     = 254;            %max rotor torque in Nm
r           = 0.315;          %wheel radius
ng          = 8.19;           %gear ratio
m           = 1645;           %vehicle mass in kg
A           = 133.3;          %coastdown parameter A
B           = 0.7094;         %coastdown parameter B
C           = 0.491;          %coastdown parameter C

Effg = 0.97;                  %Assumed gear efficiency
J = 3;                        %Assumed axle-reference MOI

wrrated     = Prrated/Trrated;   %motor base angular speed in rad/s
vrated      = wrrated*r/ng    %vehicle speed in m/s

N           = 100;            %number of steps
tend        = 12;             %end time of 12 s, say, for Nissan Leaf
dT          = tend/N;         %time step
t           = linspace(0,tend,N); %time variable from 0 to tend in
                                    N steps

v           = zeros(1,N);     %initialize speed array at zero.
Tr          = zeros(1,N);     %initialize torque array at zero.
wr          = zeros(1,N);     %initialize angular array speed
                               at zero.
Pr          = zeros(1,N);     %initialize rotor power array at zero.

for n = 1:N-1                 %looping N-1 times
   wr(n) = v(n)*ng/r;         %rotor speed
   if v(n) < vrated           %less than rated speed
      Tr(n) = Trrated;        %rotor torque array
   else                       %greater than rated speed
      Tr(n) = Prrated/wr(n); %rotor torque array
   end;
```

```
    v(n+1)      = v(n)+dT*(ng*Effg*Tr(n)-r*(A+B*v(n)+C*(v(n))^2))/
                  (r*m+J/r);   %speed equation (2.34)
    Pr(n)       = Tr(n)*wr(n);%rotor power array
end;

[hAx,hline1,hline2] = plotyy(t(1:end-1),v(1:end-1)*3.6,
t(1:end-1),Pr(1:end-1)/1000);
%title('Full power acceleration of Nissan Leaf and Power
vs. Speed');
set(hline1,'color','black','linewidth',3)
set(hline2,'color','black','linewidth',3)
set(hAx,{'ycolor'},{'black';'black'})
xlabel('Time (s)');
ylabel(hAx(1),'Vehicle Speed (km/h)');
ylabel(hAx(2),'Rotor Power (kW)');
%legend('Speed','Power','Location','northwest');
ylim([0 100]);
set(gca,'YTick', [20 40 60 80]);
grid on;
ha = findobj(gcf,'type','axes');
set(ha(1),'ytick',linspace(0,200,10));
set(hAx(1),'YLim', [0 100])
set(hAx(1),'YTick', [0:20:100])
set(hAx(2),'YLim', [0 100])
set(hAx(2),'YTick', [0:20:100])
```

Assignment: Modeling of a BEV

2.1 Research a BEV of your choosing and model the vehicle acceleration using a software of your choice. BEVs are of interest at this point as they can be charged and driven in electric-only mode and have a fixed gear ratio making them relatively easy to analyze.

Useful information on your vehicle can be found on [3]. Feel free to research other sources.

The following information can be found from the test car list data files [3]:

i) rated power in horsepower in column K
ii) test weight in pounds in column V
iii) gear ratio (=axle ratio) in column W
iv) *ABC*s in columns BC, BD, BE

The wheel radius can be calculated from the tire specification. A simple example is now used.

The tire of the Tesla Model S is specified as follows: 245/45YR19.

The standard tire for the Tesla is 245 mm wide as per the first numbers.

The aspect ratio is 0.45, and so the tire height is the width times the aspect ratio, or 245 mm × 0.45 = 110.3 mm.

The wheel rim diameter, or the internal tire diameter, is 19 inches (× 25.4 mm = 482.6 mm).

The tire diameter is twice the height plus the rim, or 2 × 110.3mm + 482.6 mm = 703.2 mm.

The tire radius is half the diameter, r = 351.6 mm.

2.2 Modify the MATLAB code of the Simplified BEV Acceleration Profile on page 65 to model the acceleration of the Tesla truck-trailer using the information presented in Problem 2.8.

Various BEVs on the EPA files [3] for model years 2011–2019

(See the Wiley website for the latest vehicles)

1) 2019 Audi e-tron 55 quattro
2) 2019 Honda Clarity
3) 2019 Hyundai Kona Electric
4) 2019 Jaguar I-Pace
5) 2019 Kia Niro Electric
6) 2019 Tesla Model 3 (multiple variations)
7) 2019 Tesla Model S (multiple variations)
8) 2019 Tesla Model X (multiple variations)
9) 2017 BMW i3
10) 2017 Chevy Bolt
11) 2017 Fiat 500e
12) 2017 Hyundai Ioniq EV
13) 2017 Kia Soul
14) 2017 Mercedes-Benz B-class
15) 2017 Mitsubishi i-MiEV
16) 2017 Nissan Leaf SV
17) 2016 BRG Bluecar
18) 2016 Ford Focus
19) 2016 Chevy Spark
20) 2016 Mercedes-Benz Smart for two
21) 2016 Tesla Model X
22) 2016 Volkswagen eGolf
23) 2014 BYD E6
24) 2014 Tesla Model S
25) 2014 Toyota RAV4
26) 2013 Azure Dynamics Transit Connect Electric
27) 2013 Coda
28) 2013 Honda Fit EV
29) 2013 Toyota Scion iQ EV
30) 2011 BMW 1 Series Active E
31) 2011 Nissan Leaf

3

Batteries

"The storage battery is, in my opinion, a catchpenny, a sensation, a mechanism for swindling the public by stock companies. The storage battery is one of those peculiar things which appeals to the imagination, and no more perfect thing could be desired by stock swindlers than that very self-same thing.... Just as soon as a man gets working on the secondary battery it brings out his latent capacity for lying.... Scientifically, storage is all right, but, commercially, as absolute a failure as one can imagine," Thomas Edison, 1883.

"Don't you know that anyone who has ever done anything significant in physics has already done it by the time he was your age?" Remark by professor to World War II veteran John Goodenough, aged 24 in 1946. John was to invent the first Li-ion battery at the age of 57 in 1980 [1].

"Cost, safety, energy density, rates of charge and discharge, and cycle life are critical for battery-driven cars to be more widely adopted. We believe that our discovery solves many of the problems that are inherent in today's batteries." John Goodenough speaking in 2017 on his latest breakthrough – a solid-state battery [2].

In this chapter, the reader is introduced to the electrochemical batteries used for the electric powertrain. Initially, the chapter provides a battery overview and concentrates on the widely used automotive batteries of lead-acid, nickel-metal hydride, and lithium-ion. The basic definitions and material considerations are presented. The critical issue of battery lifetime and related factors are then discussed. Introductory electrochemistry is presented.

3.1 Introduction to Batteries

3.1.1 Batteries Types and Battery Packs

An electric **battery** stores and converts electrochemical energy to electrical energy. The battery symbol is shown in Figure 3.1(a).

Electric Powertrain: Energy Systems, Power Electronics and Drives for Hybrid, Electric and Fuel Cell Vehicles, First Edition. John G. Hayes and G. Abas Goodarzi.

Companion website: www.wiley.com/go/hayes/electricpowertrain

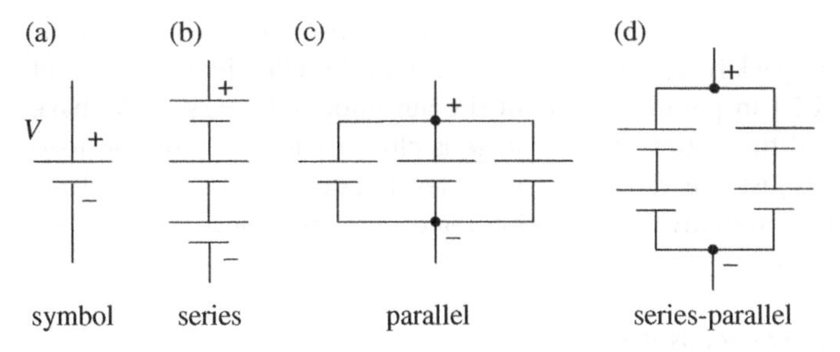

Figure 3.1 (a) Battery symbol; batteries in (b) series, (c) parallel, and (d) series-parallel.

The batteries used for automotive powertrain applications are secondary batteries. A **secondary** battery is a battery that can be repeatedly charged and discharged. Examples of secondary batteries are **lead-acid** (with symbol **PbA**), **nickel-metal hydride (NiMH),** and **lithium-ion (Li-ion).**

Smartphones typically use a **lithium-ion polymer** (**LiPo**) battery. The LiPo battery operates similarly to a standard Li-ion battery. The LiPo battery has a gel-like electrolyte which makes the cell more malleable and easier to shape. The standard Li-ion cell electrolytes are liquid with a polypropylene separator.

The other principal type of battery is a **primary** battery. The primary battery cannot be recharged after discharge. Examples of primary batteries are the **alkaline-manganese** batteries we commonly use in our TV remote controls and other common appliances.

A battery **cell** is the basic battery unit. A **battery pack** comprises multiple cells. Multiple cells may be required to increase the energy storage, the pack voltage, or the battery pack power. Many electrical devices require higher voltages than the basic cell voltage for operation. For example, the speed of a dc electric motor, powered directly by a battery, is approximately proportional to the battery voltage. Many electronic devices require battery voltages in excess of certain minimum values for the electronics to work. For example, modern electronic phones have a processor optimized for operation at 1.5 V, which is easily powered by a Li-ion cell using a power converter to step down the voltage.

The battery cells can be arranged in series or in parallel. Cells are arranged in **series**, as shown in Figure 3.1(b), in order to generate a higher voltage, and higher power, as the battery pack voltage is simply the sum of the individual cell voltages. A typical TV remote control has two 1.5 V alkaline batteries in series for an overall voltage of 3 V. A commonly encountered problem for consumers is when the two 1.5 V alkaline primary cells are replaced with two rechargeable NiMH cells. The cell voltage of NiMH is only 1.2 V, and the reduced cell voltage from 1.5 V to 1.2 V can cause operating problems for devices if the device is designed for operation using a cell voltage of 1.5 V. The 12 V lead-acid battery on a standard vehicle comprises six 2 V lead-acid cells in series.

Cells can also be arranged in **parallel**, as shown in Figure 3.1(c), in order to generate a higher output current and power. The stored energy, lifetime, and voltage of a battery are dependent on the current or power pulled from the battery. Adding more cells in parallel increases the energy, lifetime, and voltage for a given power.

Automotive EV batteries are typically arranged in **series-parallel**, as shown in Figure 3.1(d), in order to obtain higher voltage, current, power, energy, and lifetime. The 2012 Tesla Model S with an 85 kWh battery pack has 16 modules in series.

Each module has six strings of submodules in series with each sub-module having 74 cells in parallel. Thus, the battery pack has a total number of 7104 cells, effectively a matrix of cells with 96 in series and 74 in parallel. Many of the automotive battery packs have 96 Li-ion cells in series, and the battery pack voltage is close to 400 V as the no-load voltage on each Li-ion cell is just over 4 V. The voltage level of 400 V is a common level for many power converters, and many related technologies have been developed to efficiently and safely convert power at this voltage level.

3.1.1.1 Recent EVs and Battery Chemistries

There are many different types of battery chemistries. Batteries can also be optimized for a particular application as the usage can dictate the material selection. A battery electric vehicle (BEV) battery is optimized for a wide operating range, while a hybrid electric vehicle (HEV) battery is optimized for a narrow operating range in order to maximize the number of discharge cycles.

The General Motors (GM) EV1 electric car came to market in 1996 with a lead-acid battery pack. The lead in the lead-acid battery makes it a very heavy battery, thus limiting the on-vehicle energy storage. The chemistry has a relatively short lifetime, further reducing the energy storage as the capacity degrades. While lead-acid was the only available chemistry at that time for the vehicle, NiMH technology was developing rapidly, and the second-generation GM EV1, launched in 1999, featured a NiMH battery pack. NiMH has a greater energy capacity and longer lifetime than PbA. NiMH was the chemistry of choice for Toyota for the 1997 Toyota Prius.

In the meantime, Li-ion was being developed as the chemistry of choice for cellular phones and laptops due to its large capacity and lifetime. Li-ion chemistry has been the basis for the EV renaissance which began with the launch of the Tesla Roadster in 2008. Lithium is the lightest metal, and the Li-ion battery has many advantages over the other technologies, such as a higher energy density, a higher cell voltage, and a longer life.

There are a number of variations of Li-ion chemistry that have been developed by various manufacturers. Earlier Li-ion chemistries featured cobalt and manganese as the main metals. The mix and choice of materials can significantly influence the energy density, lifetime, safety, and cost. The cathode, to be defined in the next section, is a mix of lithium, various metals, and oxygen, and is termed lithium metal oxide. It is common to use the chemical formula to describe a battery. For example, $LiCoO_2$ is the chemical formula for lithium cobalt oxide.

The Panasonic battery used in the Tesla Model S has a high nickel content. In a typical nickel-cobalt-aluminum (NCA) battery, nickel can make up 85% of the cathode with aluminum and cobalt making up 10% and 5%, respectively. The high nickel content results in a battery with a very high energy density and a long life.

The LG Chem battery used in the 2011 Chevy Volt features an NMC cathode. NMC stands for nickel, manganese, and cobalt with the formula $LiNiMnCoO_2$. The metals are mixed at approximately 1/3 each. Although the NMC chemistry has a lower energy density than NCA, NMC has a lower cost and a longer life. The energy density improved from 87 Wh/kg in the 2011 Chevy Volt to 101 Wh/kg in the 2016 Chevy Volt.

The AESC batteries in the 2011 Nissan Leaf feature a blend of manganese and nickel. In chemical terms, the cathode is a mix of $LiMn_2O_4$ and $LiNiO_2$.

It is useful to review the published parameters on selected vehicles from the 1996 GM EV1 on, and the developments in battery technology will become apparent. The key parameters are presented in Table 3.1. The 1996 GM EV1 had a vehicle weight of

Table 3.1 Representative battery cell and pack parameters for various vehicles.

Vehicle	Vehicle weight (kg)	Battery weight (kg)	Battery manufacturer	Chemistry	Rated energy (kWh)	Specific energy (Wh/kg)	Cell/pack nominal Voltage (V)	Rated power (kW)	Specific Power (W/kg)	P/E
1996 GM EV1	1400	500	Delphi	PbA	17	34	2/312	100	200	6
1999 GM EV1	1290	480	Ovonics	NiMH	29	60	1.2/343	100	208	3
1997 Toyota Prius	1240	53	Panasonic	NiMH	1.8	34	1.2/274	20	377	11
2008 Tesla Roadster	1300	450	Panasonic	Li-ion	53	118		185	411	3
2011 Nissan Leaf	1520	294	AESC	Li-ion	24	82	3.75/360	80	272	3
2011 ChevyVolt	1720	196	LG Chem	Li-ion	17	87	3.75/360	110	560	6
2012 Tesla Model S	2100	540	Panasonic	Li-ion	85	157		270	500	3
2017 Chevy Bolt	1624	440	LG Chem	Li-ion	60	136	3.75/360	150	341	3

1400 kg, of which over 500 kg was the battery pack. The rated energy and power of the battery pack were about 17 kWh and 100 kW, respectively. The **specific energy**, defined as energy per kilogram (Wh/kg), was relatively low at 34 Wh/kg, while the **specific power**, defined as watts per kilogram (W/kg), was 200 W/kg. A common metric to compare batteries is to calculate the **power to energy ratio**, P/E, by dividing the specific power by the specific energy. The P/E ratio tends to be low for EVs and higher for HEVs. This consideration of optimizing batteries for EV versus HEV operation can affect the ultimate battery chemistry as a particular battery can be optimized for energy or power. The 1996 GM EV1 lead-acid battery had a P/E of 6, a relatively high number for an EV pack compared to the next generation of EVs.

The introduction of NiMH technology to the GM EV1 in 1999 almost doubled the energy of the battery pack while reducing the weight slightly. Applying similar NiMH technology to the 1997 Toyota Prius HEV resulted in a battery with half the specific energy of the 1999 GM EV1 NiMH battery but with a far higher specific power as the battery is optimized for long-life hybrid applications. In regular operation, the 1997 Toyota Prius uses only 20% of the available energy capacity in order to maximize its lifetime. The 1997 Toyota Prius P/E ratio is close to 11 versus 3 for the 1999 GM EV1 NiMH.

The Tesla Roadster introduced the high-specific-energy Li-ion technology and achieved a specific energy three to four times greater that of the 1996 GM EV1 lead-acid battery and twice that of the 1999 GM EV1 NiMH battery, while having a high specific power and a long lifetime.

The 2012 Tesla Model S increased the specific energy and power to higher values with a pack size close to five times greater than the 1996 GM EV1 but with a similar weight.

The 2011 Chevy Volt features a HEV battery with a P/E of 9 and has over double the specific energy of the 1996 GM EV1. The 2011 Nissan Leaf features a P/E close to 3 and a specific energy similar to the Chevy Volt.

The 2017 Chevy Bolt BEV features a higher-power-density lower-cost battery pack than the 2011 Chevy Volt. The pack is shown in Figure 3.2. This GM vehicle illustrates

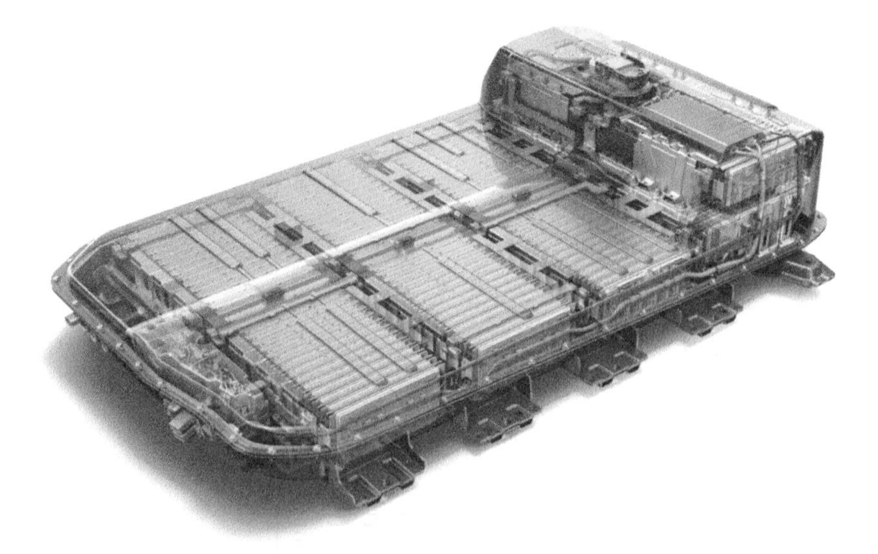

Figure 3.2 Battery pack of Chevy Bolt. (Courtesy of General Motors.)

the progress of battery technology as the 2017 Chevy Bolt has approximately four times the energy density of the 1996 GM EV1 at a fraction of the original cost.

3.1.2 Basic Battery Operation

The discovery of the electrochemical battery is credited to **Alessandro Volta**, an Italian who lived from 1745 to 1827. The measure and unit of electric potential are **voltage** and the **volt**, respectively, and are named in his honor. Volta discovered that an electrical potential is generated when two dissimilar metals, known as **electrodes**, are submerged in an **electrolyte**. The voltage is dependent on many factors, including the type and size of the electrodes and the electrolyte, pressure, and temperature. The basic structure of a cell is as shown in Figure 3.3(a). The cathode is typically made from a metal oxide, while the anode is made from a conductive metal or graphite carbon. The electrodes are connected to conductive current collectors which connect the electrodes to the battery terminals. A **separator** is required in order to eliminate the possibility of shorting between the electrodes and to facilitate the flow of ions.

The positive voltage of the cell is at the cathode, and the negative voltage is at the anode. During charging of the cell, the electrical charger is connected between the terminals, and the current *I* flows from the charger into the cathode and out of the anode. During charging, electrons, with the symbol e^-, flow externally in the opposite direction to the current from the cathode to the anode. The positive ions travel through the electrolyte from the cathode to the anode.

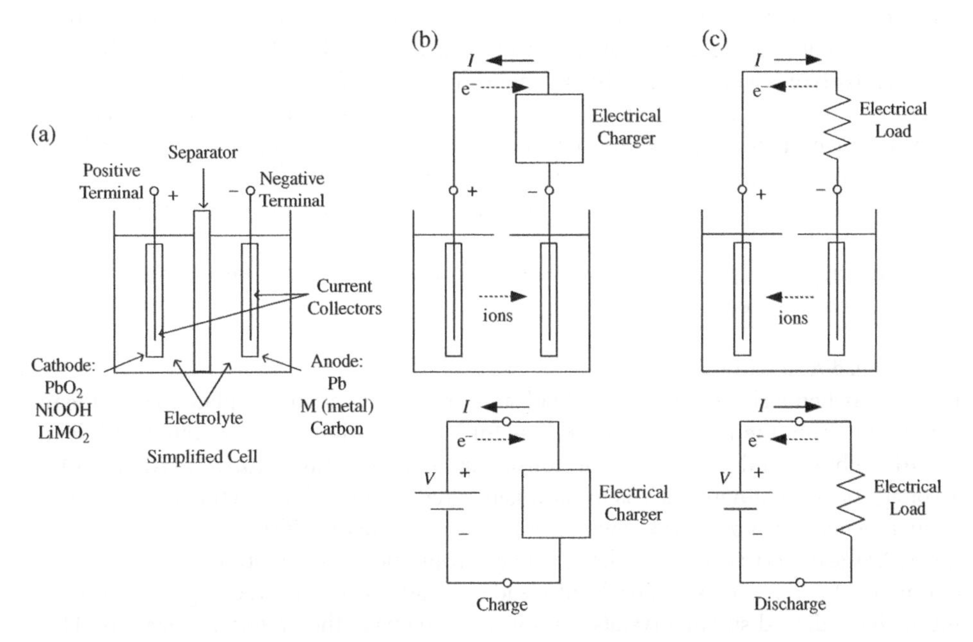

Figure 3.3 Simplified battery cell.

During discharge, when an external electrical load is connected to the battery cell, the current flows from the cathode through the load and back into the anode. The electron flow is external to the cell and in the opposite direction to the current. The ions travel internally through the electrolyte from the anode to the cathode.

3.1.3 Basic Electrochemistry

Electrochemistry is the study of the relationships between electricity and chemical reactions. Oxidation and reduction are the basic chemical reactions in an electric battery.

Oxidation is the loss of electrons in a chemical reaction. This is a common phenomenon in nature, especially when a material combines with oxygen, for example, the rusting of iron.

Reduction is the gain of electrons in a chemical reaction.

An **oxidation-reduction** reaction, commonly termed a **redox** reaction, occurs when different materials lose and gain electrons in a shared reaction. The **reducing agent**, or **reductant**, loses electrons and is **oxidized**, while the **oxidizing agent**, or **oxidant**, gains electrons and is **reduced**. The energy released in a redox reaction can be used to generate or store electricity by enabling the flow of electrons externally. This is the basis of operation of an **electrochemical battery cell** or **voltaic cell**. A simple electrochemical cell comprises two electrodes and an electrolyte.

A voltaic cell is composed of two **half-cells**. One half-cell is the site of the oxidation reaction and the other half-cell is the site of the reduction reaction.

The **anode** is the solid metal connection or electrode within the battery at which oxidation occurs during discharge. The anode is at the **negative** terminal of the battery. By definition, an anode is an electrode through which current flows into a device (during discharge in the case of the battery).

The **cathode** is the solid metal connection or electrode within the battery at which reduction occurs during discharge. The cathode is the **positive** electrode of the battery.

An **electrolyte** is a substance which contains ions and allows the flow of ionic charge. A **cation** is an ion with a positive charge, and an **anion** is an ion with a negative charge.

An electric **battery** comprises one or more of these electrochemical cells and produces dc current by the conversion of chemical energy into electrical energy.

A **primary** battery is a battery with one or more cells in which an **irreversible** chemical reaction produces electricity.

A **secondary** battery is a battery with one or more cells in which a **reversible** chemical reaction produces electricity.

3.1.3.1 Lead-Acid Battery

The electrochemical elements of the **lead-acid** battery are an anode plate made of **lead** (symbol Pb, from the Latin word *plumbum* for lead), a cathode plate covered with **lead dioxide** (PbO_2), and an electrolyte consisting of water and **sulfuric acid** (H_2SO_4). A voltage of approximately 2 V is initially generated by the redox reactions. As the electricity is generated, a process known as sulfation causes **lead sulfate** ($PbSO_4$) to coat the plates. As the battery is discharged, the voltage drops due to the sulfation. Lead sulfate is a soft material and is converted back into lead and sulfuric acid on recharge. However, if left uncharged, lead sulfate crystals can form and damage the battery irreversibly. This is a common source of lead-acid battery failure. The battery can lose capacity due to

Table 3.2 Cell parameters for various batteries.

Chemistry	Electrode	Reaction	Equation	Voltage (V)
Lead-acid charge	Cathode	Oxidation	$PbSO_4 + 2H_2O \rightarrow PbO_2 + H_2SO_4 + 2H^+ + 2e^-$	1.69
	Anode	Reduction	$PbSO_4 + 2H^+ + 2e^- \rightarrow Pb + H_2SO_4$	−0.36
	Cell		$2PbSO_4 + 2H_2O \rightarrow Pb + PbO_2 + 2H_2SO_4$	2.05
Lead-acid discharge	Cathode	Reduction	$PbO_2 + H_2SO_4 + 2H^+ + 2e^- \rightarrow PbSO_4 + 2H_2O$	1.69
	Anode	Oxidation	$Pb + H_2SO_4 \rightarrow PbSO_4 + 2H^+ + 2e^-$	−0.36
	Cell		$Pb + PbO_2 + 2H_2SO_4 \rightarrow 2PbSO_4 + 2H_2O$	2.05
NiMH charge	Cathode	Oxidation	$Ni(OH)_2 + OH^- \rightarrow NiOOH + H_2O + e^-$	0.45
	Anode	Reduction	$MH + OH^- + e^- \rightarrow M + H_2O$	−0.83
	Cell		$Ni(OH)_2 + M \rightarrow NiOOH + MH$	1.28
NiMH discharge	Cathode	Reduction	$NiOOH + H_2O + e^- \rightarrow Ni(OH)_2 + OH^-$	0.45
	Anode	Oxidation	$M + H_2O \rightarrow MH + OH^- + e^-$	−0.83
	Cell		$NiOOH + MH \rightarrow Ni(OH)_2 + M$	1.28
Li-ion charge	Cathode	Oxidation	$LiMO_2 \rightarrow MO_2 + Li^+ + e^-$	1
	Anode	Reduction	$C_6 + Li^+ + e^- \rightarrow LiC_6$	−3
	Cell		$LiMO_2 + C_6 \rightarrow MO_2 + LiC_6$	4
Li-ion discharge	Cathode	Reduction	$MO_2 + Li^+ + e^- \rightarrow LiMO_2$	1
	Anode	Oxidation	$LiC_6 \rightarrow C_6 + Li^+ + e^-$	−3
	Cell		$MO_2 + LiC_6 \rightarrow LiMO_2 + C_6$	4

sulfation and the battery should periodically undergo **charge equalization**, a process to ensure that the charge in each cell is brought to similar levels.

The half-cell and cell reactions for lead-acid, NiMH, and Li-ion batteries are shown in Table 3.2. During charging of the lead-acid cell, the lead sulfate and water at the cathode are converted to lead dioxide and sulfuric acid, releasing electrons and ions to flow to the anode. A half-cell voltage of 1.69 V is generated at the cathode. The electrons and ions combine with the lead sulfate at the anode to create lead and sulfuric acid and a half-cell voltage of −0.36 V.

The reactions are reversed during discharge, and the nominal cell voltage during charge and discharge is 1.69 V minus −0.36 V, which equals 2.05 V.

Note that the voltages presented for the anode or cathode are with respect to a standard hydrogen electrode, which is the basis used to determine the voltage from a half-cell.

Lead-acid batteries remain the choice for starter batteries for conventional vehicles due to the low internal resistance, high cranking current, low costs and recyclability.

3.1.3.2 Nickel-Metal Hydride

The NiMH cell is similar to the once widely used **nickel-cadmium** (NiCad) cell. However, cadmium is toxic and a carcinogen, in addition to being expensive. The metal hydride enables the use of hydrogen and the replacement of the cadmium in the anode of the NiCad cell. A metal hydride is a compound in which hydrogen is bonded to the metal. The actual mix of metals tends to be proprietary but is a mix of common metals such as nickel and cobalt and smaller quantities of other materials. The nominal cell voltage for the NiMH cell is about 1.2 V, similar to NiCad.

While NiMH has many advantages over NiCad and the other chemistries, it has a very high self-discharge compared to the other chemistries. NiMH and NiCad batteries can suffer from "memory" effects. The **memory effect** is the loss of capacity due to multiple partial charges. Similar to lead-acid, these battery cells must be equalized periodically in order to ensure that each cell in a battery pack is equally charged.

3.1.3.3 Lithium-Ion

The Li-ion cell has a cathode formed from lithium metal oxide ($LiMO_2$) and an anode formed from carbon in the form of graphite carbon. The Li-ion cell operates differently from the lead-acid cell in that during charging the lithium itself is released as a positive ion from the lithium metal oxide, and travels from the cathode through the electrolyte to the anode, where it combines with the carbon electrode to form lithiated carbon. The electrolyte is a high-conductivity lithium salt which facilitates the movement of the Li cation.

The cathode reaction generates a half-cell voltage of approximately 1 V, while the anode half-cell reaction generates −3 V for an overall cell voltage of approximately 4 V.

Lithium metal can deposit on the anode at low temperatures. This buildup of lithium is known as a **dendrite**. The dendrite can result in a short circuit to the cathode by penetrating the separator.

The Li-ion cell does not have a memory effect, but it loses its capacity with time and cycling. The active lithium within the battery decreases with time. Higher cell voltages can accelerate the loss of capacity. The cycling of the battery results in mechanical fracturing of the electrodes and a reduction in the battery lifetime. These effects increase significantly with temperature.

Lithium-titanate (LiT) batteries feature titanium rather than graphite in the anode of the cell. This significantly improves the lifetime of the battery by eliminating the cracking of the graphite. The reduced battery internal resistance improves the power capability of the LiT battery, especially at low temperatures, making the battery an attractive option for HEVs. The disadvantages of LiT compared to Li-ion are cost, low voltage, and low specific energy.

Newer Li-ion batteries are using advanced materials for the anode. **Silicon-based alloys** are being used to replace the carbon in order to achieve a higher energy density.

3.1.4 Units of Battery Energy Storage

A battery stores energy. An object is said to possess energy if it can do work. The unit of energy and work is the **joule** with the symbol J. The unit is named after an English physicist **James Joule** (1818–1889). In his professional career, he was a beer brewer, a useful career path when considering the complexities of electrochemistry.

Power is defined as the work done per second. The unit of power is the **watt** (symbol W). In equation form:

$$\text{energy} = \text{power} \times \text{time} \tag{3.1}$$

In units:

$$1\,\text{J} = 1\,\text{W s} \tag{3.2}$$

However, a joule is a relatively small unit of energy, and it is more common to reference other energy storage units such as **watt-hours**. One watt-hour (Wh) is defined as follows:

$$1\,\text{Wh} = 3{,}600\,\text{W s} = 3{,}600\,\text{J} \tag{3.3}$$

The watt-hour is commonly used as a measure of energy for automotive applications. It can also be more useful to express energy in **kilowatt-hours** (kWh):

$$1\,\text{kWh} = 3{,}600{,}000\,\text{W s} = 3{,}600{,}000\,\text{J} \tag{3.4}$$

A common quantity of measure related to battery energy storage is **capacity**, with the unit of **ampere-hour** (Ah). The unit of current I is the **ampere** (symbol A). The unit is named after André-Marie Ampère (1775–1836), a French physicist and mathematician who lived through the French Revolution and the Napoleonic era and contributed to electromagnetism. The standard SI unit for charge Q is the **coulomb** (symbol C), named after another Frenchman, Charles-Augustin de Coulomb (1736–1806), who performed great work in the field of electrostatics and developed Coulomb's law relating the forces between electric charges. In equation form:

$$\text{charge} = \text{current} \times \text{time} \quad \text{or} \quad Q = It \tag{3.5}$$

In units,

$$1\,\text{C} = 1\,\text{A s} \tag{3.6}$$

Again, the coulomb is a small unit, and the ampere-hour is more commonly referenced:

$$1\,\text{Ah} = 3{,}600\,\text{A s} = 3{,}600\,\text{C} \tag{3.7}$$

The capacity in ampere-hours can be related to the energy in watt-hours by multiplying the capacity by the battery voltage V:

$$\text{energy} = \text{capacity} \times \text{voltage} \tag{3.8}$$

In units,

$$1\,\text{J} = 1\,\text{A s} \times \text{V} \tag{3.9}$$

3.1.5 Capacity Rate

In describing batteries, it is common to use the **C rate**. The C in this case stands for capacity rate and not the unit of coulombs. A C rate is a measure of how quickly the battery is charged or discharged relative to its maximum capacity.

For example, a **1C** rate discharges the battery pack at a fixed given current in one hour, while a 10C rate discharges the battery pack at ten times the 1C current. A C/3 rate discharges the battery pack at one third of the 1C current.

3.1.5.1 Example of the 2011 Nissan Leaf Battery Pack

The 2011 Nissan Leaf has a nominal battery capacity of 24 kWh and a nominal battery voltage of 360 V [3]. The battery pack has 192 battery cells in all. The cells are arranged into 48 series modules with 4 cells per module, as shown in Figure 3.4. The cells were manufactured by AESC Corp., a collaboration between Nissan and NEC Corp.

The cell, module, and pack parameters are shown in Table 3.3. A single cell weighs 0.787 kg and has a length and breadth of 29 cm and 21.6 cm, respectively, and an approximate thickness of 0.61 cm. The cell has an energy density of 317 Wh/L and a specific energy of 157 Wh/kg. As the cells are integrated into the modules, these values drop to 206 Wh/L and 128 Wh/kg, respectively, due to the additional packaging, interconnects, cabling, and circuitry required in the module. As the pack is made up of the 48 modules in addition to the cabling, interconnects, and circuitry, a strong chassis is required to package and protect the cells. The overall pack weight increased to 294 kg with a specific energy of approximately 82 Wh/kg.

The rated capacity of the cell is 32.5 Ah at a $C/3$ rate for 3 h when discharged. The C rate for this battery is quoted at $C/3$, which means that the rated capacity of 32.5 Ah or 121.875 Wh per cell, or 65 Ah or 23.4 kWh per pack, is available when the cell or battery is discharged at the $C/3$ rate of 10.8 A per cell, or 21.6 A per pack for 3 h. The $C/3$ rate is quoted for the battery capacity as this level is closer to the typical discharge current of the

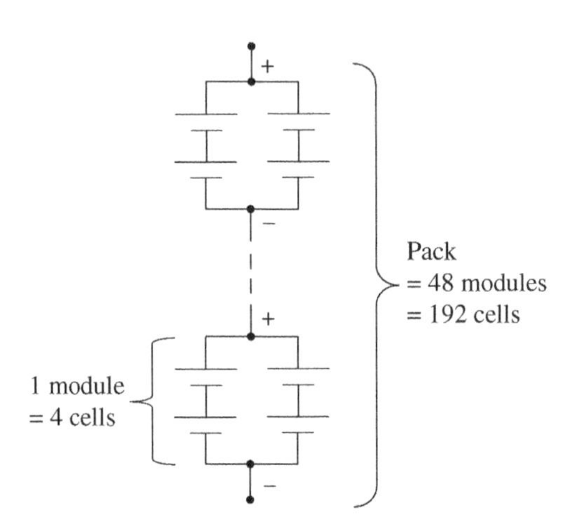

Figure 3.4 Cell, module, and battery pack for 2011 Nissan Leaf [3].

Table 3.3 Approximate cell, module, and pack parameters for the 2011 Nissan Leaf [3,4].

	Number of cells	Voltage (V)	Capacity (Ah)	Energy (Wh)	Volume (L)	Weight (kg)	Energy density (Wh/L)	Specific energy (Wh/kg)
Cell	1	3.75	32.5	121.9	0.384	0.787	317	157
Module	4	7.5	65	487.5	2.365	3.8	206	128
Pack	192	360	65	23,400–24,000	NA	294	NA	82

Nissan Leaf battery pack under normal driving conditions. The 1C rate is commonly used for other applications.

We now introduce two battery terms that are closely related.

The **state of charge** (**SOC**) is the portion of the total battery capacity that is available for discharge. It is often expressed as a percentage, and can be seen as a measure of how much energy remains in the battery.

The **depth of discharge** (**DOD**) is the portion of electrical energy stored in a battery that has been discharged. It is often expressed as a percentage.

For example, if the pack capacity is 24 kWh and 6 kWh has been discharged, the DOD is 6/24 or 25%. The remaining energy in the pack is then 18 kWh, and the SOC is 18/24 or 75%.

3.1.6 Battery Parameters and Comparisons

There are many parameters related to battery technology; at this point, we concern ourselves with the critical parameters that are of importance when selecting a battery for the automotive application. While focusing on the main batteries, we should also note that the battery chemistries can be optimized for particular automotive applications. Batteries used for HEVs typically feature smaller energy storage with a relatively narrow range of energy usage compared to the batteries for a BEV, which typically have a large storage and a very wide operating range.

3.1.6.1 Cell Voltage

The cell voltage is a function of the chemical reaction within the battery and can vary significantly with the SOC, age, temperature, and charge or discharge rate. The nominal cell voltages for the batteries of interest are shown in the third column of Table 3.4. The **rated voltage** of a battery cell is the average voltage over a full discharge cycle. For instance, the Li-ion cell has a rated voltage of 3.75 V, while the cell voltage can actually vary from about 4.2 V when fully charged to 2.5 V when fully discharged. NiMH has a rated cell voltage of 1.2 V. As automotive batteries are typically packaged in series-parallel strings, the higher the cell voltage, the fewer battery cells are required, making Li-ion more attractive for packaging in battery packs. Lead-acid has a nominal cell voltage of 2 V.

Table 3.4 Representative parameters for various batteries.

Chemistry	Symbol	Cell voltage (V)	Specific energy (Wh/kg)	Cycle life	Specific power (W/kg)	Self-discharge (% per month)
Lead-acid	PbA	2	35	≈500	250–500	5
Nickel-metal hydride	NiMH	1.2	30–100	>1000	200–600	>10
Lithium-ion	Li-ion	3.8	80–160	>1000	250–600	<2
Lithium-titanate	LiT	2.5	50–100	>20000	NA	NA
Alkaline	$ZnMnO_2$	1.5	110	NA		<0.3

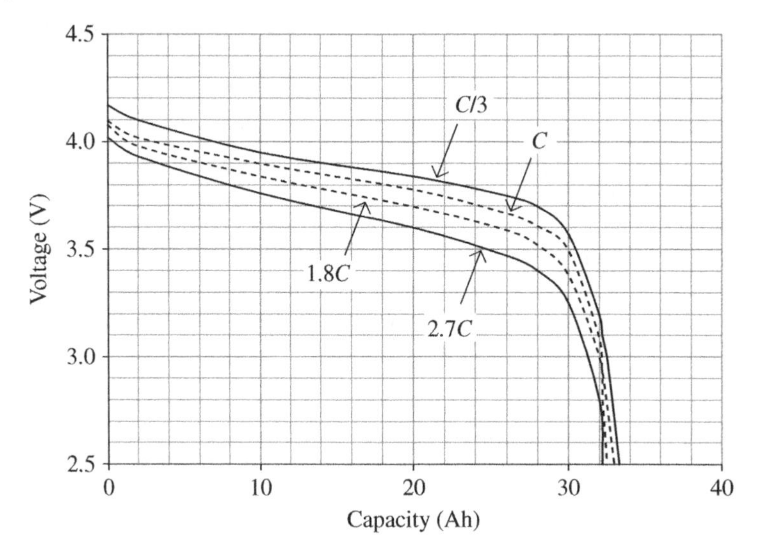

Figure 3.5 Representative discharges curves for a 33.3 Ah Li-ion cell.

Operation at high voltages can result in very significantly reduced battery lifetime. Similarly, operation at lower voltages can result in cell failure. Representative plots of the cell voltage as a function of capacity are presented in Figure 3.5 for various discharge rates, and are known as **discharge curves**. For the cell shown in Figure 3.5, the capacity of the cell is about 33.3 Ah for a *C*/3 discharge. Four levels are shown: *C*/3 or 11.1 A, 1*C* or 33.3 A, 1.8*C* or 60 A, and 2.7*C* or 90 A.

We can see one of the important characteristics of a battery. As the rate of discharge is increased, the cell terminal voltage, that is, the voltage at the output terminals of the cell, drops because the internal resistance results in voltage drops within the cell due to the flow of current from the cell. These losses result in increased internal heating within the cell and reduced available energy. For the cell shown in Figure 3.5, the capacity is about 33.3 Ah for a *C*/3 discharge. Interestingly, there is little change in capacity with increasing *C*. However, we shall see later that the available energy can drop significantly with increased *C* – the drop in useful energy from the battery can be deduced from the reduced average terminal voltage shown for each curve.

3.1.6.2 Specific Energy

The specific energy of a battery is a measure of the stored energy of a battery per unit weight. Li-Ion has the highest energy density of the batteries, as shown in column 4 of Table 3.4. The specific energy of the Li-ion battery is approximately 3 to 5 times that of the lead-acid battery.

3.1.6.3 Cycle Life

Cycle Life is a measure of the number of times a battery can be charged and discharged before it reaches its end of life. Electrochemical batteries degrade with time and usage. Factors such as temperature and cell voltage also play a critical role. Again, Li-ion has the highest cycle life and NiMH is similar. Lead-acid batteries have a significantly lower lifetime while the primary alkaline cell cannot be recharged. The nominal values quoted in Table 3.4 are representative for 100% deep discharges. Lithium-titanate (LiT) is a variation on a Li battery that eliminates the significant aging problem of Li-ion. Although LiT has a lower cell voltage and specific energy, the very significant increase in cycle life makes the battery an attractive option for HEVs.

3.1.6.4 Specific Power

Specific power is a measure of the discharge power available from a battery pack per unit weight. Lead-acid traditionally has had a high specific power and is used as the starter battery for conventional cars. The newer batteries, such as Li-ion and NiMH, have comparable specific powers.

3.1.6.5 Self-Discharge

Electrochemical cells consume energy even when not being charged or discharged. This energy usage is a parasitic use of stored energy and is termed self-discharge. Nominal self-discharge rates are shown for the various chemistries in the table. Self-discharge rates are relatively high for nickel-based batteries compared to their competitors. There can be a very high initial self-discharge of a battery in the first 24 hours after being fully charged, but this rate tapers off. The self-discharge rates can increase significantly with temperature. Note that while the self-discharge of Li-ion is less than 2%, the overall self-discharge of a battery pack may be closer to 5% as up to an additional 3% may be required by the electronic system and circuits managing the battery pack.

3.2 Lifetime and Sizing Considerations

Unlike many electrical and electronic products, an electrochemical battery can have a relatively short lifetime depending on its chemistry. Automotive products are typically designed for a 10- to 12-year lifetime. The typical semiconductor device, such as a microprocessor or switch, exhibits minor degradation with time. Electrochemical devices, such as batteries, fuel cells, and electrolytic capacitors, can degrade relatively fast as they age, especially with increased usage. Thus, a significant challenge for battery manufacturers has been to develop batteries which match the automotive product lifetime. The market has shifted from relatively short-life lead-acid batteries to longer-life NiMH and Li-ion batteries.

First, let us review some useful terms:

Time and **charge/discharge cycles**: One of the characteristics of electrochemical cells is that the ability to store charge degrades with time. Repeated deep charge/discharge cycles can even more significantly result in a reduced lifetime.

Lifetime: The lifetime of a battery can be described using time (years) or repeated cycles. Automotive batteries may have an additional requirement expressed as range in miles or kilometers.

Beginning of life (BOL): The beginning-of-life parameters are typically the values for the capacity and the internal resistance of the battery when it is initially manufactured.

End of life (EOL) The end-of-life parameters are the values of critical components once they degrade with time or usage. A typical end-of-life criterion is for the battery energy storage capacity to drop to 80% of the BOL value or for the internal resistance to increase by 50%.

The factors affecting the life of a battery are many and complex. The following are some of the related factors:

Voltage: Too high a cell voltage can result in breakdown of the electrolyte, increased effects of impurities, and an accelerated loss of lithium from the electrodes, all of which increase resistance, reduce storage capacity, and consequently reduce cycles and lifetime. Thus, while lowering the cell voltage increases the battery lifetime, there is the trade-off in that the lower voltage also reduces energy storage within the cell. A representative plot of the number of battery charge\discharge cycles and battery capacity versus the float voltage is presented in Figure 3.6. The **float voltage** is the voltage at which the cell is maintained once the battery has been fully charged in order to compensate for the self-discharge of the cell. While charging the battery to a higher voltage increases stored energy in the battery, the higher voltages can accelerate battery aging and the number of charge\discharge cycles to the end of the life can drop very significantly. Some Li-ion cells are operated at maximum values in the range of 3.6 to 3.8 V in order to optimize energy storage and cycle life. BEV and HEV Li-ion batteries typically go to approximately 4.2 V maximum under normal conditions but can go slightly higher to approximately 4.3 V on a pulsed basis.

High temperatures: In general, in chemistry and electronics, operating at high temperatures results in effects that can significantly reduce the lifetime and reliability of a component. The classic work published by the Swedish scientist **Svante Arrhenius** in 1889 relating lifetime to temperature applies very much to today's electrochemical storage devices. Note that Arrhenius was one of the first scientists to investigate the effects of increased carbon emissions on the global climate. The usual rule of thumb based on the Arrhenius equation is that the lifetime of a device halves for every 10°C rise in temperature. Thus, battery lifetime can be significantly reduced in very high-temperature climates. Many parts of the world have climates that are hostile

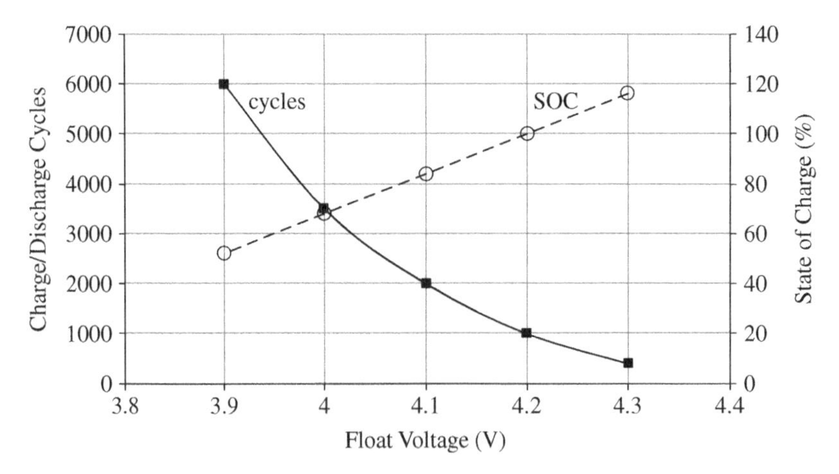

Figure 3.6 Representative charge\discharge cycles and capacity versus cell float voltage for Li-ion battery.

to battery technology. For example, a temperature of greater than 40°C is common in Phoenix, Arizona, United States, in the summer. The temperature seen by the battery can be significantly higher as the car is sitting under a hot sun absorbing solar energy in addition to sitting on a hot asphalt or tarmacadam surface. The necessary use of air-conditioning in Phoenix results in deeper discharges of the battery. Thus, the increased temperatures and deeper discharges can accelerate capacity fade and increase the internal resistance.

Low temperatures: Operation at very low temperatures can also be a problem for some battery technologies. The electrolyte can become more viscous and have decreased conductivity. Freezing of Li-ion cells at temperatures less than −10°C (14°F) reduces the amount of power and stored energy available from a battery. For this reason, manufacturers offer battery heaters for colder climates in order to ensure adequate performance. Ideally, the battery is heated while plugged in to the grid in charging mode so that the battery energy is saved for driving.

Time: One of the biggest challenges for EV manufacturers has been to develop a battery with a calendar life greater than the lifetime of the vehicle. A battery life of 8 to 12 years is necessary. As discussed already, factors such as voltage, temperature, and cycles significantly affect the lifetime. As the battery ages, there is a reduction with time in the lithium available as an active material. A lower SOC results in a lower cell voltage, which slows the degradation of the electrolyte and the loss of active lithium.

Battery life testing can be very complex and time consuming. Typical testing involves a series of partial and full charges and discharges on multiple identical battery packs to determine the number of cycles from BOL to EOL. A representative set of curves of charge/discharge versus DOD is plotted in Figure 3.7. The plots are based on the number of charge/discharge cycles until the capacity drops to the EOL capacity, typically specified as 80%.

The solid line in Figure 3.7 has been generated using a basic assumption – that the total energy converted by a battery is related to the DOD. For the solid line curve ($L = 1$), we assume that the energy throughput of the battery is constant. Thus, if the battery can

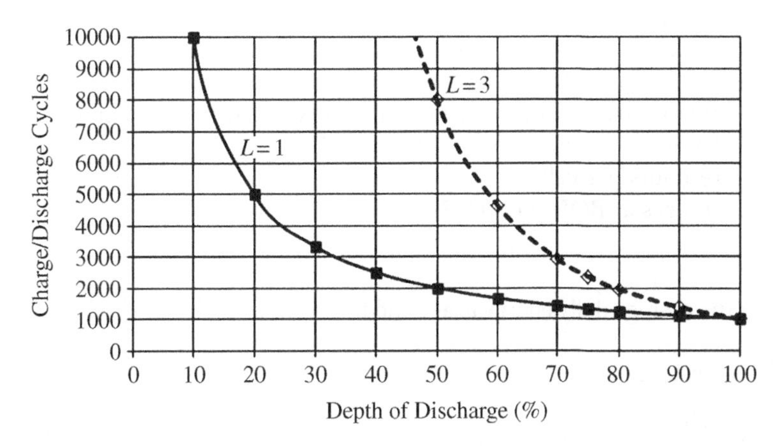

Figure 3.7 Charge\discharge cycles versus depth of discharge (DOD).

sustain 1000 cycles at 100% DOD, then it will sustain 2000 cycles for a 50% DOD and 10,000 cycles for a 10% DOD. In equation form:

$$N = N_{100\%} \times \frac{100\%}{\text{DOD}} \quad \text{or} \quad \text{DOD} = \frac{N_{100\%}}{N} \times 100\% \tag{3.10}$$

where $N_{100\%}$ is the number of cycles for a 100% DOD, and N is the number of cycles for a given DOD.

The assumption of a constant energy throughput is a reasonable starting assumption when sizing a battery for a BEV.

However, battery lifetimes can be significantly higher than the numbers just outlined for HEV batteries, which are designed for shallower discharges and a very high cycle life. Representative lifecycle numbers for HEV batteries can range from 3000 cycles for a DOD of 75% (from a 95% SOC to a 20% SOC), to 9000 cycles for a DOD of 50% (from 95% SOC to 45% SOC). Smaller charges and discharges can result in the hundreds of thousands of cycles necessary for a HEV.

Thus, a **cycle lifetime index** L is introduced here as a novel concept, and is used in this section as a parameter to quantify battery life.

Equation (3.10) is modified to include index L as follows:

$$N = N_{100\%} \left(\frac{100\%}{\text{DOD}}\right)^{L} \quad \text{or} \quad \text{DOD} = \left(\frac{N_{100\%}}{N}\right)^{1/L} \times 100\% \tag{3.11}$$

An enhanced-lifetime curve is shown in Figure 3.7 for $L = 3$. The lifetime of the battery can be enhanced by modifying the chemical composition of the cell, improving the battery cooling, reducing the battery float voltage, and reducing the DOD. Such a battery could be used in a plug-in hybrid-electric vehicle (PHEV) or hybrid-electric vehicle (HEV).

3.2.1 Examples of Battery Sizing

3.2.1.1 Example: BEV Battery Sizing

Determine the beginning-of-life kilowatt-hour storage required in a BEV battery pack based on the following requirements: eight years of operation, an average of 48 km of driving per day s_{day} over the 365 days of the year, daily charging, and an average battery output energy per kilometer E_{km} = 180 Wh/km. Assume L = 1 and $N_{100\%}$ = 1000.

Assume two parallel battery strings with 96 Li-ion cells per string, with a total number of cells N_{cell} = 192, and a nominal voltage of 3.75 V per cell.

Determine the ampere-hours per cell.

What are the vehicle ranges at BOL and EOL?

Solution:

First, let us determine the average daily energy usage and the number of charge\discharge cycles.

The total number of cycles N is

$$N = 8 \text{ years} \times 365 \frac{\text{charges}}{\text{year}} = 2920 \text{ cycles} \tag{3.12}$$

The average daily battery output energy E_{day} is

$$E_{day} = s_{day} E_{km} = 48\ \text{km} \times 0.18 \frac{\text{kWh}}{\text{km}} = 8.64\ \text{kWh} \tag{3.13}$$

The required DOD can be simply determined as follows:

$$\text{DOD} = \left(\frac{N_{100\%}}{N}\right)^{1/L} \times 100\% = \left(\frac{1000}{2920}\right)^{1/1} \times 100\% = 34.25\% \tag{3.14}$$

The BOL storage is

$$E_{BOL} = \frac{E_{day}}{\text{DOD}} = \frac{8.64\ \text{kWh}}{0.3425} = 25.23\ \text{kWh} \tag{3.15}$$

The EOL storage is

$$E_{EOL} = 0.8 \times E_{BOL} = 20.18\ \text{kWh} \tag{3.16}$$

This simple example shows that a battery pack with an initial energy storage of 25.23 kWh is required in order to supply the required average daily energy over 8 years.

The pack has a total of 192 battery cells with two strings of 96 cells in series. The battery pack voltage V_{bp} is

$$V_{bp} = \frac{N_{cell}}{2} V_b = \frac{192}{2} \times 3.75\ \text{V} = 360\ \text{V} \tag{3.17}$$

The battery pack ampere-hours Ah_{bp} is given by

$$\text{Ah}_{bp} = \frac{E_{BOL}}{V_{bp}} = \frac{25230}{360} \text{Ah} = 70.08\ \text{Ah} \tag{3.18}$$

The cell ampere-hours Ah_b is simply derived by dividing the pack Ah by the number of strings:

$$\text{Ah}_b = \frac{\text{Ah}_{bp}}{2} = \frac{70.08}{2} \text{Ah} = 35.04\ \text{Ah} \tag{3.19}$$

The vehicle range at BOL is

$$\text{Range (BOL)} = \frac{E_{BOL}}{E_{km}} = \frac{25230}{180} \text{km} = 140.2\ \text{km} \tag{3.20}$$

The vehicle range at EOL is

$$\text{Range (EOL)} = \frac{E_{EOL}}{E_{km}} = \frac{20180}{180} \text{km} = 112.1\ \text{km} \tag{3.21}$$

3.2.1.2 Example: PHEV Battery Sizing

In this example, we repeat the above exercise for an enhanced-lifetime battery for use in a PHEV with a cycle lifetime index $L = 3$. A single string of batteries is assumed.

Solution:

As before, the total number of cycles N is 2920 cycles, and the average daily energy consumption E_{day} is 8.64 kWh.

The required DOD can be simply determined as follows:

$$\mathrm{DOD} = \left(\frac{N_{100\%}}{N}\right)^{1/L} \times 100\% = \left(\frac{1000}{2920}\right)^{1/3} \times 100\% = 70\% \tag{3.22}$$

Thus, in order to have an 80% SOC after 2920 cycles, the BOL storage is

$$E_{BOL} = \frac{E_{day}}{\mathrm{DOD}} = \frac{8.64\,\mathrm{kWh}}{0.70} = 12.34\,\mathrm{kWh} \tag{3.23}$$

The EOL storage is

$$E_{EOL} = 0.8 \times E_{BOL} = 9.87\,\mathrm{kWh} \tag{3.24}$$

This simple example shows that a battery pack with initial energy storage of 12.34 kWh is required in order to supply the required average daily energy over 8 years.

The vehicle range at BOL for the PHEV is

$$\mathrm{Range\,(BOL)} = \frac{E_{BOL}}{E_{km}} = \frac{12340}{180}\mathrm{km} = 68.6\,\mathrm{km} \tag{3.25}$$

The vehicle range at EOL is

$$\mathrm{Range\,(EOL)} = \frac{E_{EOL}}{E_{km}} = \frac{9880}{180}\mathrm{km} = 54.9\,\mathrm{km} \tag{3.26}$$

The cell and battery pack ampere-hours are equal for a single string and are given by

$$\mathrm{Ah}_b = \mathrm{Ah}_{bp} = \frac{E_{BOL}}{V_b} = \frac{12340}{360}\,\mathrm{Ah} = 34.3\,\mathrm{Ah} \tag{3.27}$$

These two examples illustrate some of the challenges of sizing a battery pack. The lower-lifetime BEV battery pack has to be oversized for daily driving in order to meet the lifetime. The enhanced-lifetime PHEV battery pack can be much smaller but has less overall energy and reduced range at BOL and EOL. As the DOD is reduced, the pack cycle life can increase very significantly into hundreds of thousands of cycles. The typical HEV battery pack uses only a relatively small range of DOD, resulting in very long lifetimes over hundreds of thousands of cycles.

3.2.2 Battery Pack Discharge Curves and Aging

EVs and their batteries have been tested by the US Department of Energy using Idaho National Laboratory and the Center for Evaluation of Clean Energy Technology, an Intertek Company [4]. The test vehicles have been driven in the desert climate of Phoenix, Arizona. At various intervals, the battery packs are disconnected and discharged at a constant rate in order to characterize the performance of the battery pack with miles

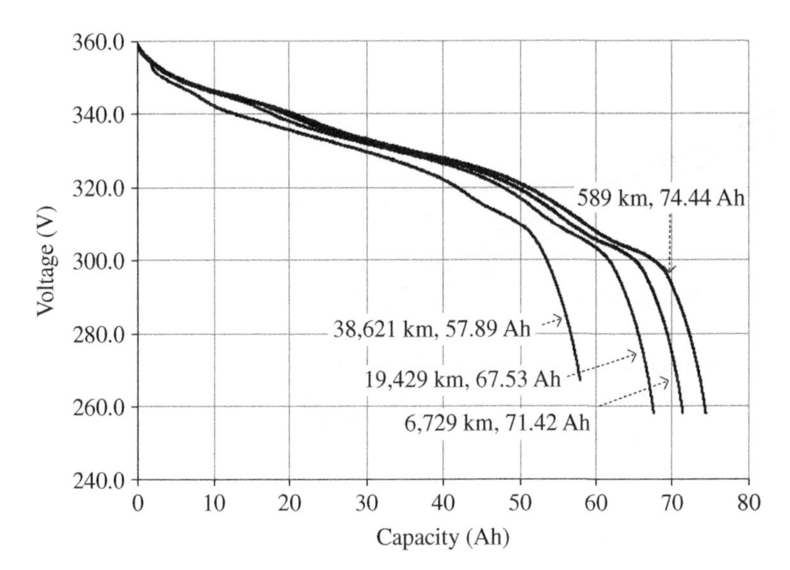

Figure 3.8 2013 Ford Focus BEV voltage versus capacity discharged during the static discharge test at a C/3 rate. (Data courtesy of Idaho National Laboratory and the Center for Evaluation of Clean Energy Technology, an Intertek Company).

driven. Test results for the 2013 Ford Focus BEV battery pack, manufactured by LG Chem, are as shown in Figure 3.8. In these tests, the vehicle battery pack is discharged at a constant rate of 8 kW. There is a significant drop in voltage from approximately 360 V to 260 V as the battery pack is discharged. The battery has been tested at various driving intervals. The baseline capacity of the battery was measured to be 74.44 Ah after 589 km on the odometer. The testing was repeated after 6,729 km, 19,429 km, and 38,621 km, with the overall testing taking 36 months. The battery capacity decreased to 71.42 Ah after 6,729 km, to 67.53 Ah after 19,429 km, and to 57.89 Ah at 38,621 km. The pack Ah degraded to 77.8% of the Ah at BOL.

The 2013 Chevy Volt PHEV features a similar LG Chem battery chemistry as the Ford Focus BEV. This pack was tested at the slightly higher 10 kW discharge rate. The results for the Chevy Volt are as shown in Figure 3.9. The battery capacity dropped from the baseline of 46.46 Ah at 7,227 km to 38 Ah at 257,541 km. Over the 44 months of testing the pack, the Ah degraded to 81.8% of the Ah at BOL. The pack storage had degraded about 10% by 195,387 km, and several percentage points more between 195,387 km and 257,541 km.

The degradation in battery capacity with kilometers driven is lower in the PHEV compared to the BEV as the battery pack DOD in the PHEV is controlled within a narrower range using the gasoline engine. The test results confirm this observation. The PHEV experienced an 18.4% degradation by 257,541 km and 44 months, whereas the BEV experienced a 22.2% degradation by 38,621 km and 36 months. The driving conditions were relatively harsh. The Chevy Volt was driven in EV mode only for about 20% of the overall distance, but the battery pack played an active role in energy management while in HEV mode. The air-conditioning was running for over 90% of the distance.

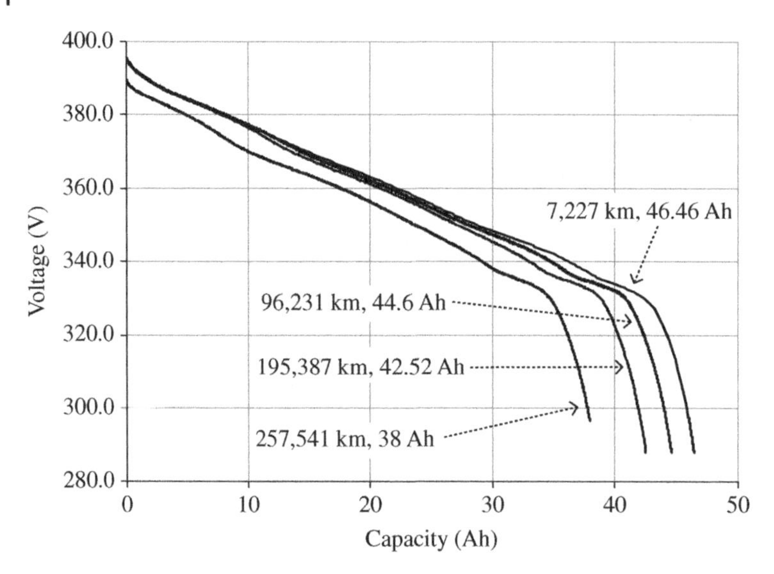

Figure 3.9 2013 Chevy Volt PHEV voltage versus capacity discharged during the static discharge test at a 10 kW rate. (Data courtesy of Idaho National Laboratory and the Center for Evaluation of Clean Energy Technology, an Intertek Company).

3.3 Battery Charging, Protection, and Management Systems

3.3.1 Battery Charging

All batteries have a similar charging profile. Generally, when discharged, the battery can accept a significant charge current. At a certain voltage or SOC, the current is reduced in order to ensure the correct charging voltage to the battery. As we have seen earlier in the chapter, the maximum cell voltage for the Li-ion battery is an important factor in the lifetime of the battery. A typical charging profile for a Li-ion battery is as shown in Figure 3.10. Initially, the battery is charged at a constant current. When the battery reaches around 80% SOC, the charge current is reduced in order to control the voltage applied to the battery. In this mode, the current is tapered in order to maintain the appropriate battery voltage. During this stage, the SOC slowly increases, and the battery can take significantly longer to go from 80% to 100% compared to going from a low SOC to 80%. Generally, high-power charging is applied to Li-ion batteries up to an SOC of 80%.

3.3.2 Battery Failure and Protection

Batteries can be a source of catastrophic failure resulting in dangerous and possible life-threatening consequences. The automotive battery pack must undergo rigorous testing in order to ensure benign failure modes. In general, the battery should not emit parts, fire, or toxic or hazardous gases. Care must additionally be taken in manufacturing, transporting, using, and recycling batteries. Many safety standards exist to define the level of danger from an automotive battery. The following are examples of safety tests.

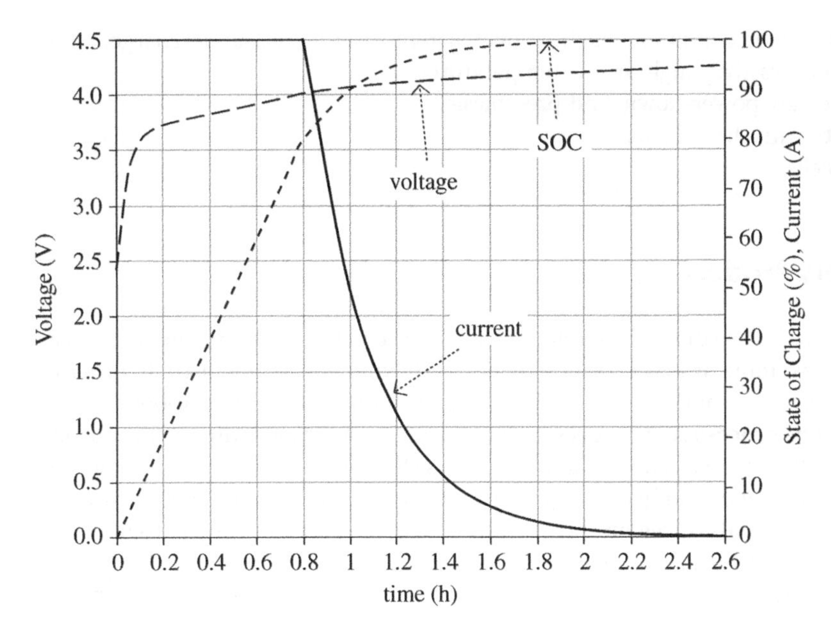

Figure 3.10 Li-ion charge profile.

Penetration: Penetration of the pack can expose the lithium-based materials to air, resulting in fire.

Crush: A battery pack must not become hazardous if it is crushed in a collision.

Thermal stability: A battery pack can be susceptible to thermal runaway if the pack temperature is too high. This was a significant problem for sodium-sulfur batteries but is generally not a problem for Li-ion. Li-ion batteries are thermally safe up to 80°C. Potentially dangerous reactions can occur at higher temperatures, and additional safety features must be implemented in order to reduce the likelihood of emissions of flames or hazardous gases. Reference the Further Reading section.

Overcharge\overdischarge: These conditions can cause internal heating with the possibility of thermal runaway.

External short: An external short can result in a very significant internal power dissipation within the battery with associated heating, elevated temperatures, and thermal runaway.

3.3.3 Battery Management System

As can be understood from the earlier material in this chapter, a battery pack is a complex energy storage system. In order to carry out the basic task of storing energy, a battery management system (BMS) is required to take all the battery and vehicle information and ensure safe operation of the battery. The typical BMS has the following tasks:

Monitor voltage, current, power, and temperature.
Estimate the SOC.
Maintain a healthy battery and **conduct** diagnostics.

Protect against fault conditions such as overcurrent, overcharge, undercharge, short circuit, and excessively high or low temperatures.
Control power-up, power-down, and pre-charge.
Communicate with the vehicle.
Balance the cells.

3.4 Battery Models

Electrochemistry is the study of the relationship between electricity and chemical reactions and is a vast topic. It is necessary to delve a little into electrochemistry in order to model the battery. The authors found [5] to be a useful introductory textbook.

A **spontaneous process** is a process which can proceed on its own without any outside assistance. A battery is an example of a spontaneous electrochemical redox reaction converting chemical energy and generating an electromotive force (emf) and performing electrical work. The **Gibbs free energy** is a measure of the reaction's ability to do work.

The open-circuit reversible voltage of an electrochemical cell V_r^0 is the no-load open-circuit voltage of the cell, and is related to the change in the Gibbs free energy ΔG^0 by the equation:

$$V_r^0 = -\frac{\Delta G^0}{nF} \tag{3.28}$$

where n represents the number of moles of electrons transferred according to the balanced equation for the redox reaction, and F is Faraday's constant (96,485 J/V-mol). Josiah Gibbs (1839–1903) was the first person to be awarded a PhD in engineering from an American university (Yale in 1863), and he contributed greatly to the development of chemical thermodynamics.

As the electrochemical cell is discharged, and the DOD increases or the SOC reduces, the reactants within the cell are consumed, and the concentrations of these reactions are reduced. A German chemist, Walther Nernst (1864–1941), also contributed significantly to electrochemistry and developed an equation relating the battery cell voltage V_b to the concentration of the reactants:

$$V_b = V_r^0 - \frac{RT}{nF}\log_e Q_R \tag{3.29}$$

where R is the ideal gas constant, T is the temperature, and Q_R is the reaction quotient, which is a function of the reactant concentrations.

The Nernst equation is important as it relates the cell voltage to the SOC or DOD of the battery. The earlier Figure 3.5, reproduced in Figure 3.11, clearly illustrates the drop in cell voltage with the SOC. However, strictly speaking, the Nernst equation only applies when there is no current in the cell. In addition, we see that the cell voltage is dependent on the cell current. This relationship is governed by a number of voltage drops within the cell. There are three main voltage drops:

Ohmic, the internal battery resistance due to the terminations and current collectors;
Activation polarization, the internal resistance due to charge-transfer reactions faced by the electrons at the electrode–electrolyte junctions;

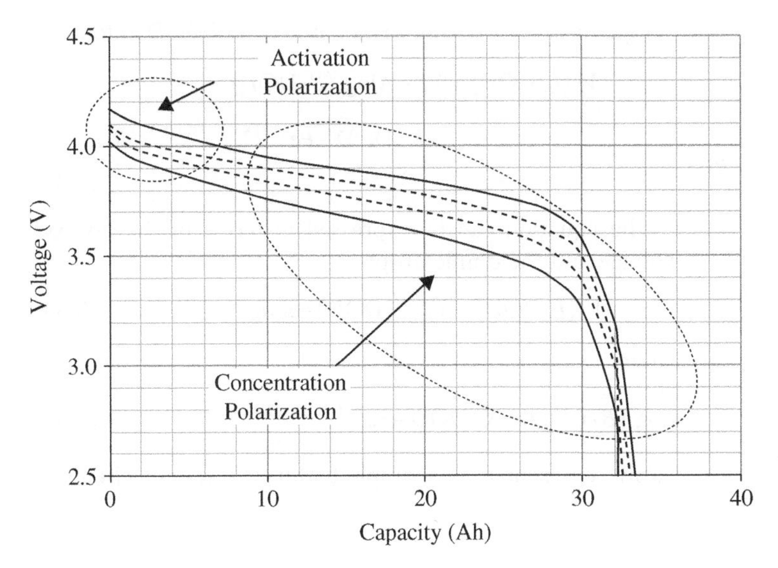

Figure 3.11 Discharge curves for Li-ion cell with polarization regions outlined.

Concentration polarization, the internal resistance faced by the ions due to the electrolyte concentration.

For simplicity, we assume in this discussion that these resistances can be modeled by a single lumped equivalent battery internal resistance R_b. While this is a simplification of the complex electrochemical resistances, it provides a useful engineering approximation.

Thus, the battery cell voltage of Equation (3.29) can be modified to include an ohmic drop:

$$V_b = V_r^0 - \frac{RT}{nF}\log_e Q_R - R_b I_b \tag{3.30}$$

where R_b is the internal resistance of the battery due to the combined ohmic and polarization drops, and I_b is the cell current. The resistance can be different between charging and discharging, but this text assumes a constant resistance for simplicity.

Finally, for this textbook, the battery equation is modified to account for two significant voltage drops. As can be seen from Figure 3.11, after the initial logarithmic drop, the battery voltage drops quasi-linearly with battery capacity followed by an exponential drop. Thus, Equation (3.30) is modified as follows to show the cell voltage equation as a function of the no-load reversible voltage; the logarithmic, linear, and exponential drops with capacity; and an ohmic drop with current:

$$V_b(I_b,y) = V_r^0 - A\log_e(B\cdot y) - K\cdot y - Fe^{G\cdot(y-y_3)} - R_b I_b \tag{3.31}$$

where y is a variable that can be related to the capacity, the DOD, the SOC, or the cell energy, y_3 is the value at which the exponential drop-off begins, and A, B, K, F, and G are values determined by curve fitting.

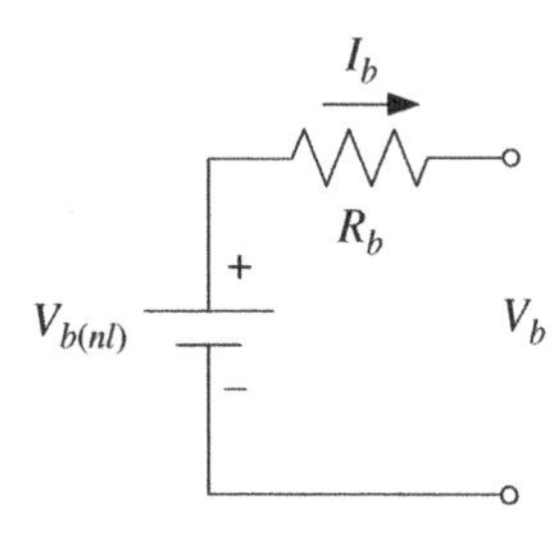

Figure 3.12 Static battery equivalent circuit model.

Equation (3.31) can be written in terms of DOD as follows:

$$V_b(I_b,\text{DOD}) = V_r^0 - A\log_e(B\cdot\text{DOD}) - K\cdot\text{DOD} - Fe^{G\cdot(\text{DOD}-\text{DOD}_3)} - R_bI_b \tag{3.32}$$

The equation can be expressed in terms of the no-load voltage and the ohmic drop:

$$V_b(I_b,\text{DOD}) = V_{b(nl)}(\text{DOD}) - R_bI_b \tag{3.33}$$

where

$$V_{b(nl)}(\text{DOD}) = V_r^0 - A\log_e(B\cdot\text{DOD}) - K\cdot\text{DOD} - Fe^{G\cdot(\text{DOD}-\text{DOD}_3)} \tag{3.34}$$

These equations can be easily represented as a simple battery equivalent circuit as shown in Figure 3.12.

3.4.1 A Simple Novel Curve Fit Model for BEV Batteries

There are many approaches to curve fitting to generate a battery model. The focus of this section is on a static or dc steady-state model rather than on a more complex dynamic model. This novel static model can be generated on the basis of the discharge curves, while additional testing is required to generate a dynamic model with capacitive components.

The static curve can be modeled in a variety of ways. It is often useful to have a closed-form equation as generated in Equation (3.32).

The following curve segments can be identified on the basis of the general characteristics of the discharge curves, as shown in Figure 3.13(a): an initial quasi-logarithmic drop and a quasi-linear drop in the middle range, followed by a quasi-exponential drop as the cell is deeply discharged. In addition, the cell voltage drops with the cell discharge current due to the internal voltage drops.

The first step is to identify the various modes within the discharge curve of Figure 3.13 (a). The characteristic is quasi-logarithmic between (Ah_1, V_1) and (Ah_2, V_2), quasi-linear between (Ah_2, V_2) and (Ah_3, V_3), and quasi-exponential between (Ah_3, V_3) and (Ah_5, V_5). An interim point is required for characterization in the exponential region, and this is (Ah_4, V_4). These data are recorded in Table 3.5 for the capacity rate of $C/3$. Points 6 and 7, as presented in Table 3.5, will be used in a later simplification. Curve-fitting programs can be used to generate a curve fit for the battery cell.

A simple curve fit is developed for the 33.3 Ah battery cell. The curve-fit approach is outlined in the appendix of this chapter. As discussed in the appendix, the values of Table 3.6 are determined for the various parameters in Equation (3.32).

Plots of cell voltage can now be easily reproduced for various discharge rates. Plots based on Equation (3.32) are shown in Figure 3.13(b). These curves are an approximation of the cell terminal voltage. Now that we have a closed-form equation for the cell voltage in terms of the current and DOD, we can plot the curves as shown in Figure 3.14(a). As expected, the cell voltage drops to 2.5 V at $C/3$, the rated capacity, at DOD = 100%. The characteristic at $4C$ is also shown.

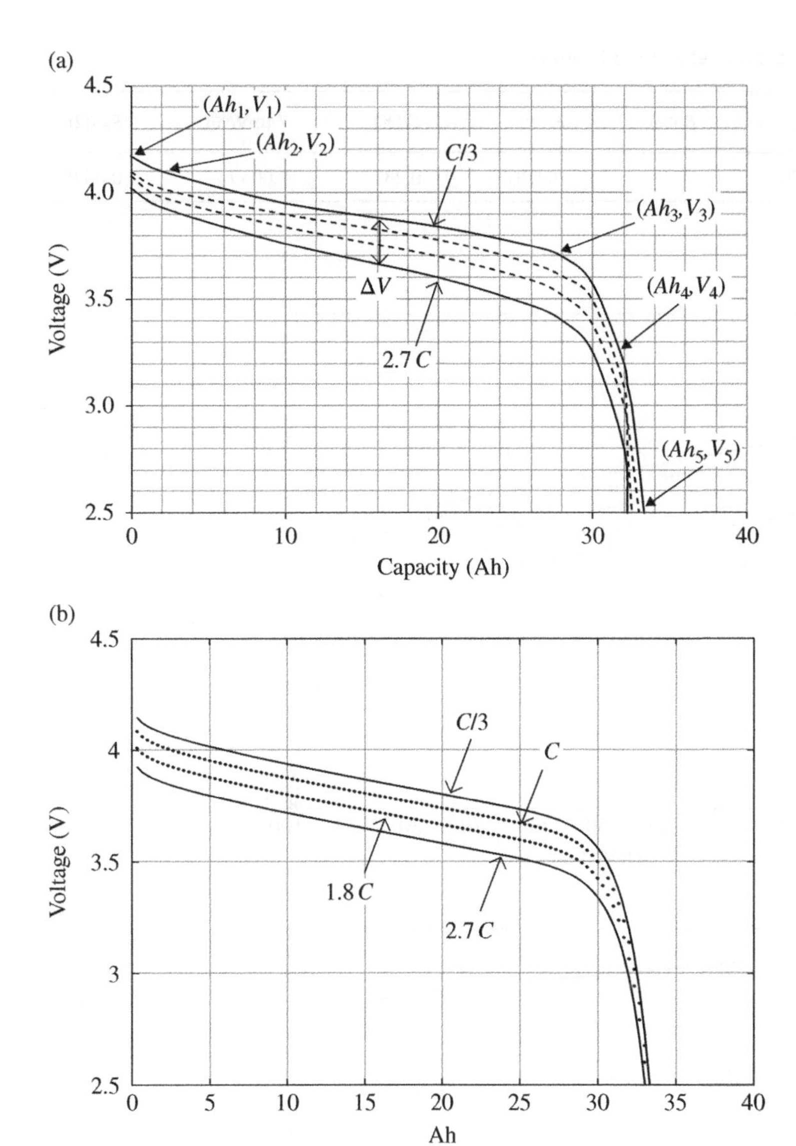

Figure 3.13 Discharge curves for (a) 33.3 Ah cell and (b) simulation.

Table 3.5 Curve-fitting coordinates for 33.3 Ah cell.

		1	2	3	4	5	6	7
Ah	Ah	0.3	2	28	32	33.3	8.3	25
Vcell	V	4.15	4.1	3.7	3.2	2.5	4	3.75
DOD	%	1	6	84	96	100	25	75
y	%	1	6	84	96	100	25	75

Table 3.6 Curve-fitting parameters for 33.3 Ah cell.

V_r^0 (V)	A (V)	B (%)	F (V)	G (%)	K (mV/%)	R_b (Ω)
4.18	0.0279	1	0.0281	0.231	0.0039	0.0028

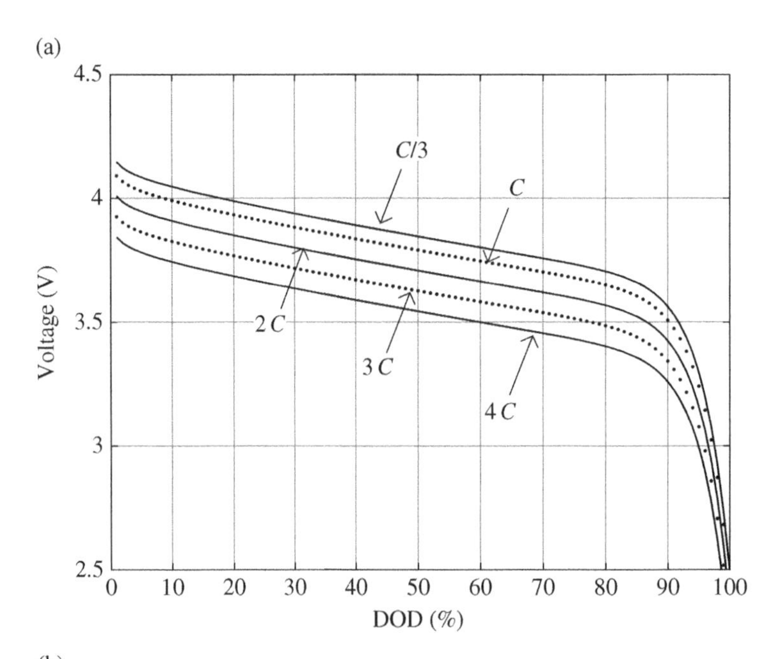

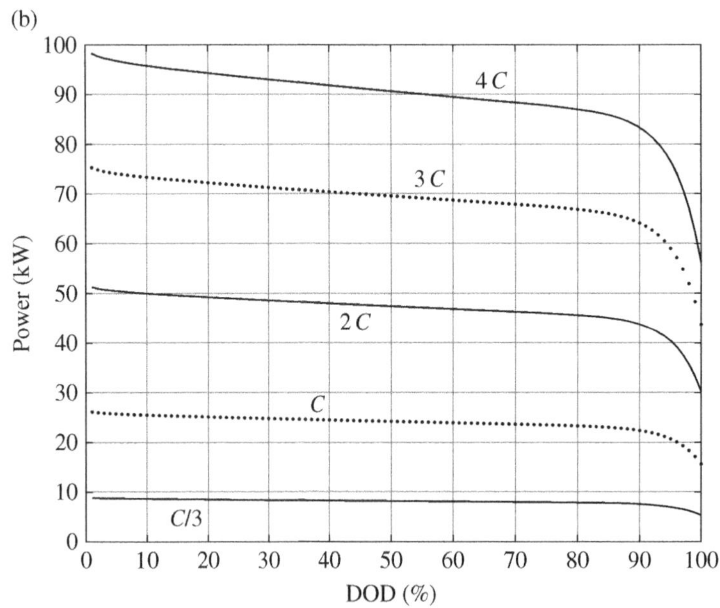

Figure 3.14 (a) 33.3 Ah cell voltage versus DOD and (b) pack power versus DOD for 192 33.3 Ah cells.

Simulated curves for battery pack power as a function of DOD for constant discharge currents are recreated in Figure 3.14(b) for a 192 cell battery pack. For example, the 3*C* discharge rate results in a peak power of about 75 kW when fully charged, dropping to about 67 kW at a DOD of 80%. A discharge current of 133.2 A or a 4*C* rate would result in output powers greater than 80 kW at 90% DOD.

3.4.2 Voltage, Current, Resistance, and Efficiency of Battery Pack

Once the cell voltage and current are known, the battery pack terminal voltage and current can be determined. If a battery pack has N_{par} strings in parallel and N_{ser} cells in series for each string, as shown in Figure 3.15, then the battery pack voltage V_{bp} and current I_{bp} are related to the cell voltage V_b and current I_b by

$$V_{bp} = N_{ser} V_b \tag{3.35}$$

and

$$I_{bp} = N_{par} I_b \tag{3.36}$$

The battery pack resistance R_{bp} is related to the cell resistance by

$$R_{bp} = \frac{N_{ser}}{N_{par}} R_b \tag{3.37}$$

The efficiency of the battery pack during discharge, η_{dis}, is the ratio of the useful output power to the sum of the useful output power and the internal ohmic loss:

$$\eta_{dis} = \frac{V_{bp} I_{bp}}{V_{bp} I_{bp} + R_{bp} I_{bp}^2} \times 100\% = \frac{V_{bp}}{V_{bp(nl)}} \times 100\% \tag{3.38}$$

where $V_{bp(nl)}$ is the no-load battery pack terminal voltage.

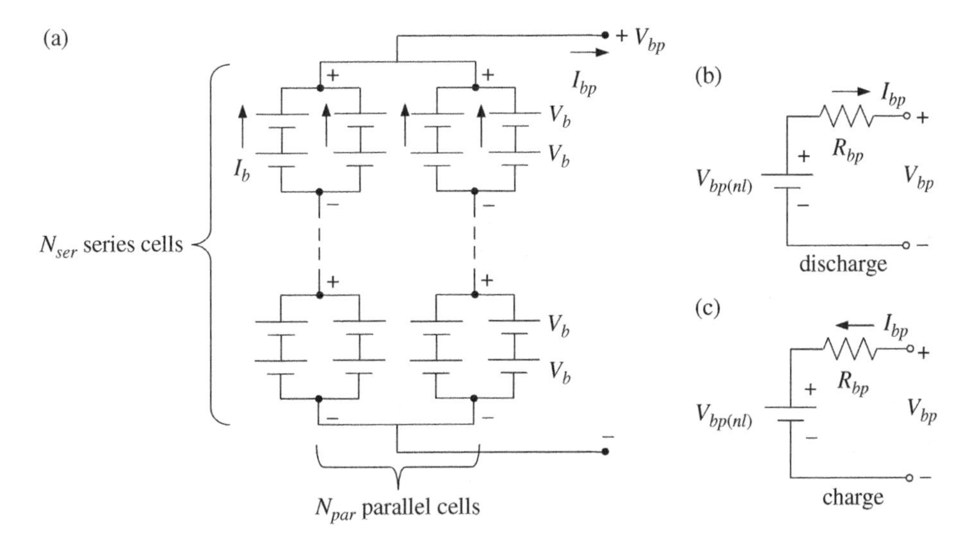

Figure 3.15 Series-parallel battery pack equivalent circuits.

The efficiency of the battery pack during charging η_{ch} is the ratio of the input power minus the internal resistive loss to the input power:

$$\eta_{ch} = \frac{V_{bp}I_{bp} - R_{bp}\,I_{bp}^2}{V_{bp}I_{bp}} \times 100\% = \frac{V_{bp(nl)}}{V_{bp}} \times 100\% \tag{3.39}$$

3.4.2.1 Example: Determining the Pack Voltage Range for a BEV

What is the voltage range for the 33.3 Li-ion Ah cell when used in a BEV application with a DOD of 0% to 100%?

What is the battery pack voltage range if 192 cells are arranged in series-parallel?

Solution:

From the earlier analysis, the cell voltage is 4.18V at 0% DOD and is 2.5 V when fully discharged.

The 192 cells arranged in series-parallel results in the equivalent of 96 cells in series. Thus, we simply multiply the cell voltage range of 2.5 V to 4.18 V by 96, yielding a pack voltage ranging from 240 V when fully discharged to 401 V when fully charged.

3.4.3 A Simple Curve-Fit Model for HEV Batteries

The DOD for a HEV is typically lower than that for a BEV. If we are planning on using only 50% of the SOC of the battery between 25% and 75% with the typical SOC being 50%, then the life of the battery will increase significantly. The modeling can also become easier as we are largely dealing with the quasi-linear portion of the discharge curve, as shown in Figure 3.16. Thus, we can model the battery cell by ignoring the logarithmic and exponential portions of the cell equation. The simplified equation is

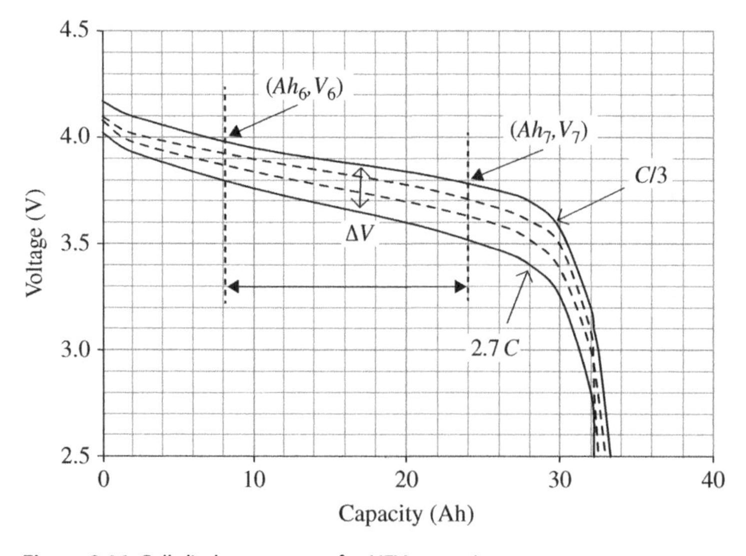

Figure 3.16 Cell discharge curves for HEV operation.

$$V_b = V_{b(nl,y6)} - K \cdot (y - y_6) - R_b I_b \tag{3.40}$$

where $V_{b(nl,y6)}$ is the no-load voltage at y_6. The equation can be expressed in terms of DOD as

$$V_b = V_{b(nl,\mathrm{DOD6})} - K \cdot (\mathrm{DOD} - \mathrm{DOD}_6) - R_b I_b \tag{3.41}$$

An estimate of the cell internal resistance R_b can be determined by measuring the cell voltage drop with increased current. Thus, the voltage drop ΔV is estimated at about 0.22 V at 50% of the rated Ah, as the discharge current increases from 11.1 A at $C/3$ to 90 A at $2.7C$.

$$R_b = \frac{\Delta V}{I_{2.7C} - I_{C/3}} = \frac{0.22}{90 - 11.1} \Omega = 2.8\ \mathrm{m}\Omega \tag{3.42}$$

The no-load voltage at a 25% DOD is obtained as

$$V_{b(nl,\mathrm{DOD6})} = V_6 + R_b I_b = 4.0\ \mathrm{V} + 0.0028 \times 11.1\ \mathrm{V} = 4.03\ \mathrm{V} \tag{3.43}$$

The voltage drop is modeled by a straight line of slope K through the quasi-linear region between (Ah_6, V_6) and (Ah_7, V_7), with values given in Table 3.5:

$$K = \frac{V_6 - V_7}{\mathrm{DOD}_7 - \mathrm{DOD}_6} = \frac{4 - 3.75}{75 - 25} \frac{\mathrm{mV}}{\%} = 5 \frac{\mathrm{mV}}{\%} \tag{3.44}$$

3.4.3.1 Example: Determining the Pack Voltage Range for a HEV

What is the voltage range for the cell when used in a HEV application with a DOD of 25% to 75% and a load ranging from no-load to a full load of $6C$?

What is the battery pack voltage if there are 192 cells arranged with 96 cells in series and two strings in parallel?

Solution:

At no-load, the cell voltage is simply $V_{b(nl,\mathrm{DOD6})} = 4.03$ V.

At full load, the $6C$ rate is 199.8 A, and the cell voltage is

$$V_b = V_{b(nl,\mathrm{DOD6})} - K \cdot (\mathrm{DOD} - \mathrm{DOD}_6) - R_b I_b$$

$$= 4.03\ \mathrm{V} - 0.005 \times (75 - 25)\ \mathrm{V} - 0.0028 \times 199.8\ \mathrm{V} = 3.22\ \mathrm{V}$$

Thus, the cell output voltage ranges from 3.22 V to 4.03 V.

The 192 cells arranged in series-parallel results in 96 cells in series. Thus, we simply multiply the cell voltage by 96, obtaining a pack voltage ranging from 387 V at no-load to 309 V at full load.

3.4.4 Charging

There is a range of available powers for charging a battery pack. A standard home charger may be able to output from 2 kW to 20 kW depending on the installation. Fast charging

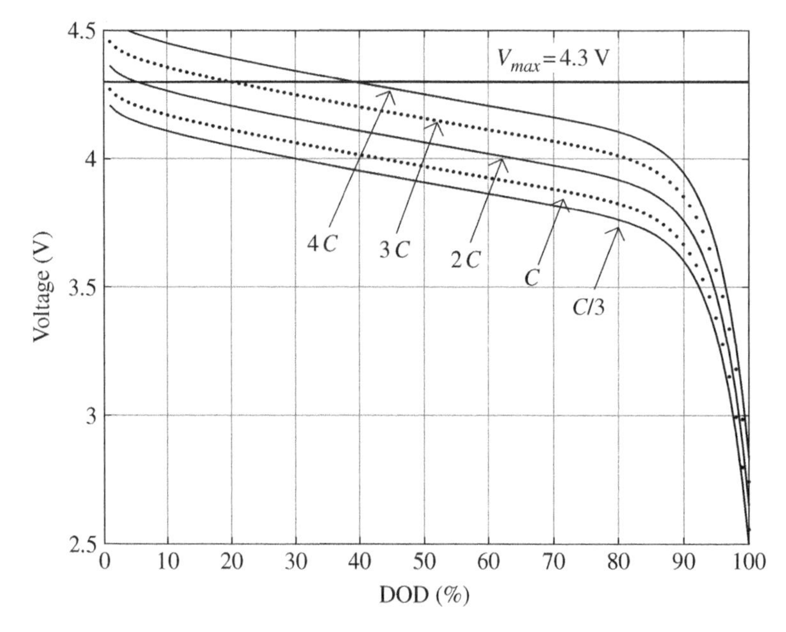

Figure 3.17 Simulated charge curves for the 33.3 Ah cell.

can run into many tens of kilowatts. The same equations can be used. All we have to do is change the sign on the battery current, ensuring that the terminal voltage is greater than the no-load cell voltage to recharge the cell.

$$\text{BEV}: \; V_b = V_r^0 - A\log_e(B \cdot y) - K \cdot y - Fe^{G \cdot (y - y_3)} + R_b I_b \tag{3.45}$$

$$\text{HEV}: \; V_b = V_{b(nl,y6)} - K \cdot (y - y_6) + R_b I_b \tag{3.46}$$

A set of characteristic curves of cell terminal voltage versus DOD are plotted in Figure 3.17 for charging. Care has to be taken when charging to not exceed the maximum cell voltage. A maximum voltage of 4.3 V is shown.

Using 4.3 V as the maximum, it is clear that charging may have to be curtailed at a low DOD (or at a high SOC). For example, at 3*C* the cell hits the limit at approximately 20% DOD or 80% SOC. The curves show that charging at 1*C* or below keeps the cell voltage below the limit.

3.4.4.1 Example: Fast Charging a Battery Pack

A 24 kWh battery pack can be fast charged from 0% to 80% SOC in 30 min. Determine the approximate charge current and power in order to achieve this charge time.

Solution:

The nominal pack size is 24 kWh, and so 80% is 19.2 kWh.

Thus, the battery has to be recharged at an average rate of approximately 2 times 19.2 kW, which equals 38.4 kW net in order to recharge the pack in 30 min.

If the nominal battery pack voltage is 360 V, say, then the average charge current is 38.4 kW/360 V or 107 A.

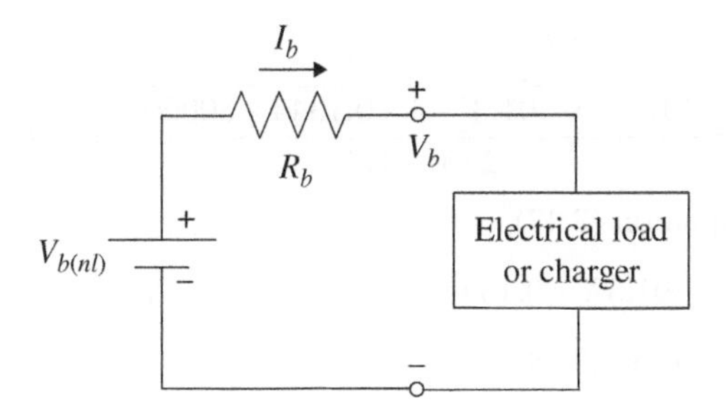

Figure 3.18 Simple battery model.

3.4.5 Determining the Cell/Pack Voltage for a Given Output\Input Power

If the DOD is known, then the cell voltage is simply

$$V_b = V_{b(nl)} - R_b I_b \tag{3.47}$$

as the no-load voltage $V_{b(nl)}$ is a function of the DOD. The current is positive for discharging and negative for charging. The basic circuit model is presented in Figure 3.18.

If the battery input or output power P_b is known, as often is the case, then the current and terminal voltage are easily determined by solving the following quadratic equation:

$$P_b = V_b I_b = V_{b(nl)} I_b - R_b I_b^2 \tag{3.48}$$

$$\Rightarrow R_b I_b^2 - V_{b(nl)} I_b + P_b = 0 \tag{3.49}$$

The solution is

$$I_b = \frac{V_{b(nl)} - \sqrt{V_{(b)nl}^2 - 4R_b P_b}}{2R_b} \tag{3.50}$$

for discharge, and

$$I_b = \frac{V_{b(nl)} - \sqrt{V_{b(nl)}^2 - 4R_b(-P_b)}}{2R_b} \tag{3.51}$$

for charge.

3.4.5.1 Example: Battery Discharge

A battery has 96 cells in series per string with two parallel strings. Each cell has a no-load voltage of 4.18 V and an internal resistance of 2.8 mΩ.

i) Determine the pack current and voltage under a 80 kW discharge if the battery is fully charged.
ii) Determine the discharge efficiency of the battery.

Solution:

The no-load voltage when fully charged is 96 times 4.18 V or $V_{bp(nl)} = 401.3$ V.

From Equation (3.37), the pack resistance is

$$R_{bp} = \frac{N_{ser}}{N_{par}} R_b = \frac{96}{2} \times 2.8\ \text{m}\Omega = 134.4\ \text{m}\Omega$$

From Equation (3.50),

$$I_{bp} = \frac{401.3 - \sqrt{401.3^2 - 4 \times 0.1344 \times 80000}}{2 \times 0.1344} \text{A} = +214.8\text{A}$$

From Equation (3.47),

$$V_{bp} = 401.3\,\text{V} - 0.1344 \times 214.8\,\text{V} = 372.4\,\text{V}$$

From Equation (3.38),

$$\eta_{dis} = \frac{V_{bp}}{V_{bp(nl)}} \times 100\% = \frac{372.4}{401.3} \times 100\% = 92.8\%$$

3.4.5.2 Example: Battery Charge

Determine the pack current and voltage under a 50 kW charge if the battery is fully discharged. The cell voltage drops to 2.5 V when fully discharged. How efficient is the charging of the battery at this power level?

Solution:

The no-load voltage when fully discharged is 96 times 2.5 V or $V_{bp(nl)} = 240$ V. The pack resistance is as before.

$$I_{bp} = \frac{240 - \sqrt{240^2 - 4 \times 0.1344 \times (-50000)}}{2 \times 0.1344} \text{A} = -188.4\text{A}$$

and the pack voltage is

$$V_{bp} = 240\text{V} - 0.1344 \times (-188.4)\,\text{V} = 265.3\,\text{V}$$

From Equation (3.39),

$$\eta_{ch} = \frac{V_{bp(nl)}}{V_{bp}} \times 100\% = \frac{240}{265.3} \times 100\% = 90.5\% \tag{3.52}$$

3.4.6 Cell Energy and Discharge Rate

The usable battery capacity drops as the C rate increases. This is basically due to the increased resistive losses at the higher discharge rates. Note that the following approach to estimating the cell energy of a battery works for the Li-ion cells presented in this book but is not generally applicable to other battery chemistries.

The Ah value for an automotive BEV battery is typically specified at a low rate, for example, the $C/3$ rate. The total energy stored within the cell prior to discharge E_{cell} is equal to the rated capacity times the rated voltage plus the resistive losses for the rated condition:

$$E_{cell} = \text{Ah}_{rated} \cdot V_{rated} + R_b I_{rated}^2 \cdot \frac{\text{h}}{C_{rated}} \tag{3.53}$$

where "h" is one hour.

The nominal voltage V_{xC} at other C rates, designated xC, can be approximated by subtracting the internal resistance voltage drop from the rated voltage:

$$V_{xC} \approx V_{rated} - R_b(I_{xC} - I_{rated}) \tag{3.54}$$

where x is the coefficient of the C rate.

The available output energy at xC is then approximated by

$$E_{xC} \approx E_{cell} - R_b I_{xC}^2 \cdot \frac{\text{h}}{x} \tag{3.55}$$

The efficiency of the discharge (and also of charge) is also a function of the C rate;

$$\eta_{dis} = \frac{E_{xc}}{E_{cell}} (\%) \text{ for discharge, } \eta_{ch} = \frac{E_{cell}}{E_{xc}} (\%) \text{ for charge} \tag{3.56}$$

3.4.6.1 Example: Cell Capacity

The capacity of the cell is approximately 33.3 Ah at $C/3$ with a rated voltage of 3.75 V.

Determine the capacity at $3C$. Assume R_b = 2.8 mΩ.

Solution:

The total energy stored within the cell prior to discharge is equal to the product of the capacity and the rated voltage plus the resistive losses at $C/3$.

$$\begin{aligned} E_{cell} &= \text{Ah}_{C/3} V_{rated,C/3} + R_b I_{C/3}^2 \times 3\,\text{h} \\ &= 33.3 \times 3.75\,\text{Wh} + 0.0028 \times 11.1^2 \times 3\,\text{Wh} \\ &= 124.88\,\text{Wh} + 1.03\,\text{Wh} \\ &= 125.91\,\text{Wh} \end{aligned} \tag{3.57}$$

At $3C$, the current increases to 99.9 A, and there is an increased internal resistive power loss that reduces the energy available from the cell.

$$E_{3C} = E_{cell} - R_b I_{3C}^2 \times \frac{\text{h}}{3} = 125.91\,\text{Wh} - 0.0028 \times 99.9^2 \times \frac{\text{h}}{3}\,\text{W} = 116.6\,\text{Wh} \tag{3.58}$$

It is clear from the example that the available output energy from the cell is reduced as the C rate increases. This occurs because the internal battery losses increase with the square of the current. Thus, if the $C/3$ rate is the baseline for a particular cell, then the available energy is reduced as the rate increases. The ampere-hour capacity may not necessarily drop as it is equal to the available energy divided by the nominal voltage, which is also dropping.

The rated voltage of the cell is specified as 3.75 V for this cell, and this is the average voltage as the cell discharges from a SOC of 100% to 0% at rated C ($C/3$ in this case). The average voltage for $3C$ is

$$V_{3C} \approx V_{rated} - R_b(I_{3C} - I_{rated}) = 3.75\,\text{V} - 0.0028 \times (99.9 - 11.1)\text{V} = 3.50\,\text{V}$$

Table 3.7 Estimations of various quantities for the 33.3 Ah Li-ion cell.

Parameter	Unit	C/3 (11.1A)	1C (33.3A)	3C (99.9 A)
x		1/3	1	3
V	V	3.75	3.69	3.50
Ah	Ah	33.3	33.3	33.3
Wh	Wh	124.88	122.8	116.6
η	%	99.2	97.5	92.6

At the 3C rate,

$$\eta_{dis} = \frac{E_{xc}}{E_{cell}}(\times 100\%) = \frac{116.6}{125.91}(\times 100\%) = 92.6\% \tag{3.59}$$

The capacity can be determined by dividing the output energy by the nominal voltage

$$Ah_{3C} = \frac{E_{3C}}{V_{3C}} = \frac{116.6}{3.50}\,\text{Ah} = 33.3\,\text{Ah} \tag{3.60}$$

a value which shows little change from the rated capacity.

These quantities are summarized in Table 3.7 for the 33.3 Ah cell for $C/3$, C, and 3C. As can be seen from the table, the capacity remains approximately constant for the various rates, whereas the average cell terminal voltage, useful output energy, and efficiency vary relatively significantly.

3.5 Example: The Fuel Economy of a BEV Vehicle with a Fixed Gear Ratio

In this example, the vehicle fuel performance is investigated for the simple drive cycle and data presented in Section 2.3. The BEV has a 60 kWh battery. It is assumed that the combined battery and electric drive efficiency η_{ed} is 85%, the charging efficiency from the plug η_{chg} is 85%, and the gearing\transmission efficiency η_g is 95%, as shown in Figure 3.19. The battery power drain during idle time is 200 W.

Consider two different countries in determining the CO_2 emissions per kilometer for the electrical energy from the plug: the United States with about 500 gCO_2/kWh (electrical), and Norway with about 10 gCO_2/kWh (electrical).

Solution:

From Section 2.3, the traction powers are P_{t1} = 3.478 kW and P_{t2} =12.050 kW. The overall traction energy for the drive cycle is E_{tC} = 20.72 MJ (5.76 kWh).

The power required from the battery P_b is the motor output power divided by the electric drive efficiency, η_{ed}:

$$P_b = \frac{P_t}{\eta_{ed}} \tag{3.61}$$

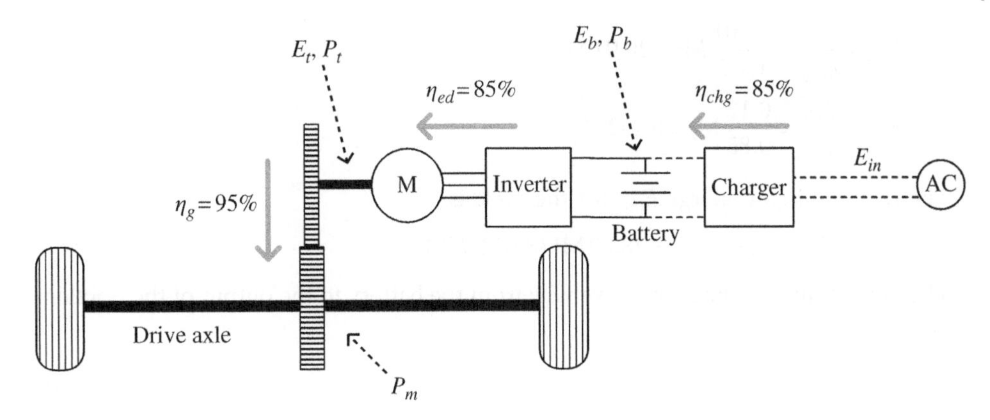

Figure 3.19 BEV architecture.

The battery powers P_{b1} and P_{b2} are

$$P_{b1} = \frac{P_{t1}}{\eta_{ed}} = \frac{3.478}{0.85}\,\text{kW} = 4.092\,\text{kW}$$

$$P_{b2} = \frac{P_{t2}}{\eta_{ed}} = \frac{12.050}{0.85}\,\text{kW} = 14.176\,\text{kW}$$

The battery power during idle time is

$$P_{b3} = 0.2\,\text{kW}$$

The energy consumed from the battery is simply the product of power and time:

$$E_b = P_b \times t \tag{3.62}$$

The battery output energies E_{b1}, E_{b2}, and E_{b3} are

$$E_{b1} = P_{b1} \times t_1 = 4.092 \times 1800\,\text{kJ} = 7.37\,\text{MJ}$$

$$E_{b2} = P_{b2} \times t_2 = 14.176 \times 1200\,\text{kJ} = 17.01\,\text{MJ}$$

$$E_{b3} = P_{b3} \times t_2 = 0.2 \times 600\,\text{kJ} = 0.120\,\text{MJ}$$

The battery output energy over the drive cycle, E_{bC}, is

$$E_{bC} = E_{b1} + E_{b2} + E_{b3} = 7.37\,\text{MJ} + 17.01\,\text{MJ} + 0.12\,\text{MJ} = 24.5\,\text{MJ}\ \ (6.81\,\text{kWh})$$

The BEV is fueled using electricity from the plug. Assuming a charging efficiency η_{chg} from the plug to the battery of 85%, the supplied energy from the grid E_{in} is

$$E_{in} = \frac{E_b}{\eta_{chg}} \tag{3.63}$$

The input energies, E_{in1}, E_{in2}, and E_{in3}, are

$$E_{in1} = \frac{E_{b1}}{\eta_{chg}} = \frac{7.37}{0.85}\,\text{MJ} = 8.67\,\text{MJ}$$

$$E_{in2} = \frac{E_{b2}}{\eta_{chg}} = \frac{17.01}{0.85}\,\text{MJ} = 20.01\,\text{MJ}$$

$$E_{in3} = \frac{E_{b3}}{\eta_{chg}} = \frac{0.12}{0.85}\,\text{MJ} = 0.14\,\text{MJ}$$

The required input energy E_{inC} for the cycle is

$$E_{inC} = E_{in1} + E_{in2} + E_{in3} = 28.82\,\text{MJ}\ (8.01\,\text{kWh})$$

The powertrain efficiency of the vehicle from the battery to the output of the traction motor is

$$\eta_{pt} = \frac{E_t}{E_b} \tag{3.64}$$

and so

$$\eta_{ptC} = \frac{E_{tC}}{E_{bC}} = \frac{20.72}{24.5} \times 100\% = 84.6\%$$

The fuel consumption (FC) of the vehicle and the related fuel economy are typically based on the input energy from the plug:

$$FC = \frac{E_{in}}{s}\left(\text{in}\,\frac{\text{Wh}}{\text{km}}\right) \tag{3.65}$$

and so the fuel economy for this cycle is

$$FC = \frac{E_{inC}}{s} = \frac{8.01\,\text{kWh}}{55\ \ \text{km}} = 146\,\frac{\text{Wh}}{\text{km}}$$

An **equivalent mpg** is estimated for BEVs on the basis of a simple calculation, and is termed **mpge**. An energy density of 33.705 kWh/gal (US) is used as the basis.

$$\text{mpge} = \frac{\dfrac{s}{1.609}}{E_{in} \times \dfrac{\text{gal}}{33.705\,\text{kWh}}} \tag{3.66}$$

Thus, the cycle mpge is

$$\text{mpge} = \frac{\dfrac{s}{1.609}}{E_{inC} \times \dfrac{\text{gal}}{33.705\,\text{kWh}}} = \frac{\dfrac{55}{1.609}\,\text{miles}}{8.01 \times \dfrac{\text{gal}}{33.705\,\text{kWh}}} = \frac{34.18\,\text{miles}}{0.238\,\text{gal}} = 144\,\text{mpge}$$

Knowing the input energy, the upstream carbon emissions due to the BEV can be estimated. Typically, the figures for carbon emissions per electrical kilowatt-hour are published for the various electrical power grids around the globe. A representative number for the United States is about 500 gCO_2/kWh (electrical) (including upstream emissions). Countries such as Norway and France can have much lower emissions, about 10 gCO_2/kWh in Norway due to hydroelectric power and about 50 gCO_2/kWh in France due to nuclear power. Other countries can have higher emissions due to the use of coal.

Table 3.8 Results for simplified drive cycle using a BEV.

Speed (km/h)	Time (s)	E_b (MJ)	E_{in} (MJ)	η_{pt} (%)	FC (Wh/km)	mpg (US) (mpge)	Range (km)	CO_2/km United States/ Norway (gCO_2/km)
50	1800	7.37	8.67	85	96	217	732	48/1
90	1200	17.01	20.01	85	185	113	380	93/1.8
Idle	600	0.12	0.14		—	—	—	—
Total	3600	24.5	28.82	84.6	146	144	485	73/1.5

$$\frac{gCO_2}{km} = \frac{E_{in} \times \frac{gCO_2}{kWh}}{s} \tag{3.67}$$

The emissions for this cycle in the United States is

$$\frac{gCO_2}{km} = \frac{E_{inC} \times \frac{gCO_2}{kWh}}{s} = \frac{8.01 \times 500\frac{gCO_2}{kWh}}{55} = 73\frac{gCO_2}{km}$$

This number is about 50 times lower for Norway at about 1.5 gCO_2/km. It is little wonder that Norway has invested heavily in BEVs as its electricity is almost free of fossil fuel.

The vehicle range for the BEV is simply the energy storage of the battery kWh_b divided by the energy consumption:

$$\text{Range} = \frac{kWh_b}{E_b/s} \tag{3.68}$$

Therefore, the range of the 60 kWh vehicle for this cycle is

$$\text{Range} = \frac{kWh_b}{E_{bC}/s} = \frac{60}{6.81/55}\text{km} = 485\,\text{km}$$

The various data are summarized in Table 3.8.

This example illustrates the advantages and disadvantages of the BEV compared to the fossil fuel and fuel cell options to be discussed in the following chapters. The BEV powertrain is far more efficient than the gasoline or diesel vehicles. However, the range is limited. The ability of the BEV to reduce carbon emissions ultimately depends on the fuel source used to power the electric grid. A major advantage of the BEV is its ability to decouple the emissions from the place of use to the grid power station.

References

1 S. Levine, *The Powerhouse, Inside the Invention of a Battery to Save the World*, Viking, 2015.

2 M. H. Braga, N. S. Grundish, A. J. Murchison, and J. B. Goodenough, "Alternative strategy for a safe rechargeable battery," *Journal of Energy and Environmental Science*, **10**, pp. 331–336, 2017.

3 Website of AESC: http://www.eco-aesc-lb.com/en/product/liion_ev

4 Battery testing at Idaho National Laboratory: http://avt.inl.gov/fsev.shtml

5 T. L. Brown, H. E. LeMay, B. E. Bursten, C. J. Murphy, and P. M. Woodward, *Chemistry: The Central Science*, 12th edition, Prentice Hall (Pearson Education).

Further Reading

1 T. B. Reddy, *Linden's Handbook of Batteries*, 4th edition, McGraw-Hill, 2011.
2 O. Gross, "Introduction to advanced automotive batteries," seminar, *IEEE Vehicle Power & Propulsion Conference*, 2011.
3 F. Hoffart, "Proper care extends Li-ion battery life," *Power Electronics Technology*, pp. 24–28, April 2007.
4 M. Petersen, "Probe blames Boeing and FAA for 787 Dreamliner battery fire," *Los Angeles Times*, 1 December 2014.
5 E. Hu, "Samsung Pins Blame On Batteries For Galaxy Note 7 Fires," *National Public Radio*, 22 January 2017.
6 J. E. Harlow, et al., "A Wide Range of Testing Results on an Excellent Lithium-Ion Cell Chemistry to be used as Benchmarks for New Battery Technologies," *Journal of the Electrochemical Society*, 166 (13), pp. A3031–A3044, September 2019.

Problems

3.1 A BEV has the following requirements: eight years of operation at an average of 24,000 km per year, averaged out over 365 days per year. Assume an average battery output of 204 Wh/km and a rated cell voltage of 3.6 V, a capacity of 3.4 Ah, and a lifetime index of $L = 1$. Assume $N_{100\%} = 1000$.

i) Determine the BOL kWh storage.
ii) How many cells do you need and what is the BOL range?
iii) What is the BOL storage and how many cells are required for a larger pack in order to increase the BOL range to 425 km?
iv) How many parallel strings are required if the pack has 96 cells in series?
v) What is the battery pack mass, assuming a battery with a pack density of 150 Wh/kg?
vi) If the peak power is 325 kW, what is the P/E ratio of the battery for the larger pack?

[Ans. 39.16 kWh, 3,200, 192 km, 86.7 kWh, 7,083, 74, 578 kg, 3.5]

3.2 A PHEV battery pack has the following requirements: ten years of operation at an average of 50 km per day, an average battery output of 5 km/kWh, and a 14.6 Ah cell with a rated voltage of 3.65 V and an index of $L = 3$. Assume $N_{100\%} = 1000$.

i) What is the BOL battery pack energy storage?
ii) What is the total number of cells required, to the nearest factor of 3?
iii) What is the pack voltage if the pack is in series-parallel with three cells in parallel?
iv) If the peak power is 110 kW, what is the P/E ratio of the battery?
v) What is the battery pack mass, assuming a with a pack density of 150 Wh/kg?

[Ans. 15.4 kWh, 288 (to nearest multiple of 3), 350 V, 7.1, 103 kg]

3.3 A NiMH HEV battery pack is sized based on the following requirements: 10,000 cycles of 60 Wh per year for ten years, a 6.5 Ah cell with a rated voltage of 1.2 V and an index of $L = 1.5$. Assume $N_{100\%} = 1000$.

i) What is the BOL battery pack energy storage?
ii) What is the total number of cells required?
iii) What is the pack voltage if the cells are all in series?
iv) If the peak power is 30 kW, what is the P/E ratio of the battery?

[Ans. 1.29 kWh, 166, 199.2 V, 23]

3.4 A PHEV battery has a 100% SOC of 36 kWh. The battery DOD is maintained within a range of 20% to 80%. The simplified HEV battery pack model has the following parameters: open-circuit voltage is $V_{bp(nl)}$ = 360 V at 20% DOD, with an internal ohmic resistance of R_{bp} = 200 mΩ and a constant of K_{bp} = 0.5 V/(% DOD).
i) Determine the battery terminal voltage if the battery is discharged to 80% at a constant 100 A.
ii) Determine the battery terminal voltage, current, and efficiency at DOD = 80% for (a) +80 kW discharge and (b) −80 kW charge.
iii) If the battery has 96 cells in series, at which charging power does the battery hit a 4.3 V per cell limit at 20% DOD?

[Ans. 310 V, 270.9 V, 295.3 A, 82.1 %, −214.5 A, 372.9 V, 88.5%, 109 kW]

3.5 The LiT cell of Figure 3.20 is rated at 2.5 V, 2 Ah at 1*C*.

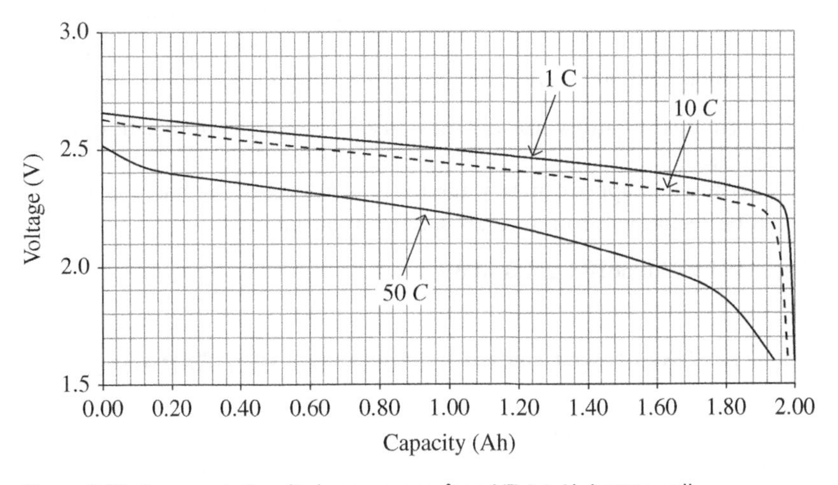

Figure 3.20 Representative discharge curves for a LiT 2.0 Ah battery cell.

i) Determine an approximate HEV cell voltage model as a function of y = DOD and I_b for the cell.
ii) Determine the approximate cell Wh, Ah, and efficiency for the 50 *C* rate.

[Ans. R_b = 2.8 mΩ, $V_{b(nl)}$ = 2.61 V at 20%, K = 3.3 mV/(%DOD), 4.45 Wh, 2 Ah, 88.6%]

3.6 A Li-ion cell is rated at 3.6 V, 3.4 Ah at 0.2*C* and has an internal resistance of 65 mΩ. Determine the cell Wh, Ah, and efficiency for the 4*C* rate.

[Ans. 9.38 Wh, 3.4 Ah, 75.7%]

3.7 An EV battery has a 100% SOC of 85 kWh. The battery can be charged at high power when the battery DOD is maintained within a range of 20% to 100%.

The pack has 96 cells in series per string with 74 parallel strings. Each cell has an average no-load cell voltage during charge of 3.64 V and an internal resistance of 65 mΩ.

i) Determine the battery terminal voltage, current, and efficiency for a 120 kW charge.
ii) What approximate time is required to charge the battery from a DOD of 100% to 20%?

[Ans. −318.9 A, 376.3 V, 92.9%, 30–40 min]

3.8 For the vehicle in Example 3.5 determine the carbon emissions and range when traveling at a constant speed of 120 km/h. See Chapter 2, Problem 2.7.

[Ans. 142 gCO2/km, 248 km]

3.9 The drive cycle of Example 3.5 is modified such that the idle period is replaced by a highway cruise at 120 km/h for 600 s.

Determine the carbon emissions and range for the revised drive cycle.

[Ans. 91 gCO2/km, 388 km]

3.10 The Tesla truck-trailer in Problem 2.8 is specified to have a battery life of 1,609,000 km (1,000,000 miles). Assume a battery pack with $L = 3$ and $N_{100\%} = 1000$, and 804.5 km (500 miles) per charge.

i) Determine the battery size and mass in kWh and kg if the specific energy is 0.18 kWh/kg.
ii) Determine an approximate value for power level of the charger if the vehicle can be recharged from 0 to 80 % SOC in 30 minutes.

[Ans. 1192 kWh, 6618 kg, 1.5 MW]

Appendix: A Simplified Curve-Fit Model for BEV Batteries

In this section, a curve fit is done by hand by making some simplifying engineering assumptions about the dominating components of Figure 3.13(a) as the cell is discharged. Initially, the cell resistance R_b is determined on the basis of the ohmic drop with current. It is then assumed that the initial voltage drop is dominated by a $\log_e$ component, then followed by a quasi-ohmic drop, and finally followed by a quasi-exponential drop. For this study, it is assumed that y = DOD, as DOD is a commonly used variable.

An estimate of the cell internal resistance R_b can be obtained by measuring the cell voltage drop with increased current. Thus, the voltage drop ΔV is estimated at about 0.22 V as the discharge current increases from 11.1 A ($C/3$) to 90 A ($2.7C$) at about 50% of the rated Ah.

$$R_b = \frac{\Delta V}{I_{2.7C} - I_{C/3}} = \frac{0.22}{90 - 11.1}\Omega = 2.8\,\text{m}\Omega \tag{A3.1}$$

In order to simplify the mathematics, it is assumed that the V_r^0 is the voltage at a DOD of 1% rather than at 0%. Thus, the fully charged open-circuit cell voltage V_r^0 is determined by accounting for the resistive drop.

$$V_r^0 \approx V_1 + R_b I_b = 4.15\,\text{V} + 0.0028 \times 11.1\,\text{V} = 4.18\,\text{V} \tag{A3.2}$$

When discharged, the fully charged cell voltage initially drops logarithmically from V_1 to V_2. For this portion, we assume that the voltage drops are as follows:

$$V_b = V_r^0 - A\log_e(B \cdot y) - R_b I_b \tag{A3.3}$$

It can be shown that

$$A = \frac{V_1 - V_2}{\log_e\left({}^{y_2}/_{y_1}\right)} \quad \text{and} \quad B = \frac{e^{\left(\frac{V_r^0 - V_1 - R_b I_b}{A}\right)}}{y_1} \tag{A3.4}$$

Thus, for the 33.3 Ah cell:

$$A = \frac{4.15 - 4.1}{\log_e({}^6/_1)} = 0.0279, \text{ and } B = \frac{e^{\left(\frac{4.18 - 4.15 - 0.028 \times 11.1}{0.0279}\right)}}{1} \approx \frac{e^0}{1} = 1$$

Through the quasi-linear region, the voltage continues to drop due to the $\log_e$ function. It is assumed that the balance of the drop in the linear region can be modeled with a straight line of slope K:

$$V_3 = V_r^0 - A\log_e(B \cdot y_3) - R_b I_b - K \cdot y_3 \tag{A3.5}$$

or

$$K = \frac{V_r^0 - A\log_e(B \cdot y_3) - R_b I_b - V_3}{y_3} \tag{A3.6}$$

For this cell:

$$K = \frac{4.18 - 0.0279\log_e(1 \times 84) - 0.0028 \times 11.1 - 3.7}{84} = 3.9\frac{\text{mV}}{\%}$$

Thus, the cell voltage drops 3.9 mV for each percentage rise in DOD. Finally, the voltage drops exponentially from V_3 to V_5. An intermediate point V_4 is required in order to solve for the F and G coefficients.

Let

$$V_4 = V_r^0 - A\log_e(B \cdot y_4) - K \cdot y_4 - Fe^{G \cdot (y_4 - y_3)} - R_b I_b \tag{A3.7}$$

$$\Rightarrow Fe^{G \cdot (y_4 - y_3)} = V_r^0 - A\log_e(B \cdot y_4) - K \cdot y_4 - R_b I_b - V_4 \tag{A3.8}$$

and

$$V_5 = V_r^0 - A\log_e(B\cdot y_5) - K\cdot y_5 - Fe^{G\cdot(y_5-y_3)} - R_bI_b \tag{A3.9}$$

$$\Rightarrow Fe^{G\cdot(y_5-y_3)} = V_r^0 - A\log_e(B\cdot y_5) - K\cdot y_5 - R_bI_b - V_5 \tag{A3.10}$$

It can be shown that

$$G = \frac{\log_e\left(\dfrac{V_r^0 - A\log_e(B\cdot y_5) - K\cdot y_5 - R_bI_b - V_5}{V_r^0 - A\log_e(B\cdot y_4) - K\cdot y_4 - R_bI_b - V_4}\right)}{y_5 - y_4} \tag{A3.11}$$

and

$$F = \left[V_r^0 - A\log_e(B\cdot y_4) - K\cdot y_4 - R_bI_b - V_4\right]e^{G\cdot(y_3-y_4)} \tag{A3.12}$$

resulting in the following values for the cell:

$$G = \frac{\log_e\left(\dfrac{4.18 - 0.0279\times\log_e(1\times 100) - 0.0039\times 100 - 0.0028\times 11.1 - 2.5}{4.18 - 0.0279\times\log_e(1\times 96) - 0.0039\cdot 96 - 0.0028\times 11.1 - 3.2}\right)}{100-96}$$

$$= 0.231$$

and

$$F = \left[4.18 - 0.0279\times\log_e(1\times 96) - 0.0039\cdot 96 - 0.0028\times 11.1 - 3.2\right]e^{0.231\times(84-96)}$$

$$= 0.0281$$

4

Fuel Cells

""Yes, but water decomposed into its primitive elements,"" replied Cyrus Harding, ""and decomposed doubtless, by electricity, which will then have become a powerful and manageable force, for all great discoveries, by some inexplicable laws, appear to agree and become complete at the same time. Yes, my friends, I believe that water will one day be employed as fuel, that hydrogen and oxygen which constitute it, used singly or together, will furnish an inexhaustible source of heat and light, of an intensity of which coal is not capable ... I believe, then, that when the deposits of coal are exhausted we shall heat and warm ourselves with water. Water will be the coal of the future.""
"I would like to see that," observed the sailor.
"You were born too soon, Pencroft," returned Neb.

From the The Mysterious Island, Jules Verne, 1874.

In this chapter, the reader is introduced to electrochemical fuel cells. The fuel cell is explained by building on the understanding of electrochemical storage developed in Chapter 3. The focus of this chapter is on the polymer electrolyte membrane fuel cell, which is the technology of choice for automotive applications. The relative efficiency and energy density of the fuel cell compared to the IC engine make the fuel cell a competitive technology for mobile applications. Topics such as fuel cell sizing and the sourcing of the hydrogen fuel are also considered.

4.1 Introduction to Fuel Cells

A fuel cell is an electrochemical device which converts the chemical energy of a fuel, for example, hydrogen, and an oxidant, for example, air, to electrical energy and heat. Similar to the electrochemical battery, the electrical energy is output in the form of dc power. Unlike a battery, the fuel and the oxidant are stored outside of the cell and transferred into the cell as the reactants are consumed. The fuel cell converts energy rather than storing it and can continuously provide power as long as fuel is provided.

Electric Powertrain: Energy Systems, Power Electronics and Drives for Hybrid, Electric and Fuel Cell Vehicles, First Edition. John G. Hayes and G. Abas Goodarzi.

Companion website: www.wiley.com/go/hayes/electricpowertrain

Fuel cells have been around for a long time. The first recorded development of a fuel cell was in England by William Grove in 1838. However, the first commercial development of a fuel cell was in the 1950s for the National Aeronautics and Space Administration (NASA) in the United States. Fuel cells are now regularly used for primary power production on space craft. For example, fuel cells power the International Space Station. Given that hydrogen gas is the rocket fuel for spacecraft, it makes sense to also use that fuel for on-board power generation. The **proton exchange membrane** or **polymer electrolyte membrane** (PEM) fuel cell of interest for today's vehicles was originally invented at General Electric (GE) in the United States by Willard Thomas Grubb and Leonard Niedrach in the 1950s. There are a number of different types of fuel cells. For example, **solid-oxide** fuel cells have been productized for stationary power backup applications, while **alkaline** fuel cells have been commonly used for space craft. All of these fuel cells can consume hydrogen fuel.

The PEM fuel cell is the technology of choice for automotive applications due to its low operating temperature range (less than 100 °C), small size, high efficiency, and wide operating range. Challenges for the fuel cell have been the costs of platinum for the electrodes, lifetime, sensitivity to impurities, and the significant accessory system, known as **balance of plant**, required to manage the fuel cell. The voltage of the PEM fuel cell is quite low, in the range of 0.5 to 1 V over the load range. Thus, automotive fuel cells are typically arranged in stacks with hundreds of cells in series. Stacks can be paralleled in order to increase power capability.

There have been many approaches over the decades at developing fuel cell vehicles for the automotive industry. Several car companies experimented with fuel cell vehicles in the 1960s and 1970s before a resurgence in the twenty-first century. In an environmentally- sensitive age, the fuel cell vehicle is attractive as its only emission at the point of use is water vapor. Fuel cell vehicles are particularly attractive as they combine the energy density of a fossil fuel with the powertrain efficiency of an electric vehicle. However, the development of fuel cells has been relatively slow due the hurdles of cost, size, manufacturability, and lifetime. In recent years, these challenges are being addressed, and new vehicles are coming onto the market. Thus, the Hyundai Tucson was introduced in 2014, the Toyota Mirai in 2015, and the Honda Clarity in 2017. New and improved vehicles are forthcoming from many different manufacturers.

The fuel cell is of particular interest for heavy-duty vehicles because of the high energy density of hydrogen and the fast rate of refueling. These two factors are major challenges for battery-electric vehicles. The primary competitors for long-range heavy-duty fuel cell vehicles are diesel and compressed natural gas (CNG) vehicles. However, these fossil-fuel-burning vehicles can generate significant particulate matter, hydrocarbons, and NO_x at the point of use, emissions which are experiencing greater curtailment, especially within urban environments. In addition, the diesel-fueled vehicle is relatively heavy in weight and requires significant maintenance. The fuel cell can achieve the required energy and power densities without the use of on-board fossil fuels. A fuel cell can be viewed as being a "combustion-less engine" as it produces electrical energy through the electrochemical reaction of hydrogen and oxygen.

4.1.1 Fuel Cell Vehicle Emissions and Upstream Emissions

Presently, hydrogen is generated in large quantities by the petroleum industry via the reformation of hydrocarbon gas. The **reforming** process is as follows: CNG (mostly methane CH_4) reacts with steam at very high temperatures to produce hydrogen and carbon monoxide (CO). The chemical reaction is

$$CH_4 + H_2O + \text{heat} \rightarrow CO + 3H_2 \tag{4.1}$$

The carbon monoxide is captured during the production and is widely used for various industrial operations.

Natural gas has become the fuel of choice in many countries for electricity generation due to its lower carbon content compared to oil or coal. Abundant resources of shale gas have expanded the role of gas in electricity generation, especially in the United States. Significant concerns exist regarding fracking and methane leakage.

Per the GREET model [1], the efficiency of producing and converting natural gas to hydrogen by steam reformation, transporting the compressed fuel, and finally fueling the vehicle is about 60%.

The powertrain energy efficiency of the fuel cell electric vehicle (FCEV) is about 45%, as per the discussion in Chapter 1, Section 1.7, and the overall well-to-wheel efficiency of about 27% is comparable to electric vehicles which are powered from a conventional grid.

The well-to-wheel emissions for the FCEV can also be improved by reforming the hydrogen from fermented biomass or by producing hydrogen using electrolysis, either at nuclear plants, where steam heat is also available, or by renewable sources.

While the efficiency of charging a battery electric vehicle (BEV) from renewables is higher than for the FCEV, the density of energy storage makes the FCEV an extremely attractive option for heavy-duty vehicles, such as large buses and trucks, which have to travel long distances and move heavy cargoes. The refueling infrastructure is more easily established for these types of vehicles as they often operate from centralized locations. The existing options are based on diesel and CNG – both of which result in significant pollutants at the point of use.

4.1.2 Hydrogen Safety Factors

Since the crash of the flaming fireball that was the LZ 129 *Hindenburg* airship on May 6, 1937, in a New Jersey field, perceived safety in particular has been a factor for hydrogen in the public eye. Even now, after all these years, there have been many theories as to the causes of the fire but no confirmed hypothesis or indication that the use of hydrogen for buoyancy was a significant contributing factor. Hydrogen has been widely used for a long time, and, as mentioned, is the fuel commonly used for spacecraft.

There are many factors which must be considered in assessing the safety of hydrogen compared to the competitive fuels. A number of these factors are presented in Table 4.1 for hydrogen, natural gas, and gasoline. Hydrogen and natural gas are colorless, odorless gases, whereas gasoline is a liquid. Natural gas and gasoline can both be toxic.

Hydrogen is significantly lighter than natural gas or gasoline vapor, and so the hydrogen gas tends to rise and disperse rapidly when it leaks in open space, as measured by the diffusion coefficient. Leakages of hydrogen and CNG are both easily detected.

Table 4.1 Various safety factors for different fuels.

	Hydrogen	Natural gas	Gasoline
Physical state at 25°C and 1 atm	Gas	Gas	Liquid
Color	None	None	Clear to amber
Odor	None	None	Yes
Toxicity	None	Some	High
Buoyancy relative to air	14.4× lighter	1.6× lighter	3.7× heavier
Diffusion coefficient in air (cm^2/s)	0.61	0.16	0.05
Detection	Yes	Yes	Limited
NPFA 704 Diamond 0 = none, 4 = Severe Red (top): Flammability Blue (left): Health Yellow (right): Reactivity White (bottom): Special hazard	4 (top), 0 (left), 0 (right)	4 (top), 2 (left), 0 (right)	3 (top), 1 (left), 0 (right)

The National Fire Protection Association (NFPA) in the United States provides a simple easily recognized standard for labeling hazardous materials. The applicable standard is NFPA 704, and the required labeling identifies on a scale of 0 (minimal hazard) to 4 (most severe) the safety hazards posed by the material for health (blue on left), flammability (red on top), reactivity (yellow on right), and special hazards (white on bottom). As noted in Table 4.1, both CNG and hydrogen are highly flammable, while hydrogen poses less of a hazard to health than CNG or gasoline.

4.2 Basic Operation

An oxidation-reduction reaction, commonly termed a **redox** reaction, is the basis of operation of an electrochemical fuel cell, and also of the electrochemical battery cell. As with the battery, a simple electrochemical fuel cell comprises two electrodes and an electrolyte, as shown in Figure 4.1.

The fuel cell is composed of two half-cells. One half-cell is the site of the oxidation reaction, and the other half-cell is the site of the reduction reaction.

The **anode** is the solid metal connection or electrode within the fuel cell at which oxidation occurs. The anode is at the negative terminal of the fuel cell.

The **cathode** is the solid metal connection or electrode within the fuel cell at which reduction occurs. The cathode is the positive electrode of the fuel cell.

An **electrolyte** is a substance which contains ions and allows the flow of ionic charge. The fuel cell features a polymer electrolyte. **PEM** is the commonly used name for this type of fuel cell, and refers to the **proton exchange membrane** or **polymer electrolyte membrane** which is designed to conduct the positive charges and insulate the electrodes. Platinum is typically included as a catalyst for both the anode and the cathode in order to

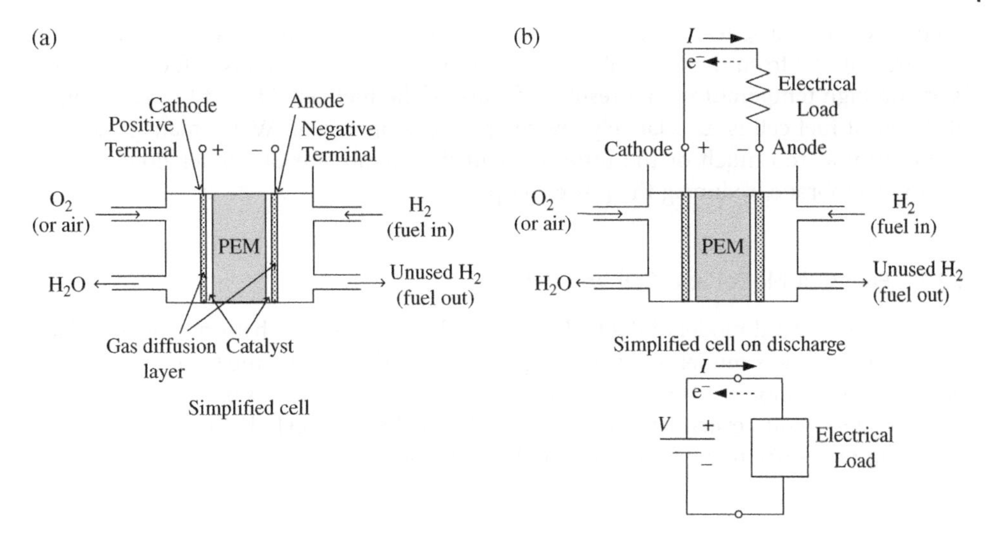

Figure 4.1 Electrochemical fuel cell.

split the hydrogen molecules into ions and electrons at the anode and to facilitate combination of the hydrogen and oxygen at the cathode.

The hydrogen and oxygen are absorbed into the **gas diffusion layer** (**GDL**), which acts as an electrode and allows the reactants to diffuse along the membrane, and also helps remove the water.

The full assembly of the electrodes and membrane is known as the **membrane electrode assembly** (**MEA**).

The half-cell and cell reactions for the PEM fuel cell are shown in Table 4.2. During discharging of the cell, the hydrogen fuel is converted to hydrogen ions and electrons at the anode. The electrons flow externally to power the load, and the hydrogen ions flow through the electrolyte to the cathode. The oxygen molecules, hydrogen ions, and electrons combine at the cathode to generate water molecules. A nominal half-cell voltage of 1 V is generated at the cathode. The half-cell voltage at the anode is 0 V. Thus, the overall cell voltage is 1 V.

Note that the voltages presented for the anode or cathode are with respect to a standard hydrogen electrode, the commonly used basis for determining the voltage from a half-cell.

Table 4.2 Cell parameters for a PEM fuel cell.

Chemistry	Electrode	Reaction	Equation	Voltage (V)
PEM discharge	Cathode	Reduction	$O_2 + 4H^+ + 4e^- \rightarrow 2H_2O$	1
	Anode	Oxidation	$2H_2 \rightarrow 4H^+ + 4e^-$	0
	Cell		$2H_2 + O_2 \rightarrow 2H_2O$	1

The fuel cell must feature a water and thermal management system as water and heat are outputs from the fuel cell, in addition to electricity and unused fuel. Excessive heat and high temperatures can result in failure of the fuel cell. One of the advantages of the PEM fuel cell is its relatively low temperature operation. Water management is also critical as too much or too little water in the membrane can either flood or dry out the membrane, reducing the power output.

4.2.1 Fuel Cell Model and Cell Voltage

The electrochemical model of the fuel cell is similar to the earlier battery model, and so a related cell voltage model is now developed. As with the battery, the focus here is on a simplified static model rather than on a dynamic model featuring capacitive effects.

The open-circuit reversible voltage of an electrochemical cell V_r^0 is related to the change in the Gibbs free energy ΔG^0 by the equation:

$$V_r^0 = -\frac{\Delta G^0}{2F} \tag{4.2}$$

where the number 2 represents the number of moles of electrons transferred according to the balanced equation for the redox reaction, and F is Faraday's constant (96,485 J/V-mol). The no-load cell voltage is typically significantly lower than the reversible voltage due to internal cell reactions. The voltage drop due to these internal reactions is known as the **rest-voltage drop,** ΔV_0.

Similar to the battery, the voltage of the fuel cell V_{fc} can be modified to include the effects of the reaction quotient as described by the Nernst equation:

$$V_{fc} = V_r^0 - \frac{RT}{2F}\log_e Q_R \tag{4.3}$$

where R is the ideal gas constant, T is the temperature, and Q_R is the reaction quotient. Again, the Nernst equation strictly only applies when there is no current in the cell. The cell terminal voltage is dependent on the cell current. This relationship is further governed by a number of voltage drops within the cell. There are three main voltage drops: ohmic, activation polarization, and concentration polarization.

Ohmic drop is the internal battery resistance composed of ohmic drops due to the terminations and current collectors.

The ohmic voltage drop ΔV_Ω can be simply modeled by

$$\Delta V_\Omega = R_\Omega i_{fc} \tag{4.4}$$

where R_Ω is the fuel-cell-specific resistance in Ωm^2, and i_{fc} is the fuel cell current density in A/m^2.

Activation polarization is the internal resistance due to charge-transfer reactions faced by the electrons at the electrode–electrolyte junctions. This effect is dominant from low-power to full-power operation of the fuel cell.

The activation voltage drop ΔV_a is governed by the equation

$$\Delta V_a = A\log_e\left(\frac{i_{fc}}{i_0}\right),\ i_{fc} \ge i_0 \tag{4.5}$$

where i_0 is known as the fuel cell exchange current density, and A is the activation loss coefficient. This equation is often known as the Tafel equation, named for the Swiss chemist Julian Tafel (1862–1918).

Concentration polarization is the internal resistance faced by the ions due to the electrolyte concentration. This effect is dominant at high currents beyond the maximum power condition.

The concentration voltage drop ΔV_c is governed by

$$\Delta V_c = me^{ni_{fc}} \tag{4.6}$$

where m and n are the concentration loss coefficient and exponent, respectively.

Thus, the fuel cell voltage of Equation (4.3) can be modified to include the various voltage drops:

$$V_{fc} = V_r^0 - \Delta V_0 - \Delta V_\Omega - \Delta V_a - \Delta V_c \tag{4.7}$$

4.2.1.1 Example: No-Load and Load Voltages of a PEM Fuel Cell

The parameters of a fuel cell are provided in Table 4.3. For simplicity, the effects of temperature and pressure are ignored.

Determine (i) the no-load voltage and (ii) the full-load voltage at 15,000 A/m^2.

Solution:

i) The no-load voltage is

$$V_{fc} = V_r^0 - \Delta V_0 = -\frac{\Delta G^0}{2F} - \Delta V_0 = -\frac{-240000}{2 \times 96485}\,\text{V} - 0.311\,\text{V} = 1.244\,\text{V} - 0.311\,\text{V} = 0.933\,\text{V}$$

ii) The voltage drops under load are as follows:
   The ohmic voltage drop ΔV_Ω is

$$\Delta V_\Omega = R_\Omega i_{fc} = 1.5 \times 10^{-5} \times 15{,}000\,\text{V} = 0.225\,\text{V} \tag{4.8}$$

   The activation voltage drop ΔV_a is

Table 4.3 Fuel cell parameters.

Description	Parameter	Value	Units
Change in Gibbs free energy	ΔG^0	-240×10^3	J/mole
Faraday's constant	F	96485	J/V-mol
Rest-voltage drop	ΔV_0	0.311	V
Fuel cell area specific resistance	R_Ω	1.5×10^{-5}	Ωm^2
Activation loss coefficient	A	5×10^{-3}	V
Exchange current density	i_0	20	A/m^2
Concentration loss coefficient	m	3×10^{-5}	V
Concentration loss exponent	n	0.5×10^{-3}	m^2/A
Cell thickness	t_{fc}	1.34	mm

$$\Delta V_a = A\log_e\left(\frac{i_{fc}}{i_0}\right) = 0.005 \times \log_e\left(\frac{15{,}000}{20}\right)\text{V} = 0.033\,\text{V} \tag{4.9}$$

The concentration voltage drop ΔV_c is

$$\Delta V_c = me^{ni_{fc}} = 3 \times 10^{-5} \times e^{0.0005 \times 15{,}000}\text{V} = 0.054\text{V} \tag{4.10}$$

The fuel cell voltage under full load is

$$V_{fc} = V_r^0 - \Delta V_0 - \Delta V_\Omega - \Delta V_a - \Delta V_c = 1.244\text{V} - 0.311\text{V} - 0.225\text{V} - 0.033\text{V} - 0.054\text{V} = 0.621\,\text{V}$$

4.2.2 Power and Efficiency of Fuel Cell and Fuel Cell Power Plant System

The specific power density of the fuel cell P_{sfc}, with units of W/m^2, is given by

$$P_{sfc} = V_{fc} i_{fc} \tag{4.11}$$

The efficiency of the elementary fuel cell η_{fc} is nominally the ratio of the output voltage to the no-load reversible voltage:

$$\eta_{fc} = \frac{V_{fc}}{V_r^0} \times 100\% \tag{4.12}$$

However, a significant portion of the output power of the fuel cell is required to power the balance of plant (which is discussed in the next section), and it is more useful to consider the efficiency of the overall fuel cell system or power plant.

Let η_{bop} be the efficiency of the balance of plant. Thus, the fuel cell power plant efficiency η_{fcp} is given by

$$\eta_{fcp} = \eta_{bop} \times \eta_{fc} = \eta_{bop}\frac{V_{fc}}{V_r^0} \tag{4.13}$$

4.2.2.1 Example: Full-Load Power and Efficiency of PEM Fuel Cell Stack

Determine the power density and the efficiencies of the fuel cell and plant at full load if the balance of plant consumes 20% of the fuel cell output power (η_{bop} equals 80%).

Solution:

The power density of the fuel cell is

$$P_{sfc} = V_{fc} i_{fc} = 0.621 \times 15{,}000\,\text{W/m}^2 = 9315\,\text{W/m}^2$$

The efficiency of the fuel cell is

$$\eta_{fc} = \frac{V_{fc}}{V_r^0} \times 100\% = \frac{0.621}{1.244} \times 100\% = 49.9\% \tag{4.14}$$

The fuel cell power plant efficiency is

$$\eta_{fcp} = \eta_{bop} \times \eta_{fcp} = 0.80 \times 49.9\% = 39.9\% \tag{4.15}$$

4.2.3 Fuel Cell Characteristic Curves

The plot of the fuel cell voltage as a function of the specific current is known as the **polarization curve** of the fuel cell, and is shown in Figure 4.2, for the parameters of Table 4.3. The fuel-cell-specific power is also plotted. The specific power peaks at about 15,000 A/m^2 before dropping significantly. The no-load and short-circuit operating points are also noted in the diagram.

Plots of the fuel cell and fuel cell plant efficiencies are presented in Figure 4.3. It is assumed in these curves that 20% of the fuel cell power is required to power the balance of plant. A couple of points should be noted for the practical fuel cell power plant. The fuel cell power plant typically does not operate at very light loads (lower than 5%–10%

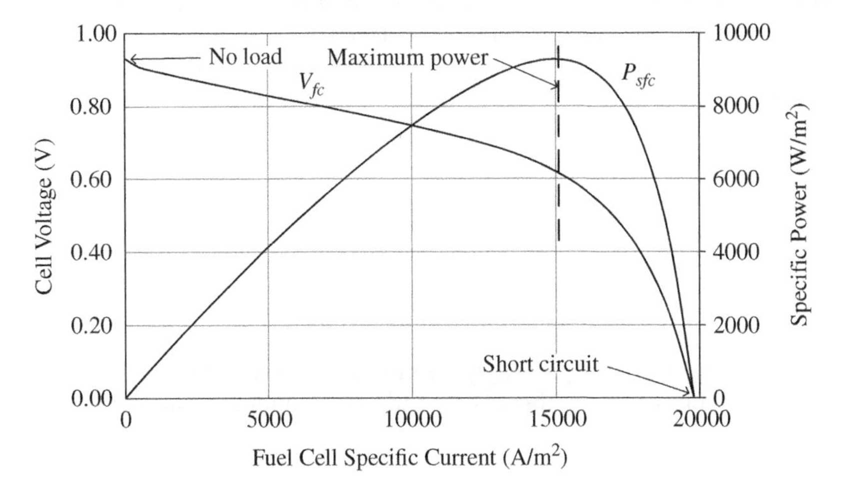

Figure 4.2 Fuel cell polarization curve and power curve.

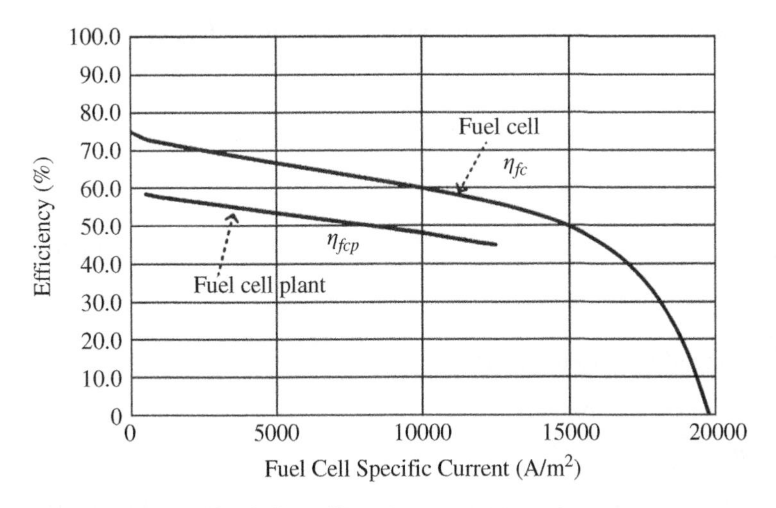

Figure 4.3 Fuel cell and plant efficiencies.

of rated power) due to the light-load losses. The plant also does not typically operate beyond the maximum power point as the system power and efficiency both drop significantly.

4.3 Sizing the Fuel Cell Plant

Multiple fuel cells are typically arranged in series as **stacks** in order to achieve the voltage and power levels required for the automotive powertrain. The fuel cell stack voltage should be as high as possible in order to maximize the overall efficiency of the electric powertrain. Hundreds of individual cells are typically required in series.

The current and power of the fuel cell are proportional to the area of the fuel cell. Thus, the area of the individual fuel cell is sized in order to output the required power.

The stack output voltage from the fuel cell power plant V_{fcp} is given by

$$V_{fcp} = N_{fc} \times V_{fc} \tag{4.16}$$

where N_{fc} is the number of cells in a fuel cell stack.

The fuel cell plant output power P_{fcp} is determined by multiplying the fuel cell stack output power P_{fc} by the efficiency of the BOP:

$$P_{fcp} = \eta_{bop} P_{fc} \tag{4.17}$$

The area of an individual fuel cell A_{fc} is simply the ratio of the maximum output power of the stack P_{fc} to the product of the number of cells and the maximum specific power P_{sfc} of the fuel cell:

$$A_{fc} = \frac{P_{fc}}{N_{fc} P_{sfc}} \tag{4.18}$$

If the thickness of the fuel cell t_{fc} is known, then the volume of the fuel cell stack v_{fc} is given by

$$v_{fc} = N_{fc} t_{fc} A_{fc} \tag{4.19}$$

The energy input to the fuel cell E_{in} is

$$E_{in} = \frac{P_{fcp}}{\eta_{fcp}} \times t \tag{4.20}$$

where t is the time.

The fuel mass flow rate of hydrogen $\dot{m}_{H2}$ is

$$\dot{m}_{H2} = \frac{E_{in}}{Q \times t} = \frac{P_{fcp} \times t}{\eta_{fcp} \times Q \times t} = \frac{P_{fcp}}{\eta_{fcp} Q} \tag{4.21}$$

where Q =120 MJ/kg (33.33 kWh) is the specific energy or lower heating value (LHV) of hydrogen.

4.3.1 Example: Sizing a Fuel Cell

A fuel cell power plant outputs 114 kW and has 370 cells in series. The parameters of the fuel cell are provided in Table 4.3. The cell thickness t_{fc} is 1.34 mm.

Determine the area of each fuel cell, the volume of the stack, the mass flow rate of fuel, and the stack voltage at full power for the full-load condition of 15,000 A/m^2. Reference the previous examples.

Solution:

The fuel cell power must factor in the balance of plant per Equation (4.17):

$$P_{fc} = \frac{P_{fcp}}{\eta_{bop}} = \frac{114}{0.8}\,\text{kW} = 142.5\,\text{kW} \tag{4.22}$$

From Equation (4.18), the area of the fuel cell stack is

$$A_{fc} = \frac{P_{fc}}{N_{fc}P_{sfc}} = \frac{142{,}500}{370 \times 9315}\,\text{m}^2 = 0.04135\,\text{m}^2\,(= 20.3\,\text{cm} \times 20.3\,\text{cm})$$

From Equation (4.19), the volume of the fuel cell stack is

$$v_{fc} = N_{fc}t_{fc}A_{fc} = 370 \times 1.34 \times 10^{-3} \times 0.04135\,\text{m}^3 = 0.0205\,\text{m}^3\,(= 20.5\,\text{litres})$$

According to Equation (4.21), the mass flow rate of hydrogen is

$$\dot{m}_{H2} = \frac{P_{fcp}}{\eta_{fcp}Q} = \frac{114\,\text{kW}}{0.5325 \times 120\,\dfrac{\text{kJ}}{\text{g}}} = 1.784\,\frac{\text{g}}{\text{s}}$$

From Equation (4.16), the stack voltage is

$$V_{fcp} = N_{fc}\,V_{fc} = 370 \times 0.621\,\text{V} = 230\,\text{V}$$

4.3.2 Toyota Mirai

The 2016 Toyota Mirai has 370 fuel cells in a single stack with a cell thickness of 1.34 mm each [2]. The vehicle stores 5 kg of hydrogen in a carbon-fiber-reinforced high-pressure tank which weighs 87.5 kg. The vehicle can be refueled in about 3 min. The fuel cell stack outputs 114 kW maximum. The energy system features a NiMH battery.

4.3.3 Balance of Plant

Power plants in general require significant support components and ancillary equipment in order to output power. These additional components and equipment are known as the balance of plant, as shown in Figure 4.4. There are several functions for the fuel cell balance of plant as follows:

i) Fuel processing
ii) Air processing
iii) Thermal and water management
iv) Electrical controls
v) Protection

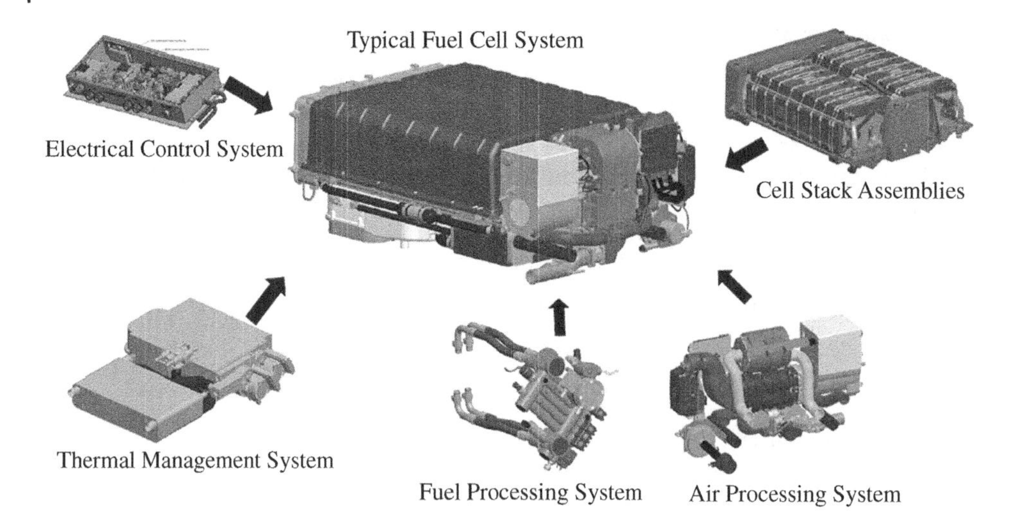

Figure 4.4 Automotive fuel cell and balance of plant. (Courtesy of US Hybrid.)

vi) Dc-dc converter
vii) Vehicle level cooling

As discussed earlier, the energy requirements for the balance of plant can be quite high, typically consuming about 20% of the fuel cell output power at full load for high-pressure fuel cell systems (see Problem 4.1), and about 10% or less for low-pressure fuel cell systems, to power the required fans, pumps, compressors, and so on.

4.3.4 Boost DC-DC Converter

A boost dc-dc power converter is required to interface the fuel cell to the high-voltage dc link. The fuel cell of the Toyota Mirai is interfaced to the electric drive by a four-phase boost converter with a maximum output voltage of 650 V. Boost converters for this application are discussed in depth in Chapter 11.

4.4 Fuel Cell Aging

Fuel cell performance decays with time as the GDLs and membrane degrade. **Lifetime** was quite a significant challenge for the earlier generations of fuel cells. The latest generation of fuel cells demonstrate lifetimes and operating hours which meet the exceedingly tough heavy-duty automotive requirements. While a car may be designed for 5000 hours of operation over its lifetime, heavy-duty commercial vehicles are designed for 20,000 hours of operation. The powertrain components of heavy-duty vehicles are typically replaced after 20,000 hours.

Polarization curves are presented in Figure 4.5 for 100 kW fuel cell power plants, which have been manufactured by US Hybrid [3] and field-tested in a fleet of heavy-duty commercial buses. The curves are based on data from regular field tests to characterize the fuel cells.

The cell voltage ranges from 0.88 V at light load to 0.7 V at full load at the beginning of life (BOL). After 20,000 hours, or at end of life (EOL), the cell voltages drops to 0.85 V at light load and to 0.57 V at full load.

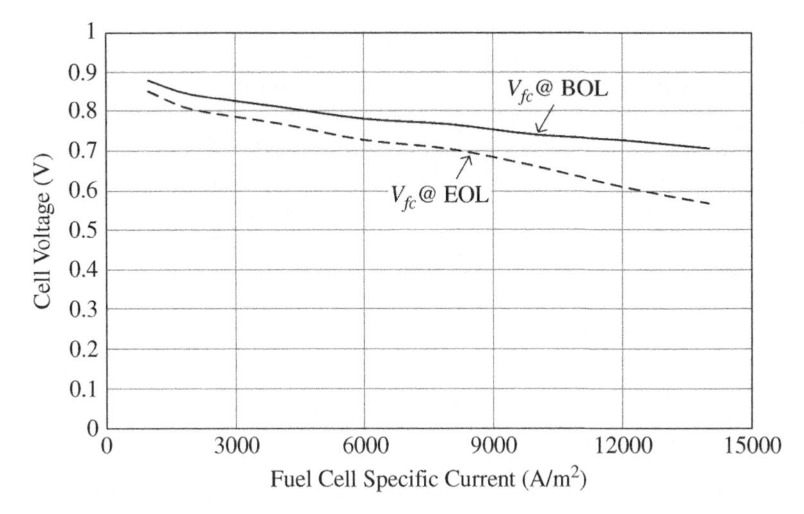

Figure 4.5 Polarization curve for heavy-duty automotive PEM fuel cell. (Courtesy of US Hybrid.)

This heavy-duty vehicle stack consists of 540 fuel cells in series. The fuel cell power plant output voltage and power are as shown in Figure 4.6.

The BOL fuel cell plant voltage ranges from 475 V at minimum load to 370 V at a full load of about 118 kW. The EOL fuel cell plant voltage ranges from 460 V at minimum load to 310 V at a full load of about 95 kW. Note that the slight oscillatory patterns shown on the voltage curves are also characteristic of the battery discharges with state of charge as shown in Chapter 3, Figure 3.8.

The fuel cell power plant output power and efficiency are plotted together in Figure 4.7.

The light-load and full-load efficiencies drop from 64% at 10 kW and 49% at a full load of 118 kW, respectively, at BOL, to 54% at 10 kW and 46% at a full load of 95 kW, respectively, at EOL. Thus, the average efficiency is about 59% at light load and about 47.5% at full load over the lifetime of the fuel cell.

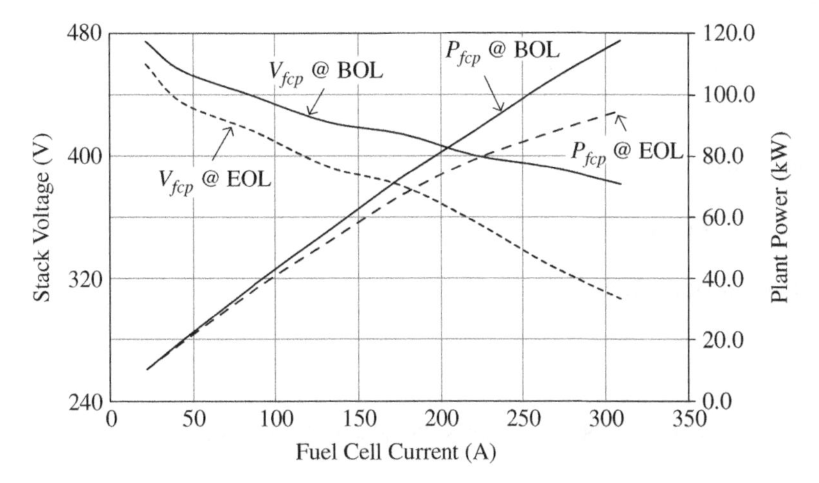

Figure 4.6 Fuel cell power plant voltage and power versus output current. (Courtesy of US Hybrid.)

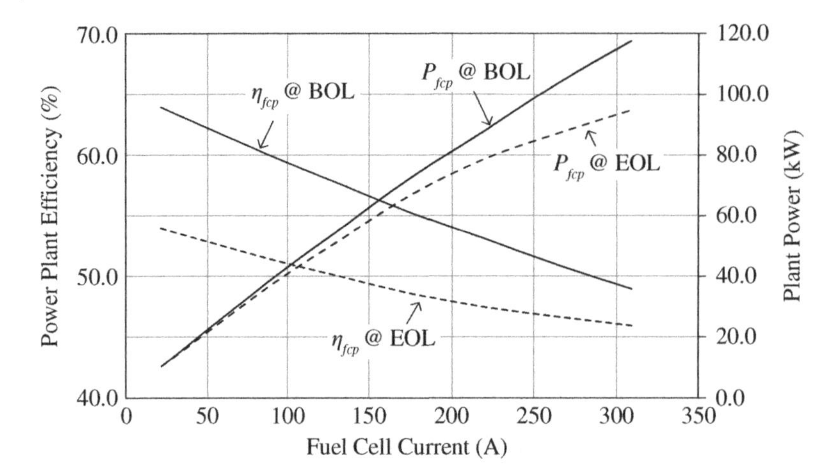

Figure 4.7 Fuel cell power plant efficiency and power versus output current. (Courtesy of US Hybrid.)

4.5 Example: Sizing Fuel Cell System for Heavy Goods Tractor–Trailer Combination

In this example, a fuel cell system is sized for a tractor–trailer carrying goods from a shipping port. The maximum gross weight for the tractor–trailer combination is 36,280 kg (80,000 lbs), and the vehicle operates at an average speed of 64 km/h. The vehicle has the following parameters: gross vehicle weight m = 36,280 kg when fully loaded, or 13,600 kg when unloaded, drag coefficient $C_D = 0.7$, vehicle cross section $A = 10\ \text{m}^2$, and coefficient of rolling resistance $C_R = 0.008$. The nominal efficiency of the powertrain and transmission $\eta_{pt} = 85\%$, and the auxiliary load $P_{aux} = 2$ kW. Let the density of air $\rho_{\text{air}} = 1.2\ \text{kg m}^{-3}$.

The overall mass of the fuel cell system (including the fuel cell, balance of plant, storage tanks, and mechanical bracketing) is 400 kg plus 80 kg per 5 kg of stored hydrogen.

i) Determine the mpge and the overall mass of the fuel cell if the vehicle is to travel at a constant speed of 64 km/h for a full work shift of 8 h, or 512 km. The fuel cell plant efficiency is 48% for this operating condition.
ii) Determine the mass of battery which would be required if the battery pack specific energy SE_b is 0.15 kWh/kg. What are challenges facing the battery for this application?

Solution:

The speed of the vehicle is

$$v = \frac{64}{3.6}\text{m/s} = 17.78\text{m/s}$$

The road load power of the fully-loaded vehicle is

$$P_v = \frac{1}{2}\rho C_D A v^3 + C_R mgv$$

$$= \frac{1}{2} \times 1.2 \times 0.7 \times 10 \times 17.78^3\,\text{W} + 0.008 \times 36280 \times 9.81 \times 17.78\,\text{W}$$

$$= 74.23\,\text{kW}$$

The fuel cell plant output power is the sum of the power to the powertrain plus the auxiliary power P_{aux}:

$$P_{fcp} = \frac{P_v}{\eta_{pt}} + P_{aux} = \frac{74.23}{0.85}\text{kW} + 2\text{kW} = 89.33\,\text{kW}$$

The overall driving time is 8 h, and so the energy output by the fuel cell plant is

$$E_{fcp} = P_{fcp}t = 89.33 \times 8\text{kWh} = 714.6\text{kWh}$$

The input energy carried by the hydrogen fuel is

$$E_{in} = \frac{E_{fcp}}{\eta_{fcp}} = \frac{714.6}{0.48}\text{kWh} = 1489\text{kWh}$$

The mass of hydrogen required is

$$m_{H2} = \frac{E_{in}}{Q} = \frac{1489}{33.33}\text{kg} = 44.67\text{kg}$$

The mpge is

$$\text{mpge} = \frac{\dfrac{s}{1.609}}{E_{in} \times \dfrac{\text{gal}}{33.705\text{kWh}}} = \frac{\dfrac{512}{1.609}\text{miles}}{1489 \times \dfrac{\text{gal}}{33.705\text{kWh}}} = \frac{318.2\text{miles}}{44.17\text{gal}} = 7.2\text{mpge}$$

The overall weight of the fuel cell system is

$$m_{fcp} = 400\text{kg} + \frac{80}{5}m_{H2} = 400\text{kg} + \frac{80}{5} \times 44.67\text{kg} = 1115\text{kg}$$

The equivalent battery output energy must equal the fuel cell output energy. Thus, the battery mass m_b is

$$m_b = \frac{E_{fcp}}{SE_b} = \frac{714.6}{0.15}\text{kg} = 4764\text{kg}$$

There are significant challenges for the battery to compete with the fuel cell in an application such as this. First, the weight of the battery can be excessive for the tractor axles. Second, the vehicle can be rapidly refueled in minutes using hydrogen, while the time for battery charging is likely to be significant and result in a proportionate drop in operating hours for the vehicle.

4.6 Example: Fuel Economy of Fuel Cell Electric Vehicle

In this example, the performance of a fuel cell vehicle is investigated for the simple drive cycle introduced in Chapter 2, Section 2.3. The following assumptions are made: electric drive efficiency η_{ed} of 85%, boost converter efficiency η_{boost} of 98%, an equivalent idling load of 1 kW, and an efficiency of steam reforming, transmission, distribution, compression, and fueling for hydrogen η_{H2} of 64%. The vehicle architecture is shown in Figure 4.8.

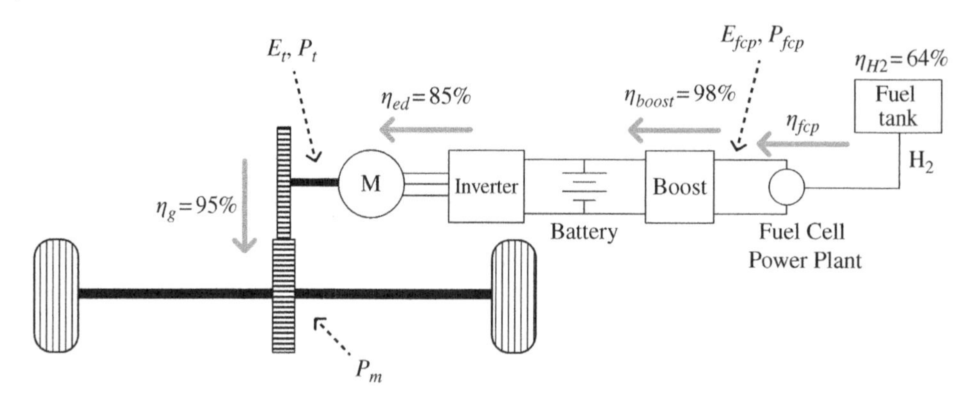

Figure 4.8 Fuel cell vehicle architecture.

It is noted that the fuel cell should be run at a minimum load to charge the battery or power vehicle auxiliaries and then shut down. For simplicity, in this problem it is assumed that the equivalent load on the fuel cell is 1 kW for time period t_3.

Let the light-load power plant efficiency $\eta_{fcp(ll)}$ be 59% at the light-load power $P_{fcp(ll)}$ of 4 kW, and the full-load power plant efficiency $\eta_{fcp(fl)}$ be 47.5% at the full-load power $P_{fcp(fl)}$ of 80 kW output. The efficiency varies approximately linearly with output power over the range between light and full load, and can be represented by

$$\eta_{fcp} = \eta_{fcp(ll)} + \frac{\left(\eta_{fcp(fl)} - \eta_{fcp(ll)}\right)}{\left(P_{fcp(fl)} - P_{fcp(ll)}\right)} \times \left(P_{fcp} - P_{fcp(ll)}\right) \tag{4.23}$$

Solution:

The power required from the fuel cell power plant P_{fcp} is the motor output power divided by the electric drive efficiency η_{ed} and the boost converter efficiency η_{boost}:

$$P_{fcp} = \frac{P_t}{\eta_{ed}\eta_{boost}} \tag{4.24}$$

The traction powers are $P_{t1} = 3.478$ kW and $P_{t2} = 12.050$ kW from Chapter 2, Section 2.3. The fuel cell plant powers, P_{fcp1}, P_{fcp2}, and P_{fcp3} are

$$P_{fcp1} = \frac{P_{t1}}{\eta_{ed}\eta_{boost}} = \frac{3.478}{0.85 \times 0.98}\,\text{kW} = 4.175\,\text{kW}$$

$$P_{fcp2} = \frac{P_{t2}}{\eta_{ed}\eta_{boost}} = \frac{12.050}{0.85 \times 0.98}\,\text{kW} = 14.466\,\text{kW}$$

and

$$P_{fcp3} = 1\,\text{kW}$$

The energy consumed from the fuel cell plant is simply the product of power and time:

$$E_{fcp} = P_{fcp} \times t \tag{4.25}$$

The fuel cell output energies E_{fcp1}, E_{fcp2}, and E_{fcp3} are

$$E_{fcp1} = P_{fcp1} \times t_1 = 4.175 \times 1800\,\text{kJ} = 7.515\,\text{MJ}$$

$$E_{fcp2} = P_{fcp2} \times t_2 = 14.466 \times 1200\,\text{kJ} = 17.359\,\text{MJ}$$

$$E_{fcp3} = P_{fcp3} \times t_3 = 1.0 \times 600\,\text{kJ} = 0.6\,\text{MJ}$$

The plant output energy over the drive cycle E_{fcpC} is

$$E_{fcpC} = E_{fcp1} + E_{fcp2} + E_{fcp3} = 7.515\text{MJ} + 17.359\,\text{MJ} + 0.6\,\text{MJ}$$

$$= 25.474\,\text{MJ}\ (= 7.08\,\text{kWh})$$

The energy supplied in the hydrogen E_{in} is

$$E_{in} = \frac{E_{fcp}}{\eta_{fcp}} \tag{4.26}$$

The plant efficiency is given by

$$\eta_{fcp} = \eta_{fcp(ll)} + \frac{\left(\eta_{fcp(fl)} - \eta_{fcp(ll)}\right)}{\left(P_{fcp(fl)} - P_{fcp(ll)}\right)} \times \left(P_{fcp} - P_{fcp(ll)}\right)$$

$$= 59\% + \frac{(47.5\% - 59\%)}{(80\text{kW} - 4\text{kW})} \times \left(P_{fcp} - 4\text{kW}\right) = 59\% - 0.1513\% \times \left(P_{fcp} - 4\text{kW}\right)$$

It is assumed that vehicle powers below the light load level can ultimately be supplied by the plant at the light-load efficiency level. The various efficiencies are then calculated as

$$\eta_{fcp1} = 59\% - 0.1513\% \times (4.175\text{kW} - 4\text{kW}) \approx 59\%$$

$$\eta_{fcp2} = 59\% - 0.1513\% \times (14.466\,\text{kW} - 4\,\text{kW}) \approx 57.4\%$$

$$\eta_{fcp3} = 59\%$$

The input energies, E_{in1}, E_{in2}, and E_{in3} are

$$E_{in1} = \frac{E_{fcp1}}{\eta_{fcp1}} = \frac{7.515}{0.59}\,\text{MJ} = 12.737\,\text{MJ}$$

$$E_{in2} = \frac{E_{fcp2}}{\eta_{fcp2}} = \frac{17.359}{0.574}\,\text{MJ} = 30.242\,\text{MJ}$$

$$E_{in3} = \frac{E_{fcp3}}{\eta_{fcp3}} = \frac{0.6}{0.59}\,\text{MJ} = 1.017\,\text{MJ}$$

The input energy E_{inC} over the cycle is

$$E_{inC} = E_{in1} + E_{in2} + E_{in3} = 12.737\,\text{MJ} + 30.242\,\text{MJ} + 1.017\,\text{MJ}$$

$$= 43.996\,\text{MJ}\ (12.22\,\text{kWh})$$

The mass of hydrogen used per interval and over the cycle, m_1, m_2, m_3, and m_C, are

$$m_1 = \frac{E_{in1}}{Q} = \frac{12.737\,\text{MJ}}{120\,\text{MJ}}\text{kg} = 0.106\text{kg} \qquad m_2 = \frac{E_{in2}}{Q} = \frac{30.242\,\text{MJ}}{120\,\text{MJ}}\text{kg} = 0.252\text{kg}$$

$$m_3 = \frac{E_{in3}}{Q} = \frac{1.017\,\text{MJ}}{120\,\text{MJ}}\text{kg} = 0.008\text{kg} \ \text{ and } \ m_C = \frac{E_{inC}}{Q} = \frac{43.996\,\text{MJ}}{120\,\text{MJ}}\text{kg} = 0.366\text{kg}$$

The powertrain conversion efficiency η_{pt} from the fuel tank to the traction machine output is

$$\eta_{pt} = \frac{E_{tC}}{E_{inC}} \times 100\% = \frac{20.72}{43.996} \times 100\% = 47.1\%$$

where $E_{tC} = 20.72$ MJ was determined in Chapter 2, Section 2.3.

The fuel consumption of the vehicle and the related fuel economy are based on the input energy from the fuel.

$$FC = \frac{E_{in}}{s} \left(\text{in} \frac{\text{Wh}}{\text{km}} \right) \tag{4.27}$$

and so the fuel consumption for this cycle is

$$FC = \frac{E_{inC}}{s} = \frac{12.22\ \text{kWh}}{55\ \ \text{km}} = 222 \frac{\text{Wh}}{\text{km}}$$

The **equivalent mpg (mpge)** for FCEVs is based on a simple calculation. The estimate is termed mpge. Note that the energy of 1 kg of hydrogen at 33.33 kWh/kg is close in value to the energy of 1 US gallon of gasoline at 33.705 kWh/gal.

$$\text{mpge} = \frac{\dfrac{s}{1.609}}{E_{in} \times \dfrac{\text{gal}}{33.705\,\text{kWh}}} \tag{4.28}$$

Thus, the cycle mpge is

$$\text{mpge} = \frac{\dfrac{s}{1.609}}{E_{inC} \times \dfrac{\text{gal}}{33.705\,\text{kWh}}} = \frac{\dfrac{55}{1.609}\text{miles}}{12.22 \times \dfrac{\text{gal}}{33.705\,\text{kWh}}} = \frac{34.18\,\text{miles}}{0.363\,\text{gal}} = 94\,\text{mpge}$$

Knowing the input energy, the upstream carbon emissions due to the FCEV can be estimated. While the carbon can be captured as carbon monoxide from the steam reformation, we will simplify the analysis by assuming 198 gCO_2/kWh (from Chapter 1, Table 1.1) for the methane input to the reformer.

$$\frac{gCO_2}{\text{km}} = \frac{E_{in}}{\eta_{H2}} \times \frac{198\,gCO_2}{\text{kWh}} \times \frac{1}{s} \tag{4.29}$$

which for this cycle is

$$\frac{gCO_2}{\text{km}} = \frac{E_{inC}}{\eta_{H2}} \times \frac{198\,gCO_2}{\text{kWh}} \times \frac{1}{s} = \frac{12.22\,\text{kWh}}{0.64} \times \frac{198\,gCO_2}{\text{kWh}} \times \frac{1}{55\,\text{km}} = 69\frac{gCO_2}{\text{km}}$$

The vehicle range for the FCEV is simply the tank stored energy divided by the fuel consumption:

$$\text{Range} = \frac{m \times Q}{E_{in}/s} \tag{4.30}$$

Thus, the range of the vehicle for this cycle is

$$\text{Range} = \frac{m \times Q}{E_{inC}/s} = \frac{5 \times 33.33}{12.22 \times 10^6/55}\text{km} = 750\text{km}$$

Table 4.4 Results for drive cycle using an FCEV.

Speed (km/h)	Time (s)	m (kg)	E_{in} (MJ)	η_{pt} (%)	FC (Wh/km)	mpg (US) (mpge)	Range (km)	CO_2 (gCO_2/km)
50	1800	0.106	12.737	49.1	142	148	1177	44
90	1200	0.252	30.242	47.8	280	75	595	87
Idle	600	0.008	1.017	—	—	—	—	—
Total	3600	0.366	43.996	47.1	222	94	750	69

The drive cycle results for the FCEV are summarized in Table 4.4.

This example illustrates the advantages and disadvantages of the FCEV compared to the fossil fuel and BEV options. The FCEV powertrain is far more efficient over the drive cycle than the gasoline or diesel vehicles, due to the light-to-medium load efficiency, and the range is significant compared to the BEV.

References

1 US Department of Energy, *Natural Gas for Cars*, DOE/GO-102015-4685, December 2015.
2 http://www.toyota-global.com/innovation/environmental_technology/fuelcell_vehicle/
3 http://www.ushybrid.com/
4 Thames and Kosmos, http://www.thamesandkosmos.com/
5 Horizon Fuel Cell Technologies, http://www.horizonfuelcell.com/
6 H. Lohse-Busch, M. Duoba, K. Stutenberg, S. Iliev, M. Kem, B. Richards, M. Christenson, and A. Loiselle-Lapointe, *Technology Assessment of a Fuel Cell Vehicle: 2017 Toyota Mirai*, Argonne National Laboratory report, 2018.

Problems

4.1 A fuel cell has the parameters of Table 4.3. Determine the fuel cell voltage, cell efficiency, and plant efficiency under (i) light load at 1500 A/m^2 with η_{bop} equals 96.5%, (ii) medium load at 7,500 A/m^2 with η_{bop} equals 89%, and (iii) overload of 17,500 A/m^2 with η_{bop} equals 77%.

Note that these BOP efficiencies are representative of a high-pressure fuel cell system, such as the Toyota Mirai [6].

[Ans. (i) 0.889 V, 71.5%, 69%, (ii) 0.79 V, 63.5%, 56.5%, (iii) 0.447 V, 35.9%, 27.6%]

4.2 A fuel cell has the parameters of Table 4.3 and has a fuel cell area of 400 cm^2. The fuel cell power plant outputs 80 kW at 15,000 A/m^2. Assume η_{bop} equals 80%. Reference examples 4.2.1.1 and 4.2.2.1.

i) Determine the number of fuel cells and the volume of the stack.
ii) Determine the plant output voltage, efficiency, and mass rate of fuel at the full-load power of 80 kW.
iii) What is the stack no-load voltage?

[Ans. 268, 14.4 L, 166.4 V, 39.9%, 1.67 g/s, 250 V]

4.3 The vehicle of Section 4.6 is traveling at a constant speed of 120 km/h and requires a traction power of 24.64 kW. See Chapter 2, Problem 2.7.

Determine the carbon emissions and range at this speed if the tank has 5 kg of hydrogen.

[Ans. 138 gCO_2/km, 373 km]

4.4 The drive cycle of Section 4.6 is modified such that the idle period is replaced by a highway cruise at 120 km/h and a required traction power of 24.64 kW (as per the previous problem) for 600 s.

Determine the carbon emissions and range for the revised drive cycle.

[Ans. 86 gCO_2/km, 599 km]

4.5 Determine the mpge and range of the vehicle in Section 4.5 if the vehicle is unloaded. The efficiency of the fuel cell plant is 52% for this condition.

[Ans. 13.4 mpge, 951 km]

4.6 A fully loaded bus has the following parameters: mass m = 20,000 kg, drag coefficient $C_D = 0.7$, vehicle cross section $A = 10$ m^2, and coefficient of rolling resistance $C_R = 0.008$. The nominal efficiency of the powertrain and transmission $\eta_{pt} = 85\%$, and the auxiliary load is 2 kW. Let the density of air $\rho_{air} = 1.2$ kg m^{-3}. The overall mass of the fuel cell system (including the fuel cell, balance of plant, storage tanks, and mechanical bracketing) is 400 kg plus 80 kg per 5 kg of stored hydrogen.

i) Determine the mpge, the hydrogen mass, and the overall mass of the fuel cell system if the vehicle is to travel at a constant speed of 64 km/h for two work shifts of 16 h total, or 1024 km. The fuel cell plant efficiency is 50% for this operating condition.
ii) Determine the mass of battery which would be required if the specific energy is 0.15 kWh/kg.
iii) What is the range of the vehicle without passengers if the weight drops by 1/3? The fuel cell plant efficiency is 53% for this operating condition.

[Ans. 10.7 mpge, 60.1 kg, 1362 kg, 6,667 kg, 1315 km]

4.7 A fuel cell energy storage system is to be sized to match the requirements of the Tesla truck-trailer of Problems 2.8 and 3.10. Assume a fuel cell plant efficiency of 59% and a boost efficiency of 98%. The overall mass of the fuel cell system (including the fuel cell, balance of plant, storage tanks, and mechanical bracketing) is 400 kg plus 80 kg per 5 kg of stored hydrogen. A hybrid battery weighing 400 kg is also part of the storage system.

Determine the mass of the combined fuel cell and battery system.

[Ans. 1585 kg]

Assignments

4.1 Experiment with a fuel cell kit from suppliers such as [4] or [5].

5

Conventional and Hybrid Powertrains

"Be ahead of the times through endless creativity, inquisitiveness and pursuit of improvement."
One of the five precepts compiled by Kiichiro Toyoda (1894–1952), founder of Toyota Motor Company. The precepts are based on the convictions of his father, Sakichi Toyoda (1867–1930), who founded Toyota Industries as a textile loom maker, and is often referred to as the father of the Japanese industrial revolution.

In this chapter, we investigate the efficiency, fuel consumption, and range factors motivating the shift to electrification and the hybridization of the automotive powertrain. Initially the brake specific fuel consumption of the internal combustion engine is introduced. The emissions, fuel economy, and range of conventional gasoline and diesel vehicles are compared with the series and series-parallel variants of the hybrid electric vehicle (HEV). A key enabling technology for the series-parallel HEV is the continuously variable transmission (CVT), a power-split device using epicyclic gears.

5.1 Introduction to HEVs

Vehicles based on the spark-ignition (SI) and compression-ignition (CI) internal combustion (IC) engines have dominated transportation for over a century. However, as discussed in Chapter 1, the proliferation of the IC engine has generated associated problems of emissions, including greenhouse gases and pollutants, while operating relatively inefficiently. Hybridization provides a pathway to mitigating these effects by providing series and/or parallel energy paths in order to run the IC engine in a high-efficiency mode and to maximize the fuel economy while minimizing the harmful emissions. This chapter provides an overview of these powertrains. Readers are referred to in-depth textbooks such as [1] in order to study the fundamentals of the IC engine.

It is worth noting again the main advantages of HEVs compared to conventional vehicles:

1) The elimination of vehicle idling losses;
2) The use of energy-recovery braking systems;

Electric Powertrain: Energy Systems, Power Electronics and Drives for Hybrid, Electric and Fuel Cell Vehicles, First Edition. John G. Hayes and G. Abas Goodarzi.

Companion website: www.wiley.com/go/hayes/electricpowertrain

3) Efficient and optimized energy management;
4) The use of downsized and more efficient engines.

These factors can double the fuel economy of a HEV compared to a conventional vehicle.

In general, the IC engine is very inefficient at low torque across the speed range and hits peak efficiency at medium-to-high torques and medium speeds. Idling operation occurs when the engine is generating no motive power, and can result in inefficient operation over the drive cycle. A typical IC engine operates at several hundred revolutions per minute (rpm) while idling.

The conventional SI IC engine is based on the Otto cycle. Nicolaus Otto (1832–1891), a German engineer, invented the first successful IC engine. The Otto-cycle machine features the four strokes of intake, compression, expansion, and exhaust. The mixture of fuel and air are ignited by a spark to release the energy to provide motive power.

The CI diesel IC engine was invented by Rudolf Diesel (1858–1913) in 1882. Diesel, also a German, discovered that the engine efficiency could be significantly increased by increasing the compression ratio and temperature within the engine cylinders in order to ignite the fuel–air mixture by compression, rather than ignition.

In general, as we shall see, these engines are most efficient at medium-to-high torque and medium speeds. However, much of the vehicle driving, especially in city driving, is at low torque and low speeds, and results in inefficient energy conversion.

The engine used in the Toyota hybrid system is a variant on the gasoline SI IC engine known as the Atkinson-cycle engine. Englishman James Atkinson (1846–1914) invented an engine with a longer expansion than compression stroke than is usual for a SI engine. The Atkinson-cycle engine has a relatively high efficiency because of this expansion of the cycle, but in general the engine was not used in automobiles due to its relatively low peak torque characteristic with speed. This weakness of the Atkinson-cycle engine was overcome by integrating the engine into the hybrid system and using the electric drive to provide the additional torque over the speed range.

Hybridizing the powertrain enables the engine to operate closer to the maximum efficiency in order to minimize emissions and fuel consumption, while operating over the full ranges of speed and torque required for vehicle propulsion.

It is useful to consider the available traction machine (motor or engine) torques and powertrain efficiencies for the various technologies. Representative plots of peak torque and efficiency are plotted in Figure 5.1(a) and (b), respectively, for a conventional gasoline engine, a diesel engine, an Atkinson-cycle engine, and electric powertrains using a fuel cell and a battery. The EV curve plotted in Figure 5.1(a) is appropriate for the battery electric vehicle (BEV) and the fuel cell electric vehicle (FCEV).

The peak torque available from the 1.5 L Atkinson-cycle gasoline engine is relatively flat over the speed range and increases from 77 Nm at 1000 rpm to 102 Nm at 4000 rpm. The peak efficiency varies between about 36% and 38% at peak torque over the speed range.

The peak torque from the diesel engine is high compared to the gasoline engine, and the percentage efficiency over the speed range is in the mid to high 30s. Newer diesel engines can have efficiencies in the 40s.

The standard gasoline engine has a wider speed range than the diesel and Atkinson-cycle engines, and has a maximum torque that is lower than its diesel equivalent but is

(a)

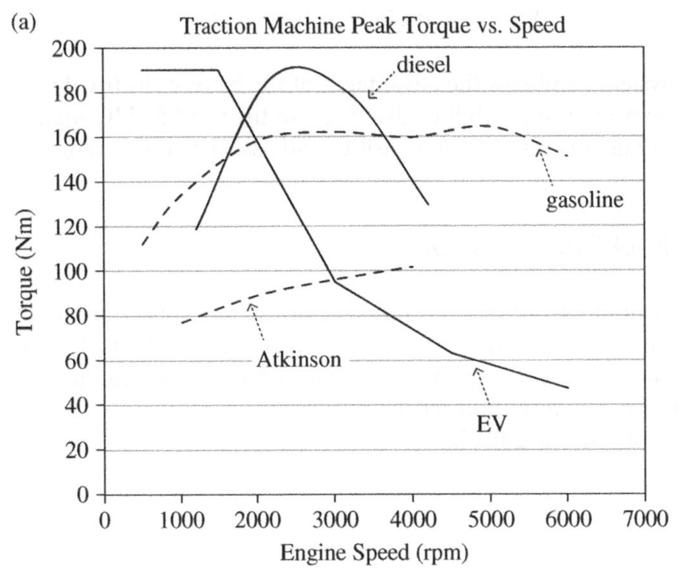

(b)

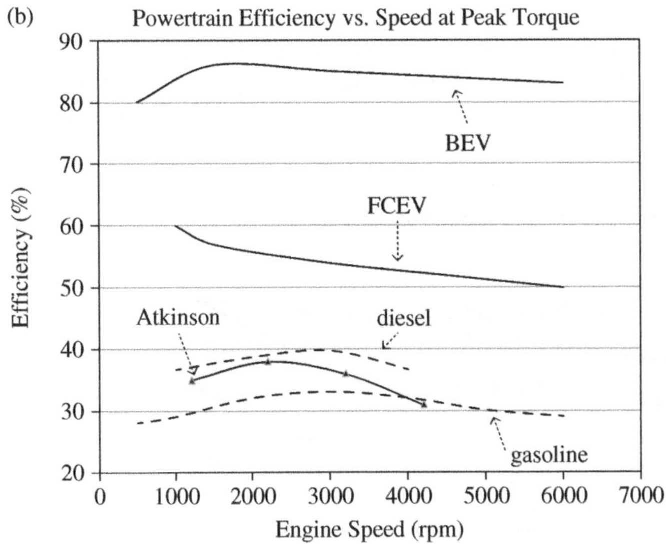

Figure 5.1 Representative torque and efficiency characteristics for various powertrains.

significantly higher than the Atkinson-cycle engine. The efficiency ranges from the high 20s to the low 30s over the speed range.

Now let us bring in the electric powertrain. The typical EV has a very high low-speed torque available all the way down to zero speed, and this torque drops inversely with speed beyond the rated speed as the machine outputs a constant power. This torque characteristic is available for any electric vehicle, whether battery, fuel cell, or hybrid. The efficiency of the BEV from the battery to the output of the electric motor can be very high – a range in the 80s is reasonable, including battery losses. The fuel cell vehicle has a much lower efficiency than the BEV, but at close to 50% to 60% is much higher than the IC engine options.

A well-designed hybrid system combines the advantages of the battery electric drive, which is range limited but very efficient, with the advantages of the fossil fuel IC engine or hydrogen FC in realizing high energy-storage densities and associated long range.

5.2 Brake Specific Fuel Consumption

The brake specific fuel consumption (BSFC) is a measure of the fuel consumption of an engine and is inversely related to the efficiency. The BSFC is equal to the fuel mass consumption rate $\dot{m}$, with units of grams/second (g/s), adjusted to 1 h by multiplying by 3600, and divided by the engine output power P_{eng}.

The commonly used unit of BSFC is g/kWh.

$$BSFC = \frac{3600\,\dot{m}}{P_{eng}} \tag{5.1}$$

The fuel consumption of an engine is often characterized by a BSFC map. The BSFC map plots the engine fuel consumption with speed and output torque when the engine is tested on a dynamometer. A sample BSFC map is shown in Figure 5.2 for the 1997 Toyota Prius engine using data from [2]. The fuel consumption of the engine shown

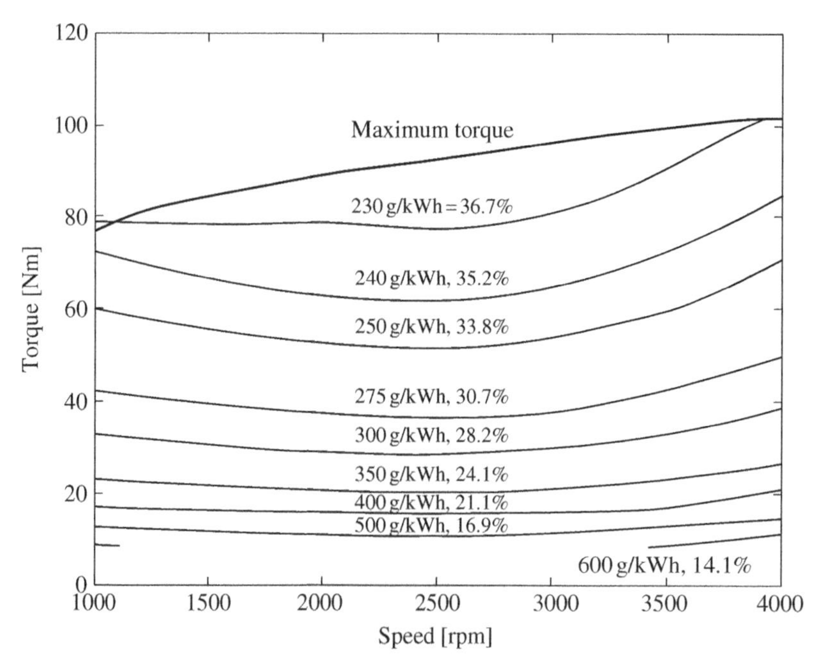

Figure 5.2 BSFC map for 1997 Toyota Prius using data from [2].

in Figure 5.2 can range from a low of about 230 g/kWh at 80 Nm and 2000 rpm to a high of about 600 g/kWh at 10 Nm and 4000 rpm.

Knowing the specific energy (Q) of a fuel, the engine efficiency η_{eng} can be calculated from

$$\eta_{eng} = \frac{P_{eng}}{\dot{m}Q} = \frac{3600}{BSFC \times Q} \tag{5.2}$$

Thus, the efficiency of an engine is inversely proportional to the BSFC. The engine efficiency values are also shown in Figure 5.2.

The Q value is dependent on the specific blends of fuel. Values of 42.6 kJ/g and 42.9 kJ/g are typically used in this particular chapter for gasoline and diesel, respectively, on the basis of the test values in [2].

5.2.1 Example: Energy Consumption, Power Output, Efficiency, and BSFC

Determine the fuel mass flow rate, output power, and engine efficiency for the 1997 Toyota Prius engine at 80 Nm and 2000 rpm. Assume that the gasoline fuel has a specific energy of 42.6 kJ/g and a density of 0.749 kg/L.

Solution:

At the operating point of 80 Nm and 2000 rpm, the engine is operating within the contour of the lowest BSFC of 230 g/kWh, which is the most efficient operating region.

The output power is

$$P_{eng} = T_{eng} \times \omega_{eng} = 80 \times 2000 \times \frac{2\pi}{60}\,\text{W} = 16.755\,\text{kW} \tag{5.3}$$

The fuel mass flow rate is

$$\dot{m} = \frac{BSFC}{3600} \times P_{eng} = \frac{230}{3600} \times 16.755\,\frac{\text{g}}{\text{s}} = 1.07\,\frac{\text{g}}{\text{s}} \tag{5.4}$$

The efficiency is given by

$$\eta_{eng} = \frac{3600}{BSFC \times Q} = \frac{3600}{230 \times 42.6} \times 100\% = 36.74\% \tag{5.5}$$

If we take the raw BSFC data and replot the data, we can gain some additional insights about the engine. Plots of fuel consumption and engine efficiency versus torque for various speeds in revolutions per minute are shown in Figure 5.3(a) and (b), respectively. Clearly, for a particular speed for this Atkinson-cycle engine, the fuel consumption increases approximately linearly with torque (and power, because power is torque times speed). The efficiency of the engine increases with torque. At high torque, the engine efficiency is quite high, reaching a peak of about 38% at 93 Nm and 3000 rpm. However, at lower torque levels, the engine efficiency can drop very significantly and is as low as 12% at 9 Nm and 4000 rpm. Similarly, the engine efficiency drops at high speed compared to the medium speed. For example, the engine efficiency at 93 Nm and 4000 rpm is about 35% compared to 38% at 3000 rpm.

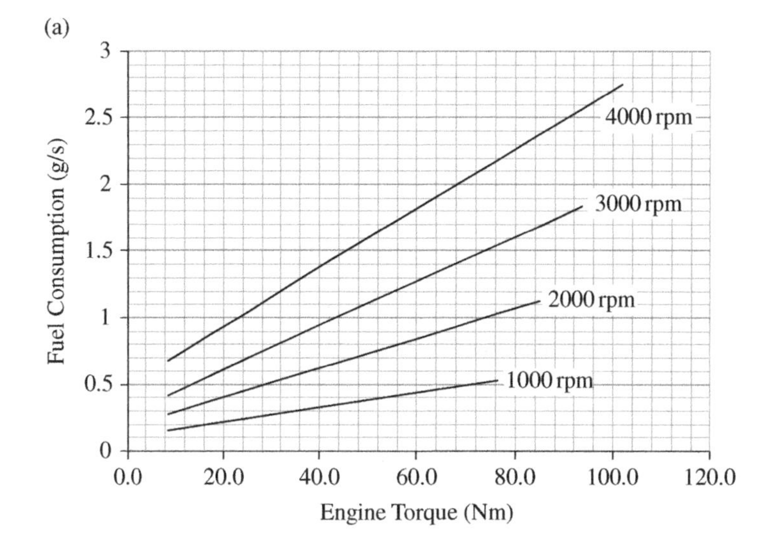

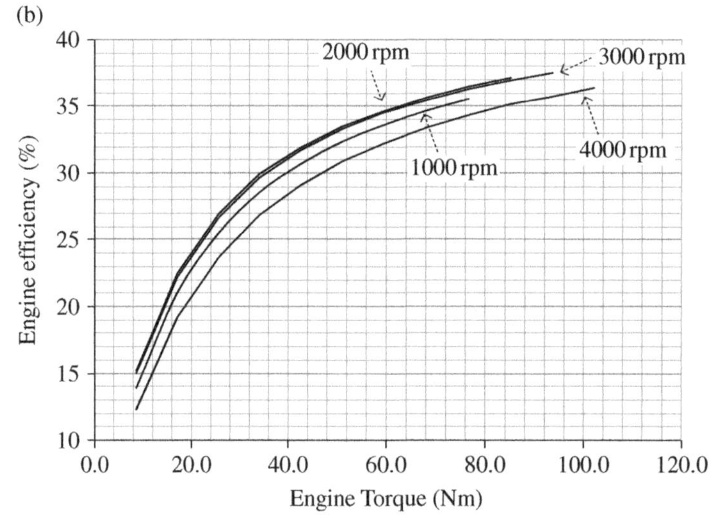

Figure 5.3 1997 Toyota Prius Atkinson-cycle SI engine characteristics for (a) fuel consumption and (b) efficiency.

Representative BSFC maps for a 1.9 L gasoline engine and a 1.7 L diesel engine are presented in Figure 5.4 and Figure 5.5, respectively, on the basis of data from [2]. The efficiency for the gasoline engine peaks at about 33.8% at 125 Nm and 2500 rpm. At a high speed of 6000 rpm, the efficiency is 11% at 16 Nm and rises to a peak of 29% at 125 Nm.

The diesel engine has a peak efficiency of about 37.8% at 180 Nm and 2200 rpm and maintains an engine efficiency in excess of 30% for much of the torque range between 2000 and 3000 rpm.

In the next section, a simple drivetrain example is used to illustrate the relative merits of the various vehicle powertrains.

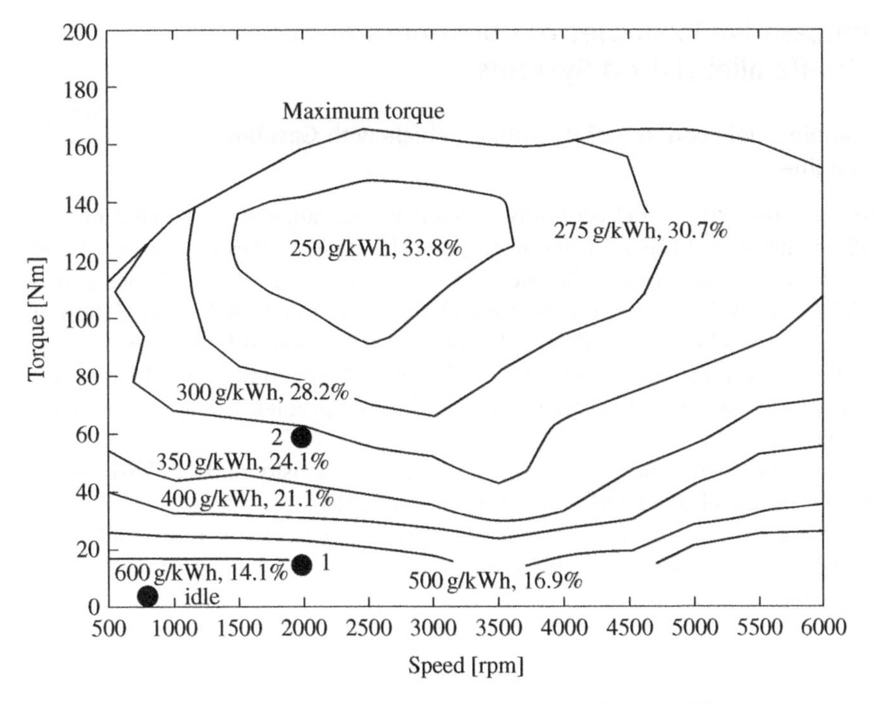

Figure 5.4 BSFC map for GM Saturn 1.9 L SI IC engine using data from [2].

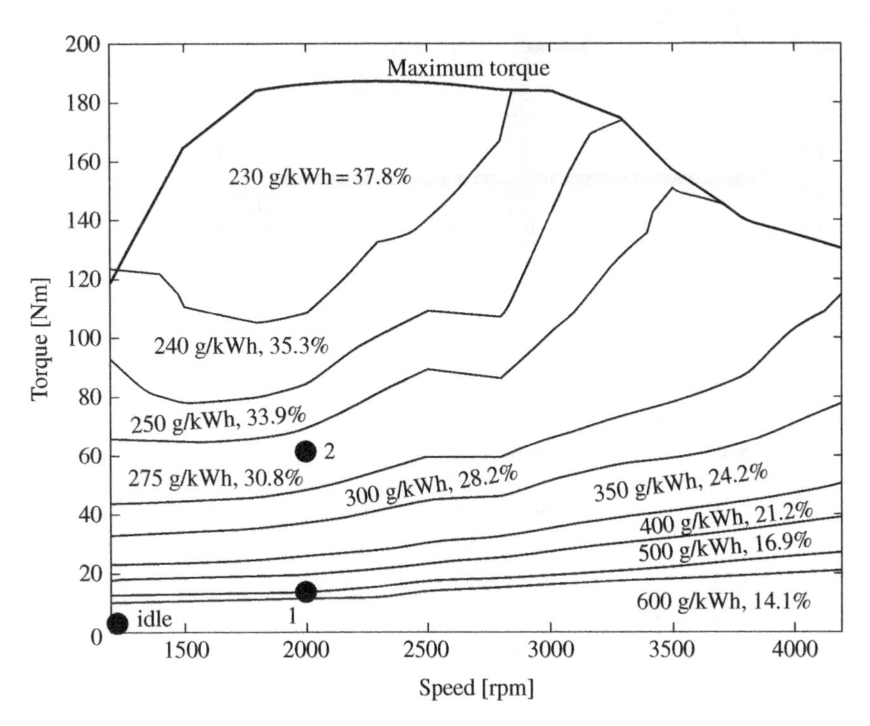

Figure 5.5 BSFC map for Mercedes 1.7 L CI IC engine using data from [2].

5.3 Comparative Examples of Conventional, Series, and Series-Parallel Hybrid Systems

5.3.1 Example: Fuel Economy of IC Engine Vehicle with Gasoline or Diesel Engine

In this example, the vehicle fuel economy, efficiency, and range are estimated for the 1.9 L gasoline engine and the 1.7 L diesel engine. The BSFC curves of Figure 5.4 and Figure 5.5 are used to determine the fuel consumption for gasoline and diesel, respectively. The BSFC values are approximate and are based on a visual interpolation of the operating points. The basic vehicle architecture is presented in Figure 5.6.

The idle fuel consumption for both the gasoline and diesel engines is 0.15 g/s with idling at a speed of 700 rpm (this value is not used in the problem). The vehicle has a 40 L fuel tank.

Assume a specific energy for gasoline of 42.6 kJ/g and a density of 0.749 kg/L, and a specific energy for diesel of 42.9 kJ/g and a density of 0.843 kg/L.

The vehicle transmission is assumed to have the representative gear ratios and operating range presented in Table 5.1. A **differential** is a torque-splitting gearing mechanism which enables the wheels to spin at different speeds when turning.

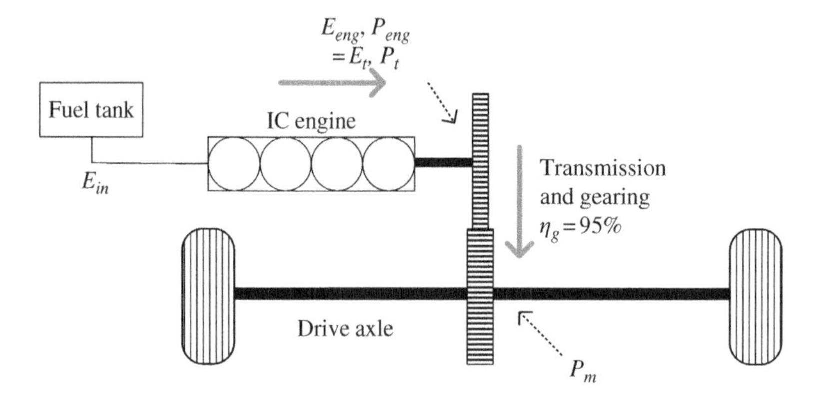

Figure 5.6 Conventional vehicle power flow.

Table 5.1 Test vehicle gear ratios.

Gear	Ratio	Operating range
1	3.72	0 < km/h < 11
2	2.13	11 < km/h < 37
3	1.30	37 < km/h < 53
4	0.89	53 < km/h < 72
5	0.69	72 < km/h
Reverse	3.31	
Differential	3.70	

Determine the fuel economy, range, and carbon emissions for the vehicle of Chapter 2, Section 2.3 using (a) gasoline and (b) diesel engines.

Solution:
First, the engine speed is estimated on the basis of the vehicle speed and the gear ratio.

The gear ratio from the engine shaft to the drive axle is the product of the particular gear ratio and the differential gear ratio n_{diff}. From Table 5.1, the vehicle operates in 3rd gear, with ratio n_{g3}, at 50 km/h and in 5th gear, with ratio n_{g5}, at 90 km/h. The linear speeds are $v_1 = 13.89$ m/s and $v_2 = 25$ m/s from Chapter 2, Section 2.3.

The engine angular speeds ω_{eng1} and ω_{eng2} are given by

$$\omega_{eng1} = \frac{v_1}{r} \times n_{g3} n_{diff} = \frac{13.89}{0.315} \times 1.30 \times 3.70 \, \text{rad/s} = 212.1 \, \text{rad/s}$$

and

$$\omega_{eng2} = \frac{v_2}{r} \times n_{g5} n_{diff} = \frac{25}{0.315} \times 0.69 \times 3.70 \, \text{rad/s} = 202.6 \, \text{rad/s}$$

where r is the wheel radius from Chapter 2, Table 2.1.

The engine speeds N_{eng1} and N_{eng2} are given by

$$N_{eng1} = \omega_{eng1} \times \frac{60}{2\pi} = 212.1 \times \frac{60}{2\pi} \, \text{rpm} = 2025 \, \text{rpm}$$

and

$$N_{eng2} = \omega_{eng2} \times \frac{60}{2\pi} = 202.6 \times \frac{60}{2\pi} \, \text{rpm} = 1935 \, \text{rpm}$$

The engine shaft torques T_{eng1} and T_{eng2} are given by

$$T_{eng1} = \frac{P_{eng1}}{\omega_{eng1}} = \frac{P_{t1}}{\omega_{eng1}} = \frac{3478}{212.1} \, \text{Nm} = 16.4 \, \text{Nm}$$

and

$$T_{eng2} = \frac{P_{eng2}}{\omega_{eng2}} = \frac{P_{t2}}{\omega_{eng2}} = \frac{12050}{202.6} \, \text{Nm} = 59.5 \, \text{Nm}$$

where P_{t1} and P_{t2} are provided in Chapter 2, Section 2.3.

The data generated so far are summarized in Table 5.2.

Table 5.2 Drive cycle operation for SI and CI engines.

Time	Duration (s)	Speed (km/h)	Gear	Speed (rpm)	T_{eng} (Nm)	P_{eng} (kW)	d (km)	E_t (kWh)
1	1800	50	3	2025	16.4	3.478	25	1.74
2	1200	90	5	1935	59.5	12.050	30	4.02
3	600	idle	neutral	700	—	—	0	0
Cycle	3600						55	5.76

Given that the engine torque and speed are known for the two cruising speeds, the fuel consumption of the vehicle can be determined. The closest values from the BSFC curves for the 50 km/h data points of 16.4 Nm at 2025 rpm are approximately 600 g/kWh for gasoline and 500 g/kWh for diesel. Values from the curves for the 90 km/h data points of 59.5 Nm at 1935 rpm are approximately 310 g/kWh for gasoline and 260 g/kWh for diesel.

Knowing the fuel consumption rate, the mass and volume of the fuel can be calculated in order to complete the required distance. The example is explained as follows for the gasoline engine, and the results are summarized in Table 5.3. The results for the diesel engine are summarized in Table 5.4.

The fuel mass consumption rate $\dot{m}$ is given by

$$\dot{m} = \frac{BSFC \times P_{eng}}{3600} \tag{5.6}$$

The fuel mass consumption rates $\dot{m}_1$, $\dot{m}_2$, and $\dot{m}_3$ are

$$\dot{m}_1 = \frac{BSFC \times P_{eng}}{3600} = \frac{600 \times 3.478}{3600}\frac{\text{g}}{\text{s}} = 0.58\,\frac{\text{g}}{\text{s}}$$

$$\dot{m}_2 = \frac{310 \times 12.05}{3600}\frac{\text{g}}{\text{s}} = 1.038\,\frac{\text{g}}{\text{s}}$$

and the fuel consumption rate during idle $\dot{m}_3$ is

$$\dot{m}_3 = 0.15\,\frac{\text{g}}{\text{s}}$$

The fuel mass m consumed over time is

$$m = \dot{m} \times t \tag{5.7}$$

The fuel masses m_1, m_2, and m_3 are given by

$$m_1 = \dot{m}_1 \times t_1 = 0.58 \times 1800\,\text{g} = 1.044\,\text{kg}$$

$$m_2 = \dot{m}_2 \times t_2 = 1.038 \times 1200\,\text{g} = 1.246\,\text{kg}$$

and

$$m_3 = \dot{m}_3 \times t_3 = 0.15 \times 600\,\text{g} = 0.09\,\text{kg}$$

The total mass of fuel consumed over the drive cycle m_C is the sum of the fuel masses consumed during each interval:

$$m_C = m_1 + m_2 + m_2 = 1.044\,\text{kg} + 1.246\,\text{kg} + 0.090\,\text{kg} = 2.38\,\text{kg}$$

The volume of fuel is given by

$$V = \frac{m}{\rho} \tag{5.8}$$

where the fuel density ρ is 0.749 kg/L for gasoline and 0.843 kg/L for diesel.

The fuel volumes V_1, V_2, and V_3 are given by

$$V_1 = \frac{m_1}{\rho} = \frac{1.044}{0.749}\text{L} = 1.394\,\text{L},\ V_2 = \frac{m_2}{\rho} = \frac{1.246}{0.749}\text{L} = 1.664\,\text{L},\ \text{and}\ V_3 = \frac{m_3}{\rho} = \frac{0.090}{0.749}\text{L} = 0.12\,\text{L}$$

Table 5.3 Results for drive cycle using gasoline engine.

Speed (km/h)	Time (s)	BSFC (g/kWh)	$\dot{m}$ (g/s)	Fuel (g)	Fuel (L)	E_{in} (MJ)	η_{eng} (%)	FC		mpg (US) (mpg)	CO_2 (gCO_2/km)	Range (km)
								(Wh/km)	(L/100km)			
50	1800	600	0.58	1044	1.394	44.47	14.1	494	5.58	42.2	129	717
90	1200	310	1.038	1246	1.664	53.08	27.2	491	5.55	42.4	128	725
Idle	600	—	0.15	90	0.12	3.83	—	—	—	0	—	—
Total	3600	—		2380	3.178	101.38	20.4	512	5.78	40.7	134	692

Table 5.4 Results for drive cycle using diesel engine.

Speed (km/h)	Time (s)	BSFC (g/kWh)	$\dot{m}$ (g/s)	Fuel (g)	Fuel (L)	E_{in} (MJ)	η_{eng} (%)	FC		mpg (US) (mpg)	CO_2 (gCO_2/km)	Range (km)
								(Wh/km)	(L/100km)			
50	1800	500	0.483	869	1.03	37.28	16.8	414	4.12	57	110	971
90	1200	260	0.870	1044	1.24	44.79	32.3	415	4.13	57	110	969
Idle	600	—	0.15	90	0.11	3.86	0	—	—	0	—	—
Total	3600			2003	2.38	85.93	24.1	434	4.32	54.4	115	926

The total volume of fuel consumed over the drive cycle V_C is the sum of the fuel volumes consumed during each interval:

$$V_C = V_1 + V_2 + V_3 = 1.394\,\text{L} + 1.664\,\text{L} + 0.12\,\text{L} = 3.178\,\text{L}$$

The energy input to the IC engine E_{in} is the product of the specific energy and the mass:

$$E_{in} = m \times Q \tag{5.9}$$

The energy inputs E_{in1}, E_{in2}, and E_{in3} are

$$E_{in1} = m_1 \times Q = 1.044 \times 42.6\,\text{MJ} = 44.47\,\text{MJ}$$

$$E_{in2} = m_2 \times Q = 1.246 \times 42.6\,\text{MJ} = 53.08\,\text{MJ}$$

$$E_{in3} = m_3 \times Q = 0.090 \times 42.6\,\text{kJ} = 3.83\,\text{MJ}$$

The total energy consumed over the drive cycle E_{inC} is the sum of the fuel volumes consumed during each interval:

$$E_{inC} = E_{in1} + E_{in2} + E_{in3} = 44.47\,\text{MJ} + 53.08\,\text{MJ} + 3.83\,\text{MJ} = 101.38\,\text{MJ}\,[= 28.16\,\text{kWh}]$$

The engine efficiency η_{eng} is

$$\eta_{eng} = \frac{E_t}{E_{in}} \times 100\% \tag{5.10}$$

The engine efficiencies are

$$\eta_{eng1} = \frac{E_{t1}}{E_{in1}} \times 100\% = \frac{6.26}{44.47} \times 100\% = 14.1\%$$

$$\eta_{eng2} = \frac{E_{t2}}{E_{in2}} \times 100\% = \frac{14.46}{53.08} \times 100\% = 27.2\%$$

$$\eta_{engC} = \frac{E_{tC}}{E_{inC}} \times 100\% = \frac{20.72}{101.38} \times 100\% = 20.4\%$$

where E_{t1}, E_{t2}, and E_{tC} are calculated in Chapter 2, Section 2.3.

These simple calculations illustrate the challenge for the IC engine. While the efficiency at medium-to-high torques can be relatively high, the efficiency drops significantly at the lower torques. Thus, the engine efficiency drops from 27.2% at 90 km/h to 14.1% at 50 km/h. The efficiency for the overall cycle factors in the idling condition, when no motive energy is generated, resulting in an overall efficiency for the cycle of 20.4%.

Given that we know the input and output energies and distances traveled, we can calculate additional useful information.

The fuel consumption of the vehicle and the related fuel economy are based on the input energy from the fuel:

$$FC = \frac{E_{in}}{s} \left(\text{in } \frac{\text{Wh}}{\text{km}}\right) \tag{5.11}$$

The values for fuel consumption in L/100 km, a commonly used measure, and fuel economy in mpg are given by

$$FC = \frac{V}{s} \times 100 \frac{\text{L}}{100\,\text{km}} \tag{5.12}$$

$$\text{mpg} = \frac{\frac{s}{1.609}}{\frac{V}{3.785}} \frac{\text{miles}}{\text{gal}} \tag{5.13}$$

The carbon emissions are estimated on the basis of the CO_2 emissions of 3.09 kg per kg of fuel, as per Chapter 1, Table 1.1 for gasoline. The mass of CO_2 m_{CO2} released by the combustion is

$$m_{CO2} = 3.09\,m \tag{5.14}$$

and the emissions per kilometer are given by

$$\frac{\text{gCO}_2}{\text{km}} = \frac{m_{CO2}}{s} \tag{5.15}$$

The vehicle range is simply the volume of the vehicle fuel tank divided by the fuel consumption:

$$\text{Range} = \frac{V_{tank}}{FC} \tag{5.16}$$

The various values for the overall drive cycle using the gasoline engine are

$$FC = \frac{E_{inC}}{s} = \frac{28160\,\text{Wh}}{55\ \ \text{km}} = 512 \frac{\text{Wh}}{\text{km}}$$

$$FC = \frac{V}{s} \times 100 \frac{\text{L}}{100\,\text{km}} = \frac{3.178}{55} \times 100 \frac{\text{L}}{100\,\text{km}} = 5.78 \frac{\text{L}}{100\,\text{km}}$$

$$\text{mpg} = \frac{\frac{s}{1.609}}{\frac{V}{3.785}} \frac{\text{miles}}{\text{gal}} = \frac{\frac{55}{1.609}}{\frac{3.178}{3.785}} \frac{\text{miles}}{\text{gal}} = \frac{34.18}{0.84} \text{mpg} = 40.7\,\text{mpg}$$

$$\frac{\text{gCO}_2}{\text{km}} = \frac{3.09\,m_C}{s} = \frac{3.09 \times 2.38}{55} \frac{\text{gCO}_2}{\text{km}} = 134 \frac{\text{gCO}_2}{\text{km}}$$

$$\text{Range} = \frac{V_{Tank}}{FC} = \frac{40\,\text{L}}{5.78\,{}^{\text{L}}/_{100}\ \text{km}} = 692\,\text{km}$$

The results for the full cycle for the gasoline engine are summarized in Table 5.3.

A similar exercise is conducted for the diesel engine – see Table 5.4. The fuel economy in mpg is significantly higher for the diesel engine than for the gasoline engine for two basic reasons: the increased energy density of diesel and the improved combustion efficiency. As the specific energies of diesel and gasoline are very close, the estimated carbon emissions reflect the difference in efficiencies for the two engine types.

5.3.2 Example: Fuel Economy of Series HEV

In this example, the electric drive is used to propel the vehicle. The Atkinson-cycle IC engine is only used to recharge the battery, as shown in Figure 5.7. For simplicity, a generating efficiency of 85% is also assumed for the engine output through the generator and inverter to recharge the battery. The engine is operated at its optimum efficiency point in order to supply the required energy to the battery.

Refer to the related BEV example in Chapter 3, Section 3.5.

Solution:

From the previous BEV example, we know the energy required from the battery for the drive cycle E_{bC}. The output energy required from the IC engine E_{eng} is the energy required to recharge the battery and is the battery output energy divided by the generating efficiency:

$$E_{eng} = \frac{E_b}{\eta_{gen}} \tag{5.17}$$

Thus, the engine output energy required is

$$E_{eng} = \frac{E_{bC}}{\eta_{gen}} = \frac{24.5}{0.85}\,\text{MJ} = 28.82\,\text{MJ}\,(8.01\,\text{kWh})$$

The optimum efficiency operating point for the Atkinson-cycle engine is at a torque T_{eng} of about 80 Nm at an engine speed N_{eng} of 2000 rpm. The output power P_{eng} at this operating point is 16.755 kW.

The time required to charge the battery t_{gen} is the output energy E_{eng} divided by the output power P_{eng}:

$$t_{gen} = \frac{E_{eng}}{P_{eng}} = \frac{28.82 \times 10^6}{16755}\,\text{s} = 1720\,\text{s}$$

The value of the BSFC is 230 g/kWh from Figure 5.2. The fuel mass flow rate $\dot{m}$ is

$$\dot{m} = \frac{BSFC \times P_{eng}}{3600} = \frac{230 \times 16.755}{3600}\frac{\text{g}}{\text{s}} = 1.07\,\frac{\text{g}}{\text{s}} \tag{5.18}$$

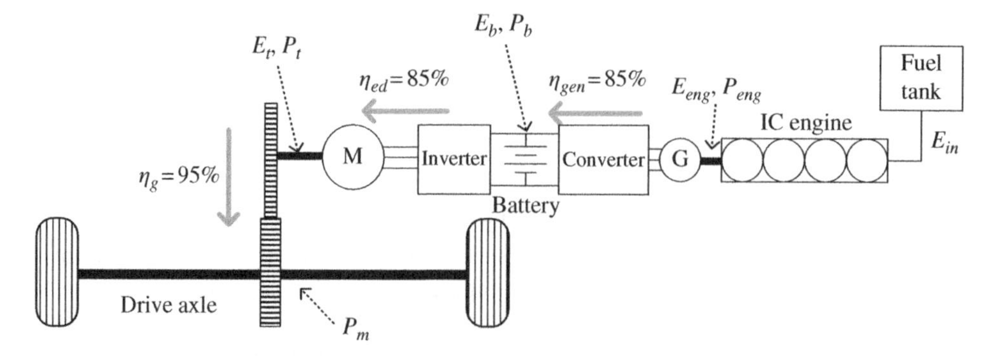

Figure 5.7 Series HEV power flow.

The mass m_C, volume V_C, and input energy E_{inC} of the fuel to the engine to provide the energy to power the vehicle over the full drive cycle are

$$m_C = \dot{m} \times t_{gen} = 1.07 \times 1720\,\text{g} = 1.84\,\text{kg}$$

$$V_C = \frac{m_C}{\rho} = \frac{1.84}{0.749}\text{L} = 2.457\,\text{L}$$

and

$$E_{inC} = m_C \times Q = 1840 \times 42.6\,\text{kJ} = 78.38\,\text{MJ}\ (21.77\,\text{kWh})$$

The powertrain efficiency is simply the traction energy divided by the input energy, expressed as a percentage:

$$\eta_{pt} = \frac{E_{tC}}{E_{inC}} \times 100\% = \frac{20.72}{78.38} \times 100\% = 26.4\%$$

where the cycle traction energy E_{tC} was determined in Chapter 2, Section 2.3.

Note that the engine is operating very efficiently at this point.

$$\eta_{eng} = \frac{E_{eng}}{E_{inC}} \times 100\% = \frac{28.82}{78.38} \times 100\% = 36.7\%$$

The various other data are calculated on the basis of the earlier example for the IC engine:

$$FC = \frac{E_{inC}}{s} = \frac{21770\,\text{Wh}}{55\ \ \text{km}} = 396\frac{\text{Wh}}{\text{km}}$$

$$FC = \frac{V}{s} \times 100\frac{\text{L}}{100\,\text{km}} = \frac{2.457}{55} \times 100\frac{\text{L}}{100\,\text{km}} = 4.47\frac{\text{L}}{100\,\text{km}}$$

$$\text{mpg} = \frac{\frac{s}{1.609}}{\frac{V}{3.785}}\frac{\text{miles}}{\text{gal}} = \frac{\frac{55}{1.609}}{\frac{2.457}{3.785}}\frac{\text{miles}}{\text{gal}} = \frac{34.18}{0.649}\text{mpg} = 52.7\,\text{mpg}$$

$$\frac{\text{gCO}_2}{\text{km}} = \frac{3.09\,m_C}{s} = \frac{3.09 \times 1.84}{55}\frac{\text{gCO}_2}{\text{km}} = 103\frac{\text{gCO}_2}{\text{km}}$$

$$\text{Range} = \frac{V_{tank}}{FC} = \frac{40\,\text{L}}{4.47\,\text{L}/100\,\text{km}} = 895\,\text{km}$$

The results for the series HEV are summarized in Table 5.5.

Table 5.5 Series HEV simplified drive cycle results.

Time (s)	BSFC (g/kWh)	$\dot{m}$ (g/s)	Fuel (g)	Fuel (L)	E_{in} (MJ)	η_{pt} (%)	FC (Wh/km)	FC (L/100km)	mpg (US) (mpg)	CO_2 (gCO_2/km)	Range (km)
Cycle	230	1.07	1840	2.457	78.38	26.4	396	4.47	52.7	103	895

The advantage of hybridization can be understood from this example. The engine is operated at close to its maximum efficiency, and the efficient electric drive train is used to propel the wheels. The carbon emissions and fuel consumption improve on the conventional vehicles, and the range is improved on the BEV. However, a disadvantage of the series HEV is that the engine energy has to pass through two electric stages in series to propel the vehicle.

5.3.3 Example: Fuel Economy of Series-Parallel HEV

The basic architecture of the series-parallel HEV is shown in Figure 5.8. In this example, the electric drive is used to propel the vehicle for only a portion of the drive cycle.

The vehicle is powered by the electric drive and battery for the 50 km/h portion during t_1 and when idling in t_3. The engine is optimally controlled to provide a short burst of energy, as just happened for the series HEV.

The engine is then used to directly power the vehicle at the higher speed of 90 km/h for time duration t_2 when it is more efficient to directly power the vehicle from the IC engine.

Solution:

We know from the earlier BEV example in Chapter 3, Section 3.5 that the energies consumed from the battery during periods t_1 and t_3 are E_{b1} and E_{b3}, respectively. Thus, the energy required from the IC engine to power the vehicle for this duration is

$$E_{eng} = \frac{E_{b1} + E_{b3}}{\eta_{gen}} = \frac{7.37 + 0.12}{0.85} \text{MJ} = 8.81 \text{ MJ}$$

where the battery energies E_{b1} and E_{b3} were determined in Chapter 2, Section 2.3.

As before for the series HEV, the optimum efficiency operating point for the Atkinson-cycle engine is about 80 Nm at 2000 rpm for an engine efficiency of 36.7%.

$$P_{eng} = T_{eng}\omega_{eng} = 80 \times 2000 \times \frac{2\pi}{60} \text{kW} = 16.755 \text{ kW}$$

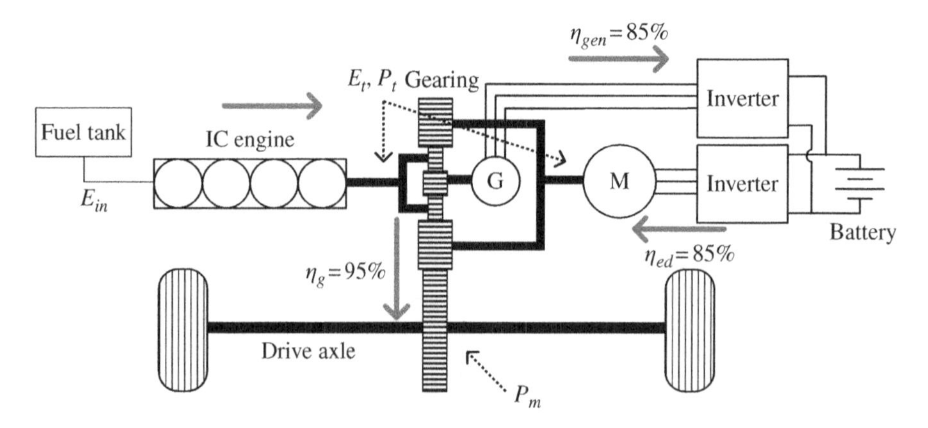

Figure 5.8 Series-parallel HEV.

The time required to charge the battery is

$$t_{gen} = \frac{E_{eng}}{P_{eng}} = \frac{8.81 \times 10^6}{16755}\,\text{s} = 526\,\text{s}$$

Let subscripts 1,3 represent the combined values for intervals 1 and 3.

Again, the rate of fuel consumption at this power is 1.07 g/s. Thus, the fuel mass, volume, and input energy required to provide energy for periods t_1 and t_3 are

$$m_{1,3} = \dot{m}_{1,3} \times t_{gen} = 1.07 \times 526\,\text{g} = 563\,\text{g}$$

$$V_{1,3} = \frac{m_{1,3}}{\rho} = \frac{0.563}{0.749}\,\text{L} = 0.752\,\text{L}$$

$$E_{in1,3} = m_{1,3} \times Q = 563 \times 42.6\,\text{kJ} = 23.98\,\text{MJ}\ (= 6.66\,\text{kWh})$$

During interval t_2, the engine directly powers the drive gear at 90 km/h. The power required at the output of the engine is 12.05 kW. If the engine is run at 1500 rpm, say, then the engine torque is given by

$$T_{eng} = \frac{P_{eng}}{\omega_{eng}} = \frac{12050}{1500 \times \frac{2\pi}{60}}\,\text{Nm} = 76.7\,\text{Nm}$$

The value of the BSFC is about 230 g/kWh from Figure 5.2. The mass flow rate $\dot{m}$ is

$$\dot{m}_2 = \frac{BSFC \times P_{eng}}{3600} = \frac{230 \times 12.050}{3600}\,\frac{\text{g}}{\text{s}} = 0.77\,\frac{\text{g}}{\text{s}} \tag{5.19}$$

and the remaining values are

$$m_2 = \dot{m}_2 \times t_2 = 0.77 \times 1200\,\text{g} = 924\,\text{g}$$

$$V_2 = \frac{m_2}{\rho} = \frac{0.924}{0.749}\,\text{L} = 1.234\,\text{L}$$

$$E_{in2} = m_2 \times Q = 924 \times 42.6\,\text{kJ} = 39.36\,\text{MJ}\ (= 10.93\,\text{kWh})$$

Knowing the energy consumptions required for the motive intervals, the overall energy consumption can be determined for the cycle. The results are summarized in Table 5.6. The table is arranged such that the results are presented for the vehicle when operating in the series mode, as happens during intervals t_1 and t_3, and in the parallel mode, as happens during interval t_2. The advantage of the series-parallel hybrid is clear from this example.

Table 5.6 Series-parallel HEV simplified drive cycle results.

						FC				
Interval	**Mode**	**Fuel (kg)**	**Fuel (L)**	E_{in} **(MJ)**	η_{pt} **(%)**	**(Wh/km)**	**(L/100km)**	**mpg (US) (mpg)**	CO_2 **(gCO$_2$/km)**	**Range (km)**
$t_1 + t_3$	Series	0.563	0.752	23.98	26.5	266	3.0	78.2	70	1331
t_2	Parallel	0.924	1.234	39.36	36.7	364	4.11	57.2	95	973
Total	—	1.487	1.986	63.34	32.7	320	3.6	65.2	84	1108

Table 5.7 Drive train comparison.

	Fuel	Fuel (kg)	Fuel (L)	E_{in} (MJ)	η_{pt} (%)	FC_{in} (Wh/km)	(L/100km)	mpge (US) (mpg)	CO_2 (gCO_2/km)	Range (km)
SI	Gasoline	2.38	3.178	101.38	20.4	512	5.78	40.7	134	692
CI	Diesel	2.003	2.38	85.93	24.1	434	4.32	54.4	115	926
BEV	Electricity			28.82	84.6	146		144	73/1.5	485
Series HEV	Gasoline	1.840	2.457	78.38	26.4	396	4.47	52.7	103	895
Series-Parallel HEV	Gasoline	1.487	1.986	63.34	32.7	320	3.6	65.2	84	1108
FCEV	Hydrogen	0.366		43.996	47.1	222		94	69	750

The vehicle's operating system can run the vehicle in the series or parallel modes depending on which mode provides the optimum efficiency or performance.

5.3.4 Summary of Comparisons

A summary for the various vehicles analyzed in Chapters 2 to 5 is presented in Table 5.7. The BEV of Chapter 3 requires the least amount of energy and is followed by the FCEV of Chapter 4. The series-parallel HEV outperforms the series HEV, and the conventional diesel and gasoline vehicles, all of this chapter. The series-parallel HEV has the longest range, while the BEV has the shortest range. Note that the fossil-fueled vehicles have 40 L of fuel storage, the FCEV has 5 kg of hydrogen, and the BEV has a 60 kWh battery.

The numbers on the carbon emissions are interesting. In general, the numbers reflect the efficiency of the energy conversion for the conventional and hybrid vehicles, and the blend of power generation for the BEV. The BEV, FCEV, and series-parallel HEV all have emissions less than 100 gCO_2/km. Two representative numbers are estimated for the BEV. The first number of 73 gCO_2/km is based on an average of about 500 gCO_2/kWh(electrical) in the United States in 2012 with a fuel mix of about 70% fossil fuel and 30% nuclear and renewables. With the bulk of the electrical power in Norway being sourced from renewable hydropower, the carbon emissions drop to about 10 gCO_2/kWh (electrical), resulting in less than 1.5 gCO_2/km for the BEV in Norway.

5.4 The Planetary Gears as a Power-Split Device

The splitting of the engine power in a series-parallel HEV is usually achieved using epicyclic gears arranged in a ring, sun, and planetary set. This type of gearing is often used to implement a CVT on a vehicle. The CVT enables a variable gearing system in contrast to the discrete gearing system used in the earlier example. HEVs, such as the Toyota Prius and Chevy Volt, implement some form of CVT based on epicyclic gears. Vehicles such as

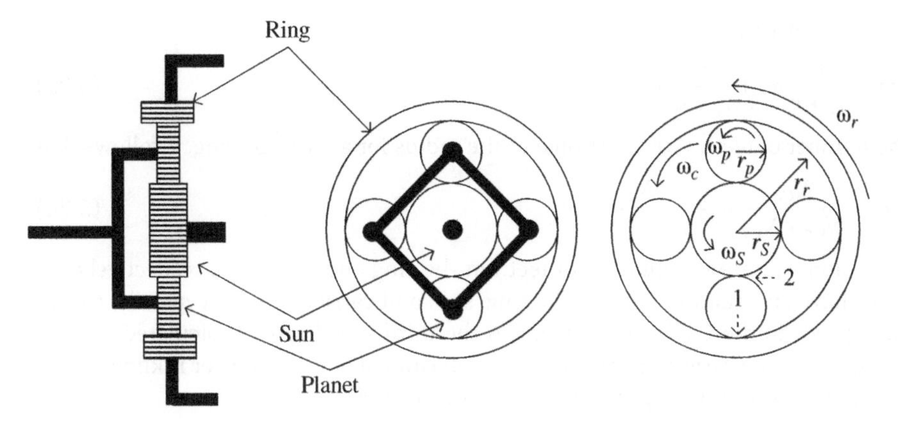

Figure 5.9 Sun and planetary gearing system.

the Hyundai Ioniq and Honda Fit use a competitive dual-clutch multigear system in a parallel hybrid configuration.

An illustration of epicyclic gears featuring the ring, sun, and four planetary gears is presented in Figure 5.9. There are three sets of gears and four relative motions, as shown in Figure 5.9. The outer gear is known as the **ring gear**, and the ring gear has the teeth on the inside surface. In the 2004 Toyota Prius, the ring gear is directly coupled to the rotor of the electrical motor and to the drive train. The innermost gear is known as the **sun gear**, with its gear teeth on the outside surface. The sun gear is directly coupled to the electrical generator in the 2004 Toyota Prius. The intermediate gears are known as the **planetary gears**. The planetary gears, four in this example, are directly coupled together, and they spin together as the **carrier gear**. The carrier gear is directly coupled to the IC engine in the 2004 Toyota Prius.

The ring gear has an internal radius r_r and a rotational speed ω_r. The planetary gear has a radius r_p and a rotational speed ω_p. The sun gear has a radius r_s and a rotational speed ω_s. The carrier gear has a rotational speed ω_c.

A simple way to characterize the gearing operation is by considering the relative gear velocities of points 1 and 2 shown in Figure 5.9.

The velocity of the two gears is v_1 at the point of contact at point 1, at the intersection of the planetary and ring gear. Thus

$$v_1 = r_r\omega_r = r_p\omega_p + r_r\omega_c \tag{5.20}$$

The velocity at this point is therefore the sum of the velocities due to the planetary gear and the coupled carrier mechanism.

There is a similar relationship for the velocity v_2 at the second point of contact between the planetary and sun gears, shown as point 2:

$$v_2 = r_s\omega_s = -r_p\omega_p + r_s\omega_c \tag{5.21}$$

Adding Equation (5.20) and Equation (5.21) yields the following relationship:

$$r_r\omega_r + r_s\omega_s = (r_r + r_s)\omega_c \tag{5.22}$$

which can be rewritten as

$$\omega_c = \frac{r_r}{r_r + r_s}\omega_r + \frac{r_s}{r_r + r_s}\omega_s \tag{5.23}$$

As the number of teeth is proportional to the radius for a given gearing, it follows that

$$\omega_c = \frac{N_{Gr}}{N_{Gr} + N_{Gs}}\omega_r + \frac{N_{Gs}}{N_{Gr} + N_{Gs}}\omega_s \tag{5.24}$$

where N_{Gr} and N_{Gs} are the number of teeth in the ring and sun gears, respectively.

We now have an equation relating the ring, sun, and carrier speeds. Typically, two of the three speeds in the CVT are known, and the third speed can be calculated.

The torques for the various gearings can be determined once the power is known. The powers supplied to the three gearings must sum to zero:

$$P_r + P_c + P_s = 0 \tag{5.25}$$

where P_r, P_c, and P_s are the powers supplied to the ring, carrier and sun gears, respectively.

In the next section, a particular application of the epicyclic gears in a CVT is considered.

5.4.1 Powertrain of 2004 Toyota Prius

The 2004 Toyota Prius is used to illustrate the operation of the CVT and the power split. Pictures of various CVTs and gearings are shown in Figure 5.10. Various other series-

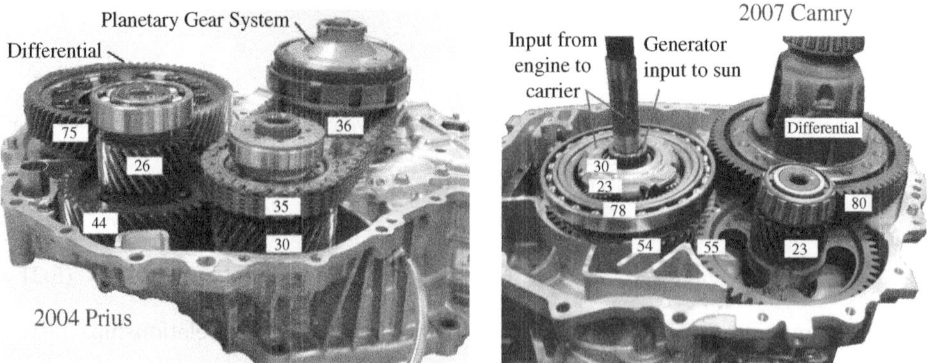

Figure 5.10 CVTs and gearings of 2010 Toyota Prius, 2004 Toyota Prius and 2007 Toyota Camry [3]. (Courtesy of Oak Ridge National Laboratory, US Department of Energy.)

Table 5.8 Gearing for CVT of 2004 Toyota Prius.

Gear	Coupling	No. of teeth
Ring	Motor & Drive train	78
Planetary	Engine	23
Sun	Generator	30

parallel HEVs use some form of power split. The operation is similar, but the detail and complexity can vary depending on the vehicle requirements.

The ring gear of the CVT is directly coupled to the motor shaft and to the drive train using a chain. The CVT is coupled to the drive axle via the chain drive and a number of gears. The final gear ratio from the ring gear to the drive axle is 4.113 [3].

The coupling configurations and teeth numbers of the gearing of the CVT of the 2004 Toyota Prius are presented in Table 5.8.

Thus, the earlier equation can be written as

$$\omega_{eng} = \frac{N_{Gr}}{N_{Gr} + N_{Gs}}\omega_{mot} + \frac{N_{Gs}}{N_{Gr} + N_{Gs}}\omega_{gen} \tag{5.26}$$

or

$$\omega_{eng} = \frac{78}{78+30}\omega_{mot} + \frac{30}{78+30}\omega_{gen} = 0.722\,\omega_{mot} + 0.278\,\omega_{gen}$$

where ω_{eng}, ω_{mot}, and ω_{gen} are the angular speeds of the engine, electric motor, and electric generator, respectively. The motor speed ω_{mot} is related to the axle speed ω_{axle} and linear speed v by

$$\omega_{mot} = \omega_{axle} \times n_{diff} = \frac{v}{r} \times n_{diff} \tag{5.27}$$

where $n_{diff} = 4.113$ is the differential gear ratio.

If we consider the engine and motor as supplying powers P_{eng} and P_{mot}, respectively, to the CVT, then the output power from the CVT creates generator power P_{gen}, or provides traction power P_t to the drivetrain:

$$P_{eng} + P_{mot} = P_{gen} + P_t \tag{5.28}$$

There are a number of modes of operation, and the main modes of operation are next explored using some examples.

5.4.2 Example: CVT Operating in Electric Drive Mode (Vehicle Launch and Low Speeds)

Determine the speeds and torques of the motor, generator, and engine if the vehicle is operating in electric mode, with the engine and generator off, as shown in Figure 5.11. Assume the same vehicle as in the earlier examples. The vehicle is traveling at 50 km with traction power P_t = 3.478 kW. Let r = 0.315 m and n_{diff} = 4.113.

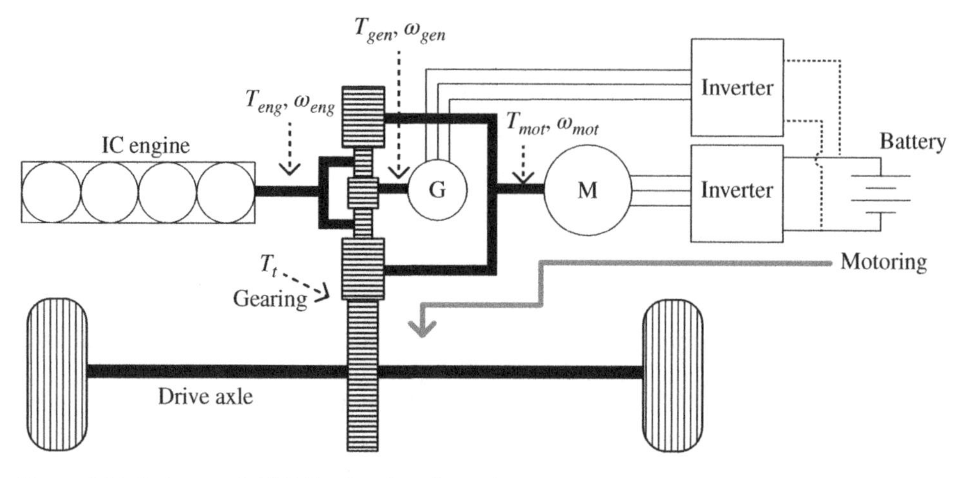

Figure 5.11 Series-parallel HEV in electric-only mode.

Solution:

This is the simplest mode of operation. In this mode, the motive power is sourced from the battery alone. The electric motor provides the traction torque, and no power is supplied by the engine. The engine shaft is locked and is not spinning, and so the resulting speed of the carrier gear is zero. The generator speed will be estimated.

The motor speed is

$$\omega_{mot} = \frac{v}{r} \times n_{diff} = \frac{50/3.6}{0.315} \times 4.113\,\text{rad/s} = 181.4\,\text{rad/s}$$

or

$$N_{mot} = \omega_{mot} \times \frac{60}{2\pi} = 181.4 \times \frac{60}{2\pi}\,\text{rpm} = 1732\,\text{rpm}$$

The motor torque T_{mot} is

$$T_{mot} = \frac{P_t}{\omega_{mot}} = \frac{3478}{181.4}\,\text{Nm} = 19.2\,\text{Nm}$$

The generator speed for this condition can be determined by rearranging Equation (5.26):

$$\omega_{gen} = \frac{N_{Gr} + N_{Gs}}{N_{Gs}}\omega_{eng} - \frac{N_{Gr}}{N_{Gs}}\omega_{mot} \tag{5.29}$$

For this example, the engine speed is zero, and so the generator gear speed is

$$\omega_{gen} = \frac{N_{Gr} + N_{Gs}}{N_{Gs}}\omega_{eng} - \frac{N_{Gr}}{N_{Gs}}\omega_{mot} = \frac{78+30}{30}\omega_{eng} - \frac{78}{30}\omega_{mot} = 3.6\,\omega_{eng} - 2.6\,\omega_{mot} \tag{5.30}$$

$$\Rightarrow \omega_{gen} = 3.6 \times 0\,\text{rad/s} - 2.6 \times 181.4\,\text{rad/s} = -471.6\,\text{rad/s}$$

and

$$N_{gen} = \omega_{gen} \times \frac{60}{2\pi} = -471.6 \times \frac{60}{2\pi}\,\text{rpm} = -4503\,\text{rpm}$$

Thus, for this condition, the motor develops 19.2 Nm at 1732 rpm, the generator spins at −4503 rpm, and the engine develops zero torque and does not spin.

5.4.3 Example: CVT Operating in Full-Power Mode

Determine the motor, generator, and engine speeds and torques if the vehicle is operating in full-power mode, with the engine on and maximum torque going to the drive axle from both the engine and the motor, as shown in Figure 5.12. Use the same vehicle as in the previous example. Assume that the vehicle requires 82 kW of traction power at 85 km/h in order to accelerate with the engine operating at its maximum power condition of 57 kW at 5000 rpm.

Solution:

In this mode, the motive power is sourced from both the battery and the engine. A minor torque is developed in the generator. While the electric motor is directly geared to the drive axle, the generator plays the important role of regulating the engine speed such that the engine can operate at its maximum power.

The engine angular speed ω_{eng} is based on its maximum power operating point of 57 kW at 5000 rpm.

$$\omega_{eng} = N_{eng} \times \frac{2\pi}{60}\,\text{rad/s} = 5000 \times \frac{2\pi}{60}\,\text{rad/s} = 523.6\,\text{rad/s}$$

The engine shaft torque T_{eng} is

$$T_{eng} = \frac{P_{eng}}{\omega_{eng}} = \frac{57000}{523.6}\,\text{Nm} = 108.9\,\text{Nm}$$

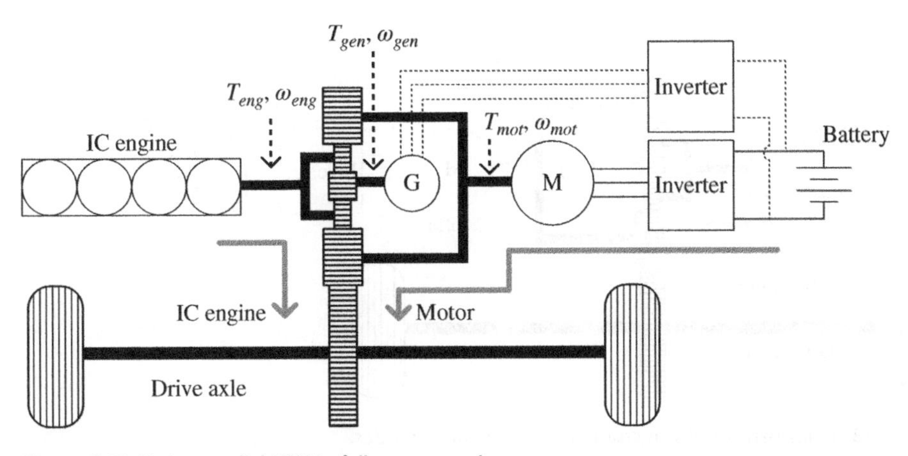

Figure 5.12 Series-parallel HEV in full-power mode.

The motor angular speed ω_{mot} is directly related to the drive axle speed:

$$\omega_{mot} = \frac{v}{r} \times n_{diff} = \frac{85/3.6}{0.315} \times 4.113\,\text{rad/s} = 308.3\,\text{rad/s}$$

The motor speed N_{mot} is

$$N_{mot} = \omega_{mot} \times \frac{60}{2\pi} = 308.3 \times \frac{60}{2\pi}\,\text{rpm} = 2944\,\text{rpm}$$

If the engine is outputting 57 kW and 82 kW is required, then the balance of 25 kW is sourced from the battery through the electric motor.

The motor torque T_{mot} is

$$T_{mot} = \frac{P_{mot}}{\omega_{mot}} = \frac{25000}{308.3}\,\text{Nm} = 81.1\,\text{Nm}$$

The generator speed is determined as follows:

$$\omega_{gen} = 3.6\,\omega_{eng} - 2.6\omega_{mot} = 3.6 \times 523.6\,\text{rad/s} - 2.6 \times 308.3\,\text{rad/s} = 1083.4\,\text{rad/s}$$

and

$$N_{gen} = \omega_{gen} \times \frac{60}{2\pi} = 1083.4 \times \frac{60}{2\pi}\,\text{rpm} = 10346\,\text{rpm}$$

In this example, it is assumed that the generator torque is minimal and that the generator is spinning at 10346 rpm in order to govern the engine speed while the engine generates 108.9 Nm at 5000 rpm, and the motor generates 81.1 Nm at 2944 rpm.

5.4.4 Example: CVT Operating in Cruising and Generating Mode

Determine the motor, generator, and engine speeds and torques if the vehicle is operating in cruise mode at a speed of 85 km/h, with the engine on and generating, as shown in Figure 5.13. The vehicle requires 57 kW at 5000 rpm, and the generator develops 33 kW. Assume negligible torque on the traction motor.

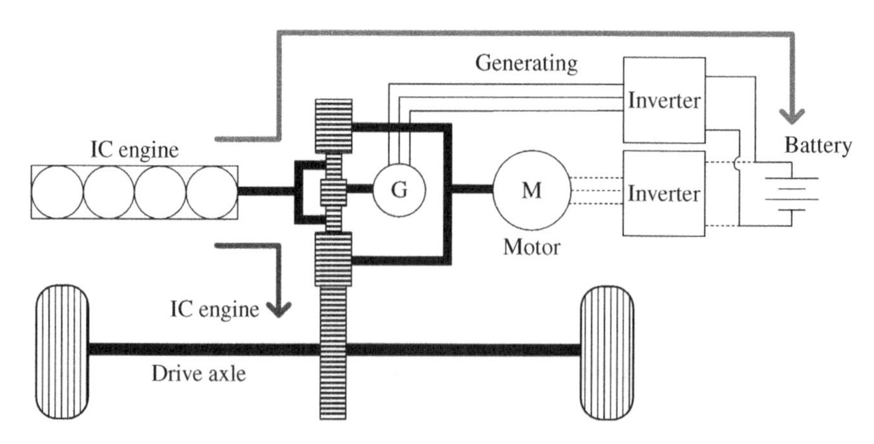

Figure 5.13 Series-parallel HEV in cruise and regenerative modes.

Solution:

In this mode, the motive power is sourced from the engine while the battery is being recharged. Negligible torque is developed in the motor.

The motor, generator, and engine speeds and engine torque are all as calculated in the previous example.

The generator shaft torque T_{gen} is

$$T_{gen} = \frac{P_{gen}}{\omega_{gen}} = \frac{-33000}{1083.4}\,\text{Nm} = -30.5\,\text{Nm}$$

In this mode, the generator generates 33 kW at 10,346 rpm while the engine runs at full power at 57 kW and 5000 rpm. The motor runs at 2944 rpm.

References

1 John Heywood, *Internal Combustion Engine Fundamentals,* 1st edition, McGraw Hill, 1988.

2 ADVISOR 2.0, National Renewable Energy Laboratory, 1999.

3 R. H. Staunton, C. W. Ayers, L. D. Marlino, J. N. Chiasson, and T. A. Burress, *Evaluation of 2004 Toyota Prius Hybrid Electric Drive System,* Oak Ridge National Laboratory report, May 2006.

Problems

5.1 Verify the data of Table 5.4 for the diesel engine.

5.2 Determine the fuel consumption, carbon emissions and range when traveling at a constant speed of 120 km/h for 600 s for the gasoline, diesel, SHEV, and SPHEV vehicles used in the examples in this chapter. Refer to Chapter 2 Problem 2.7.

[Ans. gasoline: 6.85 L per 100 km, 159 gCO_2/km, 584 km; diesel: 5.97 L per 100 km, 159 gCO_2/km, 670 km; series: 8.71 L per 100 km, 202 gCO_2/km, 459 km; series-parallel: 6.3 L per 100 km, 146 gCO_2/km, 635 km]

5.3 The drive cycle of this chapter is modified such that the idle period is replaced by a highway cruise at 120 km/h for 600 s. Determine the mpg, carbon emissions, and range for the revised drive cycle.

[Ans. gasoline: 5.9 L per 100 km, 137 gCO_2/km, 678 km; diesel: 4.62 L per 100 km, 123 gCO_2/km, 866 km; series: 5.59 L per 100 km, 129 gCO_2/km, 716 km; series-parallel: 4.29 L per 100 km, 99 gCO_2/km, 932 km]

5.4 The HEV with a CVT, of the examples used in this chapter, cruises at 45 km/h in EV-only mode. The engine is off, and the vehicle is regenerating to the battery. Determine the speeds of the motor and generator.

[Ans. N_{mot} = 1,558 rpm, N_{gen} = −4,051 rpm]

5.5 The HEV of the preceding problem reverses at 10 km/h with the engine off. Determine the speeds of the motor and generator.

[Ans. N_{mot} = −347 rpm, N_{gen} = +902 rpm]

5.6 Determine the motor, generator, and engine speeds if the vehicle of the preceding problem operates in cruise mode at a speed of 180 km/h, with the engine on and generating. The engine outputs 57 kW at 110 Nm.

[Ans. N_{eng} = 4,948 rpm, N_{mot} = 6,234 rpm, N_{gen} = +1,604 rpm]

Assignments

Advisor 2.0

5.1 Download Advisor 2.0, and simulate the various vehicle models.

MATLAB Code for BSFC Map for 1997 Toyota Prius

The MATLAB code modifies the original data file from ADVISOR 2.0 and generates a BSFC map similar to Figure 5.2.

```
%Brake specific fuel consumption for 1.5L Toyota Prius(Atkinson
cycle)engine
% with a maximum power of 43kW @4000rpm and max Torque 102 Nm @ 4000
rpm. (Kevin Davis and John Hayes)
% Data source: ADVISOR script file "FC_PRIUS_JPN.m"
Spd_rpm=[1000 1250 1500 1750 2000 2250 2500 2750 3000 3250
3500 4000];
Trq_Nm=[8.54 16.95 25.49 34.04 42.44 50.99 59.53 67.94 76.48 85.02
93.43 101.97];
% FUEL consumption table in (g/s),vertical index=Speed, horizontal
index=Torque
Fuel_cons = [
0.1513 0.1984 0.2455 0.2925 0.3396 0.3867 0.4338 0.4808 0.5279 0.5279 0.5279 0.5279
0.1834 0.2423 0.3011 0.3599 0.4188 0.4776 0.5365 0.5953 0.6541 0.6689 0.6689 0.6689
0.2145 0.2851 0.3557 0.4263 0.4969 0.5675 0.6381 0.7087 0.7793 0.8146 0.8146 0.8146
0.2451 0.3274 0.4098 0.4922 0.5746 0.6570 0.7393 0.8217 0.9041 0.9659 0.9659 0.9659
0.2759 0.3700 0.4642 0.5583 0.6525 0.7466 0.8408 0.9349 1.0291 1.1232 1.1232 1.1232
0.3076 0.4135 0.5194 0.6253 0.7312 0.8371 0.9430 1.0490 1.1549 1.2608 1.2873 1.2873
0.3407 0.4584 0.5761 0.6937 0.8114 0.9291 1.0468 1.1645 1.2822 1.3998 1.4587 1.4587
0.3773 0.5068 0.6362 0.7657 0.8951 1.0246 1.1540 1.2835 1.4129 1.5424 1.6395 1.6395
0.4200 0.5612 0.7024 0.8436 0.9849 1.1261 1.2673 1.4085 1.5497 1.6910 1.8322 1.8322
0.4701 0.6231 0.7761 0.9290 1.0820 1.2350 1.3880 1.5410 1.6940 1.8470 1.9999 2.0382
0.5290 0.6938 0.8585 1.0233 1.1880 1.3528 1.5175 1.6823 1.8470 2.0118 2.1766 2.2589
0.6789 0.8672 1.0555 1.2438 1.4321 1.6204 1.8087 1.9970 2.1852 2.3735 2.5618 2.7501
];
[Trq,spd]=meshgrid(Trq_Nm, Spd_rpm);% develop torque-speed
table of operating points
```

```
Engine_kW=Trq.*spd*(pi/30)/1000;% Calculate engine power table
in kW using P=Tw
bsfc=(Fuel_cons./Engine_kW).*3600; % bsfc=(g/s)/Power (kW)
*(seconds per hour) to get (g/kWh)
%Plotting efficiency as a contour plot
[C,h]=contour(spd,Trq,bsfc,[225 230 240 250 275 300 350 400
500],'k','linewidth',1);
clabel(C,h); % displays the selected contour values
xlabel('Speed [rpm]');ylabel('Torque [Nm]');zlabel ('BSFC
[g/kWh]');
title ('BSFC (g/kWh) for Prius 1.5L Engine');
axis([1000,4000,0,120]); %
hold on
%LIMITS Plot of maximum torque (Nm) for each operating speed (RPM)
Trq_max=[77.29 82.04 84.75 86.78 89.36 91.12 92.89 94.65 96.41
98.17 99.94 101.97];
plot(Spd_rpm,Trq_max,'--kd','linewidth',3);
% end
```

Part 2

Electrical Machines

6

Introduction to Traction Machines

"If I had asked people what they wanted, they would have said faster horses."
Henry Ford (1863–1947).

This chapter introduces the electrical machines used for traction in electric drives. Electric traction is the key contributor to efficient electric and hybrid propulsion. The very simple key relationships between the mechanical quantities of torque and speed and the electrical quantities of current and voltage are presented. The various dc and ac machines are introduced together with their advantages and disadvantages for use in the automotive powertrain. The chapter discusses the key specifications for electrical machines that can be generally applied regardless of the machine type.

6.1 Propulsion Machine Overview

Electrical machines have played a vital role in the development of modern industrial society and are a significant enabling technology for electric propulsion. The torque, or turning force, developed by an electric machine is due to the interaction between a current and a magnetic field. The magnetic field can be generated by either a permanent magnet (PM) or an electromagnet, whose field flux is generated using current.

Let us note that there are two main physical parts of an electrical machine. The **stator** is the stationary part of the machine. The **rotor** is the spinning or rotating part of the machine.

There are two broad classes of machines: alternating current or ac, and direct current or dc. Earlier electric vehicles featured dc machines for use in propulsion, while modern electrically propelled vehicles use ac machines. Ac machines are very efficient and have a higher power density, lower cost, and better reliability than dc machines. Thus, ac machines are used in applications as diverse as propulsion for electric vehicles and electricity generation in wind turbines and power stations. However, although ac machines have achieved widespread dominance for medium-to-high power applications, dc machines are ubiquitous in low-power applications due to their ease of control and cost. Dc machines utilize mechanical commutators to adjust the rotating field. The advent of

Electric Powertrain: Energy Systems, Power Electronics and Drives for Hybrid, Electric and Fuel Cell Vehicles,
First Edition. John G. Hayes and G. Abas Goodarzi.

Companion website: www.wiley.com/go/hayes/electricpowertrain

power semiconductors and low-cost microprocessors has made the electronic control of ac motor drives more cost-effective, more power dense, and inherently more reliable than dc drives.

Dc and ac machines play a significant role on the modern internal combustion (IC) engine vehicle. A series-wound dc motor is the starter motor to crank the shaft for a gasoline engine, while an ac machine is used for the alternator. The alternator generates ac voltages and uses a rectification stage to convert the ac waveforms to the dc voltages required on the vehicle.

Ac and dc machines operate on similar principles, and their operation can be described by similar speed and torque relationships with voltage, current, and magnetic flux. The operation of electromagnetic machines is based on some basic relationships that are summarized by two simple equations.

First, (as derived in Chapter 16, Section 16.7), the electromagnetic torque T_{em} developed by a machine is directly related to the current I supplied to the machine:

$$T_{em} = kI \tag{6.1}$$

where k is the machine constant and is a function of the physical and magnetic properties of the machine.

A similar relationship exists between the back-emf voltage E induced onto the machine windings and the angular speed ω_r of the machine rotor:

$$E = k\omega_r \tag{6.2}$$

6.1.1 DC Machines

For many people, the first contact with an electric propulsion system is the brushed PM dc motor, which propels the toy car and is energized by a primary alkaline battery. The ease of use of a **brushed PM dc motor** makes it the machine of choice for many applications. The current flowing in the brushed PM machine flows from the stationary part of the machine (stator) to the rotating part of the machine (rotor) via mechanical brushes and commutators, thus enabling the machine to optimally generate torque. However, the brushes wear away with time, which limits the lifetime of brushed dc motors. An example of a brushed PM dc machine is shown in Figure 6.1. This particular machine is a windshield wiper motor. The rotor comprises the drive axle, the output shaft, the commutator, the winding, and the iron core. The stator comprises the housing, the PMs, the brushes, and the flange. Note that the stator does not have an iron core to conduct the flux as the rotor does. The flux actually returns to the magnets through the metal housing of the machine.

The **wound-field dc machine** uses the field current of an electromagnet, rather than a PM, to generate the field flux. This machine also features brushes, but it has a significant advantage over its PM sibling in that the field current can be controlled to weaken the magnetic field, an attribute that enables high-speed operation of the machine.

Brushed dc machines have limitations on operating voltage, durability, size, and lifetime due to the brushes. However, there are many uses on the vehicle for which they are the optimum drive motor: windshield wiper, seat adjustment, blower motor, headlamp adjusters, electric steering-column lock, mirror adjusters, electronic throttle control, exhaust-gas recirculation, window-lift drives, electric pumps, lumbar support, power-lift gear, and more.

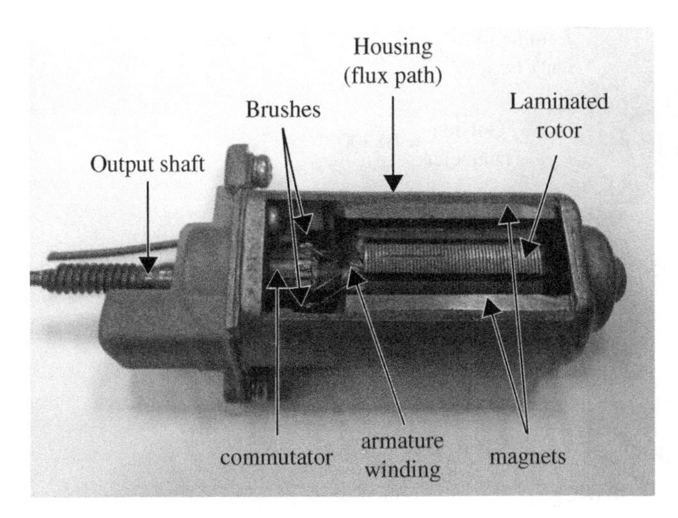

Figure 6.1 Toyota windshield wiper motor as an example of a conventional dc motor.

There are two broad classes of dc machine: the **brushed** motor with lifetime limitations, and the **electronically commutated** (EC) dc motor with a significantly greater lifetime. The EC dc machine is closer in operation to a PM ac motor in that the PMs are located on the rotor and the current-carrying conductors are located on the stator. This machine structure eliminates the need for the brushes and commutator and significantly increases the lifetime of the machine, albeit with the penalty of additional associated sensing and power circuits that are required for electronic commutation. The EC dc machine is often known as the **trapezoidal-waveform brushless dc machine** and is optimum for many electric drives. On the vehicle, the EC dc machine is often the preferred machine for fuel pumps, exhaust-gas recirculation, dual-clutch transmission, and electric power steering. EC dc motors are also commonly used for long-life applications such as fans for computers, an application where the lifetime requirements far exceed the capabilities of a brushed machine.

6.1.2 AC Machines

The ac machine is based on the wonderful inventions of Nikola Tesla, a great contributor to the industrial age. Tesla envisioned a machine being fed by three-phase ac voltages and currents with the phases being distributed such that a spinning magnetic field is generated within the machine *without* the need for the brushes and commutator. This incredible invention was a key part of the technology revolution that created the modern electrical grid. An example of the stator of an ac machine is shown in Figure 6.2.

The genius of Nikola Tesla was to visualize that by spacing three winding coils, a, b, and c, 120° apart physically within the stator of the machine, as shown in Figure 6.3(a), and by supplying these three windings with three currents which are 120° displaced in time, as we presently have in our electrical grid, a spinning magnetic field is created. Note that the coils consist of pairs of conductors carrying current in a closed path within the machine; for example, coil a is shown to have conductors a^+ and a^- carrying current out

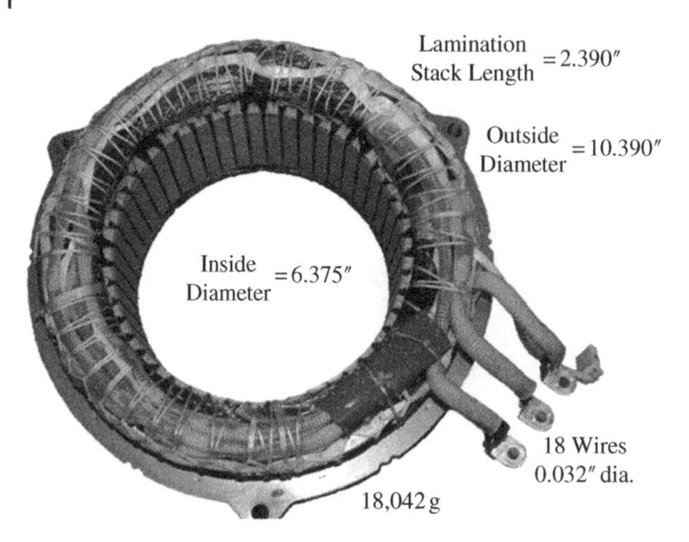

Figure 6.2 Stator of 2007 Toyota Camry hybrid electric vehicle motor [1]. (Courtesy of Oak Ridge National Laboratory, US Dept. of Energy.)

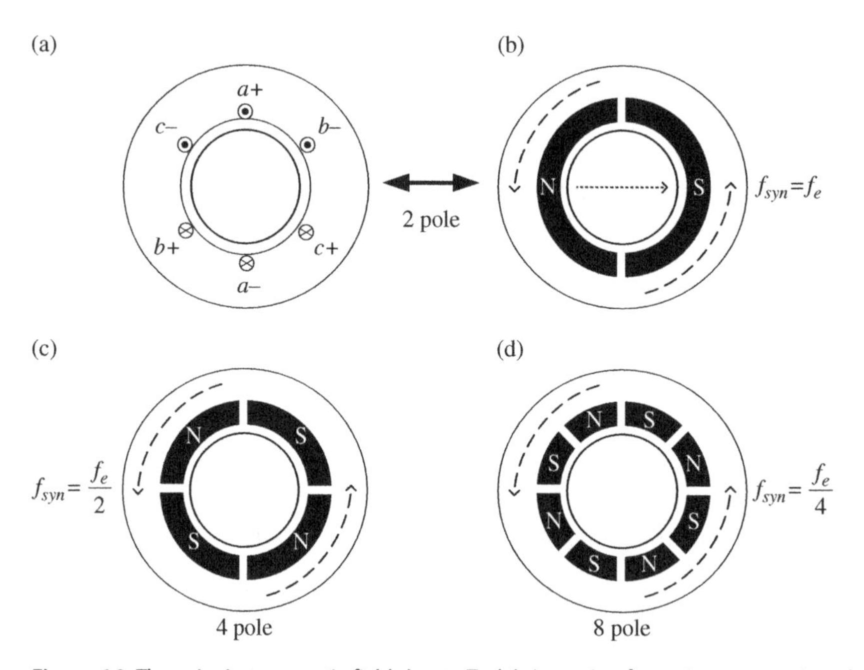

Figure 6.3 The spinning magnetic field due to Tesla's invention for various magnetic pole counts.

of and into the page, respectively. The spinning magnetic field can be modeled by a spinning magnet, as shown in Figure 6.3(b). This spinning magnetic field created on the stator can then interact with a rotor magnetic field, generated by a PM or current to develop electromechanical torque. The spinning magnetic field can even induce the rotor current with which it will interact – the basis of the induction machine. Tesla's great invention was to enable generators based in Niagara Falls to generate three-phase ac voltages for transformation and transmission to New York City for distribution and use. At the point of use, the ac voltages can be supplied to an ac motor and reconverted to mechanical power using the same electromechanical technology as for generating.

The machines are commonly designed to have more than one magnetic pole pair. The conductors can be rearranged in the stator slots to make multiple pole machines. Four- and eight-pole machines, with their stator-generated equivalent magnetic fields, are shown in Figure 6.3(c) and (d), respectively. The four-pole machine has two pole pairs, and the rotating magnetic field takes two electrical cycles to complete a full 360° mechanically. The eight-pole machine has four pole pairs, and the rotating magnetic field takes four electrical cycles to complete a full 360° mechanically.

The frequency of the rotating magnetic field is known as the **synchronous frequency** f_{syn}. The synchronous frequency can be determined by dividing the input electrical frequency by the number of pole pairs:

$$f_{syn} = \frac{f_e}{p/2} \tag{6.3}$$

where p is the number of poles, and $p/2$ is the number of pole pairs.

One of the significant advantages of increasing the pole pairs in a machine is that it runs at a higher electrical frequency for a given mechanical speed, which reduces the machine size and weight. The induction motor used in the 1996 GM EV1 was a four-pole machine. The PM motors used in the Nissan Leaf and Toyota hybrids are eight-pole.

There are two broad classes of ac machines: **asynchronous** and **synchronous**. Both types of machine have a similar stator but have different rotor constructions.

In an asynchronous machine, the spinning magnetic field operates at a different frequency from the current with which it interacts to generate torque. In a synchronous machine, the spinning magnetic field and current operate at the same frequency.

An example of an asynchronous machine is the **squirrel-cage induction motor** invented by Nikola Tesla. In the induction motor, sketched in Figure 6.4(a), with an actual stator and rotor as shown in Figure 6.5, the current supplied to the stator generates a spinning magnetic field. The rotor features a conductor structure similar to a hamster

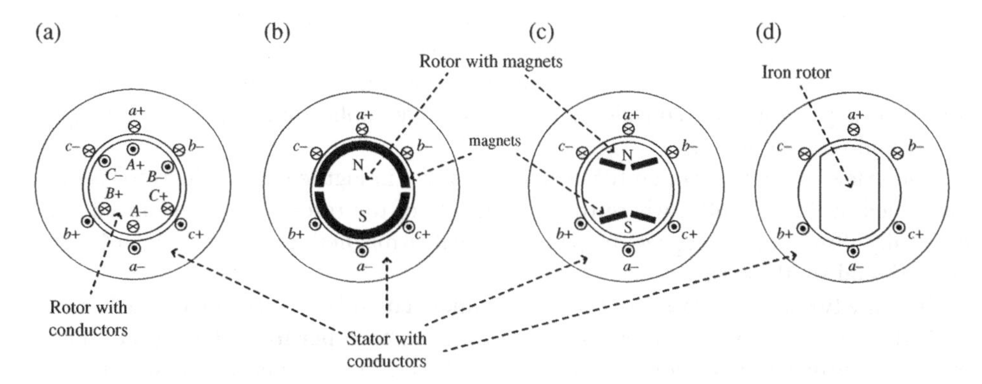

Figure 6.4 Sketches of (a) induction, (b) surface permanent-magnet, (c) interior permanent-magnet, and (d) reluctance machines.

(a)

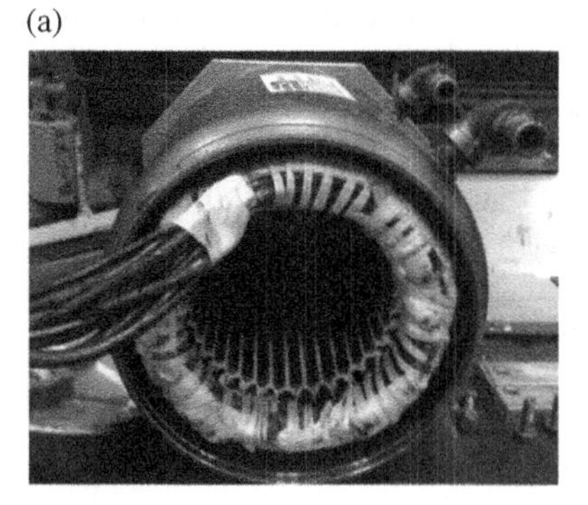

(b)

Figure 6.5 Views of automotive powertrain induction motor (a) stator and (b) rotor. (Courtesy of US Hybrid.)

or squirrel cage, with long bars running through the rotor and shorted on both ends. The spinning magnetic field induces voltages and currents on to the rotor by Faraday's law. The induced current then interacts with the spinning magnetic field, causing it to generate torque. The induction motor is commonly known as the workhorse of industry. The significant disadvantage of the induction motor is the heating loss generated on the rotor due to the induced current.

In recent years, synchronous machines based on PMs have become very popular and more than competitive with induction motors. In a **PM ac machine**, the rotor contains from one to several pairs of PM poles. The current fed to the stator interacts with the PMs on the rotor to generate torque. The technological advances of PMs and power electronics enable very high-power-density machines. The magnets may be on the surface or embedded in the interior of the rotor, as are shown in Figure 6.4(b) and (c). The **interior-permanent-magnet ac motor** is widely used for electric vehicles. Torque is generated in the interior-permanent-magnet (IPM) ac machine by two mechanisms: (1) the interaction between the PM flux and the supplied current to create **magnet torque** and (2) the interaction between the supplied currents and the iron material to create a **reluctance torque**.

The reluctance motor is another variation of the ac motor. An example of an elementary **reluctance machine** is shown in Figure 6.4(d). The spinning magnetic interacts with the salient-pole iron rotor to exert a reluctance torque on the iron, similar to the actions of a magnet and a piece of iron, and without the need for rotor magnets or conductors. Although its cost is low, the reluctance motor has had limited use in industrial applications and is generally not considered for automotive applications due to size, torque ripple, and acoustic noise.

A side view of the rotor of an IPM motor is shown in Figure 6.6(a). The rotor is disassembled in Figure 6.6(b), where the 16 magnets of the rotor have been moved and physically offset. The 16 magnets are paired to create 8 magnetic poles resulting in 4 pairs of north and south poles.

There are two general classes of PM ac machines. The IPM ac machine is widely used for higher-speed automotive propulsion, while the **surface-permanent-magnet** (SPM) machine is widely used for lower-speed electric drives. The magnets of the SPM motor are bonded to the surface of the rotor, as shown in Figure 6.4(b). However, the surface mounting of the magnets is not suitable for high-speed applications, and so the magnets

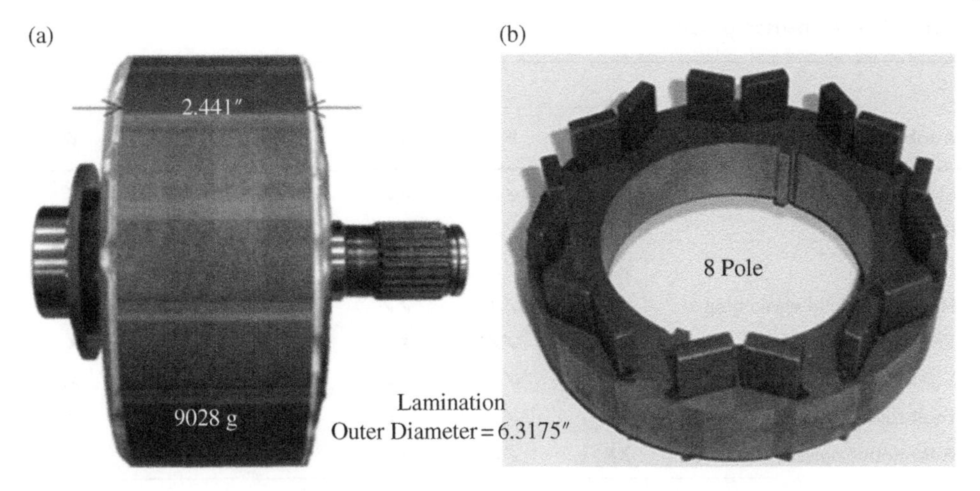

Figure 6.6 Views of the IPM rotor from the 2007 Toyota Camry HEV motor [1]. (Courtesy of Oak Ridge National Laboratory, US Dept. of Energy.)

Figure 6.7 Stator and rotor laminations of 2004 Toyota Prius motor.

are embedded in the rotor, resulting in the IPM structure, as shown in Figure 6.4(c) and Figure 6.6. The IPM structure makes the machine more robust for high-speed operation. The stator and rotor laminations of the 2004 Toyota Prius motor are shown in Figure 6.7. The Prius propulsion machine is an eight-pole IPM with each pole consisting of a pair of magnets arranged in a V shape. The other Toyota products have similar shapes, while the Nissan Leaf features three magnets per pole arranged in a $\bar{\text{V}}$ shape [2].

6.1.3 Comparison of Traction Machines

The first-generation GM EV1, brought to market in 1996, and the first- and second-generations of Tesla vehicles, the Tesla Roadster from 2007 and the Tesla Model S from 2012 all feature high-performance induction motors. The IPM ac motor is

Table 6.1 A comparison of various traction machines.

	ac				dc	
Attribute	**SPM**	**IPM**	**SC IM**	**RM**	**PM**	**WF**
Size	✓✓	✓✓	✓✓	✗✗	✗	✗
Cost	✓	✓✓	✓✓	✓✓	✗	✗
Efficiency	✓✓	✓✓	✓	✓✓	✗	✗
High-speed field weakening	✓✓	✓✓	✓✓	✓	✗✗	✓✓
Rotor cooling	✓✓	✓✓	✗	✓✓	✗	✗
Service and maintenance	✓✓	✓✓	✓✓	✓✓	✗✗	✗✗
Fault tolerance	✗✗	✓	✓✓	✓	✗✗	✓
Automotive powertrain	✗✗	✓✓	✓✓	✗✗	✗✗	✗✗

preferred by many other manufacturers for electric propulsion and is the machine of choice for the Toyota hybrid traction motor and generator, the Nissan Leaf, the Chevy Volt, the Mitsubishi iMiEV, and more.

A comparison of the SPM ac motor, the IPM ac motor, the squirrel-cage induction motor (SCIM), the reluctance motor (RM), the brushed permanent-magnet dc (PM dc) machine, and the brushed wound-field dc machine (WF dc) is presented in Table 6.1 for size, cost, efficiency, high-speed field weakening, rotor cooling, service and maintenance, fault tolerance, and overall suitability. Positive characteristics of a machine earn from one ✓ to two ✓✓. Negative characteristics earn a ✗. A double ✗✗ means that the machine is unsuitable for the automotive powertrain application.

In general, the dc machines do not compare well with the ac machines and are not suitable for the electric powertrain. The PM dc machine has a basic size limitation issue that will be explored in more detail in Chapter 7, where it is shown that the WF dc machine outperforms the PM dc machine due to its ability to control the field flux. Both the PM and WF dc machines are unsuitable for the automotive powertrain application due to the necessity of providing regular service and replacing the brushes.

The SPM and IPM machines have a size advantage over SCIMs, while the SCIM has advantages of cost and ruggedness over the PM machines. An advantage of the SPM and IPM machines over the induction motor is that the synchronous machine can be excited by dc current to generate torque at standstill.

The SCIM is more rugged because it is essentially made from iron and copper, while the PMs of the PM machines run the risk of being demagnetized or of generating excessive voltages under certain fault conditions. Both the SPM ac and PM dc machines can generate excessive voltages in the case of certain fault conditions.

The permanent magnets themselves can make up a significant portion of the machine cost, and so the SCIM has a cost advantage over the PM machines. Note that the IPM has a lower cost than the SPM and does not require containment rings to bind the magnets for high-speed operation.

The RM has potential for being cost-effective as it lacks both magnets and windings on the rotor, but it has not been used for the automotive powertrain due to its size, efficiency, and acoustic noise.

The IPM has an inherent efficiency advantage over the SCIM because the SCIM experiences rotor copper loss. Advanced cooling methods can result in high-power-density induction machines.

The PM dc, WF dc, and SCIM are all handicapped by significant copper losses on the rotor. Rotor heating is more difficult to manage compared to heating on the stator, with the result that machines with rotor copper losses are physically larger machines.

The most suitable cost-effective rugged machines for the mass-produced automotive electric powertrain are the IPM and the SCIM.

6.1.4 Case Study – Mars Rover Traction Motor

The Mars rover is an interesting case study for propulsion machines in general and for brushed dc motors in particular. The 2004 Mars rovers, known as the *Spirit* and *Opportunity,* are battery electric vehicles that were built to operate on a different planet! The *Spirit* and *Opportunity* featured advanced lightweight rechargeable lithium-ion batteries, which are recharged using on-vehicle solar panels.

The following is an excerpt from the Maxon Motors literature and references the Mars landing in January 2004 [3]. "This time, two identical rovers were to be sent up simultaneously: *Spirit* and *Opportunity*. They were significantly larger than their predecessors (about 185 kg) and equipped with more advanced technology. The rovers were able to take photos, brush and scrape rocks to reveal fresh surfaces for study. Their mission: find evidence that there used to be water on Mars, and thus, maybe, even life. The two vehicles exceeded all expectations. Their expected service life was three months. However, six years passed before *Spirit* sent its last signal to Earth. Its brother *Opportunity* is still going strong ... Despite sandstorms and temperature fluctuations from −120°C to +25°C, the Maxon dc motors are also still performing reliably. Each rover is equipped with 35 of these precision drives, which are responsible for driving the six wheels, the steering mechanism, the RAT (rock abrasion tool), the robotic arm, and the cameras. In addition, 8 Maxon motors were used in the lander."

The *Opportunity,* imaged and self-photographed in Figure 6.8, passed a magical milestone in March 2015 when it completed the equivalent of a marathon on Mars!

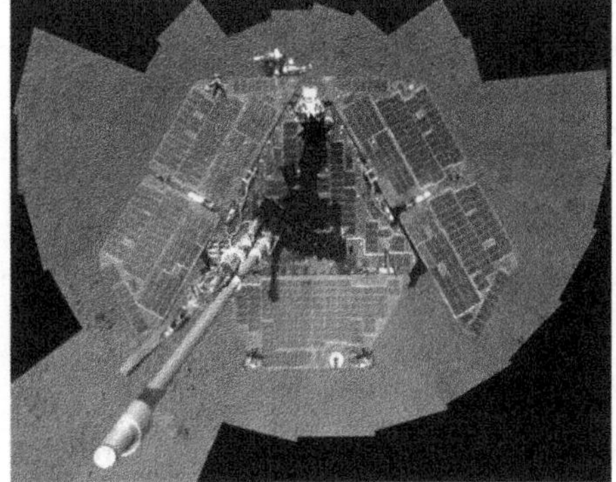

Figure 6.8 *Opportunity* Mars rover image and a self-portrait from Mars. (Courtesy of NASA/JPL-Caltech.)

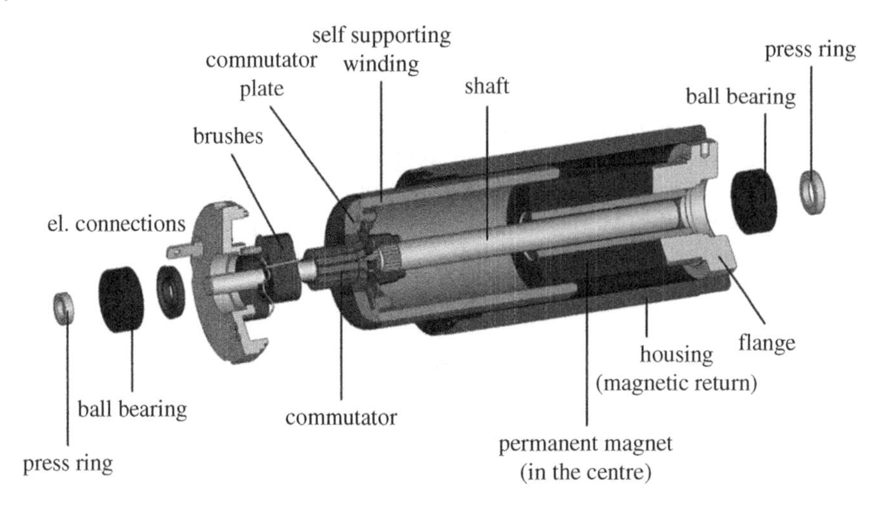

Figure 6.9 Coreless dc motor. (Courtesy of Maxon Motors.)

The traction motor used on the Mars rover, as seen in Figure 6.9, is a modified version of the standard Maxon motor, part number 339152, whose parameters are outlined in Chapter 7. This machine is now briefly discussed as an insightful example of the engineering of brushed dc machines.

The Maxon design is a "coreless" design as it features a sophisticated interleaved winding that is designed and optimized to not require iron core material on the rotor. This coreless approach has advantages of efficiency, weight, and inertia compared to the conventional construction and can also be applied to ac machines. The skewed winding approach also reduces torque cogging and operating noise. The coreless motor design is limited to low-torque applications.

The brushes on the Maxon machine are 50% graphite and 50% copper. The copper is required to reduce the contact and brush resistance. The commutator segments are also made of copper. The precious-metal brush, for example, made of silver compound, is an alternative and tends to have a constant relatively low resistance with armature current. The precious-metal brush is optimum for smaller machines operating continuously with light loads but is not optimum for stop-start or high-current loads. The graphite-copper brush was the choice for the Mars rover because of its higher current and temperature capabilities. Reference Chapter 7, Sections 7.10 and 7.11.

6.2 Machine Specifications

In this section, we review the key operating characteristics of electrical machines and identify some of the important specifications.

6.2.1 Four-Quadrant Operation

An electrical machine can operate across all four quadrants of operation, with positive and negative torque and positive and negative speed, as shown in Figure 6.10 and summarized in Table 6.2.

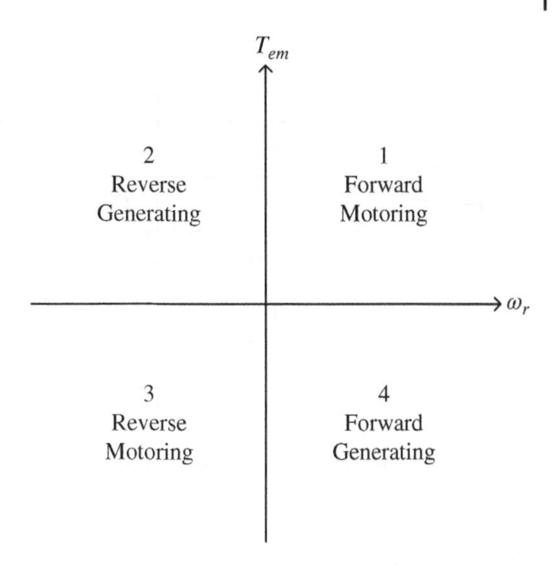

Figure 6.10 Four-quadrant torque versus speed.

Table 6.2 Machine quadrants.

Quadrant	Direction	Speed/EMF	Action	Torque/Current
1	Forward	+	Motoring	+
2	Reverse	–	Generating	+
3	Reverse	–	Motoring	–
4	Forward	+	Generating	–

The first quadrant of operation is for motoring in a forward direction with a positive torque and a positive speed. If the machine is spinning in a forward direction with a positive speed and a negative torque, then the machine is in the fourth quadrant and is forward generating/braking. If the machine is spinning in a reverse direction and motoring in the third quadrant, then the torque and speed are both negative. Finally, if the machine is reversing and generating/braking, then the speed is negative and the torque is positive and is in the second quadrant.

The practical electrical machine is limited in developing high torque as the required current can result in excessive winding heating and possible machine failure. Thus, the four-quadrant torque versus speed characteristic is as shown in Figure 6.11 for continuous and transient operation. We next discuss these considerations and the specifications for the machine.

6.2.2 Rated Parameters

The rated parameters for the machine are the parameters for which the machine has been designed. These are typically the rotor torque, rotor power, rotor speed, supply voltage, supply current, cooling conditions, ambient temperature, and time duration.

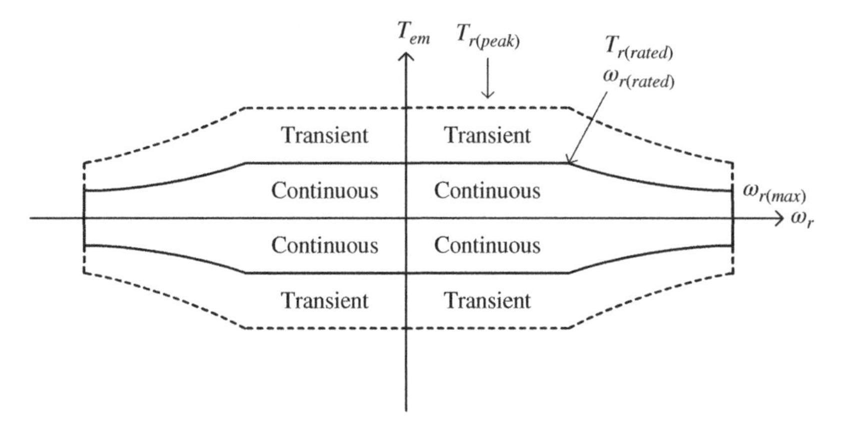

Figure 6.11 Continuous and transient operation for the practical machine over all four quadrants.

Under specified conditions, the machine outputs rated power at the rated speed when supplied by rated voltage, current, and frequency.

The machine may be specified to operate at the rated power continuously, or for a rated period of time.

6.2.3 Rated Torque

The rated torque of the machine $T_{r(rated)}$ is the rotor torque level at which the machine outputs the rated power at the rated speed. This parameter is also known as the full-load torque.

6.2.4 Rated and Base Speeds

The rated speed, also known as the **nominal** or base speed, of the machine $\omega_{r(rated)}$ is the speed at which the machine has been designed to deliver the rated power.

The rated speed is often quoted as $N_{r(rated)}$ with the units of revolutions per minute, or rpm, and is related to the angular speed by

$$N_{r(rated)} = 60 \times \frac{\omega_{r(rated)}}{2\pi} \tag{6.4}$$

The **base speed** is the minimum speed at which the machine can deliver full power. It is generally used interchangeably with the rated speed in this textbook.

6.2.5 Rated Power

The rated power of the machine $P_{r(rated)}$ is the power that the machine can output when operating at rated torque and speed:

$$P_{r(rated)} = T_{r(rated)}\omega_{r(rated)} \tag{6.5}$$

6.2.6 Peak Operation

Transient operation up to a peak torque $T_{r(peak)}$ or a maximum speed may also be specified for a time period.

The peak torque is the maximum torque at which the machine can be operated while not exceeding the current and temperature limits of the machine and inverter. The peak torque is typically for intermittent operation and not for continuous operation.

6.2.7 Starting Torque

The **starting** or **stall torque** is the maximum starting torque of the machine at zero speed, a condition that occurs at start-up or when stalled. In general, stall torque is not a significant issue for an electrical machine powered by a power inverter, as is the case for the automotive powertrain. It is a very significant issue for induction motors that are directly connected to the grid and are not buffered by a power inverter. See Chapter 7, Section 7.10 for insights into the stall torque of a dc machine.

6.3 Characteristic Curves of a Machine

We discussed the machine characteristic in Chapter 2 Section 2.2.2 when discussing vehicle acceleration. The torque versus speed and power versus speed curves for an electrical machine are shown in Figure 6.12. We can now identify different modes of operation for the electrical machines.

6.3.1 Constant-Torque Mode

In this mode, the torque is limited by the peak torque. If the vehicle is accelerating with full torque, then the machine operates in constant-torque mode, and the rotor power increases linearly with speed until the machine hits the power limit:

$$T_r = T_{r(peak)} \tag{6.6}$$

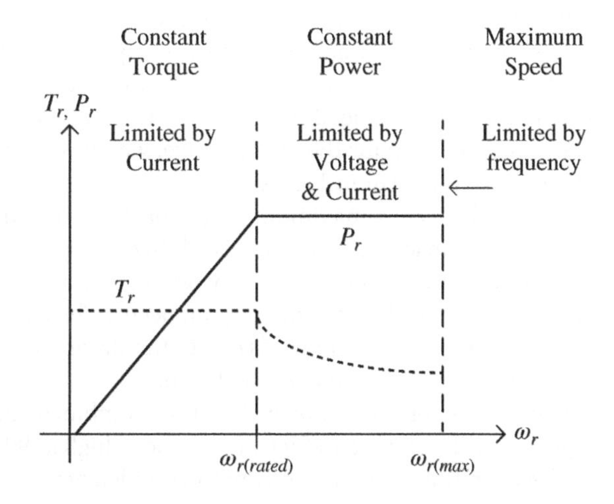

Figure 6.12 Torque and power versus speed curves.

In this mode, the torque is limited by the machine current and the related conduction loss.

6.3.2 Constant-Power Mode

In this mode, the machine's power is approximately constant, and the available rotor torque decreases inversely with rotor speed:

$$T_r = \frac{P_{r(peak)}}{\omega_r} \tag{6.7}$$

This mode is a necessary mode of operation for high speeds and is often known as the **field-weakened mode**. The machine field current is controlled in order to maximize the torque by weakening the magnetic field.

The power is limited by the machine voltage and current. The machine voltage is limited by magnetic parameters such as saturation, core loss, and related heating, and electrical parameters such as insulation thickness.

6.3.3 Maximum-Speed Mode

In this mode, the maximum operating frequency $\omega_{r(max)}$ of the power electronics inverter likely limits the maximum operating speed of the machine. The speed can also be limited by mechanical constraints such as bearing or rotor speed limits.

6.3.4 Efficiency Maps

The machine performance is often limited by the power loss within the machine. Power loss causes higher temperatures within the machine. The IPM ac machine tends to be a smaller machine than an induction machine for the same power level because the IPM does not have current-carrying conductors on its rotor to generate a magnetic field. This reduces the power loss of the machine and eliminates the rotor copper losses, which are difficult to remove from the spinning rotor.

It is common to characterize a machine by plotting efficiency isocontours as a function of the output torque and rotor speed. An example of an experimental torque–speed efficiency map for the 2010 Toyota Prius traction motor is shown in Figure 6.13. The efficiency map for the overall drive, which is the product of the motor and inverter efficiencies, is shown in Figure 6.14.

The machine has a maximum torque of about 200 Nm and a maximum power of about 60 kW. The machine efficiency is plotted using isocontours or contours of constant efficiency. The efficiency is typically very high at about 96% in the mid-speed, low-torque region, where much of the driving can occur. The efficiency can drop significantly as the torque increases at lower speeds due to the stator copper loss. For example, the efficiency is about 82% at 200 Nm and 2000 rpm.

Similar curves are shown for the combined motor–inverter electric drive in Figure 6.14. The inverter efficiency can be as high as 99% and remains over 90% for much of the operating region. The combined efficiency for the motor–inverter is simply the

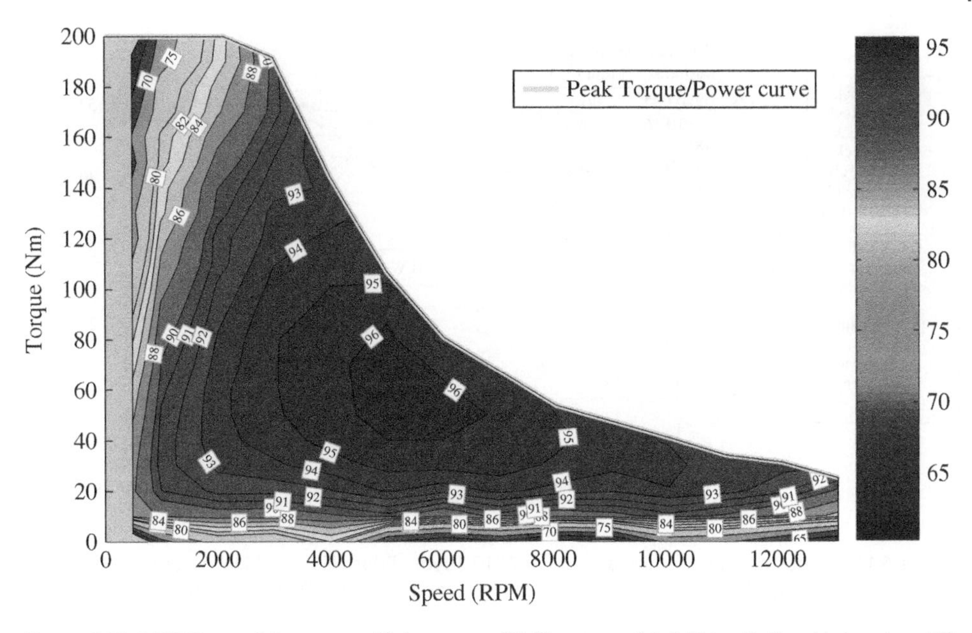

Figure 6.13 2010 Toyota Prius motor efficiency map [4]. (Courtesy of Oak Ridge National Laboratory, US Dept. of Energy.)

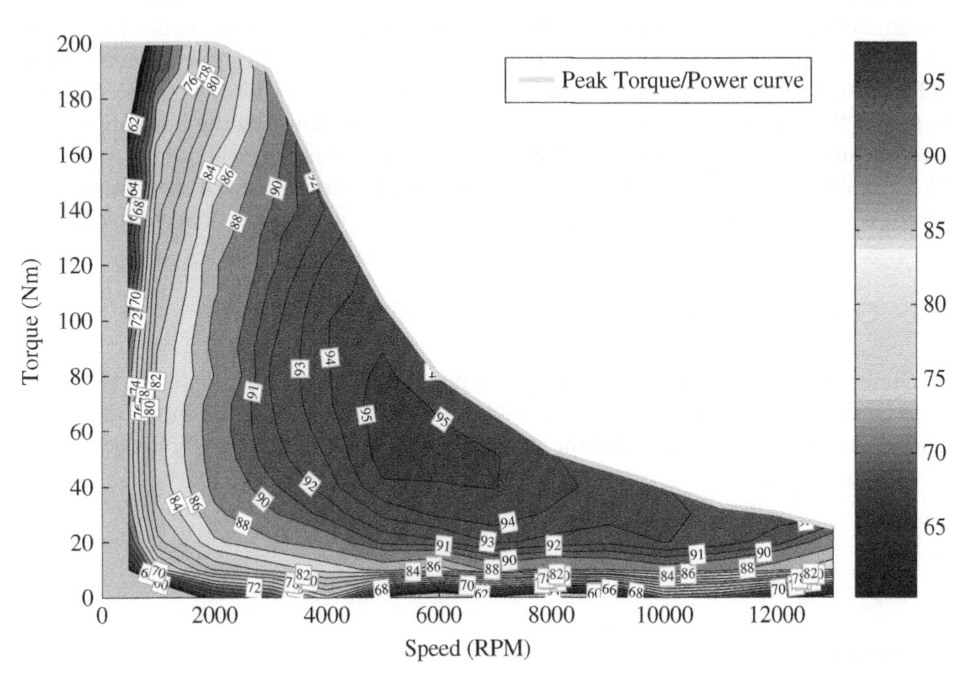

Figure 6.14 2010 Toyota Prius inverter-motor efficiency map [4]. (Courtesy of Oak Ridge National Laboratory, US Dept. of Energy.)

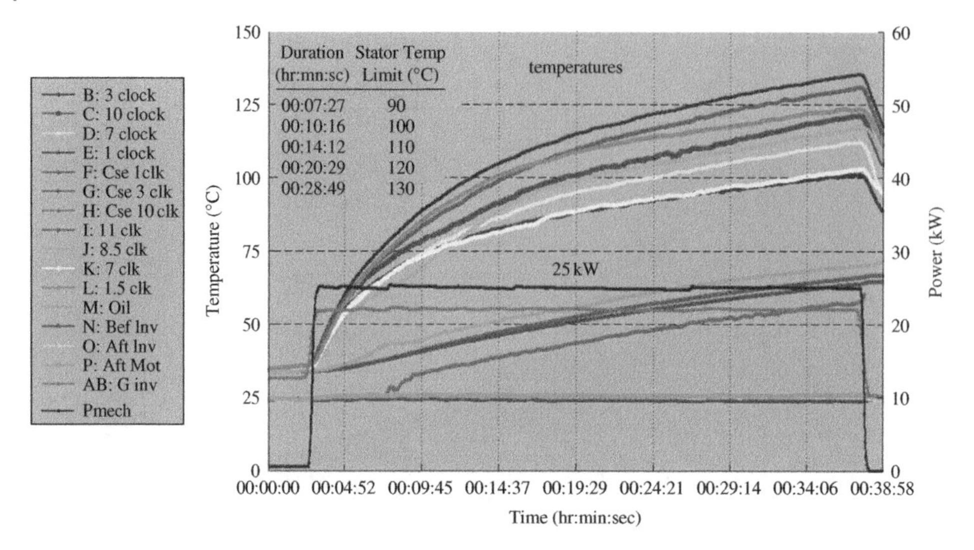

Figure 6.15 Various machine temperatures For 2010 Toyota Prius HEV at 25 kW output, 3000 rpm, and 25°C coolant [4]. (Courtesy of Oak Ridge National Laboratory, US Dept. of Energy.)

product of the two efficiencies. The motor–inverter efficiency can be as high as 95% in the mid-speed, low-torque region of operation.

Lower-efficiency operation causes higher machine and inverter power losses, which can produce excessive temperatures within the system. Typically, these conditions are not continuous operating conditions, and the system can operate in these modes for a time period as it takes time for the machine to heat up. An example of such operation is shown in Figure 6.15. In this plot, the machine is operated at 25 kW, as shown by the step waveform and the secondary vertical axis, for over 30 min, as shown by the horizontal axis. The machine temperatures are recorded using thermocouples placed about the machine. The machine heats up exponentially and takes about 10 min to reach a maximum temperature of 100°C and over 20 min to hit 125°C. Reference Chapter 7, Section 7.11.

6.4 Conversion Factors of Machine Units

Machines are commonly described using machine constants. For example, the relationships between torque and current, and voltage and speed, are often presented using the machine torque coefficient k_t and the machine voltage coefficient k_e as follows:

$$T_{em} = k_t I_a \tag{6.8}$$

and

$$E_a = k_e \omega_r \tag{6.9}$$

In the metric system:

$$k = k_e = k_t \tag{6.10}$$

Therefore,

$$T_{em} = k\, I_a \tag{6.11}$$

and

$$E_a = k\, \omega_r \tag{6.12}$$

The use of the voltage coefficient k_e and the torque coefficient k_t to describe a machine relates to machines specified in non-metric units, such as US or Imperial units. Translating coefficients for the voltage coefficient, when specified as volts/rpm, or the torque coefficient, when specified as foot-pounds force per ampere, lb_f-ft/A, back into the metric system results in the foregoing simpler metric equations.

References

1 T. A. Burress, C. L. Coomer, S. L. Campbell, L. E. Seiber, L. D. Marlino, R.H. Staunton, J. P. Cunningham, and H. T. Lin, *Evaluation of the 2007 Toyota Camry Hybrid Synergy Electric Drive System*, Oak Ridge National Laboratory report, 2008.
2 T. Nakada, S. Ishikawa, and S. Oki, "Development of an electric motor for a newly developed electric vehicle," *SAE* Technical Paper 2014-01-1879.
3 S. Rochi, "The Mars Mission – Technology for another world," *Maxon Motors Application Stories*, 2014.
4 T. A. Burress, S. L. Campbell, C. L. Coomer, C. W. Ayers, A. A. Wereszczak, J. P. Cunningham, L. D. Marlino, L. E. Seiber, and H. T. Lin, *Evaluation of the 2010 Toyota Prius Hybrid Synergy Drive System*, Oak Ridge National Laboratory report, 2011.

7

The Brushed DC Machine

"This is the first time any human enterprise has exceeded the distance of a marathon on the surface of another world," said John Callas, *Opportunity* project manager at NASA's Jet Propulsion Laboratory as NASA's Mars rover *Opportunity* completed a marathon on planet Mars in March 2015. It only took just over 11 years. The rover is an electric vehicle and is propelled by brushed dc machines.

In this chapter, the reader is introduced to the self-commutated brushed dc machine. For many people, the first contact with an electric propulsion system is the brushed permanent-magnet dc motor running a toy car. The ease of use of a brushed permanent-magnet dc motor makes it the machine of choice for many applications. However, while the machine can simply and optimally generate torque, it is limited in lifetime and durability because of its use of brushes and commutators to supply the current onto the rotating part of the machine. The wound-field dc machine uses the field current of an electromagnet to generate the field flux, rather than a permanent magnet. This machine also features brushes, but it has a significant advantage over its sibling in that the field current can be controlled to weaken the magnetic field and allow the machine to operate at high speed. This field-weakening capability is investigated here and is of much value, as we will see in the later chapters on ac machines. Wound-field machines play a wide role in vehicles. The starter motor for a conventional internal-combustion-engine vehicle is a series-wound dc machine.

This chapter serves as an introduction to machines and deals with many of the important technical considerations, such as field weakening, saturation, efficiency, power loss, and temperature rise (thermal management). The basic introductory theory of electromechanical energy conversion is presented in Chapter 16, Section 16.7.

7.1 DC Machine Structure

The dc motor has a uniform air gap and develops a rotational torque from the interaction between a stationary field flux and a rotating current. The magnetic field can be produced by a permanent magnet or by an electromagnet. Sketches of elementary permanent-magnet (PM) and wound-field (WF) brushed dc machines are shown in Figure 7.1. A top view of

Electric Powertrain: Energy Systems, Power Electronics and Drives for Hybrid, Electric and Fuel Cell Vehicles,
First Edition. John G. Hayes and G. Abas Goodarzi.

Companion website: www.wiley.com/go/hayes/electricpowertrain

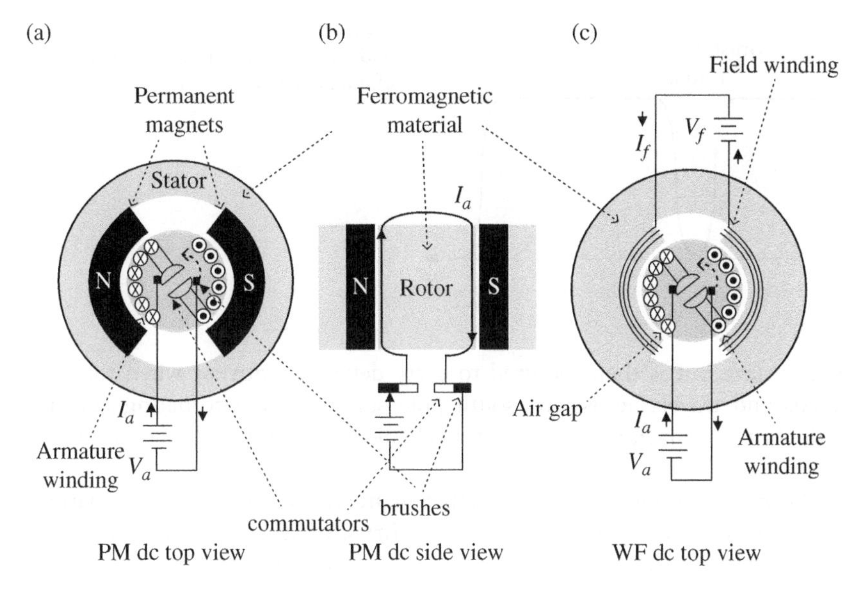

Figure 7.1 Elementary dc machines.

the PM machine is shown in Figure 7.1(a) with a side view in Figure 7.1(b). A top view of the WF brushed dc machine is shown in Figure 7.1(c). In the PM machine, the armature current I_a from the armature supply voltage source V_a interacts with the permanent magnets to generate torque.

The **stator** is the stationary part of the machine and is composed of the casing, ferromagnetic core, and the permanent magnets. The stator may have any even number of poles excited alternately north and south in sequence. An elementary two-pole machine is shown in Figure 7.1.

Note that the stator poles are curved around the rotor in order to allow a small fixed air gap between the stator and rotor. This is a very important attribute of the machine as the rotor conductors experience a constant flux density as they rotate, resulting in a tangential torque which acts to turn the rotor as desired. With this uniform air gap construction, the coils do not experience the undesired radial torque which acts to pull the coils apart.

The **rotor** is the rotating part of the motor. The **armature** is the part of the rotor which carries the current interacting with the stator flux to create torque. The rotor typically contains ferromagnetic material to provide a path for the stator-induced flux, in addition to the mechanical components such as the shaft and gearing.

Consider the elementary two-pole motor with an armature winding consisting of a single coil having coil sides at diametrically opposite points on the rotor surface, as shown in Figure 7.1. The armature winding carries an armature current I_a which is supplied by an external dc power source V_a. The magnetic field is from the north pole though the air gaps and rotor to the south pole, returning to the north pole through the outside ferromagnetic paths.

In the dc machine, the current in a given armature coil must flow in two distinct directions as the coil rotates through the magnetic field, as shown in Figure 7.1. When motoring, the conductor is required to carry approaching current while traveling across

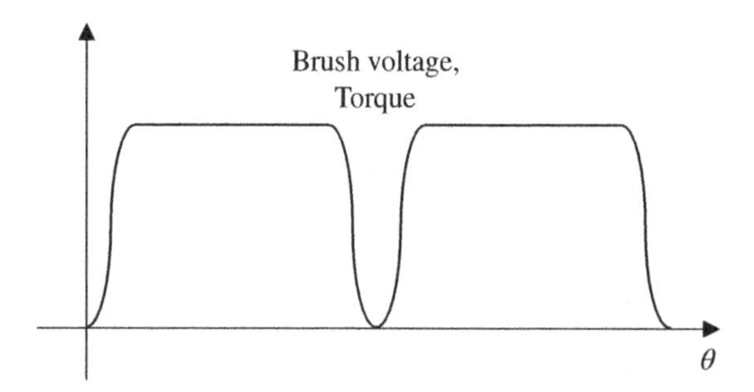

Figure 7.2 Variation of brush voltage and torque with angle θ *for a single turn* of a dc machine.

the north pole surface but is then required to carry departing current when the coil spins on its axis and travels across the south pole surface. The alternating current or voltage in the brushed machine is achieved mechanically by means of a commutator and brushes.

The **commutator** is a cylinder formed by copper segments mounted on the armature shaft but insulated from it and each other. Stationary **brushes**, directly connected to the dc power source, are held against the commutator surface, thus feeding voltage and current to the armature coils. Using this mechanism of commutator and brushes, the current direction within the coils changes direction as the rotor rotates. Thus, the coil emf is rectified by the commutator and brushes to produce a dc voltage at the brush output, as shown in Figure 7.2. The machine produces a unidirectional torque of the same wave shape.

The practical dc motor can have a number of parallel current-conduction paths in order to reduce the torque pulsations in the machine. These two classes of distributed-winding constructions are termed **lap** and **wave**. The motors feature lap or wave winding constructions because the pulsating nature of the emf and torque waveforms of the simple machine shown in Figure 7.2 is usually unacceptable.

The principal disadvantages of brushed dc machines are (1) a reduced lifetime due to the wear and tear of the brushes for commutation and (2) rotor heating and power loss due to the rotor current.

7.2 DC Machine Electrical Equivalent Circuit

The equivalent circuits for the dc machine are as shown in Figure 7.3.

The brushed motor symbol is usually shown as a voltage source of back emf E_a contacted by two brushes as shown. Components internal to the motor such as the armature winding inductance L_a and the armature equivalent series resistance R_a are often designated in circuit diagrams as discrete external components, although they are obviously integral to the motor.

It is worth noting at this point that there are additional types of separately excited brushed machine configuration, as shown in Figure 7.4. The field and armature windings can be connected in series or in parallel. The **series-wound** machine, as shown in Figure 7.4(a), is commonly used as the starter motor for an internal-combustion engine on a conventional vehicle. The **parallel** or **shunt-wound** machine, as shown in Figure 7.4 (b), is commonly used for constant-speed, variable-load applications. The **compound**

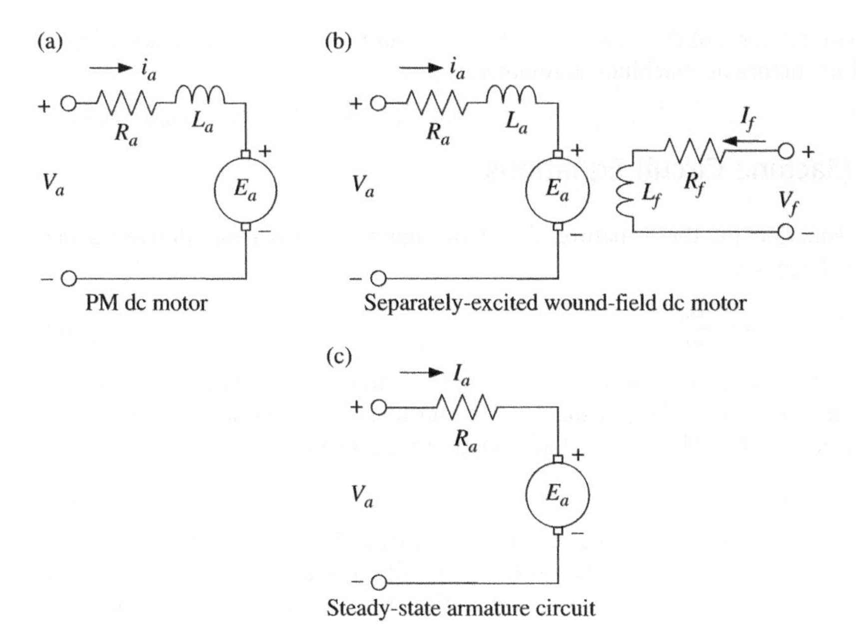

Figure 7.3 Equivalent circuit of dc motor (a) with a permanent magnet, (b) with an electromagnet, and (c) in steady state.

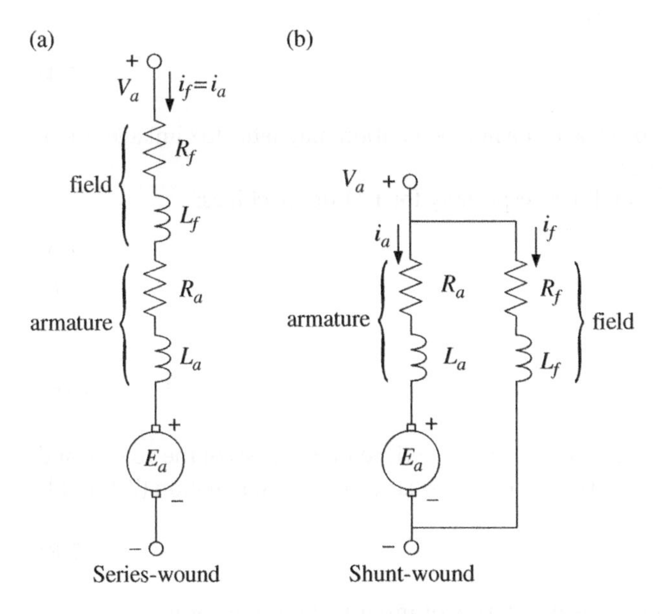

Figure 7.4 Series- and shunt-wound machines.

machine, a combination of the series and shunt configurations, was a widely used high-power machine before ac machines dominated [1].

7.3 DC Machine Circuit Equations

The circuit equation for the armature circuit of Figure 7.3(a) is easily derived using Kirchhoff's voltage law:

$$V_a = E_a + R_a i_a + L_a \frac{di_a}{dt} \tag{7.1}$$

The effects of the inductive portion of this equation can be ignored when operating in steady state in a dc circuit. Thus, in steady state, the armature equivalent circuit can be reduced to Figure 7.3(c). The steady-state equation reduces to

$$V_a = E_a + R_a I_a \tag{7.2}$$

The relationship between the generated back emf E_a and the terminal voltage V_a is significant. When $V_a > E_a$ in the forward direction, the machine acts as a motor, and I_a is positive. When $V_a < E_a$ in the forward direction, the machine acts as a generator, and I_a is negative.

The electromechanical behavior of the dc machine is described by the equations for torque T_{em} and back emf, which are derived in Chapter 16, Section 16.7:

$$T_{em} = \frac{p}{2} \lambda_f I_a \tag{7.3}$$

and

$$E_a = \frac{p}{2} \lambda_f \omega_r \tag{7.4}$$

where ω_r is the rotor angular speed, λ_f is the field-excitation magnetic flux linkage, and p is the number of magnetic poles.

Often, a constant k is used as follows, especially for PM dc machines:

$$T_{em} = k I_a \tag{7.5}$$

$$E_a = k \omega_r \tag{7.6}$$

where

$$k = \frac{p}{2} \lambda_f \tag{7.7}$$

A relationship between torque and speed can be derived on the basis of the voltage and torque equations. Substituting Equation (7.5) and Equation (7.6) into Equation (7.2) yields

$$V_a = k \omega_r + \frac{R_a}{k} T_{em} \tag{7.8}$$

Rearranging yields the angular speed in terms of input voltage and torque:

$$\omega_r = \frac{V_a}{k} - \frac{R_a}{k^2} T_{em} \tag{7.9}$$

Alternatively, the torque can be obtained in terms of the input voltage and angular speed:

$$T_{em} = \frac{k}{R_a} V_a - \frac{k^2}{R_a} \omega_r \tag{7.10}$$

which defines the **torque-speed** characteristic of the motor. Torque is plotted as a function of speed for the first quadrant of operation as shown in Figure 7.5(a) for a constant V_a. This characteristic represents a straight line.

The **electromagnetic power** developed by the machine is given by

$$P_{em} = T_{em} \omega_r = \frac{kV_a}{R_a} \omega_r - \frac{k^2}{R_a} {\omega_r}^2 \tag{7.11}$$

Power is also plotted for the first quadrant of operation as shown in Figure 7.5(a). The plot of power versus speed represents a quadratic equation. The torque drops linearly from the stall value $T_{r(stall)}$ to zero. The peak torque of the automotive machine is electronically limited to the rated value $T_{r(rated)}$ as shown in Figure 7.5 (b), in which case the available power increases linearly from zero speed to the rated speed. The rotor torque T_r equals the electromagnetic torque T_{em} minus the no-load torque T_{nl}.

$$T_{em} = T_r + T_{nl} \tag{7.12}$$

A number of key machine characteristic operating points are discussed as follows.

7.3.1 No-Load Spinning Loss

The no-load spinning loss is the power demanded by the machine to just spin the rotor without any external load. As with the vehicle itself, the rotor has aerodynamic and friction losses. This mechanical loss is commonly known as the **friction and windage** loss. The **friction loss** is largely due to the bearings and the brush–commutator assembly

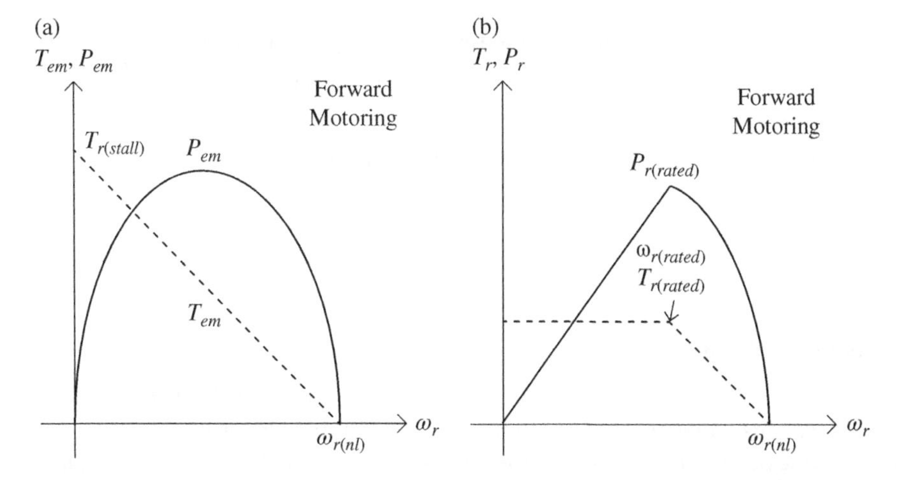

Figure 7.5 First quadrant torque and power characteristic curves for a dc machine.

while the **windage loss** is due to air resistance as the rotor spins. The **core loss** is a magnetic loss and comprises the hysteresis and eddy-current losses, as discussed in Chapter 16.

While these three losses are separate functions of speed, torque, and flux, it is common to lump the losses together and refer to the **lumped core, friction, and windage loss** P_{cfw} as the **no-load or spinning loss** of the machine. While the no-load torque varies with speed, it is usually assumed in this textbook that the loss can be modeled by a constant no-load torque T_{nl}.

The output power available on the rotor P_r is the electromagnetic power P_{em} minus the no-load spinning power loss P_{cfw}:

$$P_r = P_{em} - P_{cfw} = \frac{k V_a}{R_a}\omega_r - \frac{k^2}{R_a}\omega_r{}^2 - P_{cfw} \tag{7.13}$$

where

$$P_{cfw} = T_{nl}\omega_r \tag{7.14}$$

7.3.2 No-Load Speed

The no-load speed $\omega_{r(nl)}$ is the speed at which the output rotor torque is zero and at which the rotor free spins when supplied by voltage V_a. The no-load speed is given by

$$\omega_{r(nl)} = \frac{V_a}{k} - \frac{R_a}{k^2}T_{nl} \tag{7.15}$$

If the no-load torque T_{nl} is relatively small, then the following approximation works well for estimating the no-load speed of the machine:

$$\omega_{r(nl)} \approx \frac{V_a}{k} \tag{7.16}$$

7.3.3 Maximum Power

The power characteristic for the dc machine represents a quadratic equation with zero output power at zero speed and at the no-load speed, and a peak power at $\omega_r = \frac{\omega_{r(nl)}}{2}$, as shown in Figure 7.5(a).

The maximum electromagnetic power can be shown to be

$$P_{em(max)} = \frac{V_a{}^2}{4R_a} \tag{7.17}$$

7.3.4 Rated Conditions

As discussed in Chapter 6, the rated parameters for the machine are typically the torque, power, speed, voltage, current, cooling, temperature, and time. Under specified conditions, the machine outputs rated power at the rated speed when supplied by rated voltage and current.

7.4 Power, Losses, and Efficiency in the PM DC Machine

Consider a dc motor operating at a steady-state speed ω_r and developing a torque T_{em}, and supplied by armature voltage V_a and current I_a. There are three principal loss mechanisms: the copper loss in the armature resistance, the core loss of the rotor, and the mechanical friction and windage loss, as shown in Figure 7.6.

Multiplying Equation (7.2) by I_a gives

$$V_a I_a = (E_a + R_a I_a) I_a = E_a I_a + R_a I_a^2 \tag{7.18}$$

The power $V_a I_a$ represents the total electrical power supplied to the armature, and $R_a I_a^2$ represents the power loss due to the resistance of the armature circuit. The difference between these two quantities, namely $E_a I_a$, therefore represents the net electrical power input to the armature which is converted to electromagnetic power $T_{em}\,\omega_r$ and core loss.

The **electromechanical power** can be expressed as

$$\begin{aligned} P_{em} &= E_a I_a \\ &= \frac{p}{2}\lambda_f \omega_r \times \frac{2}{p\lambda_f} T_{em} \\ &= T_{em}\omega_r \end{aligned} \tag{7.19}$$

Thus:

$$V_a I_a = T_{em}\omega_r + R_a I_a^2 \tag{7.20}$$

The electromechanical torque developed by the rotor comprises the no-load torque T_{nl} and the useful rotor output torque T_r. Thus, the power equation can be rewritten as

$$V_a I_a = T_r \omega_r + T_{nl}\omega_r + R_a I_a^2 \tag{7.21}$$

The machine **power loss** $P_{m(loss)}$ is

$$P_{m(loss)} = T_{nl}\omega_r + R_a I_a^2 \tag{7.22}$$

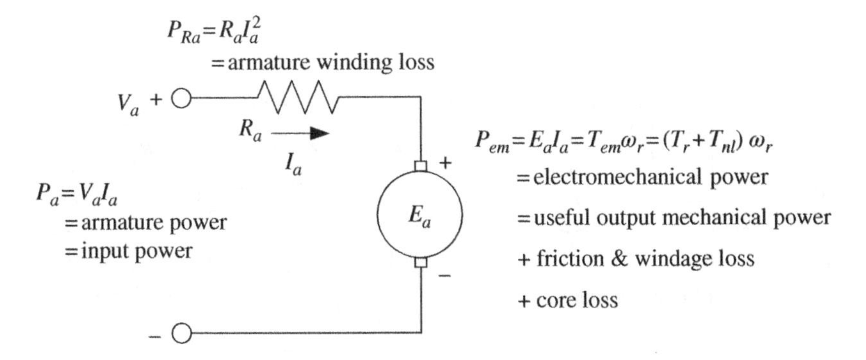

Figure 7.6 DC machine equivalent circuit motoring power flow.

The **efficiency** of the machine is the ratio of the output rotor power P_r to the input armature power P_a. When motoring,

$$\eta = \frac{P_r}{P_a} \times 100\% = \frac{T_r\omega_r}{V_aI_a} \times 100\% \text{ (motoring)} \tag{7.23}$$

When generating, the input power is the rotor mechanical power and the output power is the armature electrical power:

$$\eta = \frac{P_a}{P_r} \times 100\% = \frac{V_aI_a}{T_r\omega_r} \times 100\% \text{ (generating)} \tag{7.24}$$

7.5 Machine Control using Power Electronics

For many applications using dc machines, especially in low-power applications, it is acceptable to power the machine directly from the source without the use of a power converter to provide the optimum voltage and current to the machine. In an earlier age, an inefficient discrete variable resistance was placed between the source and the armature in order to regulate the speed or torque.

However, optimum operation of any machine for a traction application requires a switch-mode power-electronics-based power converter to provide the optimum armature voltage and current to the machine in order to operate the machine for torque or efficiency maximization. This dc machine requires a power converter that can output positive and negative voltages across the armature ranging from $+V_s$ to $-V_s$, where V_s is the dc source voltage. The dc-dc converter, as shown in Figure 7.7(a), is typically a full-bridge dc-dc converter, as shown in Figure 7.7(b). At this stage, we simply model the dc-dc converter as being a high-efficiency (typically 98%–99%) dc-dc converter. Traction power converters for dc machines are not covered any further in this textbook, but the three-phase traction inverters used for ac machines are discussed in detail in Chapter 13.

7.5.1 Example: Motoring using a PM DC Machine

In this example, we use a PM dc machine as the traction motor for a battery electric vehicle. The basic specifications for the machine are $P_{r(rated)} = 80$ kW and $T_{r(rated)} = 280$ Nm output at rated speed, a gear ratio $n_g = 8.19$, and a wheel radius $r = 0.315$ m.

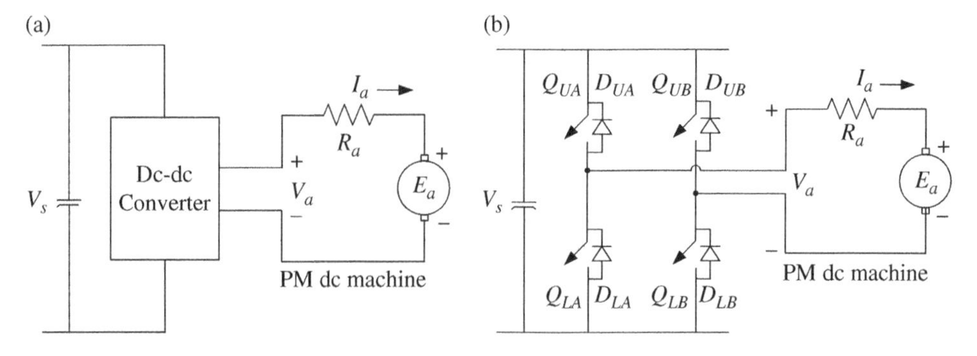

Figure 7.7 DC motor electric drive: (a) dc-dc converter and (b) full-bridge dc-dc converter.

The following assumptions are made: back emf E_a of 220 V at rated speed, armature resistance R_a = 50 mΩ, and no-load torque T_{nl} = 2 Nm.

Determine the armature voltage and current output by the dc-dc converter and the machine efficiency when the vehicle is operating under the following conditions:

a) Motoring up a hill and developing full torque at rated speed;
b) Cruising and developing 70 Nm at the rated speed;
c) Cruising and developing 70 Nm at half the rated speed.

Solution:

a) Motoring up a hill and developing full torque at rated speed.

The rated speed of the rotor is

$$\omega_r = \omega_{r(rated)} = \frac{P_{r(rated)}}{T_{r(rated)}} = \frac{80{,}000}{280}\,\text{rad}\,\text{s}^{-1} = 285.71\,\text{rad}\,\text{s}^{-1}$$

The frequency of the rotor is

$$f_r = \frac{\omega_r}{2\pi} = \frac{285.71}{2\pi}\,\text{Hz} = 45.473\,\text{Hz}$$

or

$$N_r = 60 f_r = 60 \times 45.473\,\text{rpm} = 2728.4\,\text{rpm}$$

The speed of the axle ω_{axle} is

$$\omega_{axle} = \frac{\omega_r}{n_g} = \frac{285.71}{8.19} \times \text{rad}\,\text{s}^{-1} = 34.885\,\text{rad}\,\text{s}^{-1}$$

The vehicle speed v is

$$v = r\,\omega_{axle} = 0.315 \times 34.885\,\text{rad}\,\text{s}^{-1} = 10.989\,\text{m}\,\text{s}^{-1}\,\left(\times 3.6 = 39.56\,\text{km}\text{h}^{-1}\right)$$

The machine constant can be determined knowing the back emf and angular speed at the rated condition:

$$k = \frac{E_a}{\omega_{r(rated)}} = \frac{220}{285.71}\frac{\text{V}}{\text{rad}\,\text{s}^{-1}} = 0.77\frac{\text{V}}{\text{rad}\,\text{s}^{-1}}$$

When operating in motoring mode, the electromagnetic torque is the sum of the useful output torque and the no-load torque:

$$T_{em} = T_r + T_{nl} = 280\,\text{Nm} + 2\,\text{Nm} = 282\,\text{Nm}$$

The armature current is given by

$$I_a = \frac{T_{em}}{k} = \frac{282}{0.77}\,\text{A} = 366.23\,\text{A}$$

The armature voltage output from the dc-dc converter is

$$V_a = E_a + R_a I_a = 220\,\text{V} + 0.05 \times 366.23\,\text{V} = 238.31\,\text{V}$$

The machine power loss is the sum of the ohmic and core, friction and windage losses:

$$P_{m\,(loss)} = T_{nl}\omega_r + R_a I_a^2 = 2 \times 285.71\,\text{W} + 0.05 \times 366.23^2\,\text{W} = 7.278\,\text{kW}$$

The machine input power is

$$P_a = P_r + P_{m(loss)} = 80\,\text{kW} + 7.278\,\text{kW} = 87.278\,\text{kW}$$

The machine efficiency is given by

$$\eta = \frac{P_r}{P_a} = \frac{80\,\text{kW}}{87.278\,\text{kW}} \times 100\% = 91.66\%$$

b) Cruising and developing 70 Nm at the rated speed.
The speed remains the same as in the first part:

$$\omega_r = \omega_{r(rated)} = 285.71\,\text{rad}\,\text{s}^{-1}$$

but the electromagnetic torque and rotor output power drop are

$$T_{em} = T_r + T_{nl} = 70\,\text{Nm} + 2\,\text{Nm} = 72\,\text{Nm}$$

$$P_r = T_r\omega_r = 70 \times 285.71\,\text{W} = 20\,\text{kW}$$

The armature current drops significantly:

$$I_a = \frac{T_{em}}{k} = \frac{72}{0.77}\,\text{A} = 93.51\,\text{A}$$

The armature voltage drops a little but remains high as the speed has not changed:

$$V_a = E_a + R_a I_a = 220\,\text{V} + 0.05 \times 93.51\,\text{V} = 224.68\,\text{V}$$

The machine power loss, input power, and efficiency are

$$P_{m\,(loss)} = T_{nl}\omega_r + R_a I_a^2 = 2 \times 285.71\,\text{W} + 0.05 \times 93.51^2\,\text{W} = 1.009\,\text{kW}$$

$$P_a = P_r + P_{m(loss)} = 20\,\text{kW} + 1.009\,\text{kW} = 21.009\,\text{kW}$$

and

$$\eta = \frac{P_r}{P_a} = \frac{20\,\text{kW}}{21.009\,\text{kW}} \times 100\% = 95.2\%$$

It is typical for machines to operate most efficiently at low-to-medium torque and low-to-medium speeds.

c) Cruising and developing 70 Nm at half the rated speed.
In this condition, the speed and back emf drop by half:

$$\omega_r = \frac{\omega_{r(rated)}}{2} = 142.86\,\text{rad}\,\text{s}^{-1}$$

$$E_a = k\omega_r = 0.77 \times 142.86\,\text{V} = 110.0\,\text{V}$$

The electromagnetic torque remains at

$$T_{em} = T_r + T_{nl} = 70\,\text{Nm} + 2\,\text{Nm} = 72\,\text{Nm}$$

and the rotor output power drops to

$$P_r = T_r\omega_r = 70 \times 142.86\,\text{W} = 10\,\text{kW}$$

The armature current and voltage are

$$I_a = \frac{T_{em}}{k} = \frac{72}{0.77}\,\text{A} = 93.51\,\text{A}$$

and

$$V_a = E_a + R_aI_a = 110\,\text{V} + 0.05 \times 93.51\,\text{V} = 114.68\,\text{V}$$

The machine power loss, input power and efficiency are

$$P_{m\,(loss)} = T_{nl}\omega_r + R_aI_a^2 = 2 \times 142.86\,\text{W} + 0.05 \times 93.51^2\,\text{W} = 723\,\text{W}$$

$$P_a = P_r + P_{m(loss)} = 10\,\text{kW} + 0.723\,\text{kW} = 10.723\,\text{kW}$$

and

$$\eta = \frac{P_r}{P_a} = \frac{10\,\text{kW}}{10.723\,\text{kW}} \times 100\% = 93.3\,\%$$

7.6 Machine Operating as a Motor or Generator in Forward or Reverse Modes

The electrical machine can be switched from operating as a motor to operating as a generator simply by reversing the armature current while maintaining the armature voltage positive.

As shown in Figure 7.8(a), for forward motoring the force is positive when current is flowing into the page under the north pole, and out of the page under the south pole. The force on the conductor is counterclockwise or positive and in the direction of the rotation, which is also counterclockwise.

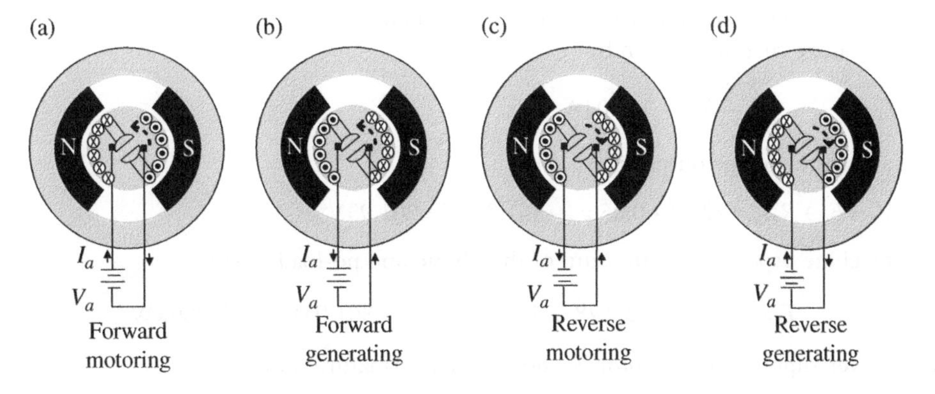

Figure 7.8 Motoring and generating modes.

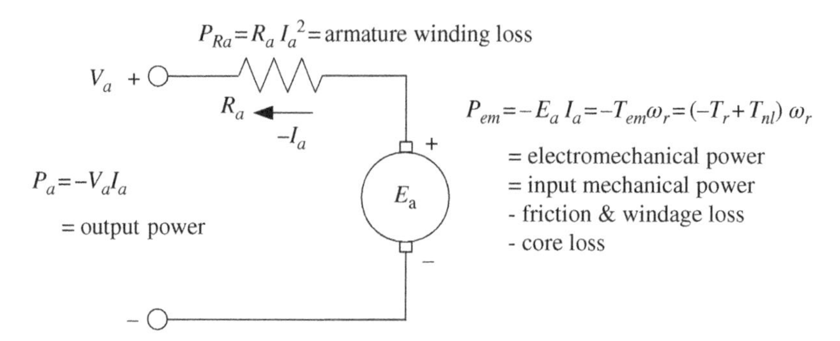

Figure 7.9 Dc machine equivalent circuit regenerating power flow.

When operating as a generator in the forward direction, as shown in Figure 7.8(b), the current in the armature winding is reversed, and the force generated is clockwise and against the direction of rotation.

The rotor can be reversed by applying a negative armature voltage, as shown in Figure 7.8(c) and (d). When motoring in reverse, the electromotive force and direction of rotation are in the same direction. While generating in reverse, the force and the direction of rotation are in opposite directions.

The forward-generating mode is shown in Figure 7.9. The input and armature currents are both negative compared to the forward-motoring condition.

7.6.1 Example: Generating/Braking using a PM DC Machine

The electric machine in the earlier example is operating in forward-generating mode and develops full torque at rated speed. Determine the armature voltage and current output by the dc-dc converter, and the machine efficiency for this operating point.

Solution:

When operating in generating or braking mode, the electromagnetic torque is the torque input at the rotor minus the no-load torque:

$$T_{em} = T_r + T_{nl} = -280\,\text{Nm} + 2\,\text{Nm} = -278\,\text{Nm}$$

The speed in hertz and rad/s are as before for the rated condition.

The armature current is given by

$$I_a = \frac{T_{em}}{k} = \frac{-278}{0.77}\,\text{A} = -361.04\,\text{A}$$

The armature voltage output is

$$V_a = E_a + R_a I_a = 220\,\text{V} + 0.05 \times (-361.04)\,\text{V} = 201.95\,\text{V}$$

The machine power loss is the sum of the ohmic and no-load losses:

$$P_{m\,(loss)} = T_{nl}\omega_r + R_a I_a^2 = 2 \times 285.71\,\text{W} + 0.05 \times (-361.04)^2\,\text{W} = 7.089\,\text{kW}$$

The power supplied to the rotor is −80 kW. The armature power is

$$P_a = P_r + P_{m(loss)} = -80\,\text{kW} + 7.089\,\text{kW} = -72.911\,\text{kW}$$

The machine efficiency is given by

$$\eta = \frac{P_a}{P_r} = \frac{-72.911}{-80} \times 100\% = 91.14\%$$

7.6.2 Example: Motoring in Reverse

The electric machine in the earlier example is operating in reverse motoring and develops 70 Nm at half the rated speed.

Determine the armature voltage and current output by the dc-dc converter and the machine efficiency for this operating point.

Solution:

When operating in reverse motoring mode, the speed is negative and the electromagnetic torque is the useful input torque plus the no-load torque, with both values also being negative.

The rotor speed, torque, and power are

$$\omega_r = -\frac{\omega_{r(rated)}}{2} = -142.86\,\text{rad s}^{-1}$$

$$T_{em} = T_r + T_{nl} = -70\,\text{Nm} - 2\,\text{Nm} = -72\,\text{Nm}$$

and

$$P_r = T_r\omega_r = -70 \times (-142.86)\,\text{W} = 10\,\text{kW}$$

The armature current, back emf, and armature voltage are

$$I_a = \frac{T_{em}}{k} = \frac{-72}{0.77}\,\text{A} = -93.51\,\text{A}$$

$$E_a = k\omega_r = 0.77 \times (-142.86)\,\text{V} = -110.0\,\text{V}$$

$$V_a = E_a + R_a I_a = -110\,\text{V} - 0.05 \times 93.51\,\text{V} = -114.68\,\text{V}$$

The machine power loss, input power, and efficiency are

$$P_{m\,(loss)} = T_{nl}\omega_r + R_a I_a^2 = -2 \times (-142.86)\,\text{W} + 0.05 \times 93.51^2\,\text{W} = 723\,\text{W}$$

$$P_a = P_r + P_{m(loss)} = 10\,\text{kW} + 0.723\,\text{kW} = 10.723\,\text{kW}$$

$$\eta = \frac{P_r}{P_a} = \frac{10\,\text{kW}}{10.723\,\text{kW}} \times 100\% = 93.3\,\%$$

When motoring in reverse, we get similar results to forward motoring as discussed in Example 7.1.4.1(c), except that the signs on some values are changed.

7.7 Saturation and Armature Reaction

Rotating machines and other electromagnetic devices are often driven into magnetic saturation during regular operation. As just derived, the machine torque and armature current are related by the magnetic flux linkage. As the current increases, the machine can be driven into magnetic saturation, effectively reducing the machine flux linkage.

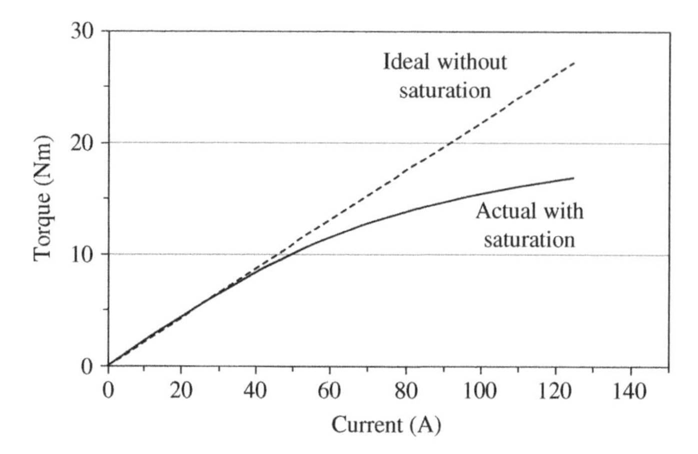

Figure 7.10 Torque versus current for an industrial PM dc machine with and without saturation.

The relationship between torque and current for an industrial PM machine is shown in Figure 7.10. The dotted line represents the theoretical torque estimation if there is no saturation, while the curved solid line is the actual experimental torque versus current plot. There are two principal saturation mechanisms. First, the rotor material can saturate due to the large armature current. Second, the saturation of the rotor increases the reluctance path for the field flux, thus, reducing the field flux. This local saturation effect is known as **armature reaction**.

The saturation introduces a field weakening with a number of knock-on effects, as can be seen from the following machine equations:

$$k = \frac{p}{2}\lambda_f = \begin{cases} \dfrac{E_a}{\omega_r} \\ \dfrac{T_{em}}{I_a} \end{cases} \tag{7.25}$$

The speed and torque are affected by saturation, as are the back emf and current. Parameters k and λ_f are similarly affected by saturation.

7.7.1 Example: Motoring using PM DC Machine and Machine Saturation

The earlier electric drive is operating in forward-motoring mode and develops full torque at the rated speed. However, the machine constant drops 30% from the nominal value at full torque due to machine saturation.

Determine the armature voltage and current output by dc-dc converter, and the efficiency for this operating point.

Solution:

Under saturation the revised machine constant drops to

$$k = 70\% \times 0.77 \frac{\text{V}}{\text{rad s}^{-1}} = 0.539 \frac{\text{V}}{\text{rad s}^{-1}}$$

The back emf drops to

$$E_a = k\omega_r = 0.539 \times 285.71\,\text{V} = 154\,\text{V}$$

The armature current increases to

$$I_a = \frac{T_{em}}{k} = \frac{282}{0.539}\,\text{A} = 523.19\,\text{A}$$

from the original value of 366.23 A without saturation.

The armature voltage output from the dc-dc converter is

$$V_a = E_a + R_a I_a = 154\,\text{V} + 0.05 \times 523.19\,\text{V} = 180.16\,\text{V}$$

The machine power loss is the sum of the ohmic and the core, friction, and windage losses:

$$P_{m\,(loss)} = T_{nl}\omega_r + R_a I_a^2 = 2 \times 285.71\,\text{W} + 0.05 \times 523.19^2\,\text{W} = 14.258\,\text{kW}$$

The machine input power is

$$P_a = P_r + P_{m(loss)} = 80\,\text{kW} + 14.258\,\text{kW} = 94.258\,\text{kW}$$

The machine efficiency is given by

$$\eta = \frac{P_r}{P_a} = \frac{80\,\text{kW}}{94.258\,\text{kW}} \times 100\% = 84.87\%$$

Thus, the machine efficiency has dropped significantly from the earlier value of 91.66% to 84.87% due to the machine saturation.

7.8 Using PM DC Machine for EV Powertrain

Generally, there are two modes of machine operation limiting the torque and speed control of an electric drive: constant-torque mode at low speeds, and constant-power or field-weakened mode at high speed. These modes are as shown in Figure 7.11(a) for the field-weakened WF dc machines.

A third mode also exists – the **power-drop-off** mode. The basic PM dc machine cannot operate in constant-power mode as there is no mechanism for weakening the field in a conventional PM dc machine (an additional field winding can be added around the magnets to field weaken the magnet, but this option is not considered here). The ac machines are easily field-weakened as discussed in Chapters 8 to 10.

Thus, the basic PM dc machine has two only operating modes: constant-torque and power-drop-off, as shown in Figure 7.11(b).

The two modes for the PM dc machine are as follows:

Constant-torque mode: In this mode, the machine can output a constant torque, and the maximum power generated by the machine increases linearly with speed. In this mode:

$$T_r = T_{r(rated)} \tag{7.26}$$

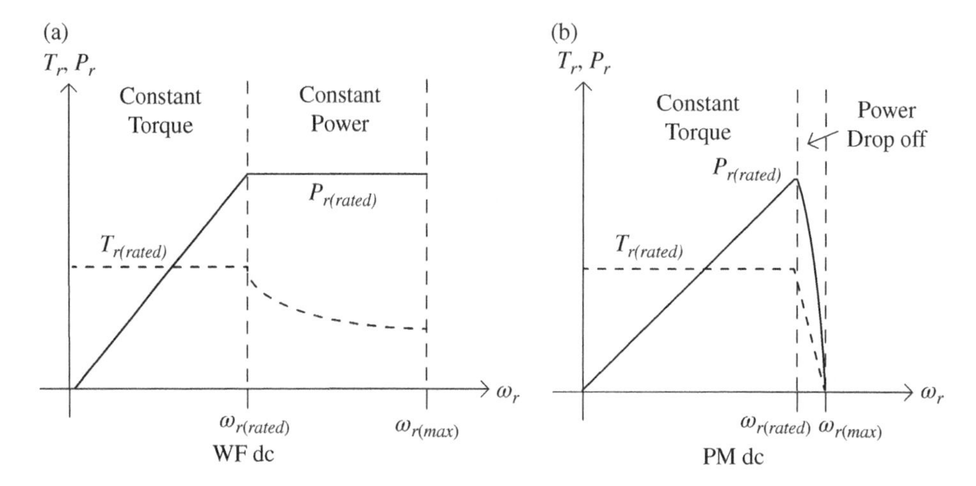

Figure 7.11 (a) Wound-field (WF) dc and (b) permanent-magnet (PM) dc machine power and torque characteristics.

Power-drop-off mode: In this mode, the machine power and torque are limited, and both tend to zero at the maximum no-load speed of the machine. It can be shown that the torque is described by the following equation:

$$T_r = T_{r(rated)} - \frac{k^2}{R_a}\left(\omega_r - \omega_{r(rated)}\right) \text{ (power-drop-off)} \tag{7.27}$$

It is a serious limitation of the PM dc machine to not have a constant-power or field-weakened mode. The machine cannot operate at higher speeds as the machine is constrained to operate with a maximum speed close to the rated speed. The PM dc machine can be used in single-gear vehicles but requires a variable gearing mechanism similar to that of a conventional internal-combustion engine vehicle in order to operate over the full speed range.

The battery electric vehicle (BEV) drivetrain is simulated using PM and WF dc machines in Figure 7.12(a) and (b), respectively.

7.8.1 Example: Maximum Speeds using PM DC Machine

Determine the motor and vehicle speeds at (i) the rated condition and at (ii) maximum speed for the PM dc machine used in the earlier examples.

Solution:

i) The results for the rated condition have already been estimated in Example 7.1.4.1. The rotor and vehicle speeds are 285.71 rad/s and 39.56 km/h, respectively.
ii) The maximum rotor speed can be determined by setting T_r to zero in Equation (7.27). Thus,

$$\omega_{r(\max)} = \omega_{r(rated)} + \frac{R_a}{k^2}T_{r(rated)} = 285.71\,\text{rad}\,\text{s}^{-1} + \frac{0.05}{0.77^2} \times 280\,\text{rad}\,\text{s}^{-1} = 309.32\,\text{rad}\,\text{s}^{-1}$$

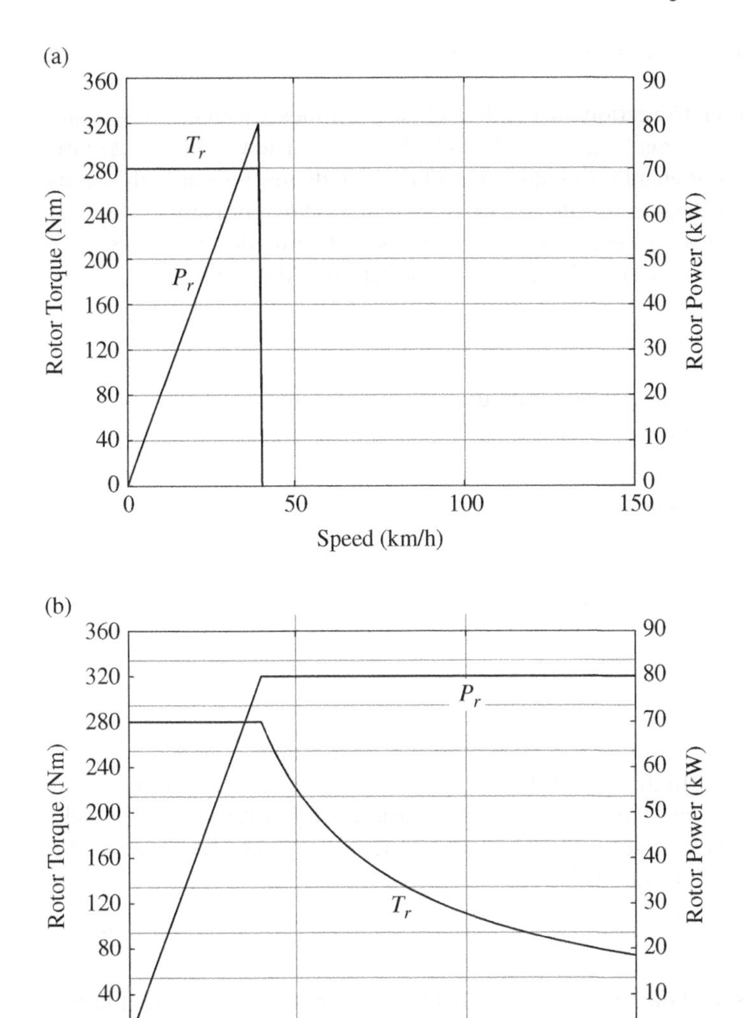

Figure 7.12 Power and torque characteristic for (a) PM dc and (b) WF dc machines.

which correlates to a vehicle speed of 42.83 km/h.

Thus, the PM machine reaches peak power at about 39.56 km/h, and the maximum vehicle speed is limited to a speed less than 42.83 km/h.

7.9 Using WF DC Machine for EV Powertrain

It is desirable, and often necessary, in motion control to operate well above the rated speed of the machine without exceeding the rated voltage. Such high-speed operation can easily be achieved in many machines by using field weakening. The field can be reduced or weakened from the rated value by reducing the field current and the resulting field flux linkage.

As discussed in Chapter 16 Section 16.4.6, the WF dc machines use an electromagnet rather than a permanent magnet to generate the field. While the field winding adds complexity and machine loss, it enables the operation of the WF dc machine at high speeds above the rated speed. A diagram of the WF drive system is shown in Figure 7.13. The armature and field currents are separately controlled using two dc-dc converters.

From Chapter 16 Section 16.4.6.1, we note that the field flux linkage is

$$\lambda_f = L_f I_f$$

where L_f and I_f are the field inductance and current, respectively.

The torque and back emf are

$$T_{em} = \frac{p}{2}\lambda_f I_a = \frac{p}{2}L_f I_f I_a = kI_a \tag{7.28}$$

and

$$E_a = \frac{p}{2}\lambda_f \omega_r = \frac{p}{2}L_f I_f \omega_r = k\omega_r \tag{7.29}$$

where k is the equivalent machine constant.

There are two principal operating modes for the WF dc machine.

Constant-torque mode: In this mode, the machine can output maximum torque, and the maximum power generated by the machine increases linearly with speed. The field current and resulting flux linkage and machine constant can be held unchanged at the rated value. The maximum torque is limited to

$$T_r = T_{r(rated)} \tag{7.30}$$

From Equation (7.28) or Equation (7.29), the field current at the rated condition is

$$I_{f(rated)} = \frac{2k}{pL_f} \tag{7.31}$$

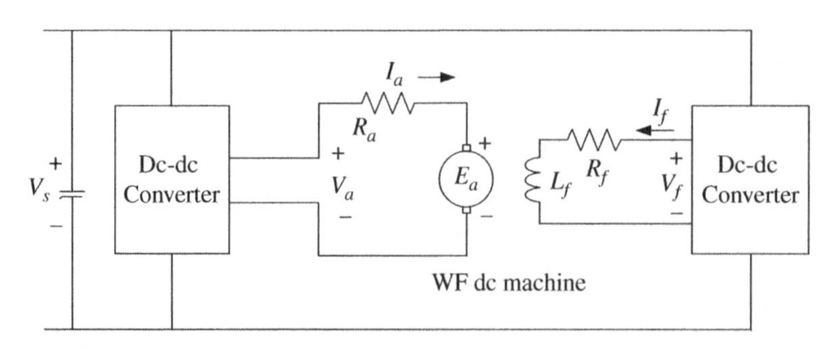

Figure 7.13 WF dc motor drive.

Constant-power mode: In this mode, the machine can output constant power and the rotor torque available decreases inversely with rotor speed, and so the maximum torque is limited to

$$T_r = \frac{P_{r(rated)}}{\omega_r} \tag{7.32}$$

The field current decreases as the speed increases above the rated value and is given by

$$I_f = \frac{\omega_{r(rated)}}{\omega_r} I_{f(rated)} \tag{7.33}$$

The equivalent machine constant becomes

$$k = \frac{\omega_{r(rated)}}{\omega_r} k_{rated} \tag{7.34}$$

where k_{rated} is the equivalent machine constant at the rated condition.

If the WF dc machine is now integrated into the EV powertrain, then we obtain the simulation results shown in Figure 7.12(b). The machine operates in field-weakened mode from approximately 40 km/h to 150 km/h. The WF dc machine meets the full specification of the test vehicle.

7.9.1 Example: Motoring using WF DC Machine

In this example, the test vehicle of the earlier examples has an eight-pole WF dc machine as the traction motor. The rated conditions are 80 kW and 280 Nm. The equivalent machine constant is 0.77 Nm/A at the rated condition. The machine has an armature resistance R_a = 50 mΩ, field resistance R_f = 40 mΩ, field inductance L_f = 1 mH, and no-load torque T_{nl} = 2 Nm.

Determine the armature and field currents output by the dc-dc converters, and the machine efficiency for the following operating points:

a) rated power at rated speed;
b) rated power at 3.75 times the rated speed;
c) a partial-load torque of 70 Nm at rated speed.

Solution:

a) The rated field current can be determined from Equation (7.31):

$$I_{f(rated)} = \frac{2k_{rated}}{pL_f} = \frac{2 \times 0.77}{8 \times 0.001} \text{A} = 192.5\,\text{A}$$

As before, the electromagnetic torque is the sum of the useful output torque and the no-load torque:

$$T_{em} = T_r + T_{nl} = 280\,\text{Nm} + 2\,\text{Nm} = 282\,\text{Nm}$$

The armature current is given by

$$I_a = \frac{T_{em}}{k} = \frac{282}{0.77} \text{A} = 366.23\,\text{A}$$

The machine power loss is the sum of the spinning loss and the armature and field copper losses:

$$P_{m\,(loss)} = T_{nl}\omega_r + R_a I_a{}^2 + R_f I_f{}^2 = 2 \times 285.71\,\text{W} + 0.05 \times 366.23^2\,\text{W}$$
$$+ 0.04 \times 192.5^2\,\text{W} = 8760\,\text{W}$$

The machine input power P is

$$P = P_{r(rated)} + P_{m(loss)} = 80\,\text{kW} + 8.76\,\text{kW} = 88.76\,\text{kW}$$

The machine efficiency is given by

$$\eta = \frac{P_r}{P} = \frac{80\,\text{kW}}{88.76\,\text{kW}} \times 100\% = 90.13\%$$

b) The vehicle is propelled on a slight incline at 3.75 times the rated speed, and 80 kW full power is developed on the rotor.

For this condition, the rotor speed is

$$\omega_r = 3.75\omega_{r(rated)} = 3.75 \times 285.71\,\text{rad s}^{-1} = 1071.4\,\text{rad s}^{-1}$$

For a constant power output in the field-weakened mode, the rotor torque must drop proportionately:

$$T_r = \frac{P_{r(rated)}}{\omega_r} = \frac{80000}{1071.4}\,\text{Nm} = 74.67\,\text{Nm}$$

Per Equation (7.33), the field current drops proportionately:

$$I_f = \frac{\omega_{r(rated)}}{\omega_r} I_{f(rated)} = \frac{1}{3.75} \times 192.5\,\text{A} = 51.33\,\text{A}$$

Similarly, per Equation (7.34), the equivalent machine constant drops proportionately with speed in order to limit the back emf:

$$k = \frac{\omega_{r(rated)}}{\omega_r} k_{rated} = \frac{1}{3.75} \times 0.77\,\text{Nm/A} = 0.2053\,\text{Nm/A}$$

As before, the electromagnetic torque is the sum of the useful output torque and the no-load torque:

$$T_{em} = T_r + T_{nl} = 74.67\,\text{Nm} + 2\,\text{Nm} = 76.67\,\text{Nm}$$

The armature current is given by

$$I_a = \frac{T_{em}}{k} = \frac{76.67}{0.2053}\,\text{A} = 373.4\,\text{A}$$

The machine power loss is the sum of the ohmic and no-load losses:

$$P_{m\,(loss)} = T_{nl}\omega_r + R_a I_a{}^2 + R_f I_f{}^2 = 2 \times 1071.4\,\text{W} + 0.05 \times 373.4^2\,\text{W}$$
$$+ 0.04 \times 51.33^2\,\text{W} = 9220\,\text{W}$$

The machine input power is

$$P = P_r + P_{m(loss)} = 80\,\text{kW} + 9.22\,\text{kW} = 89.22\,\text{kW}$$

The machine efficiency is given by

$$\eta = \frac{P_r}{P} = \frac{80\,\text{kW}}{89.22\,\text{kW}} \times 100\% = 89.67\%$$

c) The vehicle is propelled on a slope at rated speed and requires 70 Nm of rotor torque. For this condition, the rotor speed is

$$\omega_r = \omega_{r(rated)} = 285.71\,\text{rad}\,\text{s}^{-1}$$

Thus, the field current remains at 192.5 A from part (a) of this example.
The rotor output power is

$$P_r = T_r\omega_r = 70 \times 285.71\,\text{W} = 20\,\text{kW}$$

The electromagnetic torque is the sum of the useful output torque and the no-load torque:

$$T_{em} = T_r + T_{nl} = 70\,\text{Nm} + 2\,\text{Nm} = 72\,\text{Nm}$$

The armature current is given by

$$I_a = \frac{T_{em}}{k} = \frac{72}{0.77}\,\text{A} = 93.51\,\text{A}$$

The machine power loss is the sum of the ohmic and no-load losses:

$$P_{m\,(loss)} = T_{nl}\omega_r + R_a I_a{}^2 + R_f I_f{}^2 = 2 \times 285.71\,\text{W} + 0.05 \times 93.51^2\,\text{W}$$
$$+ 0.04 \times 192.5^2\,\text{W} = 2491\,\text{W}$$

The machine input power is

$$P = P_r + P_{m(loss)} = 20\,\text{kW} + 2.491\,\text{kW} = 22.491\,\text{kW}$$

The machine efficiency is given by

$$\eta = \frac{P_r}{P} = \frac{20}{22.491} \times 100\% = 88.92\%$$

7.10 Case Study – Mars Rover Traction Machine

In this final section, we review some of the specifications of an industrial dc machine. The machine of interest is the standard version of the Maxon Motors two-pole machine, part number 339152, which is similar to the traction machines used on the Mars rovers *Spirit* and *Opportunity*. Key machine specifications are shown in Table 7.1 [2,3]. The table has been modified to include some brief explanations. The parameters in italics in the first column have been calculated on the basis of the published parameters.

Plots of output torque and power are shown in Figure 7.14. Clearly, the machine can operate at far higher torque and power levels than specified for the rated conditions. The stall torque at 325 mNm is almost 10 times greater than the rated torque of 30.4 mNm. The peak power of approximately 93 W is significantly greater than the rated power of 30.8 W at 25°C.

Table 7.1 Maxon 24 V PM DC MACHINE, part no. 339152.

Parameter	Symbol	Value	Unit
Rated voltage	V_a	24	V
No-load speed at rated voltage	$N_{r(nl)}$	10900	rpm
No-load angular speed	$\omega_{r(nl)}$	1141	rad/s
No-load current	$I_{a(nl)}$	40.2	mA
No-load torque	T_{nl}	0.84	mNm
Rated speed	$N_{r(rated)}$	9690	rpm
Rated angular speed	$\omega_{r(rated)}$	1015	rad/s
Rated torque (max. continuous torque)	$T_{r(rated)}$	30.4	mNm
Rated current (max. continuous current)	$I_{a(rated)}$	1.5	A
Nominal output power (max. continuous)	$P_{r(rated)}$	30.85	W
Nominal input power (max. continuous)	$P_{a(rated)}$	36	W
Efficiency at rated condition	η	85.7	%
Stall torque	$T_{r(stall)}$	325	mNm
Stall current	$I_{a(stall)}$	15.6	A
Characteristics			
Terminal resistance	R_a	1.53	Ω
Terminal inductance	L_a	0.186	mH
Torque constant	k	0.0208	Nm/A
Mechanical time constant	τ_{mech}	5.24	ms
Rotor inertia	J	14.7	gcm^2
Thermal data			
Thermal resistance housing-ambient	$R_{\Theta(H\text{-}A)}$	14.4	°C/W
Thermal resistance winding-housing	$R_{\Theta(W\text{-}H)}$	5.1	°C/W
Thermal time constant winding	τ	29.8	s
Thermal time constant motor	τ_{motor}	543	s
Ambient temperature	T_{amb}	−30 to +100	°C
Max. winding temperature	$T_{W(max)}$	+155	°C

A sketch of the armature resistance against the armature current is presented in Figure 7.15 for graphite-copper brushes and the competitive precious-metal brushes.

The armature resistance R_a has two main components, the winding resistance of the armature winding R_{wdg} and the resistance of the brush and commutator $R_{b\text{-}c}$. As can be seen, the resistance of the graphite brush increases at light loads and may not be the optimum brush type depending on the application. The precious-metal brush is an alternative and tends to have a constant relatively low resistance with the armature current. Silver compound is a common material for a precious-metal brush. The precious-metal brush is ideal for smaller machines operating continuously with light loads.

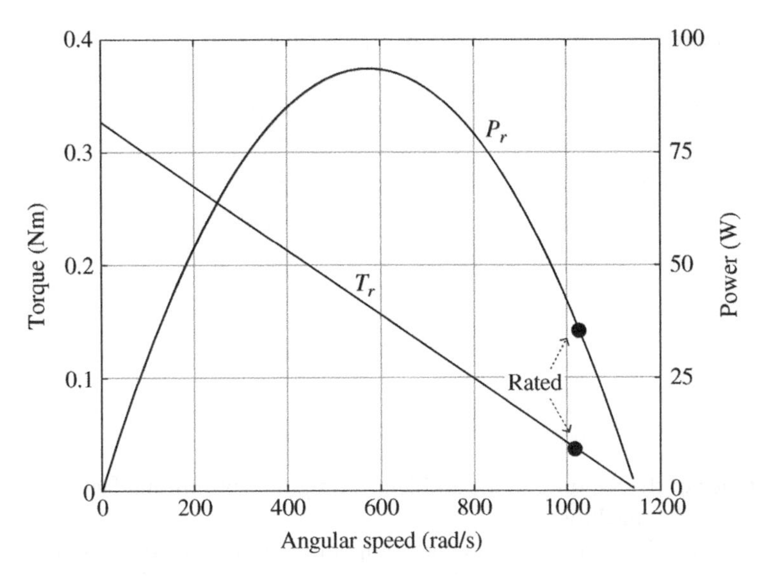

Figure 7.14 Output torque and power versus speed for Maxon 24 V motor.

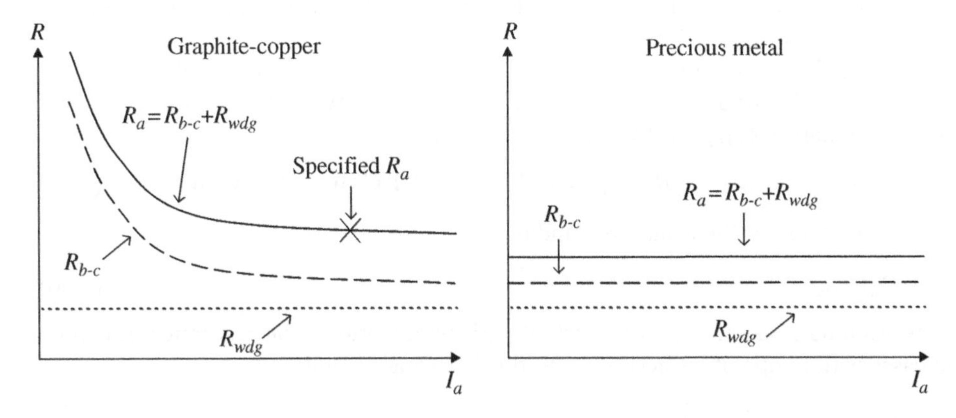

Figure 7.15 Armature resistance versus current for graphite-copper and precious-metal brushes.

The precious-metal brush is not ideal for stop-start or high-current loads but outperforms the graphite brush for audible noise, electromagnetic interference, and brush friction. The graphite-copper brush was the choice for the rover because of its higher current and temperature capabilities.

7.11 Thermal Characteristics of Machine

The thermal information for the machine is useful in allowing us to understand the machine's power capability with temperature. As the machine has a coreless construction, the losses for the Maxon machine are largely in the rotor winding, which simplifies the analysis and discussion.

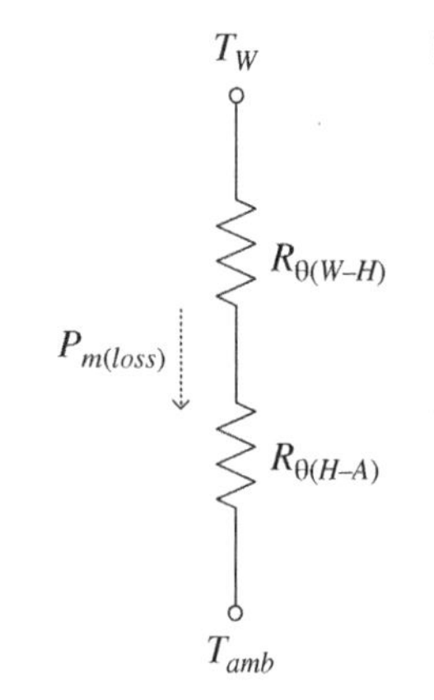

Figure 7.16 Thermal path for winding power loss.

The thermal resistance of the winding to the ambient is a significant limiting factor in the thermal capability of the machine.

The power loss in the machine causes the machine temperature to rise above the ambient temperature of the air outside the machine. The hot-spot temperature of the winding T_W can be estimated if the winding power loss and the thermal resistances to the ambient environment are known. An equivalent electrical circuit representation of the thermal circuit is shown in Figure 7.16.

The thermal resistance of the winding to the ambient $R_{\theta(W\text{-}A)}$ is the sum of the winding-housing thermal resistance $R_{\theta(W\text{-}H)}$, with a value of 5.1°C/W, and the housing-ambient thermal resistance $R_{\theta(H\text{-}A)}$, with a value of 14.4°C/W.

$$R_{\theta\,(W-A)} = R_{\theta\,(W-H)} + R_{\theta\,(H-A)} = 5.1°C/W + 14.4°C/W = 19.5°C/W \tag{7.35}$$

The copper loss in the armature winding is

$$P_{Ra} = R_a I_a^{\,2} \tag{7.36}$$

The winding resistance is not constant with temperature. The resistance of copper increases with temperature according to the following formula:

$$R_a = R_o[1 + \alpha(T - T_o)] \tag{7.37}$$

where R_o is the resistance at a standard temperature T_o, typically 25°C, and α is the temperature coefficient of resistance. The resistance of copper increases at approximately α = 0.004/°C.

7.11.1 Example of Steady-State Temperature Rise

Let us first consider the rated operating condition at 25°C. See Table 7.1 for the various values. At this condition, the winding is expected to reach a temperature of about 155°C. Thus, the resistance of the armature is approximately

$$R_a = R_o[1 + \alpha(T - T_o)] = 1.53 \times [1 + 0.004 \times (155 - 25)]\Omega = 2.33\Omega \tag{7.38}$$

The copper loss in the armature winding is

$$P_{Ra} = R_a I_{a(rated)}^{\,2} = 2.33 \times 1.5^2\Omega = 5.24\text{W} \tag{7.39}$$

The spinning friction and windage losses are

$$P_{fw} = T_{nl}\omega_{r(rated)} = 0.00084 \times 1015\,\text{W} = 0.85\,\text{W} \tag{7.40}$$

The total machine losses are

$$P_{m(loss)} = P_{fw} + P_{Ra} = 0.85\,\text{W} + 5.24\,\text{W} = 6.09\,\text{W} \tag{7.41}$$

Thus, the estimate of the winding temperature is

$$T_W = T_{amb} + R_{\theta(W-A)}P_{m(loss)} = 25°\text{C} + 19.5 \times 6.09°\text{C} = 25°\text{C} + 119°\text{C} = 144°\text{C} \tag{7.42}$$

which is a value very close to the expected winding temperature of 155°C.

7.11.2 Transient Temperature Rise

When current is fed into the armature, the machine does not instantly heat up; the temperature rises exponentially similar to the RC charging of a capacitor:

$$T_W = T_{amb} + \Delta T\left(1 - e^{-t/\tau}\right) \tag{7.43}$$

where T_{amb} is the ambient temperature. The time constant τ is the time that it takes for the temperature to rise to 63.2% of its final value. Let us consider a simple example.

7.11.3 Example of Transient Temperature Rise

Determine how long it takes for the machine winding to heat to 155°C if the ambient is at the rated temperature of 25°C and the machine torque is twice the rated torque at the rated speed.

Solution:
For simplicity, let us assume that as the torque doubles, the armature current doubles from the value of 1.5 A at the rated condition to 3.0 A, and that the armature resistance remains at 2.33 Ω, as above.

The copper loss in the armature winding is

$$P_{Ra} = R_a I_a^2 = 2.33 \times 3^2\,\Omega = 20.97\,\text{W} \tag{7.44}$$

The total machine losses are

$$P_{m(loss)} = P_{fw} + P_{Ra} = 0.85\,\text{W} + 20.97\,\text{W} = 21.82\,\text{W} \tag{7.45}$$

In the steady state, the winding temperature rises by

$$\Delta T = R_{\theta(W-A)}P_{m(loss)} = 19.5 \times 21.82\,\text{W} = 425°\text{C} \tag{7.46}$$

The time to rise from 25°C to 155°C is

$$t = -\tau\log_e\left(1 - \frac{T_W - T_{amb}}{\Delta T}\right) = -29.8 \times \log_e\left(1 - \frac{155-25}{425}\right) = 10.9\,\text{s} \tag{7.47}$$

Thus, the machine can withstand an overload torque at the rated speed, and the winding hot spot does not exceed the maximum temperature of 155°C if the time duration is less than 10.9 s.

References

1 A. E. Fitzgerald, C. Kingsley, and S. D. Umans, *Electric Machinery*, McGraw-Hill, 2002, 6th edition, ISBN 9780073660097.
2 S. Rochi, "The Mars Mission – Technology for another world," *Maxon Motors Application Stories*, 2014.
3 Maxon Motors catalog at www.maxonmotors.com.

Problems

7.1 Use a PM dc machine as the traction machine for a BEV. The machine has the following parameters: four-pole, armature resistance $R_a = 0.025\ \Omega$, machine constant $k = 0.7$ V/rad/s, and no-load torque $T_{nl} = 2$ Nm.

Determine the armature voltage and current output by the dc-dc converter, and the motor efficiency for the rated condition of 80 kW, 280 Nm in motoring mode.

[Ans. 210.1 V, 402.9 A, 94.5%]

7.2 Recalculate the above for the vehicle at +20 kW rotor power at half the rated speed.

[Ans. 105.1 V, 202.9 A, 93.8%]

7.3 Recalculate the above for the vehicle at −30 kW rotor power (regenerative mode) at the rated speed.

[Ans. 196.3 V, −147.1 A, 96.3%]

7.4 Recalculate the above for the vehicle at 10 kW motoring in reverse at 25% of the rated speed.

[Ans: −55.1 V, −202.9 A, 89.5%]

7.5 Recalculate the above for the vehicle at 80 kW, 280 Nm in motoring mode if the armature reaction and the associated saturation reduce the machine constant by 25% at the maximum current.

[Ans: 163.4 V, 537.1 A, 91.1%]

7.6 Use a WF dc machine as the traction machine for the BEV. The machine has the following parameters: eight-pole, armature resistance $R_a = 0.025\ \Omega$, field resistance $R_f = 0.020\ \Omega$, no-load torque $T_{nl} = 2$ Nm, and $L_f = 2$ mH. Control the field current I_f to emulate the machine constant of Problem 7.1.

Determine the armature voltage and current output by the dc-dc converter, the motor efficiency, and the field current for 80 kW operation in motoring mode at (i) rated speed and (ii) 3.75 times the rated speed.

[Ans. 402.9 A, 210.1 V, 94.4%, 87.5 A; 410.7 A, 210.3 V, 92.6%, 23.3 A]

7.7 Use a PM dc machine as the traction machine for a performance sports BEV. The machine has the following parameters: armature resistance $R_a = 0.02\ \Omega$, machine constant $k = 0.6$ V/rad/s, and no-load torque $T_{nl} = 2$ Nm.

For the rated condition of 300 kW, 600 Nm rotor power in motoring mode:

Determine the armature voltage and current output by the dc-dc converter, and the motor efficiency for this operating point.

[Ans. 320.1 V, 1003.3 A, 93.4%]

7.8 Recalculate the above for the vehicle of the previous problem at –300 kW, –600 Nm in generating mode.

[Ans. 280.1 V, –996.7 A, 93.0%]

7.9 Recalculate the above for the vehicle of Problem 7.7 at 300 kW, 600 Nm in motoring mode if the armature reaction and the associated saturation reduce the machine constant by 25% at the maximum current.

[Ans: 251.8 V, 1337.8 A, 89.1%]

7.10 Use a WF dc machine as the traction machine for a performance sports car. The machine has the following parameters: armature resistance R_a = 0.020 Ω, field resistance R_f = 0.015 Ω, no-load torque T_{nl} = 2 Nm, k = 0.6 Nm/A, and I_f = 300 A at the rated condition.

Determine the armature voltage and current, output by the dc-dc converter, the motor efficiency, and field current for 300 kW operation in motoring mode at (i) the rated speed and at (ii) twice the rated speed.

[Ans. 1003.3 A, 320.1 V, 93.0%, 300 A; 1006.7 A, 320.1 V, 93.0%, 150 A]

7.11 Determine how long it takes for the Maxon machine winding to heat to 110°C if the ambient is 70°C and the machine is operating at rated torque and speed. See Sections 7.11.1 and 7.11.3.

[Ans. 14 s]

7.12 Determine how long it takes for the Maxon machine winding to heat to 155°C if the ambient is at the rated temperature of 25°C and the machine is stalled at zero speed.

[Ans. 0.35 s]

8

Induction Machines

"... 1% inspiration and 99% perspiration." A famous quote by Thomas Edison explaining his success.

"His method was inefficient in the extreme ... a little theory and calculation would have saved him 90% of the labor. But he had a veritable contempt for book learning and mathematical knowledge..." Nikola Tesla commenting on his former employer Thomas Edison [1].

"Mr. Tesla may accomplish great things, but he will certainly never do this ... It is a perpetual-motion scheme, an impossible idea." Professor Pöschl at the Joanneum Polytechnic in Graz, Austria, commenting on the idea of his student Nikola Tesla [1].

"I would not give my rotating field discovery for a thousand inventions, however valuable... A thousand years hence, the telephone and the motion picture camera may be obsolete, but the principle of the rotating magnetic field will remain a vital, living thing for all time to come."

Nikola Tesla (1856–1943).

The induction motor is known as the workhorse of industry and is a great invention of Nikola Tesla. It plays a major role in our industrial economies. In recent times, as much as half the electrical power generated in the United States has been consumed by induction motors working in fans, compressors, pumps, and other equipment. In the development of electrical vehicles (EVs), the induction motor was the machine of choice for the GM EV1 and the Tesla Model S.

In this chapter, we develop the basic mathematics behind the rotating magnetic field generated by the stator winding. A squirrel-cage rotor construction is then introduced which results in the generation of torque within the machine. Electrically, the machine is operating on the principle of induction of voltages and currents onto the rotor by the spinning magnetic field of the stator – essentially, it is a transformer with a spinning secondary winding. Thus, the induction machine can be modeled using the conventional electrical equivalent circuit of a transformer. The steady-state operation of the machine

Electric Powertrain: Energy Systems, Power Electronics and Drives for Hybrid, Electric and Fuel Cell Vehicles, First Edition. John G. Hayes and G. Abas Goodarzi.

Companion website: www.wiley.com/go/hayes/electricpowertrain

is analyzed using the equivalent circuit in order to generate the electromechanical characteristics of the machine. The induction machine is characterized using the no-load and locked-rotor tests to determine its equivalent circuit parameters.

8.1 Stator Windings and the Spinning Magnetic Field

Stators for automotive interior-permanent-magnet (IPM) and induction machines (IMs) are as shown in Figure 8.1(a) and (b), respectively. Each of these stators is fed by balanced three-phase ac voltages and currents, and the result is a spinning magnetic field. The

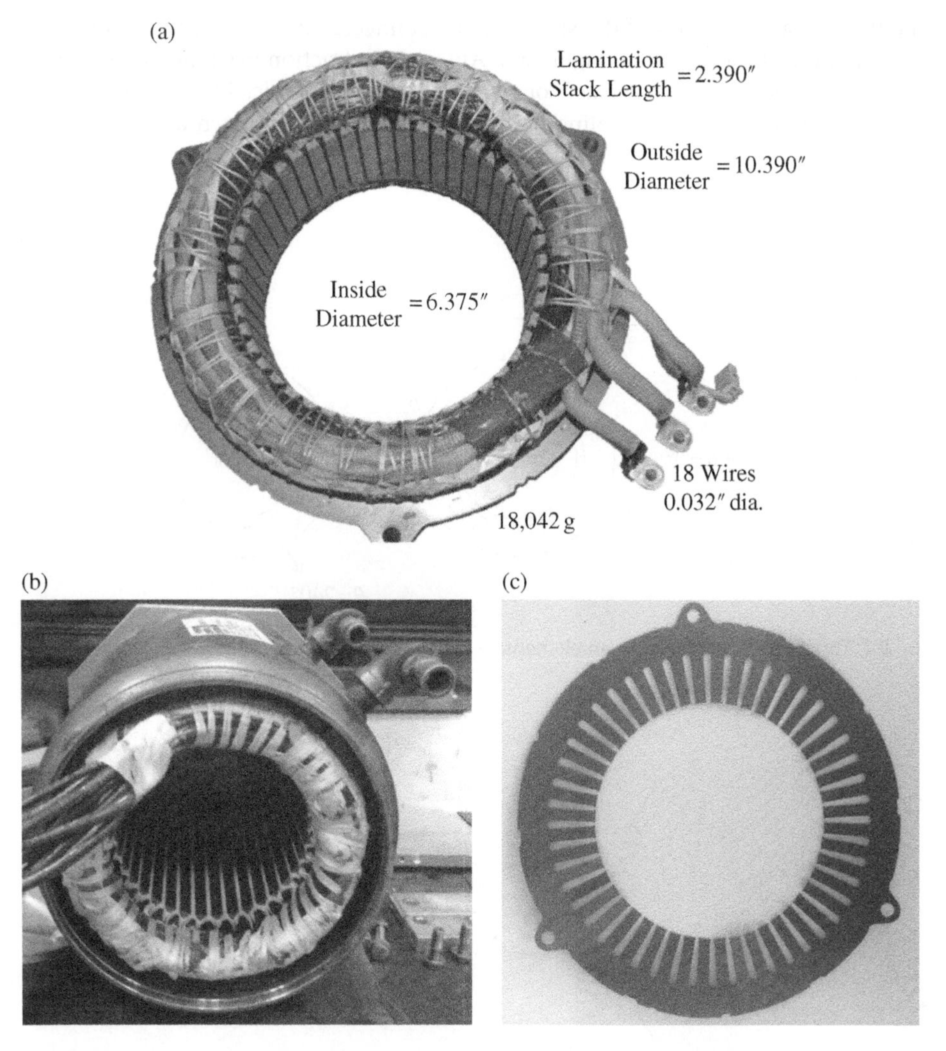

Figure 8.1 Machine stators for (a) 2007 Toyota Camry IPM motor, (b) USH induction machine, and (c) 2004 Toyota Prius stator lamination. (Images courtesy of Oak Ridge National Laboratory, US Dept. of Energy, and US Hybrid.)

stator iron is built from stacked laminations of ferromagnetic material. A photograph of a stator lamination from the 2004 Toyota Prius is shown in Figure 8.1(c). While the stators are similar for the IPM and IM, the winding construction can be different as will be discussed later.

The three-phase ac machine has three identical but separate windings which are wound into the stator slots as seen in Figure 8.1. The three-phase windings are physically wound such that they are displaced by 120° from each other. Thus, as shown in Figure 8.2 for a simple two-pole ac machine, if phase a is aligned such that its magnetic axis is at 0°, then phase b is physically offset by 120°, and phase c is offset by 240°. Hence, the winding inductances are shown in the figure as being physically offset from each other. The alignment of a magnetic axis is the direction of the magnetic field when the phase winding is energized with current.

The three-phase windings of the stator can be connected in either of two configurations: star and delta, as shown in Figure 8.3. Automotive traction machines are typically wired in **star**, which is also known as **Y** or **wye**. The phases are typically designated *abc* or *uvw*. The common point of coupling of a star configuration is known as the **neutral**.

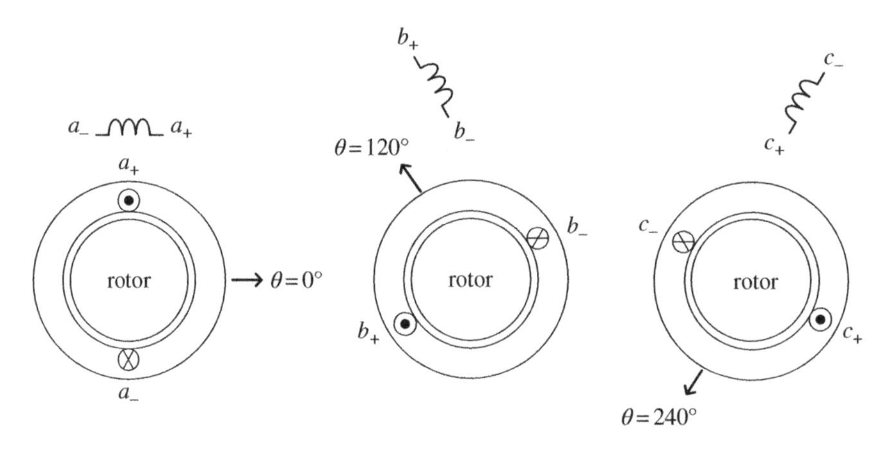

Figure 8.2 Three-phase windings for an elementary three-phase machine.

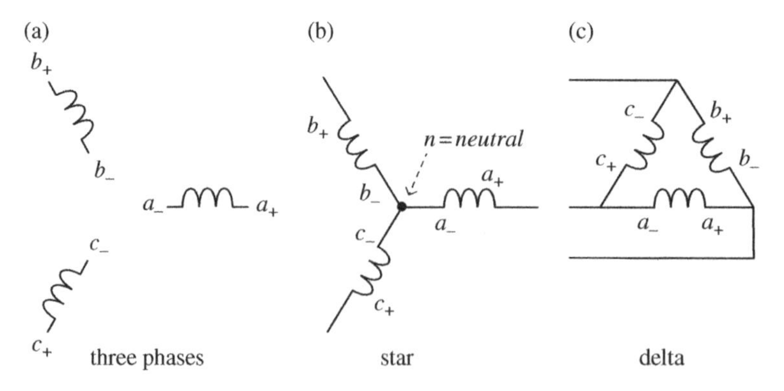

Figure 8.3 Star/wye and delta/mesh configurations of the stator windings.

Machines can also be wired in **delta**, also known as **mesh** – this is often the case for machines requiring maximum voltage across the phase winding. The voltage across the phase winding is $\sqrt{3}$ higher in the delta configuration than in the star.

The electrical symbol of the three-phase motor is shown in Figure 8.4.

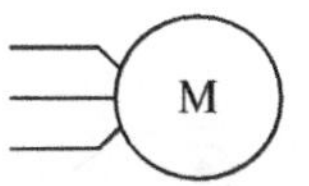

Figure 8.4 Three-phase motor electrical symbol.

8.1.1 Stator Magnetic Flux Density

In order to understand the concept of the rotating magnetic field, let us start by understanding the magnetic field due to a single coil of phase a, as shown in Figure 8.5. The current directions are as shown for the conductors of the coil, with the $a+$ conductor carrying current out of the page and the same current returning through the a- conductor going into the page. From the right-hand screw rule, a magnetic field results, and the flux lines are counterclockwise around conductor $a+$ and clockwise around conductor $a-$. The resulting magnetic flux density due to the current of phase a is as shown in Figure 8.6.

The **magnetic axis** of phase a is along the $\theta = 0°$ axis, and the flux due to phase a is symmetric around this axis. Thus, the magnetic flux density through the air gap around the rotor circumference is positive from $-\frac{\pi}{2}$ to $+\frac{\pi}{2}$ and negative from $\frac{\pi}{2}$ to $+\frac{3\pi}{2}$.

The other two phases are arranged as shown in Figure 8.7.

The magnetic flux densities for the elementary three-phase machine of Figure 8.7 are plotted in Figure 8.8. For these plots, it is assumed that the currents in phases b and c are

Figure 8.5 Magnetic field due to phase a current.

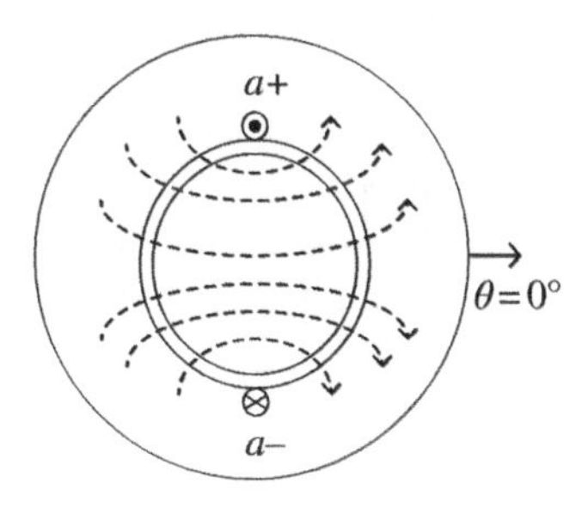

Figure 8.6 Magnetic flux density due to current of phase a.

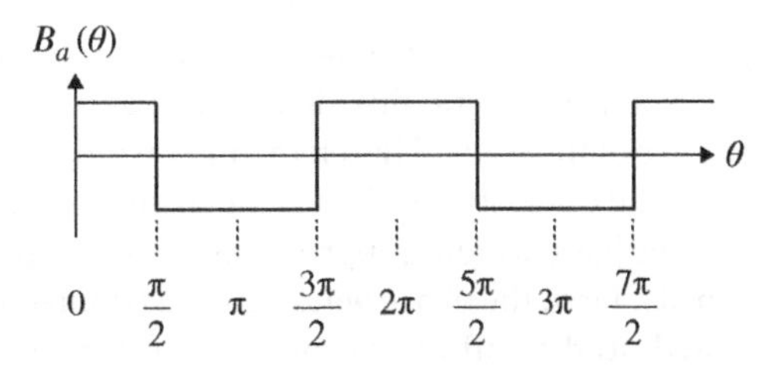

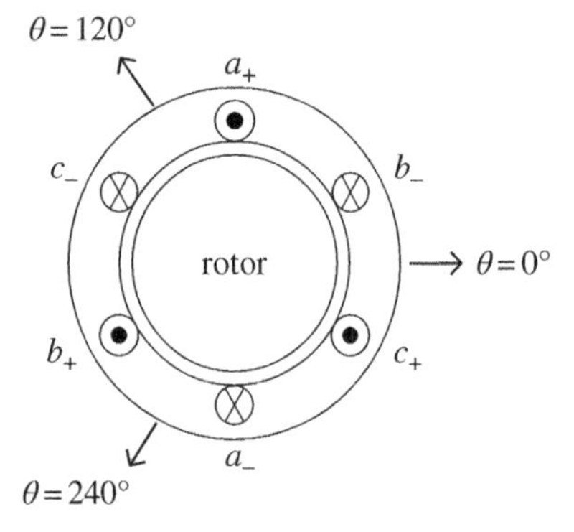

Figure 8.7 Elementary three-phase two-pole machine.

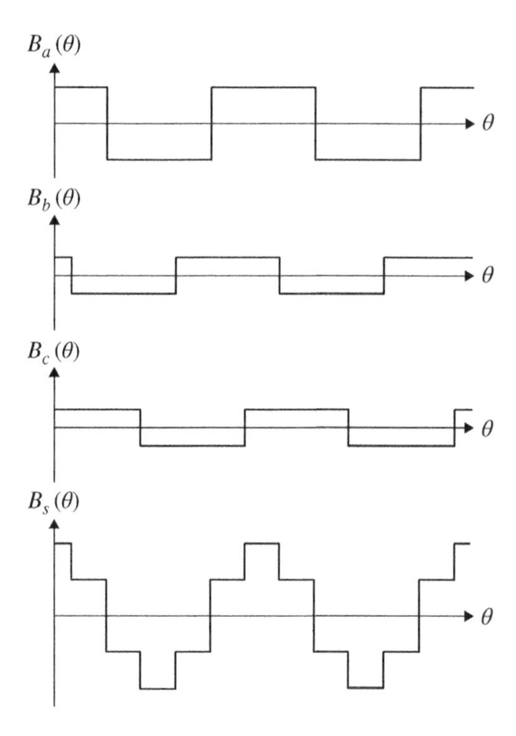

Figure 8.8 Magnetic flux densities for the phases and stator of the elementary three-phase two-pole machine.

each half and of opposite polarity to the current of phase a. The resulting magnetic flux densities created by the currents in phases b and c are offset with respect to phase a. As the magnetic flux density represents a physical vector, the flux densities from the three phases can be simply added to generate the resulting stator flux density in the air gap. The stator magnetic flux density B_s has a stepped ac characteristic around the circumference of the rotor. This elementary winding results in undesired harmonics. Ideally, the stator flux density would have a cosinusoidal waveform.

Some basic winding approaches can be employed to generate a less distorted stator flux density. First, the phase winding is distributed into more than one slot. This is known as a **distributed winding**. Second, the pitch of the winding can be varied. If the return

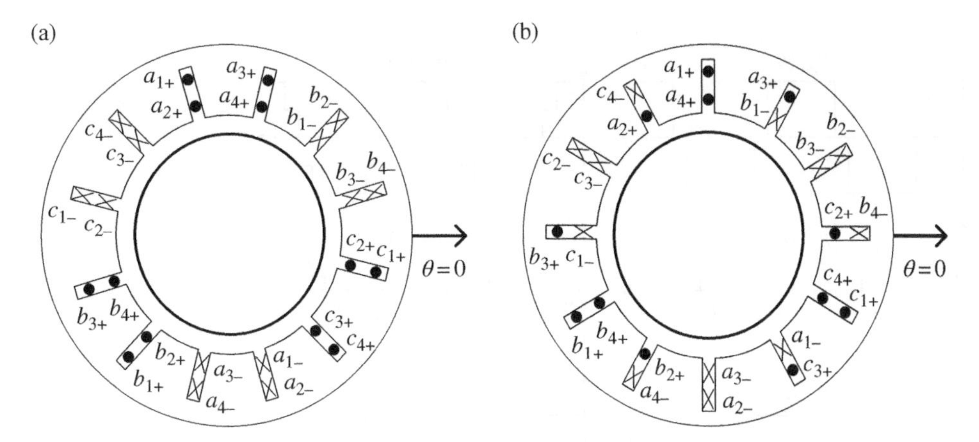

Figure 8.9 Distributed windings for three-phase two-pole machines with (a) full pitch and (b) fractional pitch.

winding for a conductor is 180° later, then such a machine winding is known as a **full-pitch** winding. If the return winding is less than a full pitch, then the winding is known as a **fractional-pitch** winding. A distributed three-phase two-pole full-pitch winding is shown in Figure 8.9(a). A variation on this type of winding is commonly used in IPM machines. A distributed three-phase two-pole fractional-pitch winding is shown in Figure 8.9(b). This type of winding is commonly used in IMs in order to reduce the distortion.

For comparison purposes, it is assumed that each phase consists of four windings in series. Flux density waveforms for the distributed three-phase two-pole machine windings with (a) full pitch and (b) fractional pitch are shown in Figure 8.10(a) and (b), respectively. The waveforms are similar. The effects of the distributed winding is to introduce a notch in the phase waveform, when compared to the earlier concentrated three-phase two-pole full-pitch winding of Figure 8.8. The effect of the fractional pitch is to shift the notch and create a wave shape with lower distortion. The harmonic distortion of the full-pitch winding is acceptable for IPM machines, while the fractional-pitch winding is necessary to further reduce the distortion in IMs. A further significant step to reduce the distortion of an induction motor is to **skew** the rotor, as will be discussed later.

Thus, it can be seen from this section that injecting balanced three-phase currents into the balanced three-phase stator of the ac machine results in a stator magnetic flux density which has an approximately sinusoidal distribution about the air gap.

8.1.2 Space-Vector Current and the Rotating Magnetic Field

If the three-phase windings are supplied by three electrical currents, which are displaced by 120° in time with respect to each other, then the stator magnetic field physically spins around the stator of the machine at a frequency determined by the frequency of the electrical currents. The basis for this fundamental operation of the spinning magnetic field is next developed.

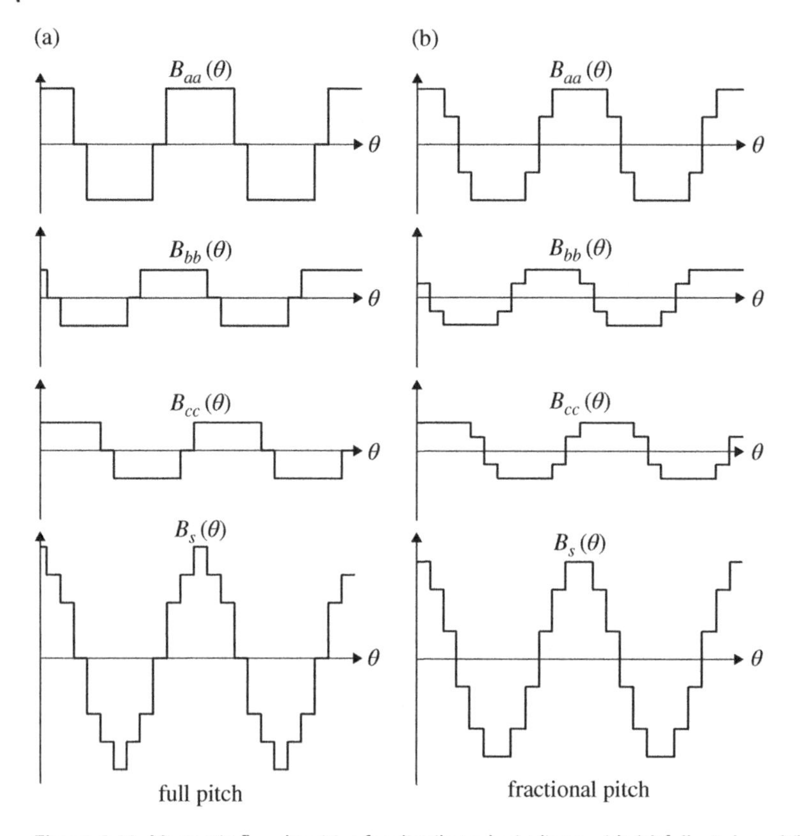

Figure 8.10 Magnetic flux densities for distributed windings with (a) full-pitch and (b) fractional pitch.

First, it is useful to identify an equivalent **space-vector current** $\vec{i}_s(t)$ which emulates or creates a similar magnetic field as the three-phase currents. The **space vector** is a virtual mathematical concept, but the concept is very useful and relates directly to the magnetic flux density vector, which is a real physical vector within the machine.

Let

$$\vec{i}_s^{\,a}(t) = i_a(t)\angle 0^\circ + i_b(t)\angle 120^\circ + i_c(t)\angle 240^\circ \tag{8.1}$$

where $\vec{i}_s^{\,a}(t)$ is the equivalent space-vector current with the phase a axis as the reference phasor, the angles relate to physical positions within the machine, and t is time.

The above equation can be rewritten in trigonometric form as

$$\vec{i}_s^{\,a}(t) = i_a(t)(\cos 0^\circ + j\sin 0^\circ) + i_b(t)(\cos 120^\circ + j\sin 120^\circ) + i_c(t)(\cos 240^\circ + j\sin 240^\circ) \tag{8.2}$$

which simplifies to

$$\vec{i}_s^{\,a}(t) = i_a(t)(1 + j0) + i_b(t)\left(-\tfrac{1}{2} + j\tfrac{\sqrt{3}}{2}\right) + i_c(t)\left(-\tfrac{1}{2} - j\tfrac{\sqrt{3}}{2}\right) \tag{8.3}$$

$$\Rightarrow \vec{i}_s^{\,a}(t) = \left\{ i_a(t) - {}^1/_2 (i_b(t) + i_c(t)) \right\} + j\,{}^{\sqrt{3}}\!/_2 (i_b(t) - i_c(t)) \tag{8.4}$$

Given that in a three-phase machine the three-phase currents sum to zero:

$$i_a(t) + i_b(t) + i_c(t) = 0 \tag{8.5}$$

Equation (8.4) simplifies to

$$\vec{i}_s^{\,a}(t) = {}^3/_2\, i_a(t) + j\,{}^{\sqrt{3}}\!/_2 \{ i_b(t) - i_c(t) \} \tag{8.6}$$

Next, we supply three cosinusoidal phase currents to the stator. The currents are displaced by 120° in time with respect to each other.

$$i_a(t) = \sqrt{2} I_{ph} \cos(2\pi f_e t) \tag{8.7}$$

$$i_b(t) = \sqrt{2} I_{ph} \cos(2\pi f_e t - 120^\circ) \tag{8.8}$$

$$i_c(t) = \sqrt{2} I_{ph} \cos(2\pi f_e t - 240^\circ) \tag{8.9}$$

where I_{ph} is the rms value of the phase current, and f_e is the electrical frequency.

These three-phase currents are substituted into Equation (8.6) as follows:

$$\vec{i}_s^{\,a}(t) = {}^3/_2 \sqrt{2} I_{ph} \cos(2\pi f_e t) + j\,{}^{\sqrt{3}}\!/_2 \left\{ \sqrt{2} I_{ph} \cos(2\pi f_e t - 120^\circ) - \sqrt{2} I_{ph} \cos(2\pi f_e t - 240^\circ) \right\} \tag{8.10}$$

which simplifies to give a simple expression for the space-vector current:

$$\vec{i}_s^{\,a}(t) = \frac{3}{2} \sqrt{2} I_{ph} \{ \cos(2\pi f_e t) + j \sin(2\pi f_e t) \} \tag{8.11}$$

which can be rewritten in polar form as

$$\vec{i}_s^{\,a}(t) = \hat{I}_s \angle 2\pi f_e t \tag{8.12}$$

or in exponential form as

$$\vec{i}_s^{\,a}(t) = \hat{I}_s e^{j2\pi f_e t} \tag{8.13}$$

where

$$\hat{I}_s = \frac{3}{2} \sqrt{2} I_{ph} \tag{8.14}$$

These complex expressions simply show that the resultant space vector, created by the vector addition of the three per-phase cosinusoidal currents, has a peak value $\hat{I}_s$ which is spinning around the machine at a frequency determined by the electrical input frequency.

For a two-pole machine, the electrical frequency is equal to the rotational frequency of the magnetic field. The rotational speed of the magnetic field is known as the **synchronous** speed or frequency f_{syn}.

For a p-pole machine, the synchronous frequency is related to the electrical frequency by

$$f_{syn} = \frac{f_e}{p/2} \tag{8.15}$$

From Ampere's circuital law the air gap magnetic field strength is approximately given by

$$\vec{H}_s^a(t) = \frac{N_{ph}}{2l_g}\vec{i}_s^a(t) \tag{8.16}$$

where N_{ph} is the number of turns per phase, and l_g is the length of the air gap between the stator and rotor. For this simple expression, the magnetic effect of the high-permeability iron core is ignored. The magnetic field strength is a physical vector which is spinning around the air gap at the synchronous frequency, which is determined by the electrical frequency.

The air gap magnetic field flux density is then given by

$$\vec{B}_s^a(t) = \frac{\mu_0 N_{ph}}{2l_g}\vec{i}_s^a(t) = \hat{B}_s \angle 2\pi f_e t \tag{8.17}$$

where μ_0 is the permeability of free space or air, and $\hat{B}_s$ is the peak value of the rotating flux density field given by

$$\hat{B}_s = \frac{3\,\mu_0\,N_{ph}\,I_{ph}}{2\sqrt{2}\,l_g} \tag{8.18}$$

Thus, the stator flux density vector is also spinning around the stator at the synchronous frequency as the flux density is directly related to the current space vector that has been created by the combination of the three time-varying currents.

This simple math essentially describes the brilliant invention by Nikola Tesla, who visualized a spinning magnetic field based on physically offset windings being fed by corresponding time-offset currents.

As a graphical illustration of the concept, consider the three-phase current waveforms shown in Figure 8.11.

The currents shown in Figure 8.11 are drawn as vectors relative to their current axes in Figure 8.12. At $\theta = 0$, the phase a current, $i_a(\theta) = i_a(\omega t)$, peaks and the phase b current, $i_b(\theta)$, and the phase c current, $i_c(\theta)$, are both at −0.5 (or cos 120°) relative to the phase a current. The resultant space vector current is aligned with the a axis.

As the currents alternate with time, three more instances are plotted for $\theta = \frac{\pi}{2}$, π and $\frac{3}{2}\pi$. For both $\theta = \frac{\pi}{2}$ and $\frac{3}{2}\pi$, the phase a current contribution is zero, and the resulting space vectors are aligned with $\frac{\pi}{2}$ and $\frac{3}{2}\pi$, respectively. As the currents change with time, the resultant current space vector and the flux density space vector rotate around the machine.

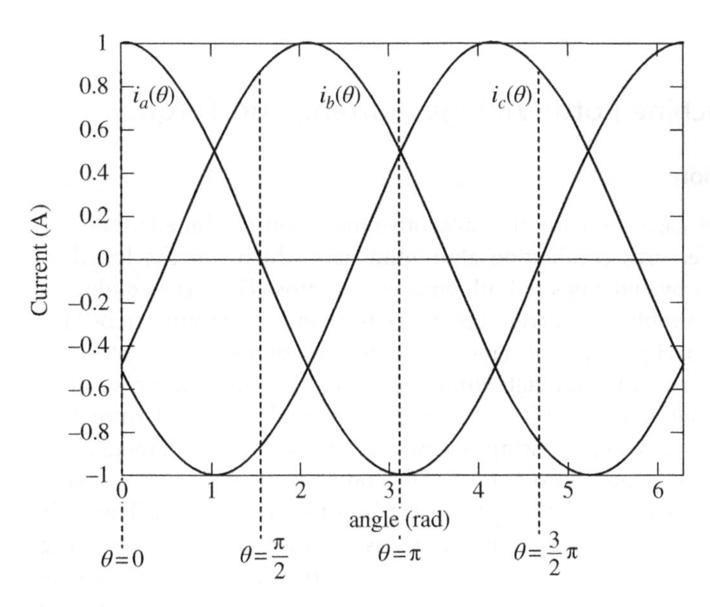

Figure 8.11 Three-phase currents.

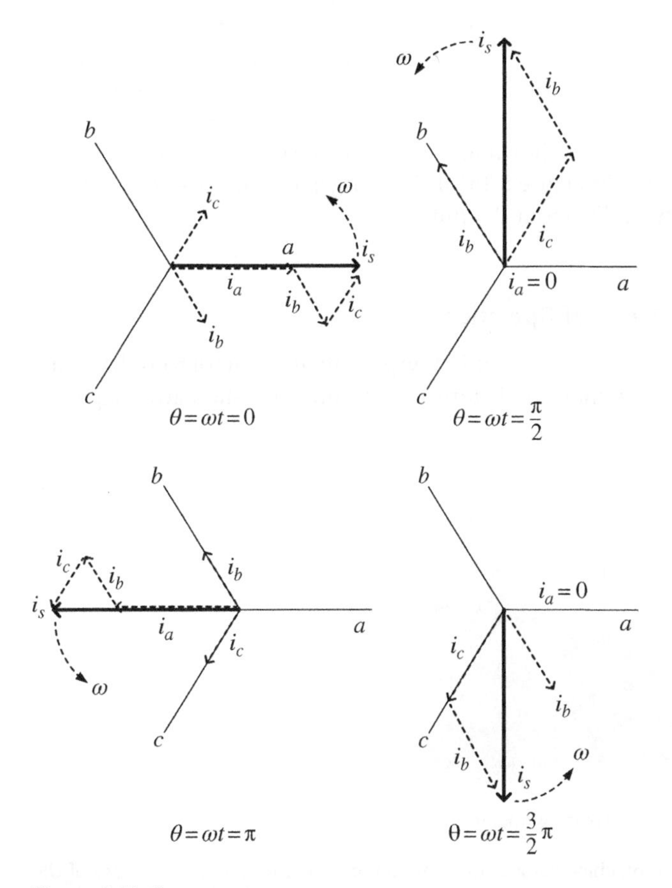

Figure 8.12 Space vectors.

8.2 Induction Machine Rotor Voltage, Current, and Torque

8.2.1 Rotor Construction

An example of a squirrel-cage rotor for an automotive induction machine is shown in Figure 8.13. The rotor features conducting aluminum bars which run the length of the rotor and are shorted by end rings at both ends of the rotor. Thus, the conductive section of the machine resembles a squirrel cage, hence the name. The rotor can be built using copper or aluminum. Higher-performance machines use copper, as the metal has a lower resistance than aluminum. Although aluminum rotors have a higher resistance, they have a manufacturing and cost advantage in that the end rings and rotor bars can be cast in a single assembly step. **Casting** is a process in which a molten metal, such as aluminum in this case, is poured into a mold. The copper rotor, on the other hand, cannot be cast and is terminated by brazing the rotor bars to the end rings. The high-temperature joining of two metals using a filler metal is commonly known as **brazing**. It can be seen from the image that the rotor bars do not run straight but rather at a skew angle from end ring to end ring. Skewing the rotor bars reduces undesirable harmonics within the machine.

The stator and rotor laminations are stamped from a single lamination. An outline of an automotive lamination is shown in Figure 8.14. The stator has 36 winding slots, and the rotor has 44 slots. The rotor bars are not only skewed but also feature more conductor slots than the stator in order to reduce harmonics. The yoke and teeth provide paths for the magnetic flux.

The rotor is placed into the stator as shown in the axial sketch of Figure 8.15(a). A side view sketch of the rotor is shown in Figure 8.15(b). The air gap between rotor and stator is relatively small in IMs – typically about 0.5 mm.

8.2.2 Induction Machine Theory of Operation

The theory of operation of the induction motor is now presented. A pair of rotor bars and the two end-ring terminations are shown in Figure 8.16. Assume that the stator magnetic

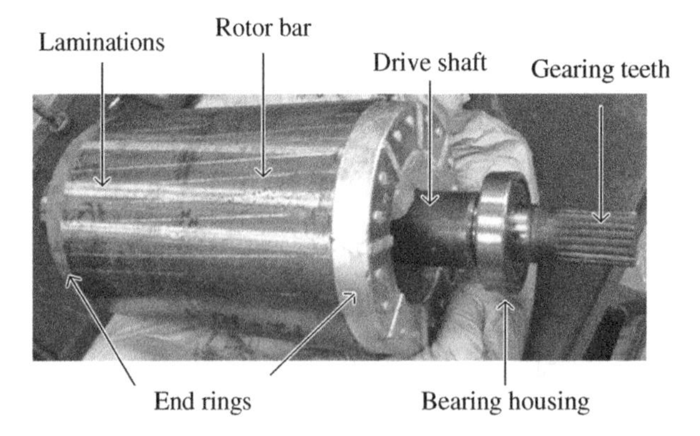

Figure 8.13 Automotive induction machine rotor with aluminum bars and end rings. (Courtesy of US Hybrid.)

Figure 8.14 Lamination outline for a three-phase four-pole induction machine showing yokes, teeth, and slots of stator and rotor.

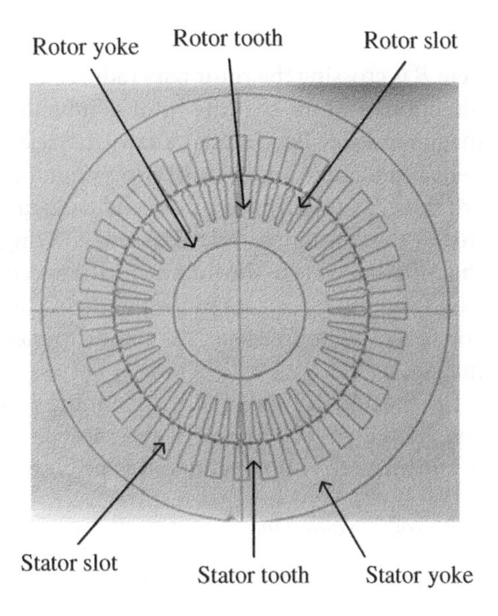

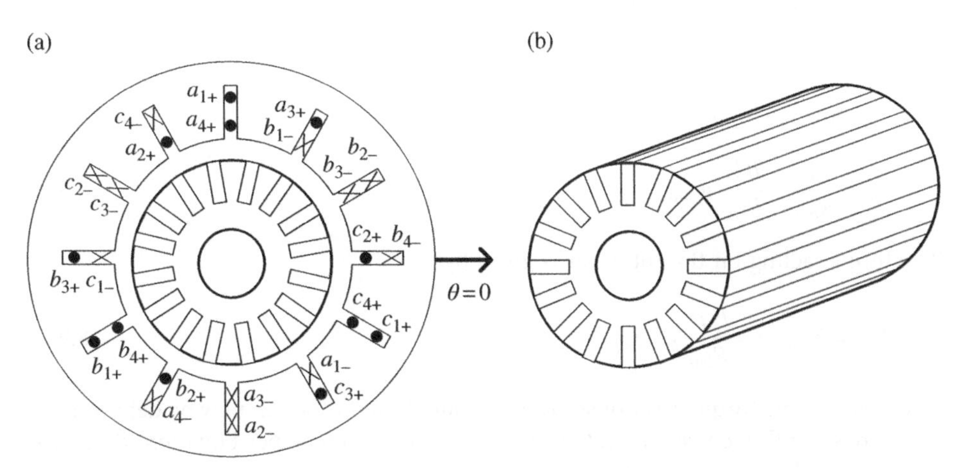

Figure 8.15 Three-phase two-pole induction motor: (a) axial view of stator and rotor and (b) side view of rotor.

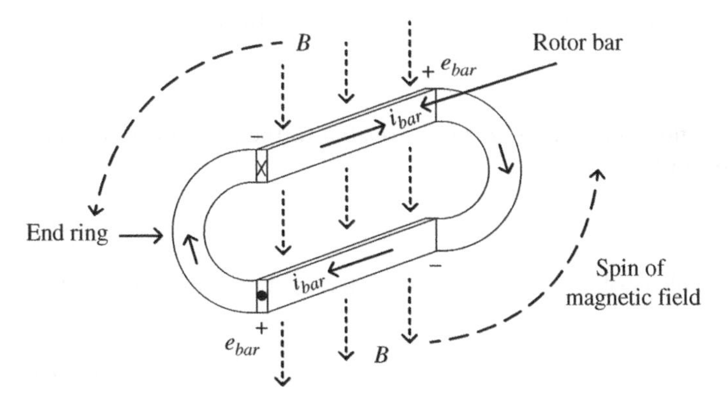

Figure 8.16 Torque generation in the induction machine rotor coil.

field B is crossing the rotor bars radially and is spinning in a positive or counterclockwise direction. Assume that the rotor conductors are stationary with respect to the spinning magnetic field. Per Faraday's law, the relative motion between the spinning magnetic field and the stationary bars causes an induced back emf e_{bar} along the bars. As the bars are shorted at both ends by the end rings, a current i_{bar} flows in the bars and is only limited by the resistance of the bars and end rings. Per the left hand rule, a force is exerted on the rotor bars due to the interaction between the current and the spinning magnetic field. The resulting force acts to torque the rotor bars around its axis of rotation. If the rotor coil is allowed to free spin, the rotor experiences a torque in the direction of the spinning magnetic field.

Let the angular speed of the spinning magnetic field be the synchronous speed ω_{syn}, and let the angular speed of the rotor be ω_r. The relative difference in angular speed is known as the **slip speed** and is given by

$$\omega_{slip} = \omega_{syn} - \omega_r \tag{8.19}$$

From Faraday's law (and as discussed in Chapter 16, Section 16.7), the back emf e_{bar} induced in a single bar ($N = 1$) is given by

$$e_{bar} = B\,l\,r\,\omega_{slip} \tag{8.20}$$

The current flowing in the rotor bar is given by

$$i_{bar} = \frac{e_{bar}}{R_{bar}} = \frac{B\,l\,r}{R_{bar}}\omega_{slip} \tag{8.21}$$

The torque acting on the rotor bar is given by

$$T_{bar} = B i_{bar} l\,r = \frac{l^2 r^2}{R_{bar}} B^2 \omega_{slip} \tag{8.22}$$

The basic simple torque relationships in an induction motor can now be determined.

For a constant flux density B and given machine parameters, per Equation (8.22) the torque is directly proportional to the slip speed:

$$T_{bar} \propto \omega_{slip} \tag{8.23}$$

Per Equation (8.22), the torque is also directly proportional to the current:

$$T_{bar} \propto i_{bar} \tag{8.24}$$

For a variable magnetic flux density, the torque is related to the product of the flux density squared and the slip frequency:

$$T_{bar} \propto B^2 \omega_{slip} \tag{8.25}$$

The classic relationship also holds – the torque is related to the product of the flux density and the current:

$$T_{bar} \propto B i_{bar} \tag{8.26}$$

The **slip** s is a commonly used term when discussing IMs. It is the ratio of the slip frequency to the synchronous frequency:

$$s = \frac{\omega_{slip}}{\omega_{syn}} \tag{8.27}$$

Slip is often expressed as a percentage.

8.3 Machine Model and Steady-State Operation

The theory of operation as just described can now be used to generate an electrical equivalent circuit based on the transformer. The transformer is an appropriate model to use, because the induction machine operates as a transformer with a spinning primary-generated magnetic field and a spinning secondary winding on the rotor.

The circuit diagram of Figure 8.17 can be used to illustrate the behavior of the induction machine on a per-phase basis. This circuit represents the per-phase transformer equivalent circuit with one change. Not only are voltage and current transformed but the frequency is also transformed, resulting in electromechanical power conversion.

When a phase voltage V_{ph} is supplied to the machine stator, a current I_{ph} flows into the stator winding. A component of the phase current magnetizes the stator-referenced magnetizing inductance L_m. This is the per-phase magnetizing current I_m. The magnetizing currents from the three phases combine to create the rotating magnetic flux density and generate a per-phase rotor back emf E_r. The magnitude of the rotor voltage depends on the equivalent turns ratio of the rotor turns N_r to the stator turns N_s and the slip. A per-phase current I_r flows in the rotor bars, with the per-phase rotor resistance R_r. A number of parasitic components can significantly affect the operation. The stator copper loss is modeled using a series resistance R_s. The core loss resulting from the magnetizing flux is modeled by placing a parallel resistance R_c across the magnetizing inductance. Stator and rotor leakage inductances are modeled using the series elements L_{ls} and L_{lr}, respectively.

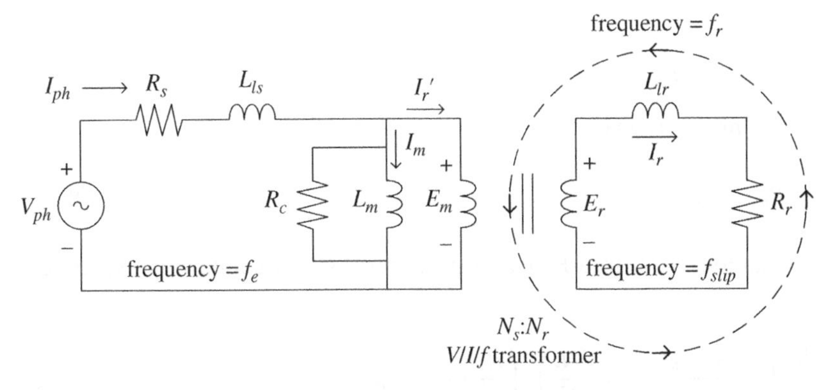

Figure 8.17 Equivalent circuit of an induction machine with a spinning rotor.

The electrical frequency of the stator is the same as the input electrical frequency f_e. The electrical frequency of the rotor is the slip frequency f_{slip}, which is usually far smaller than f_e.

The synchronous frequency can be determined by dividing the electrical frequency by the number of pole pairs:

$$f_{syn} = \frac{f_e}{p/2} \tag{8.28}$$

The rotor physically rotates at f_r. Hence, the rotor is illustrated in Figure 8.17 as a spinning secondary.

From transformer action (see Chapter 16, Section 16.5), the rotor current I_r results in a reflected current I_r' flowing in the stator. The stator and rotor currents are related by Ampere's circuital law:

$$N_r I_r = N_s I_r' \tag{8.29}$$

The stator and rotor back emfs are related by Faraday's law:

$$\frac{E_r}{N_r} = \frac{E_m}{N_s} s \tag{8.30}$$

The slip term must be included as the rotor voltage is dependent on both the equivalent machine turns ratio and the slip. If the slip is zero and the rotor is spinning at the same speed as the synchronous speed, then no voltage is induced on the rotor. If the rotor is locked or blocked from spinning, then the machine simply looks like a static transformer with a slip of 1.

The equivalent circuit can be simplified by reflecting the rotor parameters to the stator. When reflecting the rotor to the stator, it is necessary to have the same equivalent impedance Z_r', defined as

$$Z_r' = \frac{E_m}{I_r'} \tag{8.31}$$

If Equation (8.29) and Equation (8.30) are substituted into Equation (8.31), we get the following expression:

$$Z_r' = \frac{E_m}{I_r'} = \left(\frac{N_s}{N_r}\right)^2 \frac{E_r}{s\,I_r} = \left(\frac{N_s}{N_r}\right)^2 \frac{Z_r}{s} \tag{8.32}$$

The rotor impedance Z_r is given by

$$Z_r = \frac{E_r}{I_r} = R_r + j\omega_{slip} L_{lr} \tag{8.33}$$

Substituting Equation (8.33) into Equation (8.32) results in the following expression for Z_r':

$$Z_r' = \left(\frac{N_s}{N_r}\right)^2 \frac{R_r + j\omega_{slip} L_{lr}}{s} = \left(\frac{N_s}{N_r}\right)^2 \left(\frac{R_r}{s} + j\frac{\omega_e}{p/2} L_{lr}\right) \tag{8.34}$$

which is more commonly expressed as

$$Z_r' = \frac{R_r'}{s} + j\omega_e L_{lr}' \tag{8.35}$$

where the reflected-rotor parameters are related to the rotor parameters by

$$R_r' = \left(\frac{N_s}{N_r}\right)^2 R_r \text{ and } L_{lr}' = \frac{2}{p}\left(\frac{N_s}{N_r}\right)^2 L_{lr} \tag{8.36}$$

The earlier equivalent circuit can now be presented in the standard way as shown in Figure 8.18.

It is to be noted that this equivalent circuit is identical to the equivalent circuit of a transformer with a resistive load. The only difference is that the reflected rotor resistance is divided by the slip, which is, of course, a function of speed. Hence, the electrical power dissipated in the resistor must represent a certain component of mechanical power.

The rotor resistance can be split into two components in order to generate a more insightful equivalent circuit:

$$\frac{R_r'}{s} = R_r' + R_r'\left(\frac{1-s}{s}\right) \tag{8.37}$$

where the first component R_r' models the rotor copper loss, and $R_r'\left(\frac{1-s}{s}\right)$ models the electromechanical power, as shown in Figure 8.19.

The electrical power consumption of the induction machine can now be related to the power losses in the per-phase equivalent circuit model. Figure 8.20 presents the powers in the various components.

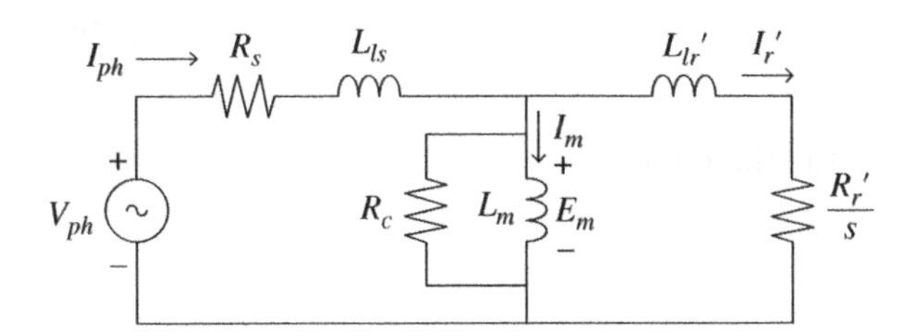

Figure 8.18 Equivalent circuit of the induction machine.

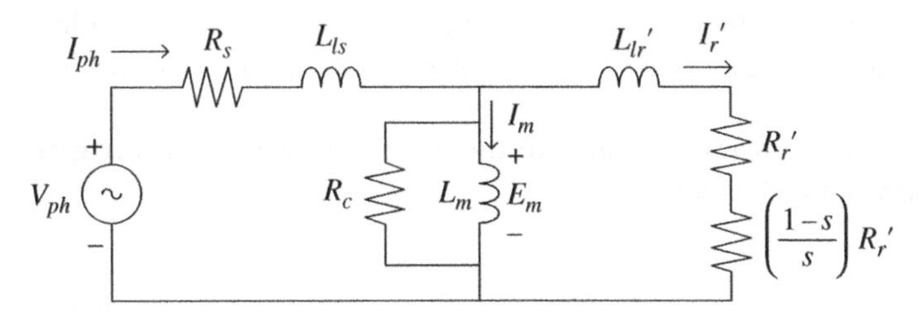

Figure 8.19 Equivalent circuit of the induction machine showing the two rotor resistance components.

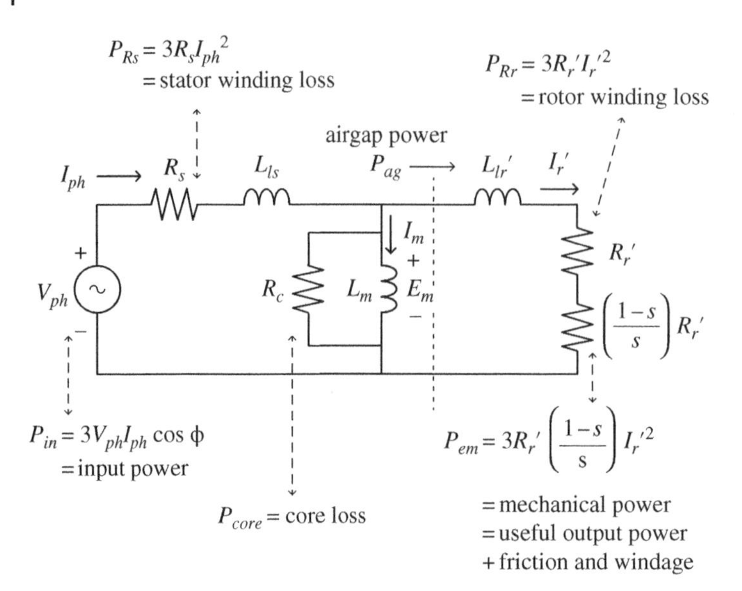

Figure 8.20 Equivalent circuit of induction machine showing power flow.

8.3.1 Power in Three-Phase Induction Machine

The **electromechanical** power P_{em} is the power converted from electrical to mechanical. The common electrical and mechanical expressions for the electromechanical power converted in an induction motor are given by

$$\begin{aligned} P_{em} &= 3R_r'\left(\frac{1-s}{s}\right)I_r'^2 \\ &= T_{em}\,\omega_r \end{aligned} \tag{8.38}$$

where the factor of "3" accounts for the three phases.

The rotor copper loss P_{Rr} is given by

$$\begin{aligned} P_{Rr} &= 3R_r'I_r'^2 \\ &= T_{em}\omega_{slip} \end{aligned} \tag{8.39}$$

Critically, we can note here that the rotor copper loss increases with the slip frequency. This can be an incentive to keep the slip frequency low in order to minimize the rotor copper loss as this heat loss can be difficult to remove from the rotor and affects machine sizing and thermal management. Note also that the slip frequency is a controlled parameter.

The power crossing the air gap from the stator to the rotor is known as the **air gap power** P_{ag}, and it is simply defined as follows:

$$\begin{aligned} P_{ag} &= \frac{3R_r'I_r'^2}{s} \\ &= T_{em}\,\omega_{syn} \end{aligned} \tag{8.40}$$

The **stator copper loss** P_{Rs} is given by

$$P_{Rs} = 3R_s I_{ph}^2 \tag{8.41}$$

The **lumped core, friction, and windage loss** P_{cfw} is given by

$$P_{cfw} = T_{nl}\omega_r \tag{8.42}$$

The total **machine loss** $P_{m(loss)}$ is the sum of the core, friction, and windage loss and the stator and rotor copper losses:

$$\begin{aligned} P_{m(loss)} &= P_{cfw} + P_{Rs} + P_{Rr} \\ &= T_{nl}\omega_r + 3R_s I_{ph}^2 + 3R_r' I_r'^2 \end{aligned} \tag{8.43}$$

The **input power** P_{ph} is the sum of the air gap power, the stator copper loss, and the core loss, and can be simply expressed as

$$P_{ph} = 3V_{ph}I_{ph}\cos\phi \tag{8.44}$$

8.3.2 Torque in Three-Phase Induction Machine

There are various expressions which can be used for the relationship between torque and current. The generation of torque in any machine is based on the orthogonality of the magnetizing flux and the current. The induction machine torque is given by

$$T_{em} = 3\frac{p}{2}\lambda_r I_r' \tag{8.45}$$

where the "3" accounts for the three phases, p is the number of poles, and λ_r is the rotor flux linkage (see Chapter 16, Section 16.7). Note that the flux linkage λ is treated in this textbook as an rms quantity for ac machines. It is very common for the peak flux linkage to be used in equations, in which case an adjustment by a factor of $\sqrt{2}$ is required.

The rotor flux linkage λ_r is equal to the stator flux linkage λ_s minus the stator and rotor leakage flux. In simpler terms, the rotor flux linkage is equal to the magnetizing flux linkage minus the rotor leakage flux linkage:

$$\lambda_r = L_m I_m - L_{lr}' I_r' \tag{8.46}$$

Thus, the electromagnetic torque is given by

$$T_{em} = 3\frac{p}{2}(L_m I_m - L_{lr}' I_r')I_r' \tag{8.47}$$

If Equation (8.45) is combined with Equation (8.38), the relationship between rotor current and slip frequency can be derived as follows:

$$I_r' = \frac{p}{2R_r'}\lambda_r\omega_{slip} \tag{8.48}$$

which results in the following expression for torque and slip:

$$T_{em} = 3\frac{p^2}{4R_r'}\lambda_r^2\omega_{slip} \tag{8.49}$$

We can now establish some basic relationships between slip, current, and torque. If λ_r is maintained at a constant value, then the following relationships hold:

$$T_{em} \propto I_r' \propto \omega_{slip} \tag{8.50}$$

This is a very important relationship in an induction machine. Basically, if the rotor flux linkage can be maintained constant (or approximately constant), then the developed electromagnetic torque is linearly proportional to both the rotor current and the rotor slip. This relationship holds in all four quadrants of operation, regardless of the electrical or mechanical frequencies, or whether motoring or generating.

A plot of torque versus speed and slip is as shown in Figure 8.21. In this curve, the rotor speed is normalized by dividing by the synchronous speed. At $s = 0$, no torque is generated. As the slip increases from 0, the torque increases approximately linearly with slip, and the machine leakage inductances have little effect. As the slip continues to increase, the torque reaches a maximum level as the leakage inductances play a larger role. As the rotor current increases, the voltage drops across stator and rotor leakage inductances increase, resulting in significant voltage drops within the machine, thus limiting the torque-generating current. The torque continues to drop as the slip increases.

If a negative slip is introduced, the machine acts as a generator. Thus, a negative torque is developed by the machine when operating in the forward mode. Again, a similar relationship exists between the torque and the negative slip. Initially the torque increases linearly with slip, reaches a maximum due to leakage, and then drops significantly.

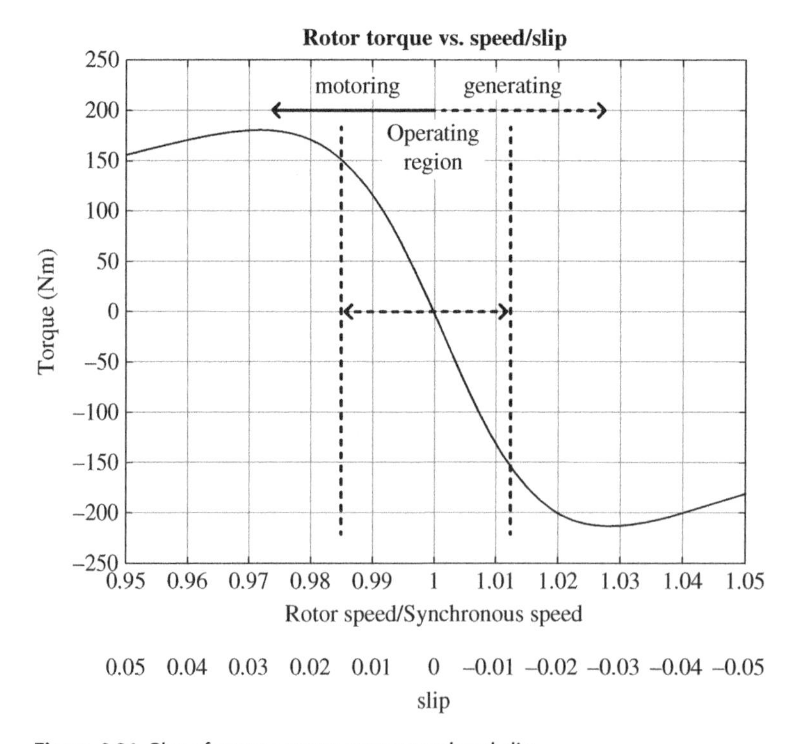

Figure 8.21 Plot of torque versus rotor speed and slip.

The automotive induction machine is operated efficiently with very low values of slip, and unlike the direct-on-line industrial machine, the inverter-fed automotive machine does not operate at higher values of slip where the torque decreases with slip.

If we focus on the operating region, as shown in Figure 8.21, it can be seen that this machine nominally operates between −150 Nm and +150 Nm, with a slip range of +1.5% to −1.3%. This curve is generated using a constant voltage on the stator. For this constant voltage-source condition, the flux linkage is higher during regeneration, resulting in slightly lower values of slip for the same torque level.

A set of characteristic curves is plotted in Figure 8.22 for a normalized rotor speed range of 0.985 to 1 times the synchronous speed, or a slip range from 0.015 to 0, for rated conditions. The torque is plotted in Figure 8.22(a). The reflected rotor and magnetizing and phase currents are plotted in Figure 8.22(b). The machine efficiency and power factor are plotted in Figure 8.22(c).

At zero slip, the machine runs at synchronous speed and develops zero torque. The rotor torque and current increase proportionately with slip. The machine is supplied with a constant electrical frequency, and so the magnetizing current remains at a relatively constant value. The phase current is composed of the reflected rotor and magnetizing currents.

As the torque increases, the machine efficiency increases to the mid 90s at medium torque with the efficiency dropping to the low-to-mid 90s at the rated torque. The power factor similarly peaks at medium torque and drops to about 0.86 at the rated condition.

8.3.3 Phasor Analysis of Induction Motor

Steady-state machine operation is easily characterized using phasor analysis. A phasor diagram for a machine operating in motoring mode is as shown in Figure 8.23. In this phasor diagram, the per-phase supply voltage phasor $\bar{V}_{ph}$ is taken as the reference. The back emf across the magnetizing inductance $\bar{E}_m$ lags $\bar{V}_{ph}$ due to the voltage drop $\bar{V}_{Zs}$ across the stator impedance $\bar{Z}_s$. The stator impedance consists of the stator resistance and leakage inductance. A magnetizing current $\bar{I}_m$ flows which lags $\bar{E}_m$ by exactly 90°. The rotor flux linkage $\bar{\lambda}_r$ lags $\bar{I}_m$ due to the leakage inductance of the rotor. The rotor current $\bar{I}_r$ lags $\bar{\lambda}_r$ by 90°. Thus, the reflected rotor current $\bar{I}_r'$ leads $\bar{\lambda}_r$ by exactly 90°. The per-phase supply current $\bar{I}_{ph}$ is the vector sum of $\bar{I}_m$ and $\bar{I}_r'$.

The phasor diagram for the induction machine, when acting as a generator, is as shown in Figure 8.24.

In order for the machine to act regeneratively in a forward direction, the back emf $\bar{E}_m$ must exceed the supply voltage $\bar{V}_{ph}$, with the voltage difference $\bar{V}_{Zs}$ being dropped across $\bar{Z}_s$. For **regeneration**, the slip is negative, and the reflected rotor current lags $\bar{\lambda}_r$ by 90°, such that it is close to −180° out of phase with $\bar{E}_m$.

8.3.4 Machine Operation When Supplied by Current Source

The automotive induction machine used for traction is supplied by a power electronics inverter which can precisely control the voltage, current, and frequency supplied to the machine. In this section, we develop the equations to characterize the machine operation. First, let us simplify the earlier equivalent circuit by combining the core loss and

(a)

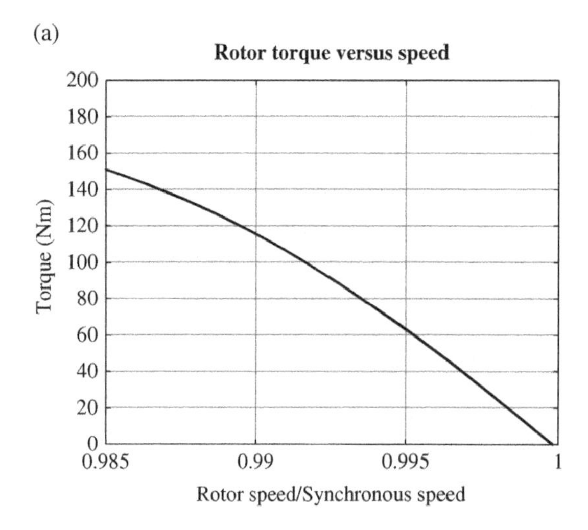

(b)

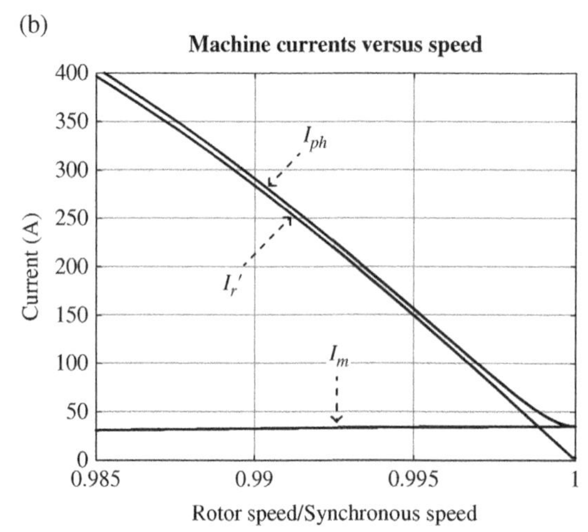

(c)

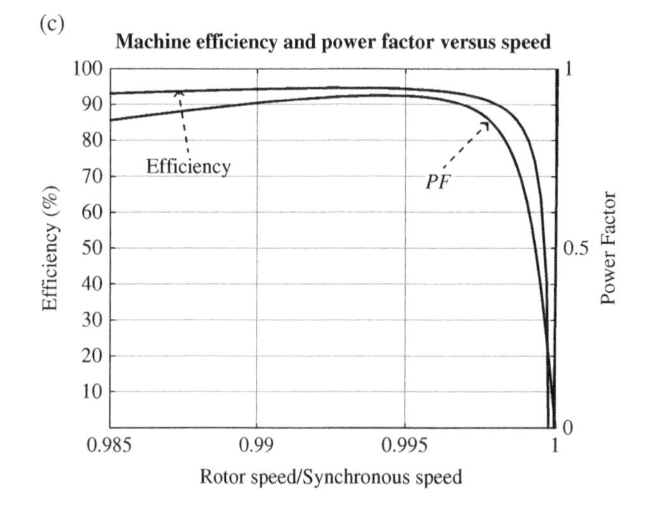

Figure 8.22 Characteristic curves for induction machine.

Figure 8.23 Phasor diagram of induction motor.

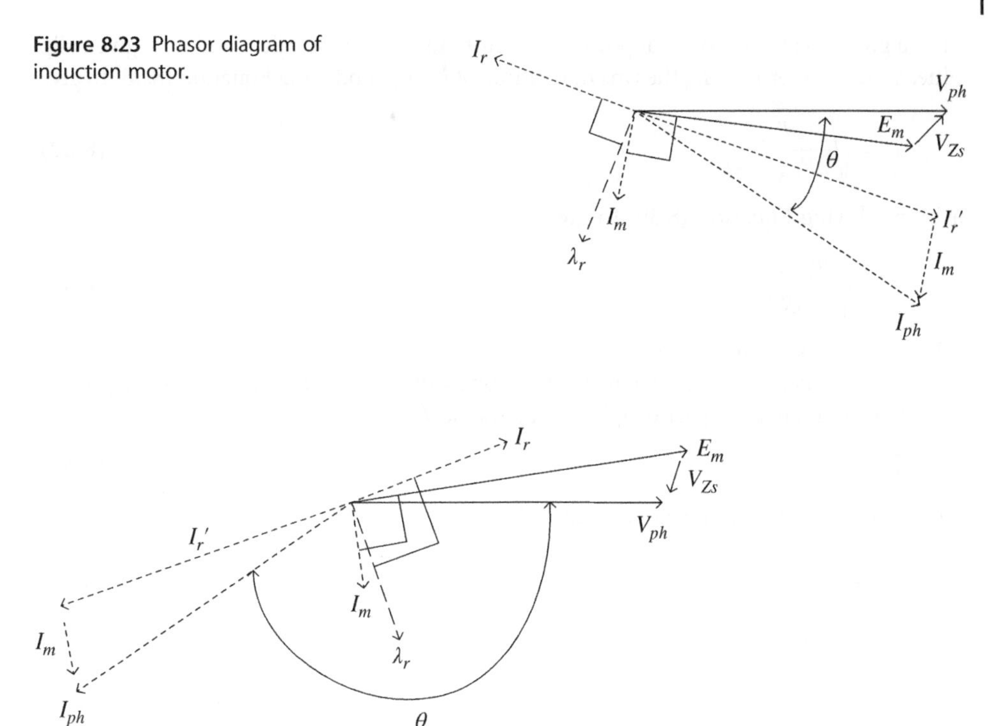

Figure 8.24 Phasor diagram of induction generator.

the friction and windage loss as P_{cfw}, as shown in Figure 8.25. Shifting the core loss eliminates the need for the discrete stator core loss resistance, and models the core loss as an electromechanical loss, making the circuit easier to analyze.

Thus, the electromechanical power consists of the useful rotor power and the core, friction, and windage power:

$$P_{em} = P_r + P_{cfw} \tag{8.51}$$

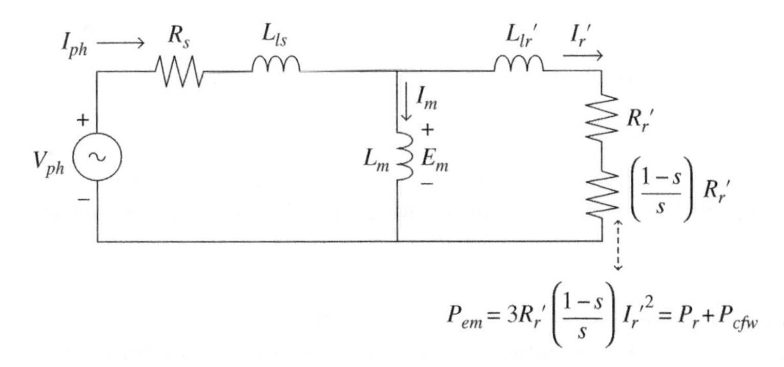

Figure 8.25 Simplified equivalent circuit model.

For a given electromechanical power, the rotor current can be calculated for a given slip value. We can determine I_r', the rms magnitude of $\bar{I}_r'$, by modifying Equation (8.38) to give

$$I_r' = \sqrt{\frac{sP_{em}}{3R_r'(1-s)}} \tag{8.52}$$

or by modifying Equation (8.39) to give

$$I_r' = \sqrt{\frac{T_{em}\omega_{slip}}{3R_r'}} \tag{8.53}$$

Both forms for I_r' are useful.

The analysis can be simplified at this time by assuming that $\bar{I}_r'$ is the reference phasor such that the imaginary part of $\bar{I}_r'$ is zero, and so $\bar{I}_r'$ is

$$\bar{I}_r' = I_r' + j0 \tag{8.54}$$

The reflected rotor impedance is given by

$$\bar{Z}_r' = \frac{R_r'}{s} + j\omega_e L_{lr}' \tag{8.55}$$

The back emf $\bar{E}_m$ is

$$\bar{E}_m = \bar{I}_r'\bar{Z}_r' \tag{8.56}$$

The magnetizing current is then given by

$$\bar{I}_m = \frac{\bar{E}_m}{j\omega_e L_m} \tag{8.57}$$

From Kirchhoff's current law, the phase current is the sum of the reflected rotor and magnetizing currents:

$$\bar{I}_{ph} = \bar{I}_r' + \bar{I}_m \tag{8.58}$$

The stator impedance is

$$\bar{Z}_s = R_s + j\omega_e L_{ls} \tag{8.59}$$

The voltage drop across the stator impedance is

$$\bar{V}_{Zs} = \bar{I}_{ph}\bar{Z}_s \tag{8.60}$$

Finally, the per-phase supply voltage is

$$\bar{V}_{ph} = \bar{E}_m + \bar{V}_{Zs} \tag{8.61}$$

Thus, the machine can be fully characterized using basic phasor analysis. Once the various currents are known, critical machine specifications such as the efficiency and power factor can be estimated.

8.3.4.1 Example: Motoring at Rated Speed using Induction Machine

In this example, we use a squirrel-cage induction machine as the traction motor for the BEV example introduced in Chapter 7. The typical automotive induction machine is four pole rather than eight pole in order to optimize the winding, and so the basic specifications for the machine are $P_{r(rated)}$ = 80 kW and $T_{r(rated)}$ = 140 Nm output at rated speed. (The mechanical gear ratio can be modified in order to achieve the same acceleration as the eight-pole machine.)

The machine has the following parameters: R_s = 12 mΩ, L_{ls} = 50 μH, L_m = 2.0 mH, L_{lr}' = 40 μH, R_r' = 4.8 mΩ, and T_{nl} = 2 Nm. The slip is 1.5% at the rated condition.

Calculate the input per-phase current, voltage, frequency, power factor, and machine efficiency. Ignore saturation and temperature rise.

Solution:

The rated rotor speed is

$$\omega_{r(rated)} = \frac{P_{r(rated)}}{T_{r(rated)}} = \frac{80000}{140}\ \text{rad/s} = 571.43\ \text{rad/s}$$

The rotor frequency is

$$f_{r(rated)} = \frac{\omega_{r(rated)}}{2\pi} = \frac{571.43}{2\pi}\ \text{Hz} = 90.95\ \text{Hz}$$

The synchronous frequency is

$$f_{syn} = \frac{f_{r(rated)}}{1-s} = \frac{90.95}{1-0.015}\ \text{Hz} = 92.33\ \text{Hz}$$

The slip frequency is

$$f_{slip} = f_{syn} - f_{r(rated)} = 92.33\ \text{Hz} - 90.95\ \text{Hz} = 1.38\text{Hz}$$

The electrical frequency is

$$f_e = \frac{p}{2} f_{syn} = \frac{4}{2} \times 92.33\ \text{Hz} = 184.66\ \text{Hz}$$

The no-load or core, friction, and windage loss is

$$P_{cfw} = T_{nl} \times \omega_r = 2 \times 571.43\ \text{W} = 1.143\ \text{kW}$$

When operating in motoring mode, the electromagnetic torque and power are the sums of the output and the no-load values:

$$T_{em} = T_r + T_{nl} = 140\ \text{Nm} + 2\ \text{Nm} = 142\ \text{Nm}$$

and

$$P_{em} = P_r + P_{cfw} = 80\ \text{kW} + 1.143\ \text{kW} = 81.143\ \text{kW}$$

Given the slip we can determine I_r' using Equation (8.52):

$$I_r' = \sqrt{\frac{sP_{em}}{3R_r'(1-s)}} = \sqrt{\frac{0.015 \times 81143}{3 \times 0.0048 \times (1-0.015)}}\ \text{A} = 292.94\ \text{A}$$

In polar form:

$$\bar{I}_r' = (292.94 + j0)\,\text{A} = 292.94\angle 0^\circ\ \text{A}$$

The reflected rotor impedance is

$$\bar{Z}_r' = \frac{R_r'}{s} + j\omega_e L_{lr}' = \left(\frac{0.0048}{0.015} + j2\pi \times 184.66 \times 40 \times 10^{-6}\right)\Omega = (0.32 + j0.0464)\Omega$$

The back emf is

$$\bar{E}_m = \bar{I}_r'\bar{Z}_r' = (292.94 + j0) \times (0.32 + j0.0464)\,\text{V} = (93.74 + j13.59)\,\text{V} = 94.72\angle + 8.25^\circ\ \text{V}$$

The magnetizing current is given by

$$\bar{I}_m = \frac{\bar{E}_m}{j\omega_e L_m} = \frac{(93.74 + j13.59)}{j2\pi \times 184.66 \times 2 \times 10^{-3}}\,\text{A} = (5.86 - j40.39)\,\text{A} = 40.81\angle - 81.74^\circ\ \text{A}$$

The rotor flux linkage is given by

$$\bar{\lambda}_r = L_m\bar{I}_m - L_{lr}'\bar{I}_r' = 2 \times 10^{-3} \times (5.86 - j40.39)\,\text{Wb-turns} - 40 \times 10^{-6} \times (292.94 + j0)\,\text{Wb-turns}$$
$$= (0 - j0.08078)\,\text{Wb-turns} = 0.08078\angle - 90^\circ\ \text{Wb-turns}$$

The phase current is the sum of the reflected rotor and magnetizing currents:

$$\bar{I}_{ph} = \bar{I}_r' + \bar{I}_m = (292.94 + j0 + 5.86 - j40.39)\,\text{A} = (298.8 - j40.39)\ \text{A}$$

or in polar form

$$\bar{I}_{ph} = 301.5\angle - 7.7^\circ\ \text{A}$$

The stator impedance is

$$\bar{Z}_s = R_s + j\omega_e L_{ls} = \left(0.012 + j2\pi \times 184.66 \times 50 \times 10^{-6}\right)\Omega = (0.012 + j0.058)\Omega$$

The voltage drop across the stator impedance is

$$\bar{V}_{Zs} = \bar{I}_{ph}\bar{Z}_s = (298.8 - j40.39)(0.012 + j0.058)\,\text{V} = (5.93 + j16.85)\ \text{V}$$

The per-phase supply voltage is

$$\bar{V}_{ph} = \bar{E}_m + \bar{V}_{Zs} = (93.74 + j13.59 + 5.93 + j16.85)\,\text{V} = (99.67 + j30.44)\ \text{V}$$

or in polar form

$$\bar{V}_{ph} = 104.21\angle + 16.98^\circ\ \text{V}$$

The machine power loss is the sum of the ohmic, core, and friction, and windage losses:

$$P_{m\,(loss)} = P_{cfw} + 3R_s I_{ph}^2 + 3R_r' I_r'^2 = 1143\ \text{W} + 3 \times 0.012 \times 301.5^2\ \text{W} + 3 \times 0.0048$$
$$\times\, 292.94^2\ \text{W} = 5651\ \text{W}$$

The machine input power is

$$P_{in} = P_r + P_{m(loss)} = 80\ \text{kW} + 5.651\ \text{kW} = 85.651\ \text{kW}$$

The machine efficiency is given by

$$\eta = \frac{P_r}{P_{in}} = \frac{80\,\text{kW}}{85.651\,\text{kW}} \times 100\% = 93.4\,\%$$

The machine power factor is given by

$$PF = \frac{P_{in}}{3V_{ph}I_{ph}} = \frac{85651}{3 \times 104.21 \times 301.5} = 0.9087$$

The various voltages and currents have been generated with the reflected rotor current as the reference phasor; it can be useful to transform the phasors such that the supply voltage is the reference phasor – in which case 16.98° is simply subtracted from all the phasors above.

8.3.4.2 Example: Motoring at Rated Speed using Induction Machine – Ignoring Leakage

Reasonable estimates of the machine currents, voltages, power loss, and efficiencies can be made by using a slightly simpler model. For this model, we assume that the leakage inductances are zero.

Recalculate the machine currents and voltages in the previous examples while neglecting the effects of leakage.

Solution:

We already know that

$$\bar{I}_r{}' = (292.94 + j0)\text{A}$$

Ignoring leakage, the reflected rotor impedance is

$$\bar{Z}_r{}' = \frac{R_r{}'}{s} = \frac{0.0048}{0.015}\Omega = 0.32\,\Omega$$

The back emf is

$$\bar{E}_m = \bar{I}_r{}'\bar{Z}_r{}' = (292.94 + j0) \times 0.32\text{V} = 93.74\text{V}$$

The magnetizing current is given by

$$\bar{I}_m = \frac{\bar{E}_m}{j\omega_e L_m} = \frac{93.74}{j2\pi \times 184.66 \times 2 \times 10^{-3}}\text{A} = -j40.39\text{A}$$

The phase current is the sum of the reflected rotor and magnetizing currents:

$$\bar{I}_{ph} = \bar{I}_r{}' + \bar{I}_m = (292.94 - j40.39)\text{A}$$

or in phasor form:

$$\bar{I}_{ph} = 295.71\angle -7.85^\circ\ \text{A}$$

Ignoring leakage, the stator impedance is

$$\bar{Z}_s = R_s = 0.012\,\Omega$$

The voltage drop across the stator impedance is

$$\bar{V}_{Zs} = \bar{I}_{ph}\bar{Z}_s = (292.94 - j40.39) \times 0.012\text{V} = (3.52 - j0.48)\text{ V}$$

The per-phase supply voltage is

$$\bar{V}_{ph} = \bar{E}_m + \bar{V}_{Zs} = (93.74 + 3.52 - j0.48)\text{V} = (97.26 - j0.48)\text{ V}$$

or in phasor form:

$$\bar{V}_{ph} = 97.26\angle -0.28^\circ \text{ V}$$

The machine power loss is the sum of the ohmic and the core, friction, and windage losses:

$$P_{m\,(loss)} = P_{cfw} + 3R_s I_{ph}^{2} + 3R_r{}' I_r{}'^{2} = 1143\text{ W} + 3 \times 0.012 \times 295.71^2\text{ W} + 3 \times 0.0048 \times 292.94^2\text{ W} = 5527\text{ W}$$

The machine input power is

$$P_{in} = P_r + P_{m(loss)} = 80\text{ kW} + 5.527\text{ kW} = 85.527\text{ kW}$$

The machine efficiency is given by

$$\eta = \frac{P_r}{P_{in}} = \frac{80\text{ kW}}{85.527\text{ kW}} \times 100\% = 93.54\,\%$$

The currents, power loss, and efficiency values calculated while ignoring the leakages compare well with the values calculated while considering the leakages.

8.3.4.3 Example: Generating at Rated Speed using Induction Machine

Using the same machine as in Section 8.3.4.1, determine the phase voltage and current, power factor, and efficiency when the machine is regenerating −80 kW and −140 Nm on the rotor at the rated speed with a slip of −1.5%. Ignore saturation and temperature effects.

Solution:

Again, the rated rotor speed is

$$\omega_{r(rated)} = \frac{P_{r(rated)}}{T_{r(rated)}} = \frac{-80{,}000}{-140}\text{ rad/s} = 571.43\text{ rad/s}$$

$$\text{or } f_{r(rated)} = \frac{\omega_{r(rated)}}{2\pi} = \frac{571.43}{2\pi}\text{ Hz} = 90.95\text{ Hz}$$

The synchronous frequency decreases due to the negative slip:

$$f_{syn} = \frac{f_{r(rated)}}{1 - s} = \frac{90.95}{1 + 0.015}\text{ Hz} = 89.6\text{ Hz}$$

The electrical frequency is

$$f_e = \frac{p}{2} f_{syn} = \frac{4}{2} \times 89.60\text{ Hz} = 179.2\text{ Hz}$$

The no-load or core, friction, and windage loss remains

$$P_{cfw} = T_{nl} \times \omega_r = 2 \times 571.43\,\text{W} = 1143\,\text{W}$$

When operating in generating mode, the electromagnetic torque is the sum of the useful output torque and the no-load torque:

$$T_{em} = T_r + T_{nl} = -140\,\text{Nm} + 2\,\text{Nm} = -138\,\text{Nm}$$

and

$$P_{em} = P_r + P_{cfw} = -80\,\text{kW} + 1.143\,\text{kW} = -78.857\,\text{kW}$$

Given the slip, we can determine I_r':

$$I_r' = \sqrt{\frac{sP_{em}}{3R_r'(1-s)}} = \sqrt{\frac{-0.015 \times -78857}{3 \times 0.0048 \times (1 + 0.015)}}\,\text{A} = 284.48\,\text{A}$$

In phasor form, the real component is negative and the imaginary component is zero:

$$\bar{I}_r' = (-284.48 + j0)\,\text{A}$$

The reflected rotor impedance is

$$\bar{Z}_r' = \frac{R_r'}{s} + j\omega_e L_{lr}' = \left(\frac{0.0048}{-0.015} + j2\pi \times 179.2 \times 40 \times 10^{-6}\right)\Omega = (-0.32 + j0.0450)\Omega$$

The back emf can be calculated from

$$\bar{E}_m = \bar{I}_r'\bar{Z}_r' = (-284.48 + j0) \times (-0.32 + j0.0450)\,\text{V} = (91.03 - j12.80)\,\text{V}$$

The magnetizing current is given by

$$\bar{I}_m = \frac{\bar{E}_m}{j\omega_e L_m} = \frac{(91.03 - j12.80)}{j2\pi \times 179.2 \times 2 \times 10^{-3}}\,\text{A} = (-5.68 - j40.42)\,\text{A}$$

The phase current is the sum of the reflected rotor and magnetizing currents:

$$\bar{I}_{ph} = \bar{I}_r' + \bar{I}_m = (-284.48 + j0 - 5.68 - j40.42)\,\text{A} = (-290.16 - j40.42)\,\text{A}$$

or in polar form:

$$\bar{I}_{ph} = 292.96\angle -172.07^\circ\,\text{A}$$

The stator impedance is

$$\bar{Z}_s = R_s + j\omega_e L_{ls} = (0.012 + j2\pi \times 179.2 \times 50 \times 10^{-6})\Omega = (0.012 + j0.0563)\Omega$$

The voltage drop across the stator impedance is

$$\bar{V}_{Zs} = \bar{I}_{ph}\bar{Z}_s = (-290.16 - j40.42)(0.012 + j0.0563)\,\text{V} = (-1.21 - j16.82)\,\text{V}$$

The per-phase supply voltage is

$$\bar{V}_{ph} = \bar{E}_m + \bar{V}_{Zs} = (91.03 - j12.80 - 1.21 - j16.82)\,\text{V} = (89.82 - j29.61)\,\text{V}$$

or in polar form:

$$\bar{V}_{ph} = 94.57\angle -18.25^\circ \text{ V}$$

The machine power loss is the sum of the ohmic and the core, friction, and windage losses:

$$P_{m\,(loss)} = P_{cfw} + 3R_s I_{ph}^2 + 3R_r' I_r'^2 = 1143\text{ W} + 3 \times 0.012 \times 292.96^2\text{ W} + 3 \times 0.0048 \times 284.48^2\text{ W} = 5398\text{ W}$$

The machine input power is

$$P_{in} = P_r + P_{m(loss)} = -80\text{ kW} + 5.398\text{ kW} = -74.602\text{ kW}$$

The machine efficiency is given by

$$\eta = \frac{P_{in}}{P_r} = \frac{-74.602\text{ kW}}{-80\text{ kW}} \times 100\% = 93.25\,\%$$

The machine power factor is given by

$$PF = \frac{P_{in}}{3V_{ph}I_{ph}} = \frac{-74602}{3 \times 94.57 \times 292.96} = -0.8975$$

8.4 Variable-Speed Operation of Induction Machine

The next stage is to be able to operate the machine across all four quadrants of torque and speed (see Chapter 6, Figure 6.11). The speed and torque range is wide; ranging from a high torque to hold the vehicle from rolling on a hill to climbing a hill at high speeds.

Full control over the torque/speed range can be obtained by controlling the input electrical frequency and the phase current. For a given input phase current, the slip magnitude determines how the phase current splits between the magnetizing and reflected-rotor components of current. This can be understood by referencing the simplified equivalent circuit of Figure 8.26, where the parasitic resistances and leakages are ignored and the machine is supplied by an inverter configured to act as a **current source**. The magnitude of the slip determines how the phase current splits between the parallel magnetizing and rotor branches.

If we review the earlier torque equations, as represented below, we can observe the basic relationships which must be controlled. By controlling the phase current and the slip, the magnetizing current can be indirectly controlled, which means that the rotor flux linkage is indirectly controlled. The torque equations can be rewritten as

$$T_{em} = 3\frac{p}{2}\lambda_r I_r' = 3\frac{p}{2}(L_m I_m - L_{lr}' I_r') I_r' \tag{8.62}$$

and

$$T_{em} = 3\frac{p^2}{4R_r'}\lambda_r^2 \omega_{slip} = 3\frac{p}{2R_r'}\lambda_r \omega_e s \tag{8.63}$$

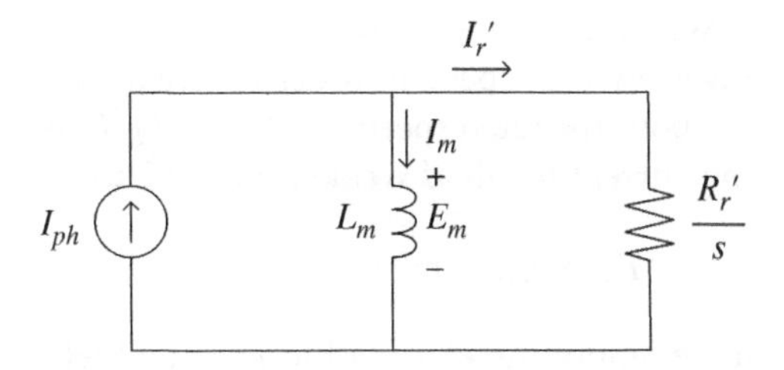

Figure 8.26 Simplified equivalent circuit model.

At or below the rated speed, maximum torque can be generated by maximizing the flux linkage. Above the rated speed, the flux linkage must be reduced as the machine is limited by the maximum flux density. Thus, at high speeds the machine operates in the field-weakening mode.

8.4.1 Constant Volts per hertz Operation

The rotor flux linkage can be held constant by controlling the magnetizing and rotor currents such that

$$\lambda_r = L_m I_m - L_{lr}' I_r' = \text{constant} \tag{8.64}$$

If the parasitic parameters are ignored, then we get the following relationship between the rotor flux linkage, per-phase supply voltage, and electrical frequency:

$$\lambda_r \approx L_m I_m \approx \frac{E_m}{\omega_e} \approx \frac{V_{ph}}{\omega_e} \propto \frac{V_{ph}}{f_e} \tag{8.65}$$

Thus, supplying a **constant volts per hertz** can provide a reasonable approximation to maintaining a constant rotor flux linkage.

8.4.1.1 Example: Maintaining a Constant Volts per Hertz

If a machine is controlled at the rated speed with a supply voltage of 100 V at 200 Hz, determine the approximate supply voltage and frequency in order to maintain a constant volts per hertz at half the rated speed.

Solution:

For this simple case, both the supply voltage and frequency are also reduced to half of their rated values, resulting in a supply of 50 V at 100 Hz to the machine.

8.4.2 Variable-Speed Operation

At or below rated speed $\lambda_r = L_m I_m - L_{lr}' I_r'$ can be held approximately constant, and so, from Equations (8.62) and (8.63):

$$T_{em} \propto I_r' \text{ and } T_{em} \propto \omega_{slip} \tag{8.66}$$

It is important to note here that operating the machine at high flux levels results in a fast dynamic response to increased torque commands but also a high core loss. If machine efficiency is the principal consideration, rather than dynamic response, as is often the case for EVs, it may be necessary to reduce the flux density in order to optimize machine efficiency at lower torque levels.

When operating above the rated speed, the magnetizing current must be reduced inversely with speed in order to weaken the magnetic field.

Above the rated speed $\lambda_r = L_m I_m - L_{lr}' I_r'$ must be reduced by decreasing the magnetizing current for **field weakening**, and so

$$T_{em} \propto I_m I_r' \text{ and } T_{em} \propto I_m^2 \omega_{slip} \tag{8.67}$$

if we ignore the effects of leakage and let $\lambda_r \approx L_m I_m$. Thus, in order to maintain rated power above the rated speed, the torque and magnetizing current drop inversely with speed while the reflected rotor current is maintained approximately constant. Similarly, for a constant power above the rated speed, the slip frequency increases with increasing frequency while the slip remains constant.

8.4.2.1 Example: Field-Weakened Motoring at Twice the Rated Speed using Induction Machine

For the earlier example in Section 8.3.4.1, calculate the input per-phase current, voltage, frequency, power factor, and machine efficiency when operating at full power at twice the rated speed. Assume a slip of 1.5%.

Solution:

When operating at twice the rated speed, the rotor torque drops in half from 140 Nm to 70 Nm in order to maintain a constant power.

The rotor speed is

$$\omega_r = \frac{P_{r(rated)}}{T_r} = \frac{80{,}000}{70} \text{ rad/s} = 1142.9 \text{ rad/s}$$

The rotor frequency is

$$f_r = \frac{\omega_r}{2\pi} = \frac{1142.9}{2\pi} \text{ Hz} = 181.89 \text{ Hz}$$

The synchronous frequency is

$$f_{syn} = \frac{f_r}{1-s} = \frac{181.89}{1-0.015} \text{ Hz} = 184.66 \text{ Hz}$$

The slip frequency is

$$f_{slip} = f_{syn} - f_r = 184.66 \text{ Hz} - 181.89 \text{ Hz} = 2.77\text{Hz}$$

The electrical frequency is

$$f_e = \frac{p}{2} f_{syn} = \frac{4}{2} \times 184.66 \text{ Hz} = 369.32 \text{ Hz}$$

The combined core, friction, and windage loss is

$$P_{cfw} = T_{nl} \times \omega_r = 2 \times 1142.9 \text{ W} = 2286 \text{ W}$$

When operating in motoring mode, the electromagnetic torque and power are the sums of the output and the no-load values:

$$T_{em} = T_r + T_{nl} = 70\,\text{Nm} + 2\,\text{Nm} = 72\,\text{Nm}$$

and

$$P_{em} = P_r + P_{cfw} = 80\,\text{kW} + 2.286\,\text{kW} = 82.286\,\text{kW}$$

Given the slip, we can determine I_r' using Equation (8.53):

$$I_r' = \sqrt{\frac{T_{em}\omega_{slip}}{3R_r'}} = \sqrt{\frac{72 \times 2\pi \times 2.77}{3 \times 0.0048}}\,\text{A} = 295.0\,\text{A}$$

In polar form:

$$\bar{I}_r' = (295 + j0)\,\text{A} = 295\angle 0^\circ\,\text{A}$$

The reflected rotor impedance is

$$\bar{Z}_r' = \frac{R_r'}{s} + j\omega_e L_{lr}' = \left(\frac{0.0048}{0.015} + j2\pi \times 369.32 \times 40 \times 10^{-6}\right)\Omega = (0.32 + j0.0928)\Omega$$

The back emf is

$$\bar{E}_m = \bar{I}_r'\bar{Z}_r' = (295 + j0) \times (0.32 + j0.0928)\,\text{V} = (94.4 + j27.38)\,\text{V} = 98.29\angle + 16.17^\circ\,\text{V}$$

The magnetizing current is given by

$$\bar{I}_m = \frac{\bar{E}_m}{j\omega_e L_m} = \frac{(94.4 + j27.38)}{j \times 2\pi \times 369.32 \times 0.002} = (5.90 - j20.34)\,\text{A} = 21.18\angle - 73.82^\circ\,\text{A}$$

The phase current is the sum of the reflected rotor and magnetizing currents:

$$\bar{I}_{ph} = \bar{I}_r' + \bar{I}_m = (295 + j0 + 5.90 - j20.34)\,\text{A} = (300.9 - j20.34)\,\text{A} = 301.59\angle - 3.87^\circ\,\text{A}$$

The stator impedance is

$$\bar{Z}_s = R_s + j\omega_e L_{ls} = (0.012 + j0.1160)\Omega$$

The voltage drop across the stator impedance is

$$\bar{V}_{Zs} = \bar{I}_{ph}\bar{Z}_s = (300.9 - j20.34) \times (0.012 + j0.1160)\,\text{V} = (5.97 + j34.66)\,\text{V}$$

The per-phase supply voltage is

$$\bar{V}_{ph} = \bar{E}_m + \bar{V}_{Zs} = (94.4 + j27.38)\,\text{V} + (5.97 + j34.66)\,\text{V} = (100.37 + 62.04)\,\text{V}$$
$$= 118.0\angle + 31.72^\circ\,\text{V}$$

The machine power loss is the sum of the ohmic, core, and friction and windage losses:

$$P_{m\,(loss)} = P_{cfw} + 3R_s I_{ph}^2 + 3R_r' I_r'^2 = 2286\,\text{W} + 3 \times 0.012 \times 301.59^2\,\text{W} + 3 \times 0.0048$$
$$\times 295^2\,\text{W} = 6814\,\text{W}$$

The machine input power is

$$P_{in} = P_r + P_{m(loss)} = 80\,\text{kW} + 6.814\,\text{kW} = 86.814\,\text{W}$$

The machine efficiency is given by

$$\eta = \frac{P_r}{P_{in}} = \frac{80\,\text{kW}}{86.814\,\text{kW}} \times 100\% = 92.15\,\%$$

The machine power factor is given by

$$PF = \frac{P_{in}}{3V_{ph}I_{ph}} = \frac{86814}{3 \times 118 \times 301.59} = 0.8131$$

The power factor has degraded compared to the rated condition due to the increased voltage drops across the leakage reactances at the higher frequency.

8.4.2.2 Example: Stall/Start-Up using Induction Machine

The induction motor can be easily controlled to develop torque at zero speed, as in the case of stall or start-up. The same analytical approach is used as for the earlier example. Equation (8.53) is used to calculate the rotor current on the basis of the desired torque. For the earlier example, calculate the input per-phase current, voltage, frequency, power factor, and machine efficiency when operating at the rated torque at zero speed. Assume the slip frequency is equal to that at the rated condition, 1.385 Hz. Ignore the core, friction, and windage loss.

Solution:

The rotor speed is

$$\omega_r = 0\,\text{rad/s}$$

The rotor frequency is

$$f_r = 0\,\text{Hz}$$

The synchronous frequency is

$$f_{syn} = f_r + f_{slip} = 0\,\text{Hz} + 1.385\,\text{Hz} = 1.385\,\text{Hz}$$

The electrical frequency is

$$f_e = \frac{p}{2} f_{syn} = \frac{4}{2} \times 1.385\,\text{Hz} = 2.77\,\text{Hz}$$

Given the slip, we can determine I_r'. In this case, we use Equation (8.53):

$$I_r' = \sqrt{\frac{T_{em}\omega_{slip}}{3R_r'}} = \sqrt{\frac{140 \times 2\pi \times 1.385}{3 \times 0.0048}}\,\text{A} = 290.9\,\text{A}$$

In polar form:

$$\bar{I}_r' = (290.9 + j0)\,\text{A} = 290.9\angle 0^\circ\,\text{A}$$

The reflected rotor impedance is

$$\bar{Z}_r' = \frac{R_r'}{s} + j\omega_e L_{lr}' = \left(\frac{0.0048}{1} + j2\pi \times 2.77 \times 40 \times 10^{-6}\right)\Omega = (0.0048 + j0.0007)\Omega$$

The back emf is

$$\bar{E}_m = \bar{I}_r'\bar{Z}_r' = (290.9 + j0) \times (0.0048 + j0.0007)\text{V} = (1.4 + j0.2)\text{V} = 1.4\angle + 8.25^\circ \text{ V}$$

The magnetizing current is

$$\bar{I}_m = \frac{\bar{E}_m}{j\omega_e L_m} = (5.82 - j40.11)\text{A} = 40.5\angle - 81.75^\circ \text{ A}$$

The rotor flux linkage is given by

$$\bar{\lambda}_r = L_m\bar{I}_m - L_{lr}'\bar{I}_r' = (0 - j0.0802)\text{Wb} - \text{turns} = 0.0802\angle - 90^\circ \text{ Wb} - \text{turns}$$

The phase current is the sum of the reflected rotor and magnetizing currents:

$$\bar{I}_{ph} = \bar{I}_r' + \bar{I}_m = (290.9 + 5.82 - j40.11)\text{ A} = (296.72 - j40.11)\text{ A} = 299.4\angle - 7.7^\circ \text{ A}$$

The stator impedance is

$$\bar{Z}_s = R_s + j\omega_e L_{ls} = \left(0.012 + j2\pi \times 2.77 \times 50 \times 10^{-6}\right)\Omega = (0.012 + j0.0009)\Omega$$

The voltage drop across the stator impedance is

$$\bar{V}_{Zs} = \bar{I}_{ph}\bar{Z}_s = (296.72 - j40.11) \times (0.012 + j0.0009)\text{V} = (3.6 - j0.22)\text{ V}$$

The per-phase supply voltage is

$$\bar{V}_{ph} = \bar{E}_m + \bar{V}_{Zs} = (1.4 + j0.2 + 3.6 - j0.22) = (5 - j0.02)\text{ V} = 5\angle - 0.23^\circ \text{ V}$$

The machine power loss is the sum of the ohmic, core, friction, and windage losses:

$$P_{m\,(loss)} = P_{cfw} + 3R_s I_{ph}^2 + 3R_r' I_r'^2 = 0\text{ W} + 3227\text{ W} + 1218\text{ W} = 4445\text{ W}$$

The machine efficiency is given by

$$\eta = \frac{P_r}{P_{in}} = \frac{0\text{ W}}{4445\text{ W}} \times 100\% = 0\%$$

The machine power factor is given by

$$PF = \frac{P_{in}}{3V_{ph}I_{ph}} = \frac{4445}{3 \times 5 \times 299.4} = 0.99$$

In this start-up or stall mode, the machine currents are limited to close to the rated values due to thermal limitations. The machine can develop full torque while the current-source inverter supplies rated current at a low voltage to the machine.

8.4.2.3 Effects of Rotor Heating

The value of the rotor resistance is critical when determining the optimum slip. However, the rotor can heat up due to the rotor power losses, and the rotor is not easily cooled. The generic formula for the relationship between conductor resistance R and temperature T is

$$R = R_0[1 + \alpha(T - T_0)] \tag{8.68}$$

where the temperature coefficient of resistance α is 0.004/°C for copper, and R_0 is the conductor resistance at the ambient temperature T_0, typically 25°C. Thus, the rotor resistance increases by approximately 50% with a rise in rotor temperature from 25°C to 150°C. Thus, some estimation of the rotor temperature is usually necessary in order to optimize the slip. Also reference Chapter 7, Section 7.11.

8.5 Machine Test

The induction machine must be experimentally characterized to determine its parameters. There are various industry test standards, for example, IEEE Std 112, which have been produced by technical standards organizations such as the IEEE, ANSI, IEC, and NEMA. In this section, we apply a simplified approach to test the machines.

There are three basic tests: the dc resistance test, the locked or blocked-rotor test, and the no-load, or free-spinning, test.

8.5.1 DC Resistance Test

In this test, the resistance of the phase winding is measured in order to determine the stator winding resistance R_s.

For a star-connected machine, as shown in Figure 8.27, the measured value of R_{dc} is twice the per-phase resistance, and R_s can be determined from

$$R_s = \frac{R_{dc}}{2} \tag{8.69}$$

8.5.2 Locked-Rotor Test

This test is similar to the short-circuit test of a transformer. In this test, the value of the slip is unity ($s = 1$) as the rotor is locked or blocked from moving. In this case, the rotor branch impedance is very low and much lower than the impedance of the magnetizing branch. Hence, the magnetizing branch can be ignored in this test. On the basis of this assumption of ignoring the magnetizing impedance, the equivalent circuit of Figure 8.28(a) can be simplified to the locked-rotor equivalent circuit of Figure 8.28(b).

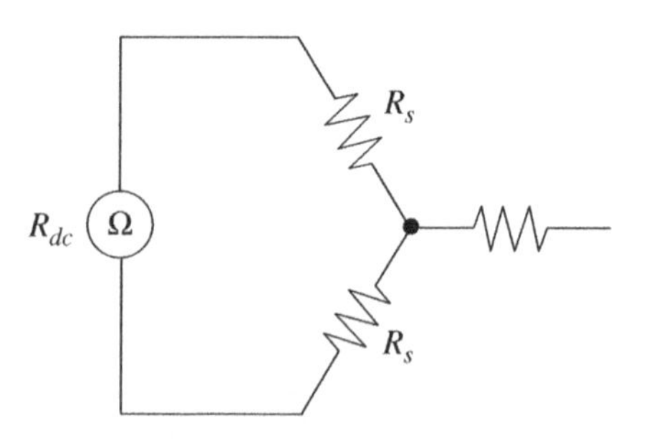

Figure 8.27 DC resistance test.

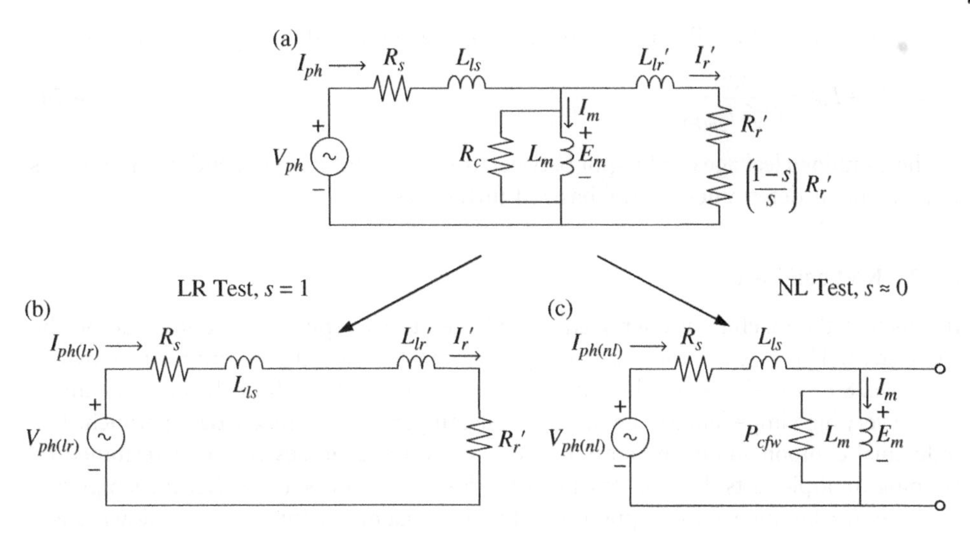

Figure 8.28 Per-phase equivalent circuits for (a) regular operation, (b) locked-rotor test, and (c) no-load test.

Note that the electrical frequency is important for this test. The rotor bars are relatively thick, and the **skin depth** of copper can significantly impact the measurement if the electrical frequency is relatively high (see Chapter 16, Section 16.3.8). Conducting the test at a high frequency can result in significant skin-depth effects for the rotor bars. A lower-frequency test is appropriate for a machine buffered by an inverter as the input electrical frequency is controlled by the inverter, and the rotor frequency is very low under regular operation.

Three measurements are required for this test. Any three of the following measurements are required: per-phase voltage $V_{ph(lr)}$, per-phase current $I_{ph(lr)}$, three-phase apparent power S_{lr}, three-phase real power P_{lr}, three-phase reactive power Q_{lr}, or power factor PF_{lr}.

The input power is

$$P_{lr} = 3(R_s + R_r')I_{ph(lr)}^2 \tag{8.70}$$

Since R_s is known from the dc resistance test, the reflected rotor resistance can be determined from

$$R_r' = \frac{P_{lr}}{3\,I_{ph(lr)}^2} - R_s \tag{8.71}$$

The three-phase reactive power is given by

$$Q_{lr} = \sqrt{\left(3V_{ph(lr)}I_{ph(lr)}\right)^2 - P_{lr}^{\;2}} \tag{8.72}$$

Ignoring the magnetizing branch:

$$Q_{lr} \approx 3 \times 2\pi f_e (L_{ls} + L_{lr}')I_{ph(lr)}^2 \tag{8.73}$$

We thus obtain the following expression for the combined leakage inductances:

$$L_{ls} + L_{lr}{}' = \frac{Q_{lr}}{6\pi f_e I_{ph(lr)}^2} \tag{8.74}$$

The combined leakage can be split between the stator and rotor depending on the class of machine or determined on the basis of further testing.

8.5.3 No-Load Test

In this test, the machine is energized and allowed to free spin without any load on the rotor shaft. This test is similar to the open-circuit test of a transformer. As shown in Figure 8.28(c), the slip is very low under no load, and the rotor branch can be assumed to be a very high impedance or an open circuit compared to the magnetizing branch. The individual components of the core, friction, and windage losses can be determined by running multiple tests. For this textbook the losses are lumped together for simplicity.

Three measurements are required for this test. Again, any three of the following will work: per-phase voltage $V_{ph(nl)}$, per-phase current $I_{ph(nl)}$, three-phase apparent power S_{nl}, three-phase real power P_{nl}, three-phase reactive power Q_{nl}, or power factor PF_{nl}.

The core, friction, and windage power P_{cfw} is determined from

$$P_{cfw} = P_{nl} - 3R_s I_{ph(nl)}^2 \tag{8.75}$$

The three-phase reactive power is given by

$$Q_{nl} = \sqrt{\left(3V_{ph(nl)} I_{ph(nl)}\right)^2 - P_{nl}{}^2} \tag{8.76}$$

If the stator leakage inductance is known from the locked-rotor test, the reactive power of the magnetizing inductance $Q_{Lm(nl)}$ is given by

$$Q_{Lm(nl)} \approx Q_{nl} - 3 \times 2\pi f_e L_{ls} I_{ph(nl)}^2 \tag{8.77}$$

As the power factor is easily measured or calculated, the per-phase current can be written in phasor form

$$\bar{I}_{ph(nl)} = I_{ph(nl)}(\cos\phi_{nl} - j\sin\phi_{nl}) \tag{8.78}$$

where

$$\phi_{nl} = \cos^{-1} PF_{nl} \tag{8.79}$$

It is assumed here that the input voltage is the reference phasor. The magnetizing inductance back-emf phasor $\bar{E}_{ph(nl)}$ is then given by

$$\begin{aligned} \bar{E}_{ph(nl)} &= E_{ph(nl)} \angle \phi_{Eph(nl)} = \bar{V}_{ph} - \bar{Z}_s \bar{I}_{ph(nl)} \\ &= V_{ph(nl)} - (R_s + j2\pi f_e L_{ls}) \times I_{ph(nl)}(\cos\phi_{nl} - j\sin\phi_{nl}) \end{aligned} \tag{8.80}$$

The per-phase magnetizing inductance is then given by

$$L_m = \frac{3\,E_{ph(nl)}^2}{2\pi f_e Q_{Lm(nl)}} \tag{8.81}$$

where $E_{ph(nl)}$ is the magnitude of the phasor.

8.5.3.1 Example of Machine Characterization

The symmetrical, four-pole, three-phase, star-connected 80 kW induction motor used in the examples in this chapter is now characterized as follows. The dc phase-to-phase resistance is measured to be 24 mΩ. A locked-rotor test with an applied per-phase voltage of 5.2 V, 10 Hz, results in a phase current of 293.1 A, and a three-phase power of 4.33 kW. A no-load test with an applied per-phase voltage of 102.6 V, 200 Hz, results in a phase current of 40 A, and a three-phase power of 1.423 kW.

Estimate the per-phase equivalent circuit parameters: R_s, L_{ls}, L_m, L_{lr}', R_r', and P_{cfw}.

Assume that $\frac{L_{ls}}{L'_{lr}} = \frac{5}{4}$ for this machine.

Solution:

Dc resistance test:

The stator resistance is given by

$$R_s = \frac{R_{dc}}{2} = \frac{24}{2}\,\text{m}\Omega = 12\,\text{m}\Omega$$

Locked-rotor test:

The rotor resistance is given by

$$R_r' = \frac{P_{lr}}{3\,I_{ph(lr)}^2} - R_s = \frac{4330}{3 \times 293.1^2}\,\text{m}\Omega - 12\,\text{m}\Omega = 4.8\,\text{m}\Omega$$

The apparent power is given by

$$S_{lr} = 3V_{ph(lr)}I_{ph(lr)} = 3 \times 5.2 \times 293.1\,\text{VA} = 4572\,\text{VA}$$

The input reactive power is given by

$$Q_{lr} = \sqrt{S_{lr}{}^2 - P_{lr}{}^2} = \sqrt{4572^2 - 4330^2}\,\text{VAr} = 1468\,\text{VAr}$$

The combined leakage inductance is

$$L_{ls} + L_{lr}' = \frac{Q_{lr}}{3 \times 2\pi f_e \times I_{ph(lr)}^2} = \frac{1468}{6\pi \times 10 \times 293.1^2}\,\text{H} = 90.7\,\mu\text{H}$$

In this problem, it is assumed that the leakage splits as $L'_{lr} = \frac{5}{4}L_{ls}$; thus

$$L_{ls} = \frac{5}{9} \times (L_{ls} + L_{lr}') = \frac{5}{9} \times 90.7\,\mu\text{H} = 50\,\mu\text{H}$$

and

$$L_{lr}' = \frac{4}{9} \times (L_{ls} + L_{lr}') = \frac{4}{9} \times 90.7\,\mu\text{H} = 40\,\mu\text{H}$$

No-load test:

The apparent power is given by

$$S_{nl} = 3V_{ph(nl)}I_{ph(nl)} = 3 \times 102.6 \times 40\,\text{VA} = 12312\,\text{VA}$$

The input power factor is given by

$$PF_{nl} = \frac{P_{nl}}{S_{nl}} = \frac{1423}{12312} = 0.1156$$

The input phase angle is

$$\phi_{nl} = \cos^{-1} PF_{nl} = \cos^{-1} 0.1156 = 83.36\,^\circ$$

The input reactive power is given by

$$Q_{nl} = \sqrt{S_{nl}{}^2 - P_{nl}{}^2} = \sqrt{12312^2 - 1423^2}\,\text{VAr} = 12229\,\text{VAr}$$

The lumped core, friction, and windage loss is

$$P_{cfw} = P_{nl} - 3R_s I^2_{ph(nl)} = 1423\,\text{W} - 3 \times 0.012 \times 40^2\,\text{W} = 1365\,\text{W}$$

The reactive power supplied to the magnetizing inductance is

$$Q_{Lm(nl)} \approx Q_{nl} - 3 \times 2\pi f_e L_{ls} I_{ph(nl)}{}^2 = 12229\,\text{VAr} - 6\pi \times 200 \times 50 \times 10^{-6} \times 40^2\,\text{VAr} = 11927\,\text{VAr}$$

Taking the input voltage as the reference, the per-phase input current is

$$\bar{I}_{ph(nl)} = I_{ph(nl)}(\cos\phi_{nl} - j\sin\phi_{nl}) = 40(\cos 83.36^\circ - j\sin 83.36^\circ)\,\text{A} = (4.62 - j39.73)\,\text{A}$$

The per-phase back emf is

$$\begin{aligned} \bar{E}_{ph(nl)} &= V_{ph(nl)} - (R_s + j2\pi f_e L_{ls}) \times I_{ph(nl)}(\cos\phi_{nl} - j\sin\phi_{nl}) \\ &= 102.6\,\text{V} - \left(0.012 + j2\pi \times 200 \times 50 \times 10^{-6}\right) \times (4.62 - j39.73)\text{V} \\ &= (100 + j0.2)\text{V} = 100\angle 0.11^\circ\,\text{V} \end{aligned}$$

The magnetizing inductance is given by

$$L_m = 3\,\frac{E^2_{ph(nl)}}{2\pi f_e Q_{Lm(nl)}} = \frac{3 \times 100^2}{2\pi \times 200 \times 11927}\text{H} = 2\,\text{mH}$$

References

1 N. Cawthorne, *Tesla vs. Edison: The Life-Long Feud That Electrified the World*, Chartwell Books, 2016.

Further Reading

1 N. Mohan, *Electric Machines and Drives: A First Course*, John Wiley & Sons, Inc., 2012, ISBN 978-1-118-07481-7.

2 A. E. Fitzgerald, C. Kingsley, and S. D. Umans, *Electric Machinery*, McGraw-Hill, 6th edition, 2002, ISBN 9780073660097.

3 P. Krause, O. Wasynczuk, S. D. Sudhoff, and S. Pekarek, *Analysis of Electrical Machinery and Drive Systems*, Wiley-IEEE Press, 3rd edition, 2013, ISBN: 978-1-118-02429-4.
4 T. A. Lipo, *Introduction to AC Machine Design*, University of Wisconsin, 3rd edition, 2007, ISBN 0-9745470-2-6.

Problems

8.1 A four-pole machine has the following specifications: $P_{r(rated)} = 100$ kW and $T_{r(rated)} = 150$ Nm.

Use the following parameters for the machine: $R_s = 8.8$ mΩ, $L_{ls} = 47$ μH, $L_m = 2.1$ mH, $L_{lr}{}' = 40$ μH, $R_r{}' = 3.4$ mΩ, and $T_{nl} = 2$ Nm.

For a slip of 1.4% at the rated condition, calculate the electrical input frequency, per-phase reflected rotor and input currents, input voltage, power factor, and machine efficiency.

[Ans. 215.22 Hz, 375.6 A, 384.11 A, 106.34 V, 0.87, 93.75%]

8.2 Using the parameters of the previous problem, recalculate the electrical input frequency, per-phase reflected rotor and input currents, input voltage, power factor, and machine efficiency when the machine is developing an output torque of 75 Nm at a slip of 0.7% at the rated speed.

[Ans. 213.7 Hz, 188.35 A, 194.66 A, 97.7 V, 0.923, 94.88%]

8.3 Recalculate the input current, voltage, and efficiency for Problem 8.1 while neglecting the effects of leakage.

[Ans. 377 A, 94.5 V, 93.9%]

8.4 The machine of Problem 8.1 is regenerating −100 kW on the rotor at the rated speed. For a slip of −1.4%, calculate the electrical input frequency, per-phase reflected rotor and input currents, input voltage, power factor, and machine efficiency.

[Ans. 209.28 Hz, 365.5 A, 373.8 A, 97.25 V, −0.858, 93.62%,]

8.5 Recalculate the input current, voltage, and efficiency in the previous generating problem while neglecting the effects of leakage.

[Ans. 366.86 A, 85.5 V, 93.75%]

8.6 Using the parameters of Problem 8.1, recalculate the electrical input frequency, per-phase reflected rotor and input currents, input voltage, power factor, and machine efficiency for the rated power at twice the base speed for a slip of 1.4%.

[Ans. 430.44 Hz, 378.0 A, 385.6 A, 132.34 V, 92.55%, 0.706]

8.7 Using the parameters of Problem 8.1, recalculate the input per-phase current, voltage, frequency, power factor, and machine power loss when operating at the rated torque at zero speed. Assume the slip frequency is equal to that at the rated condition, 1.51 Hz. Ignore the core, friction, and windage loss.

[Ans. 3.02 Hz, 373.58 A, 381.9 A, 4.66 V, 5275 W, 0.9877]

8.8 A symmetrical, four-pole, three-phase, star-connected 100 kW induction motor is characterized as follows. The dc phase-to-phase resistance is measured to be 17.6 mΩ. A locked-rotor test with an applied per-phase voltage of 5.21 V, 10 Hz, results in a per-phase current of 389.5 A and a three-phase power of 5552 W. A no-load test with an applied per-phase voltage of 97.2 V, 220 Hz, results in a per-phase current of 33.2 A and a three-phase power of 1686 W.

Estimate the per-phase equivalent circuit parameters: R_s, L_{ls}, L_m, L_{lr}', R_r', and P_{cfw}. Assume that $\frac{L_{ls}}{L'_{lr}} = \frac{4.7}{4}$ for this machine.

[Ans. $R_s = 8.8\,\text{m}\Omega, L_{ls} = 47\,\mu\text{H}, L_m = 2.1\,\text{mH}, L_{lr}' = 40\,\mu\text{H}, R_r' = 3.4\,\text{m}\Omega, P_{cfw} = 1657\,\text{W}$]

8.9 A symmetrical, four-pole, three-phase, star-connected 50 kW induction motor is characterized as follows. The dc phase-to-phase resistance is measured to be 38 mΩ. A locked-rotor test with an applied per-phase voltage of 5.57 V, 10 Hz, results in a per-phase current of 164 A and a three-phase power of 2584 W. A no-load test with an applied per-phase voltage of 99.2 V, 200 Hz, results in a per-phase current of 33.2 A and a three-phase power of 1459 W.

Estimate the per-phase equivalent circuit parameters: R_s, L_{ls}, L_m, L_{lr}', R_r', and P_{cfw}. Assume that $\frac{L_{ls}}{L'_{lr}} = \frac{5}{4}$ for this machine.

[Ans. $R_s = 19\,\text{m}\Omega, L_{ls} = 100\,\mu\text{H}, L_m = 2.3\,\text{mH}, L_{lr}' = 80\,\mu\text{H}, R_r' = 13\,\text{m}\Omega, P_{cfw} = 1396\,\text{W}$]

Sample MATLAB Code

```
% Induction Machine Torque-speed characterization (John Hayes)
%Initialisation section
clear variables;
close all;
clc;
format short g;
format compact;
% Vehicle Parameters needed
Prmax        = 80000;    % Maximum output rotor power
Trmax        = 140;      % Maximum motor positive torque range [Nm]
p            = 4;        % Number of poles

wrbase       = Prmax/Trmax;          % assume base angular speed
Nrbase       = wrbase*60/(2*pi);  % base speed in [rpm] Used for field
                                       weakening
frbase       = wrbase/(2*pi) ;     % rotor frequency in Hz
```

```
Vb          = 300;           % Battery voltage

Rs          = 12e-3          % Stator resistance
Rr          = 4.8e-3         % reflected rotor resistance
Lls         = 50e-6          % stator leakage
Llr         = 40e-6          % rotor leakage
Lm          = 2e-3           % magnetizing inductance
Tnl         = 2              % no-load torque

N           = 10000;                   %number of steps
s           = linspace(-1,+1,N); %array of values for wr
fe          = 220;           % electrical frequecy
fsyn        = fe*2/p         % synchronous frequency
fr          = fsyn*(1-s);    % rotor frequency
wr          = 2*pi*fr;       % rotor angular frequency

Vph         = Vb/(2*sqrt(2)) % max phase voltage for sinusoidal
              modulation - see Chapter 13
%Vph        = Vb/sqrt(6) % max phase voltage for SVM - see Chapter 13

Zr          = Rr./s+1i*2*pi*fe*Llr;
ZLm         = 1i*2*pi*fe*Lm;
Zpar        = Zr.*ZLm./(Zr+ZLm);
Zph         = Rs+1i*2*pi*fe*Lls+Zpar;
Iph         = Vph./Zph;
Iphmag      = sqrt(real(Iph).^2+imag(Iph).^2);
Em          = Vph-Iph.*(Rs+1i*2*pi*fe*Lls);
Ir          = Em./Zr;
Irmag       = sqrt(real(Ir).^2+imag(Ir).^2);
Im          = Em./(1i*2*pi*fe*Lm);
Immag       = sqrt(real(Im).^2+imag(Im).^2);
Lambdar     = Lm.*Im-Llr.*Ir;
Lambdarmag  = sqrt(real(Lambdar).^2+imag(Lambdar).^2);
Pem         = (3*Rr*Irmag.^2).*(1-s)./s;
Tem         = Pem./wr;
Tr          = Tem-Tnl;
Temd        = 3*p/2.*Lambdarmag.*Irmag;
Pr          = Tr.*wr;
Ploss       = 3*Rs*Iphmag.^2+3*Rr*Irmag.^2+Tnl.*wr;
Pin         = Pr+Ploss;
Eff         = 100.*Pr./Pin;
PF          = Pin./(3.*Vph.*Iphmag);

  figure(1);
    plot(1-s,Tr,'k'); hold on;
    %plot(1-s,Temd,'r'); hold on;
    title('Rotor torque vs. slip');
```

```
    axis([0,+2,-250,+250])
    xlabel('Rotor speed/Synchronous speed');
    ylabel('Torque (Nm)');
    %legend('Motor','Regen','Location','northeast');
    grid on;
 figure(2);
    plot(1-s,Tr,'k'); hold on;
    %plot(1-s,Temd,'r'); hold on;
    title('Rotor torque vs. slip');
    axis([0.95,1.05,-250,+250])
    xlabel('Rotor speed/Synchronous speed');
    ylabel('Torque (Nm)');
    grid on;
figure(3);
    plot(1-s,Tr,'k'); hold on;
    title('Rotor torque vs. speed');
    axis([0.985,1,0,+200])
    xlabel('Rotor speed/Synchronous speed');
    ylabel('Torque (Nm)');
    grid on;
figure(4);
    plot(1-s,Iphmag,'k'); hold on;
    plot(1-s,Irmag,'k'); hold on;
    plot(1-s,Immag,'k'); hold on;
    title('Machine currents vs. slip');
    axis([0.985,1,0,+400])
    xlabel('Rotor speed/Synchronous speed');
    ylabel('Current (A)');
%   legend ('Phase','Rotor','Magnetizing','Location','
     northeast');
    grid on;
figure(5);
[ax,hline1,hline2] = plotyy(1-s,Eff,1-s,PF);
title('Machine efficiency and power factor vs. Speed');
xlabel('Rotor speed/Synchronous speed');
ylabel(ax(1),'Efficiency (%)');
ylabel(ax(2),'Power Factor');
set(ax(1),'YLim',[0 100])
set(ax(2),'YLim',[0 1])
set(ax(1),'XLim',[0.985 1])
set(ax(2),'XLim',[0.985 1])
set(gca,'YTick',[10 20 30 40 50 60 70 80 90 100]);
grid on;
```

9

Surface-Permanent-Magnet AC Machines

"Thales ... said that the magnet has a soul because it moves iron." Aristotle (384–322 BC) writing in his book, commonly known as *De Anima* in Latin (translates to *On the Soul* in English).

In this chapter, we study surface-permanent-magnet (SPM) ac motors. The theory developed in the earlier chapters is used to derive relationships between speed and voltage and torque and current for the SPM ac machine. While the SPM machine is widely used in industrial applications, its sibling, the interior-permanent-magnet (IPM) machine, is the machine of choice for electric powertrains. This chapter provides the necessary background for the investigation of the IPM machine in Chapter 10.

The modern permanent-magnet machine is based on the **neodymium-iron-boron (Nd-Fe-B)** magnet, which was developed in the 1980s by General Motors in the United States and Sumitomo Corp. in Japan. The Nd-Fe-B magnets displaced other materials such as ferrite and samarium-cobalt for high-power-density industrial and automotive machines due to their high energy densities. **Ferrite magnets** continue to feature in high-power automotive machines due to concerns with the cost and availability of the rare-earth materials.

The authors found Reference [1] to be a particularly useful text in this area.

9.1 Basic Operation of SPM Machines

In our initial investigation of the SPM machine, we consider the relationship between the rotating magnet and the stationary conductor.

9.1.1 Back EMF of a Single Coil

An elementary single-coil rotating machine is shown in Figure 9.1. Consider a single-turn coil set a composed of two conductors, a_{1+} and a_{1-}, and located on the stator. The coil set is linked by a spinning magnetic field created by north and south permanent magnets, which are located on the exterior of the rotor. The rotor has radius r. The rotor

Electric Powertrain: Energy Systems, Power Electronics and Drives for Hybrid, Electric and Fuel Cell Vehicles, First Edition. John G. Hayes and G. Abas Goodarzi.

Companion website: www.wiley.com/go/hayes/electricpowertrain

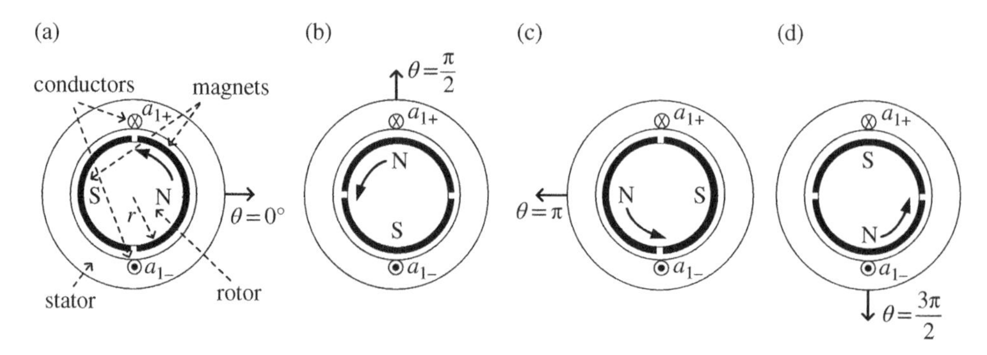

Figure 9.1 Simple single-coil machine.

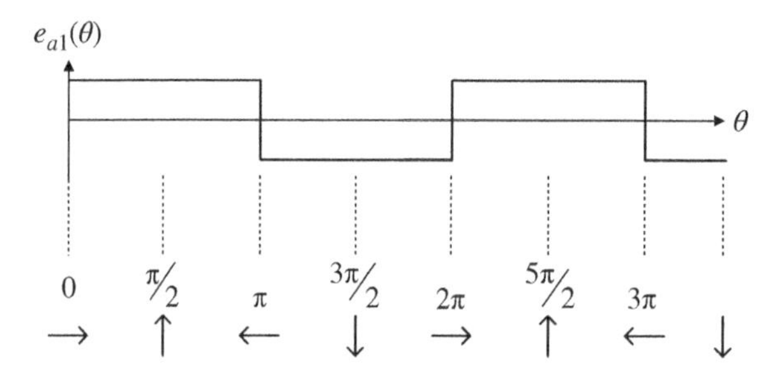

Figure 9.2 Back emf generated in coil.

position θ is aligned along the magnetic axis, as shown. As the rotor spins in the direction of the solid arrow, the magnets spin, and the rotor position changes.

When the magnet is oriented in the first position, as shown in Figure 9.1(a), at $\theta = 0\,°$, the coil links the total pole magnetic flux. As the magnets rotate past $\theta = 0\,°$, the conductor a_{1+} experiences a positive magnetic field due to the north pole, and the conductor a_{1-} experiences a negative magnetic field due to the south pole, as shown in Figure 9.1(b). Thus, per Faraday's law, a voltage is induced on the coil a_1 due to the spinning magnetic field on the rotor. The profile of the resultant back emf induced on the coil e_{a1} is shown in Figure 9.2. The coil voltage is positive when the north and south poles are rotating by the conductors a_{1+} and a_{1-}, respectively, and the voltage is negative as the poles rotate such that the north and south poles are rotating by the conductors a_{1-} and a_{1+}, respectively, as shown in Figure 9.1(b). The rotor alignment is indicated by the arrow shown along the bottom of the figure. It is seen that the back emf induced into the coil of wire $e_{a1}(t)$ has a square-wave shape.

9.1.2 Back EMF of Single Phase

The phase coils are typically distributed in multiple slots in the stator. We now add a second coil a_2, composed of two conductors, a_{2+} and a_{2-}, to the phase a winding and offset this coil by 30° with respect to a_1, as shown in Figure 9.3.

A square-wave voltage waveform is also induced on the second coil due to the spinning magnet. The waveforms for the two coil voltages, e_{a1} and e_{a2}, and the resulting per-phase

Figure 9.3 Two-coil machine.

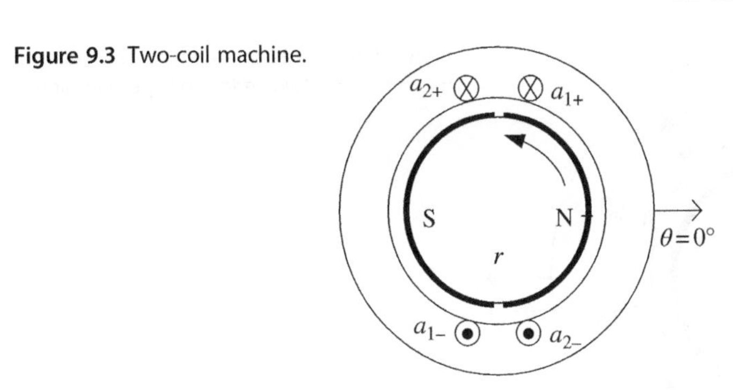

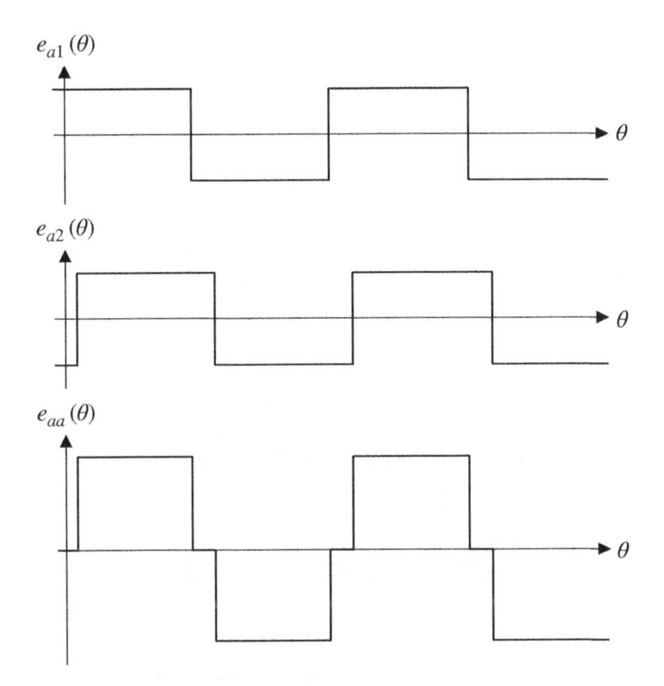

Figure 9.4 Coil and phase voltages for phase *a*.

voltage e_{aa} are shown in Figure 9.4. Assuming a series connection of the two coils, the resultant phase voltage is the sum of the two coil voltages:

$$e_{aa}(\theta) = e_{a1}(\theta) + e_{a2}(\theta) \tag{9.1}$$

Next, let us add the two other phases, *b* and *c*, with each having two coils, to create the basic three-phase machine as shown in Figure 9.5.

The three phase voltages due to the rotating magnet are as shown in Figure 9.6(a)–(c). However, the three phases are commonly coupled within the machine and so the resulting phase-to-neutral voltage is the sum of the phase back emf induced for a given phase plus the back emfs induced due to the mutual couplings of the other two phases.

There are two sources of phase back emf: (i) the back emf induced by the spinning magnets, and (ii) the back emf induced by the phase currents.

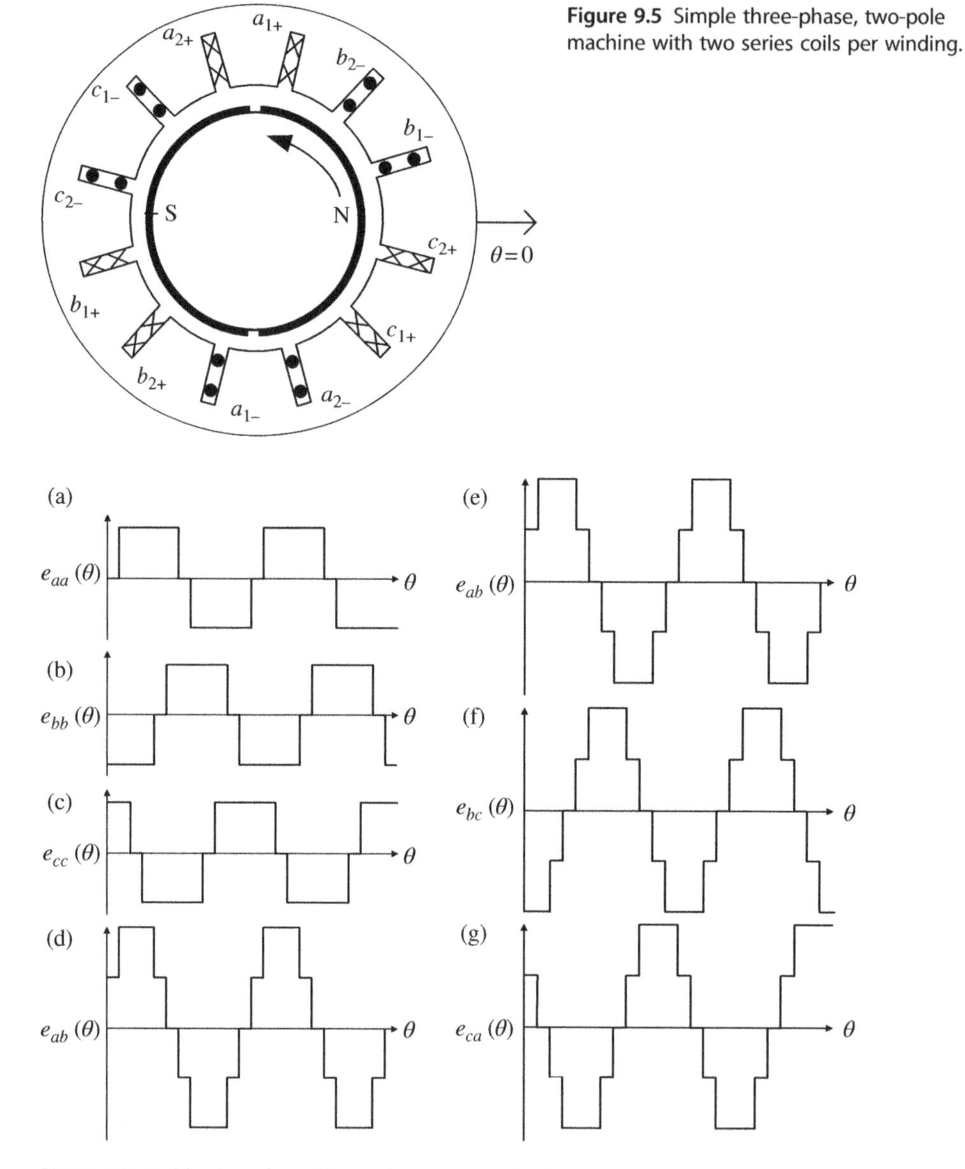

Figure 9.5 Simple three-phase, two-pole machine with two series coils per winding.

Figure 9.6 Coil back emfs and line voltages on no load.

The phase voltage back emf under load can be determined by simply summing the phase back emf due to the spinning magnets and the phase back emf due to the three phase currents.

Under no load, the phase a back emf e_{an} and the line back emf e_{ab} are simply given by

$$e_{an}(\theta) = e_{aa}(\theta) \tag{9.2}$$

and

$$e_{ab}(\theta) = e_{aa}(\theta) - e_{bb}(\theta) \tag{9.3}$$

Line back emf e_{ab} is recreated as shown in Figure 9.6 (d) by summing the coil voltages in accordance with the preceding formula. The three line back emfs for e_{ab}, e_{bc}, and e_{ca} are as shown in Figure 9.6(e)–(f). Line voltages e_{bc} and e_{ca} are delayed by 120° and 240° with respect to e_{ab}.

9.1.2.1 The Experimental Back EMF

An experimental waveform for the line back emf of the 2004 Toyota Prius traction motor is as shown in Figure 9.7. This motor has 48 stator slots and 8 rotor poles. Thus, there are six stator slots per pole resulting in 2 slots per phase per pole for the three-phase machine – similar to the elementary machine just discussed. Thus, the experimental line back emf has a similar waveform to the theoretical waveforms of Figure 9.6(e)–(f), with the difference being due to the magnetic pole arc of 120° in the Prius machine vs. 180° in the elementary machine.

9.1.2.2 Distributed Winding

Ideally, the ac machine would have a sinusoidal wave shape with little or no harmonic distortion. The waveforms of Figure 9.7 are ac waveforms with a sinusoidal fundamental component and significant harmonic distortion. The waveforms can be improved by using a number of design approaches such as distributing the winding sinusoidally, shaping the magnet, skewing the windings, and partial-pitch windings. A possible wiring scheme for a sinusoidally distributed winding is shown in Figure 9.8. In this scheme, the number of turns in each slot is determined according to the sine of the position within the machine. Phases b and c are offset 120° and 240° with respect to a.

A servo-drive industrial machine may feature a sinusoidally distributed winding. An experimental waveform from such a machine is presented in Figure 9.9. The experimental voltage shown in Figure 9.9 has far less distortion than the concentrated winding of the Toyota Prius. However, the concentrated winding costs less to manufacture and has a higher torque per ampere, making it very suitable for an automotive application.

9.1.3 SPM Machine Equations

The machine equations for the basic three-phase machine are similar to those of the dc machine of Chapter 7 and the elementary machines of Chapter 16.

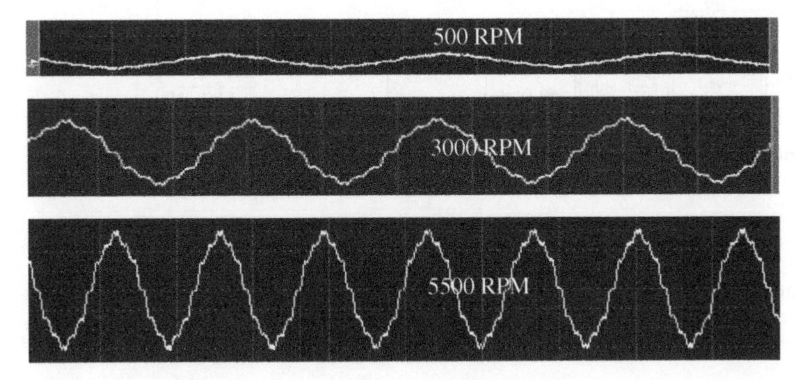

Figure 9.7 Experimental waveforms for traction motor back emf for 2004 Toyota Prius THS II [2]. (Courtesy of Oak Ridge National Laboratory, US Dept. of Energy.)

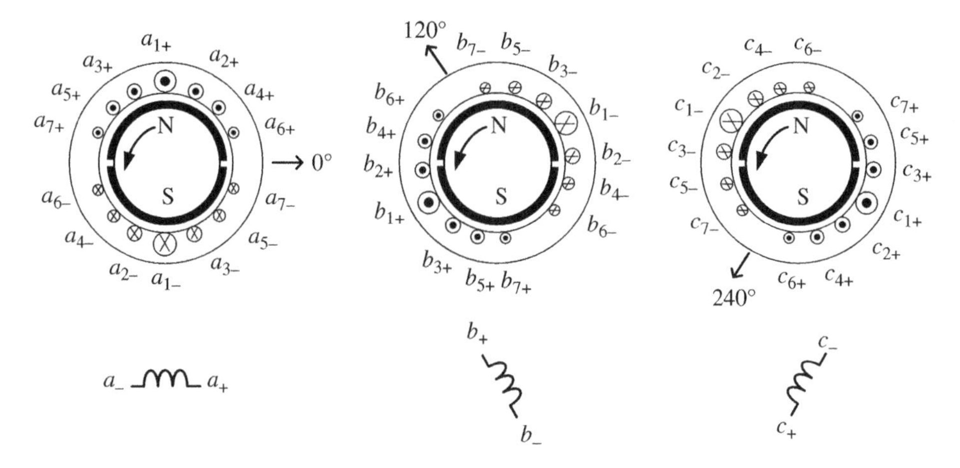

Figure 9.8 Sinusoidally distributed windings.

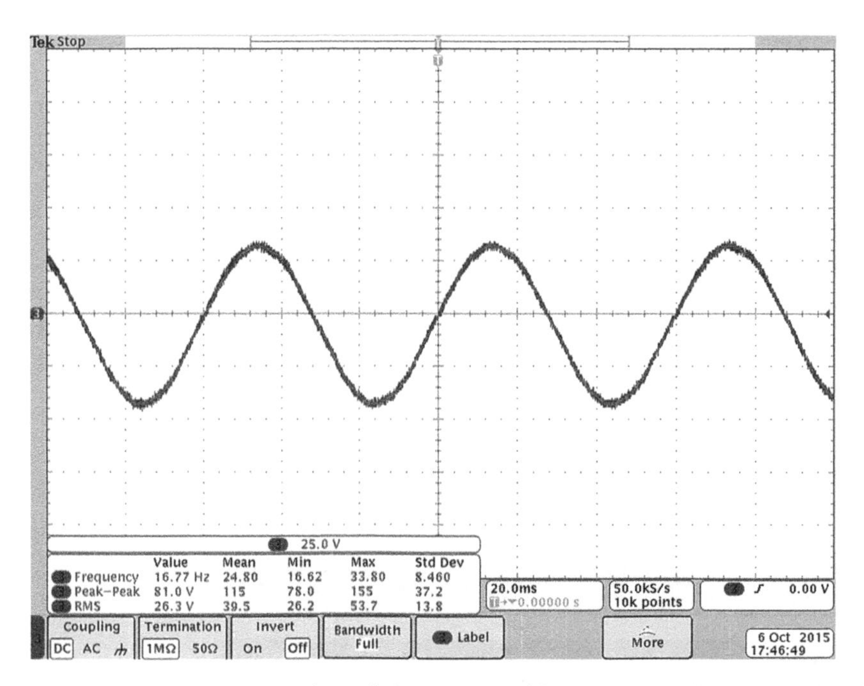

Figure 9.9 Experimental waveform of phase back emf from a sinusoidally distributed machine.

The rms value of the per-phase back emf is related to the product of the **flux linkage** λ and the angular speed ω_r:

$$E_{ph} = \frac{p}{2} \lambda \omega_r \tag{9.4}$$

Note that flux linkage λ is treated in this textbook as an rms quantity for ac machines. It is very common for peak flux linkage to be used in equations, in which case an adjustment by a factor of $\sqrt{2}$ is required.

Similar to the dc machine, it is common to express the relationship between the rms back emf E_{ph} and the angular speed ω_r using the **machine constant** k:

$$E_{ph} = k\,\omega_r \tag{9.5}$$

where k is given by

$$k = \frac{p}{2}\lambda \tag{9.6}$$

Note that k is also often expressed as a function of the line-line voltage rather than the phase voltage. In such cases, a factor of $\sqrt{3}$ is included.

If the phase current I_{ph} is vector-controlled to be in phase with the back emf, then the electromagnetic power supplied to the machine is given by

$$P_{em} = 3\,E_{ph}\,I_{ph} = T_{em}\,\omega_r \tag{9.7}$$

Thus, by combining Equation (9.7) and Equation (9.4), the electromagnetic **torque equation** of the machine is given by

$$T_{em} = 3\frac{p}{2}\lambda\,I_{ph} \tag{9.8}$$

The torque can be simply expressed in terms of the machine constant using

$$T_{em} = 3k\,I_{ph} \tag{9.9}$$

9.1.3.1 Example: Phase Voltage of SPM Machine

The eight-pole 2004 Toyota Prius traction machine is experimentally measured to have a back emf E_{ph} = 315 Vrms at N_r = 6000 rpm [2].

Determine the machine constant and flux linkage.

Solution:

The rotor frequency for this condition is

$$f_r = \frac{N_r}{60} = \frac{6000}{60}\,\text{Hz} = 100\,\text{Hz}$$

The angular frequency is then

$$\omega_r = 2\pi f_r = 2\pi \times 100\,\text{rad s}^{-1} = 628.3\,\text{rad s}^{-1}$$

The machine constant is

$$k = \frac{E_{ph}}{\omega_r} = \frac{315}{628.3\,\text{rad s}^{-1}}\frac{\text{V}}{} = 0.50\,\frac{\text{V}}{\text{rad s}^{-1}}$$

The flux linkage is

$$\lambda = \frac{2}{p}k = \frac{2}{p} \times 0.5\,\text{Wb} - \text{turns} = 0.125\,\text{Wb} - \text{turns}$$

9.2 Per-Phase Analysis of SPM Machine

The analysis of three-phase machines is usually simplified by creating an equivalent per-phase model. Results from the per-phase model are then scaled to three phase.

As is covered in Chapter 16 for the electromagnet, the **self-inductance** of a single phase of the three-phase machine L_{self} can be simply derived as

$$L_{self} = \mu_0 \frac{N^2 A_{pole}}{2l_g} = \mu_0 \frac{N^2 \pi r l}{2l_g} \tag{9.10}$$

where A_{pole} is the area of the magnetic pole, l_g is the length of the air gap, N is the number of stator turns per phase, l is the rotor length, and r is the rotor radius.

In a three-phase machine, the windings of the three phases overlap, because of which there is a mutual coupling between the phases. The flux linkage of the per-phase magnetizing inductance is the sum of the flux linkages of the self-inductance of the phase and the mutual couplings from the other two phases. In a three-phase machine, the windings are displaced by 120° with respect to each other. Thus:

$$L_m i_a(t) = L_{self}\, i_a(t) + L_{self} \cos(120^\circ)\, i_b(t) + L_{self} \cos(240^\circ)\, i_c(t) \tag{9.11}$$

This equation simplifies to

$$L_m i_a(t) = L_{self}\, i_a(t) - {}^1/_2 L_{self} \left[i_b(t) + i_c(t) \right] \tag{9.12}$$

In a balanced three-phase machine, such as the SPM, the three-phase currents sum to zero such that

$$i_a(t) + i_b(t) + i_c(t) = 0 \tag{9.13}$$

Substituting Equation (9.13) into Equation (9.12) and simplifying provides the simple relationship between the per-phase **magnetizing inductance** and the single-phase self-inductance:

$$L_m = {}^3/_2 L_{self} \tag{9.14}$$

9.2.1 Per-Phase Equivalent Circuit Model for SPM Machine

Per-phase equivalent circuit models of the SPM machine are presented in Figure 9.10. There are two inductances in the per-phase model: the **leakage inductance** L_l and the magnetizing inductance L_m, as shown in Figure 9.10(a). The combination of these two inductances is known as the **synchronous inductance** L_s. The commonly used equivalent circuit is shown in Figure 9.10(b). Under normal operation in motoring mode,

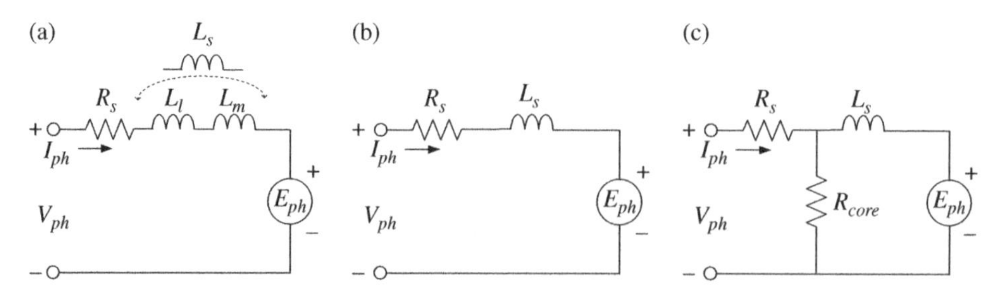

Figure 9.10 Per-phase equivalent circuit models of SPM machine.

the phase current I_{ph} in an SPM machine is controlled to be in phase with the per-phase back emf E_{ph}. The core loss can be introduced to the circuit model as shown in Figure 9.10(c). In this chapter, we take the same simplifying approach to core loss as in the earlier chapters and combine the core loss with the friction and windage losses.

9.2.2 Phasor Analysis of SPM Machine

Using Kirchhoff's voltage law, the voltages of the per-phase circuit shown in Figure 9.10 (b) can be summed as follows in phasor form:

$$\bar{V}_{ph} = \bar{E}_{ph} + R_s\bar{I}_{ph} + j\omega_e L_s\bar{I}_{ph} \tag{9.15}$$

where $\bar{V}_{ph}$, $\bar{E}_{ph}$, and $\bar{I}_{ph}$ are the per-phase supply voltage, per-phase back emf, and per-phase supply current phasors, respectively; $j\left(=\sqrt{-1}\right)$ is the complex operator; and ω_e is the electrical frequency in rad/s.

There are a number of approaches to drawing the phasor diagram which principally revolve around the choice of the reference phasor. In the discussions in this text on PM machines, the magnetic axis of the rotor permanent magnet is taken as the reference phasor. This axis is known as the **direct** or ***d* axis**, and is aligned horizontally along the reference axis at 0°. The vertical axis is known as the **quadrature** or ***q* axis** as shown in the phasor diagrams of Figure 9.11.

The per-phase current $\bar{I}_{ph}$ is resolved into two components: the quadrature current $I_{ph,q}$, and the direct component $I_{ph,d}$:

$$\bar{I}_{ph} = I_{ph,d} + jI_{ph,q} \tag{9.16}$$

When the SPM machine is motoring, the quadrature current is in phase with the back emf and is the torque-generating component of current. The direct or field component is aligned with the permanent magnet and lags the quadrature component by 90°.

It is useful to express the voltages in rectangular or Cartesian format as follows:

$$\bar{V}_{ph} = V_{ph,d} + jV_{ph,q} \tag{9.17}$$

and

$$\bar{E}_{ph} = E_{ph,d} + jE_{ph,q} \tag{9.18}$$

It is assumed in this analysis that the back emf $\bar{E}_{ph}$ can be represented by the quadrature component only, thus simplifying the analysis. Therefore:

$$\bar{E}_{ph} = jE_{ph,q} = jk\,\omega_r \tag{9.19}$$

Representative phasor diagrams are shown in Figure 9.12(a)–(d) for the motoring and generating modes of operation in forward and reverse. The phase current and back emf are represented by their quadrature components only. In these low-speed modes below the rated speed, the phase current $I_{ph,q}$ and back emf $E_{ph,q}$ are in phase for motoring and 180° out of phase for generating. The quadrature

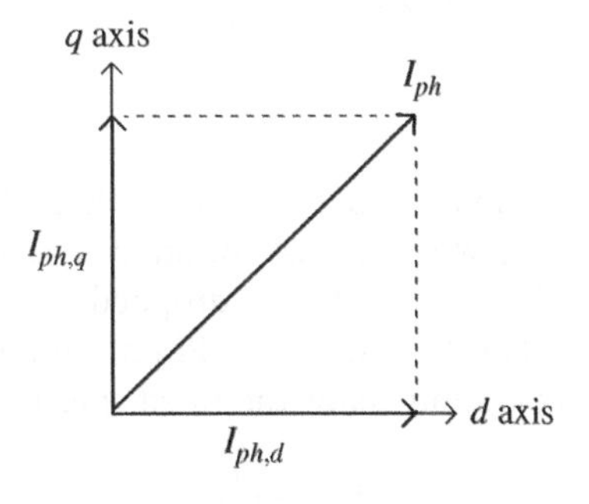

Figure 9.11 *d* and *q* axes.

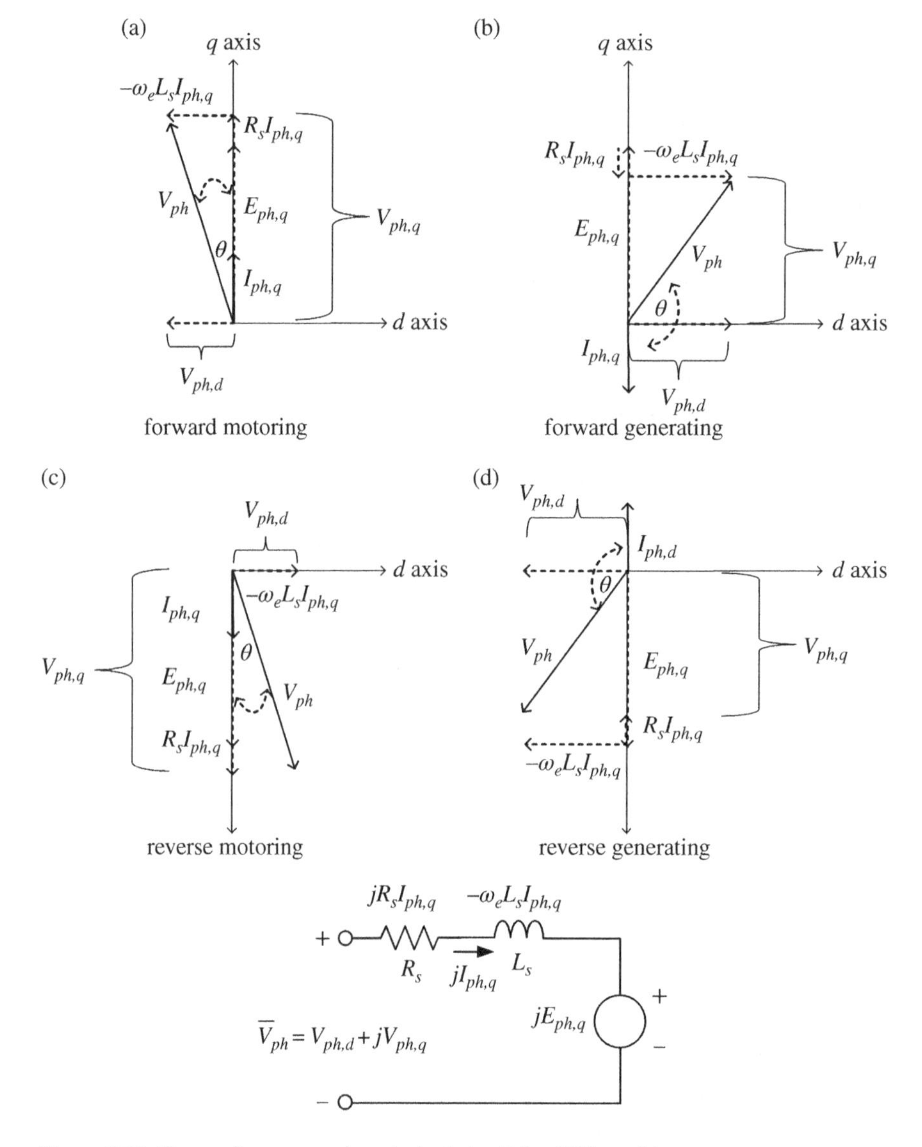

Figure 9.12 Phasor diagrams and equivalent circuit for SPM machine.

component of the supply voltage $V_{ph,q}$ equals the back emf $E_{ph,q}$ plus the ohmic voltage drop $R_s I_{ph,q}$ across the stator resistance R_s. The direct component of the supply voltage $V_{ph,d} = -\omega_e L_s I_{ph,q}$ is dropped across the synchronous inductance L_s.

For the case of the SPM machine, the direct or magnetizing component $I_{ph,d}$ is zero for operation below the rated speed. Thus:

$$\bar{I}_{ph} = jI_{ph,q} = j\frac{T_{em}}{3k} \tag{9.20}$$

and therefore Equation (9.15) can be rewritten as

$$\begin{aligned}\bar{V}_{ph} &= V_{ph,d} + jV_{ph,q} = \bar{E}_{ph} + (R_s + j\omega_e L_s)\bar{I}_{ph} \\ &= jE_{ph,q} + (R_s + j\omega_e L_s)jI_{ph,q} \\ &= -\omega_e L_s I_{ph,q} + j\left(E_{ph,q} + R_s I_{ph,q}\right)\end{aligned} \tag{9.21}$$

since $j^2 = -1$.

Thus:

$$V_{ph,d} = -\omega_e L_s I_{ph,q} \tag{9.22}$$

and

$$V_{ph,q} = E_{ph,q} + R_s I_{ph,q} \tag{9.23}$$

The phasor diagram can then be constructed as in Figure 9.12.

The **apparent power** supplied to the machine is given by

$$\bar{S} = 3\,\bar{V}_{ph}\,\bar{I}_{ph}^{\,*} = P + jQ \tag{9.24}$$

where P and Q are the **real power** and **reactive power**, respectively, and $\bar{I}_{ph}^{\,*}$ is the **complex conjugate** of $\bar{I}_{ph}$. The complex conjugate of

$$\bar{I}_{ph} = I_{ph,d} + jI_{ph,q} \tag{9.25}$$

is

$$\bar{I}_{ph}^{\,*} = I_{ph,d} - jI_{ph,q} \tag{9.26}$$

Thus, the generic equation for apparent power is

$$\begin{aligned}\bar{S} &= 3\left(V_{ph,d} + jV_{ph,q}\right)\left(I_{ph,d} - jI_{ph,q}\right) \\ &= 3\left(V_{ph,d}\,I_{ph,d} + V_{ph,q}I_{ph,q}\right) + j3\left(V_{ph,q}\,I_{ph,d} - V_{ph,d}I_{ph,q}\right)\end{aligned} \tag{9.27}$$

In the case of a SPM machine with no d-axis current, the apparent power is given by

$$\bar{S} = 3\left(V_{ph,d} + jV_{ph,q}\right)\left(-jI_{ph,q}\right) = 3\,V_{ph,q}\,I_{ph,q} - j3\,V_{ph,d}I_{ph,q} \tag{9.28}$$

The real and reactive powers supplied to the machine are given by

$$P = 3\,V_{ph,q}\,I_{ph,q} = 3E_{ph,q}\,I_{ph,q} + 3R_s\,I_{ph,q}^2 \tag{9.29}$$

and

$$Q = -3\,V_{ph,d}\,I_{ph,q} = 3\,\omega_e L_s\,I_{ph,q}^2 \tag{9.30}$$

The stator copper power loss is given by

$$P_{Rs} = 3R_s\,I_{ph,q}^2 \tag{9.31}$$

The electromagnetic power supplied to the machine is given by

$$P_{em} = 3E_{ph,q}\,I_{ph,q} \tag{9.32}$$

If the no-load torque is T_{nl}, the output rotor power is given by

$$P_r = 3E_{ph,q} I_{ph,q} - T_{nl}\,\omega_r \tag{9.33}$$

The electromagnetic and rotor torques equations are given by

$$T_{em} = 3kI_{ph,q} \tag{9.34}$$

and

$$T_r = 3kI_{ph,q} - T_{nl} \tag{9.35}$$

The machine power loss and efficiency are given by

$$P_{m(loss)} = 3R_s I_{ph,q}^2 + T_{nl}\omega_r \tag{9.36}$$

and

$$\eta = \frac{P_r}{P} \times 100\% \tag{9.37}$$

The machine **power factor** is given by

$$PF = \frac{P}{|\bar{S}|}$$

With this basic analysis, we can now work some examples.

9.2.2.1 Example: Motoring using SPM Machine

An electric drive features an eight-pole SPM machine with the following parameters: per-phase stator resistance R_s = 20 mΩ, machine constant k = 0.3 V/rad/s, synchronous inductance L_s= 0.2 mH, and no-load torque T_{nl} = 2 Nm. The rated conditions are 80 kW and 280 Nm.

Determine the applied per-phase voltage and current, machine efficiency, and power factor for the following operating points:

i) Motoring at the rated condition
ii) Generating at the rated condition
iii) Motoring at 70 Nm at half the rated speed

Ignore saturation.

Solution:

i) The rotor speed in rad/s is

$$\omega_r = \frac{P_{r(rated)}}{T_{r(rated)}} = \frac{80000}{280}\text{rad/s} = 285.71\text{ rad/s}$$

The rotor frequency in hertz is

$$f_r = \frac{\omega_r}{2\pi} = \frac{285.71}{2\pi}\text{Hz} = 45.47\text{ Hz}$$

With a pole count p = 8, the electrical input frequencies are

$$f_e = \frac{p}{2}f_r = \frac{8}{2}45.47\text{ Hz} = 181.88\text{ Hz}$$

and

$$\omega_e = \frac{p}{2}\omega_r = \frac{8}{2}285.71\,\text{rad/s} = 1142.8\,\text{rad/s}$$

When operating in motoring mode, the electromagnetic torque is the sum of the rotor output torque and the no-load torque:

$$T_{em} = T_r + T_{nl} = 280\,\text{Nm} + 2\,\text{Nm} = 282\,\text{Nm}$$

The per-phase current is given by

$$I_{ph,q} = \frac{T_{em}}{3k} = \frac{282}{3 \times 0.3}\text{A} = 313.33\,\text{A}$$

The back emf is given by

$$E_{ph,q} = k\,\omega_r = 0.3 \times 285.71\,\text{V} = 85.71\,\text{V}$$

The quadrature component of the supply voltage is

$$V_{ph,q} = E_{ph,q} + R_s I_{ph,q} = 85.71\,\text{V} + 0.020 \times 313.33\,\text{V} = 91.98\,\text{V}$$

The direct component of the supply voltage is

$$V_{ph,d} = -\omega_e L_s I_{ph,q} = -1142.8 \times 0.2 \times 10^{-3} \times 313.33\,\text{V} = -71.62\,\text{V}$$

The supplied per-phase voltage is as follows:

$$|\bar{V}_{ph}| = \sqrt{V_{ph,d}{}^2 + V_{ph,q}{}^2} = \sqrt{(-71.62)^2 + 91.98^2}\text{V} = 116.6\,\text{V}$$

The machine power loss is

$$P_{m\,(loss)} = T_{nl}\omega_r + 3R_s I_{ph,q}{}^2 = 2 \times 285.71\,\text{W} + 3 \times 0.020 \times 313.33^2\,\text{W} = 6.462\,\text{kW}$$

The input power to the machine is

$$P = P_r + P_{m\,(loss)} = 80\,\text{kW} + 6.462\,\text{kW} = 86.462\,\text{kW}$$

The machine efficiency is

$$\eta = \frac{P_r}{P} = \frac{80}{86.462} \times 100\,\% = 92.53\,\%$$

The apparent power of the machine is

$$|\bar{S}| = 3\,V_{ph} I_{ph} = 3 \times 116.6 \times 313.33\,\text{VA} = 109.6\,\text{kVA}$$

The power factor is

$$PF = \frac{P}{|\bar{S}|} = \frac{86.462}{109.6} = 0.7889\text{, lagging}$$

ii) When operating in generating or braking mode, the electromagnetic torque is the useful input torque minus the no-load torque.

The electromagnetic torque is

$$T_{em} = T_r + T_{nl} = -280\,\text{Nm} + 2\,\text{Nm} = -278\,\text{Nm}$$

The mechanical and electrical speeds in hertz and rad/s are as before.
The per-phase current is given by

$$I_{ph,q} = \frac{T_{em}}{3k} = \frac{-278}{3 \times 0.3} \text{A} = -308.89\,\text{A}$$

The supply voltages are

$$V_{ph,q} = E_{ph,q} + R_s I_{ph,q} = 85.71\,\text{V} + 0.020 \times (-308.89)\,\text{V} = 79.53\,\text{V}$$

$$V_{ph,d} = -\omega_e L_s \cdot I_{ph,q} = -1142.8 \times 0.2 \times 10^{-3} \times (-308.89)\,\text{V} = +70.6\,\text{V}$$

and

$$|\bar{V}_{ph}| = \sqrt{V_{ph,d}{}^2 + V_{ph,q}{}^2} = \sqrt{70.6^2 + 79.53^2}\text{V} = 106.35\,\text{V}$$

The machine power loss is

$$P_{m\,(loss)} = T_{nl}\omega_r + 3R_s I_{ph,q}{}^2 = 2 \times 285.71\,\text{W} + 3 \times 0.020 \times (-308.89)^2\,\text{W} = 6.296\,\text{kW}$$

The regenerative power from the machine is

$$P = P_r + P_{m\,(loss)} = -80\,\text{kW} + 6.296\,\text{kW} = -73.704\,\text{kW}$$

The machine efficiency is given by

$$\eta = \frac{P}{P_r} = \frac{-73.704}{-80} \times 100\,\% = 92.13\,\%$$

The apparent power of the machine is

$$|\bar{S}| = 3\,V_{ph} I_{ph} = 3 \times 106.35 \times 308.89\,\text{VA} = 98.551\,\text{kVA}$$

The power factor is

$$PF = \frac{P}{|\bar{S}|} = \frac{-73.704}{98.551} = -0.7479, \text{leading}$$

iii) Partial load conditions when the output torque is 70 Nm at half the rated speed.
The electromagnetic torque is

$$T_{em} = T_r + T_{nl} = 70\,\text{Nm} + 2\,\text{Nm} = 72\,\text{Nm}$$

The current is

$$I_{ph,q} = \frac{T_{em}}{3k} = \frac{72}{3 \times 0.3} \text{A} = 80\,\text{A}$$

The rotor and electrical frequencies are half of those at the rated condition:

$$\omega_r = \frac{\omega_{r(rated)}}{2} = \frac{285.71}{2} \text{rad s}^{-1} = 142.86\,\text{rad s}^{-1}$$

$$\omega_e = \frac{\omega_{e(rated)}}{2} = \frac{1142.84}{2} \text{rad s}^{-1} = 571.42\,\text{rad s}^{-1}$$

The back emf is

$$E_{ph,q} = k\,\omega_r = 0.3 \times 142.86\,\text{V} = 42.86\,\text{V}$$

The supply voltage values are

$$V_{ph,q} = E_{ph,q} + R_s I_{ph,q} = 42.86\,\text{V} + 0.020 \times 80\,\text{V} = 44.46\,\text{V}$$

$$V_{ph,d} = -\omega_e L_s I_{ph,q} = -571.42 \times 0.2 \times 10^{-3} \times 80\,\text{V} = -9.14\,\text{V}$$

$$|\bar{V}_{ph}| = \sqrt{V_{ph,d}{}^2 + V_{ph,q}{}^2} = \sqrt{(-9.14)^2 + 44.46^2}\,\text{V} = 45.39\,\text{V}$$

The machine power loss is the sum of the ohmic and no-load losses:

$$P_{m\,(loss)} = T_{nl}\omega_r + 3R_s I_{ph,q}{}^2 = 2 \times 142.86\,\text{W} + 3 \times 0.020 \times 80^2\,\text{W} = 670\,\text{W}$$

The input power to the machine is

$$P = T_r\omega_r + P_{m\,(loss)} = 70 \times 142.86\,\text{W} + 670\,\text{W} = 10.67\,\text{kW}$$

The machine efficiency is

$$\eta = \frac{P_r}{P} = \frac{10}{10.67} \times 100\,\% = 93.72\,\%$$

The apparent power of the machine is

$$|\bar{S}| = 3\,V_{ph} I_{ph} = 3 \times 45.39 \times 80\,\text{VA} = 10.89\,\text{kVA}$$

The power factor is

$$PF = \frac{P}{|\bar{S}|} = \frac{10.67}{10.89} = 0.9798,\ \text{lagging}$$

9.2.3 Machine Saturation

The earlier discussion on saturation in Chapter 7 for the brushed dc machine also applies to the SPM machine. The effects of saturation are illustrated with the following example. We see that one of the effects of saturation is a significant increase in the current required to generate the torque, resulting in increased machine power loss.

9.2.3.1 Example: Motoring using SPM Machine

Recalculate the example in Section 9.2.2.1 if the machine constant k and the synchronous inductance L_s drop by 25% due to machine saturation at the rated torque.

Solution:

Parameters k and L_s are proportionally reduced to 0.225 Nm/A and 0.15 mH, respectively, due to saturation.

The per-phase current is

$$I_{ph,q} = \frac{T_{em}}{3k} = \frac{282}{3 \times 0.225}\,\text{A} = 417.78\,\text{A}$$

The back emf is

$$E_{ph,q} = k\,\omega_r = 0.225 \times 285.71\,\text{V} = 64.28\,\text{V}$$

The supply voltage values are

$$V_{ph,q} = E_{ph,q} + R_s I_{ph,q} = 64.28\,\text{V} + 0.020 \times 417.78\,\text{V} = 72.64\,\text{V}$$

$$V_{ph,d} = -\omega_e L_s I_{ph,q} = -1142.8 \times 0.15 \times 10^{-3} \times 417.78\,\text{V} = -71.62\,\text{V}$$

$$\left|\bar{V}_{ph}\right| = \sqrt{{V_{ph,d}}^2 + {V_{ph,q}}^2} = \sqrt{(-71.62)^2 + 72.64^2}\text{V} = 102\,\text{V}$$

The machine power loss is

$$P_{m\,(loss)} = T_{nl}\omega_r + 3R_s {I_{ph,q}}^2 = 2 \times 285.71\,\text{W} + 3 \times 0.020 \times 417.78^2\,\text{W} = 11.044\,\text{kW}$$

The input power to the machine is

$$P = P_r + P_{m\,(loss)} = 80\,\text{kW} + 11.044\,\text{kW} = 91.044\,\text{kW}$$

The machine efficiency is

$$\eta = \frac{P_r}{P} = \frac{80}{91.044} \times 100\,\% = 87.87\,\%$$

The apparent power of the machine is

$$|\bar{S}| = 3\,V_{ph} I_{ph} = 3 \times 102 \times 417.78\,\text{VA} = 127.84\,\text{kVA}$$

The power factor is

$$PF = \frac{P}{|\bar{S}|} = \frac{91.044}{127.84} = 0.7122,\ \text{lagging}$$

9.2.4 SPM Torque–Speed Characteristics

The torque–speed characteristic of the SPM machine can be easily generated by modifying the earlier equations. From Equation (9.21), the magnitude of the per-phase voltage is given by

$$V_{ph} = \sqrt{\left(\omega_e L_s I_{ph,q}\right)^2 + \left(E_{ph,q} + R_s I_{ph,q}\right)^2} \tag{9.38}$$

The machine equations

$$\omega_r = \frac{2}{p}\omega_e,\ I_{ph,q} = \frac{T_{em}}{3k},\ \text{and}\ E_{ph,q} = k\omega_r$$

are substituted into Equation (9.38) to obtain the quadratic equation:

$${V_{ph}}^2 = \left(\frac{p}{2}\omega_r L_s \frac{T_{em}}{3k}\right)^2 + \left(k\omega_r + R_s \frac{T_{em}}{3k}\right)^2 \tag{9.39}$$

which is rearranged as follows:

$$\left\{\left(\frac{p}{2}L_s\cdot\frac{T_{em}}{3k}\right)^2+k^2\right\}\omega_r^{\,2}+\left(2R_s\frac{T_{em}}{3}\right)\omega_r+\left(R_s\frac{T_{em}}{3k}\right)^2-V_{ph}^{\,2}=0 \tag{9.40}$$

This quadratic equation can be solved in order to determine the maximum frequency at which we can achieve rated torque.

The maximum **no-load speed** $\omega_{r(nl)}$ for a given supply voltage can also be easily estimated. If R_s and T_{nl} are both assumed to negligibly small, as is the case for the automotive traction machine, then we get

$$\omega_{r(nl)}\approx\frac{V_{ph}}{k} \tag{9.41}$$

9.2.4.1 Example: Determining No-Load Speed

Determine the no-load speed for the earlier machine of Section 9.2.2.1 when the per-phase supply voltage is 122.5 V (the maximum phase voltage based on a 300 V battery, per Chapter 13, Section 13.2). Ignore the battery and converter losses.

Solution:

As just discussed, the no-load speed can be calculated relatively accurately as

$$\omega_{r(nl)}=\frac{V_{ph}}{k}=\frac{122.5}{0.3}\,\text{rad s}^{-1}=408.3\,\text{rad s}^{-1}\;(3899\,\text{rpm})$$

The torque–speed characteristic for the 280 Nm, 80 kW machine used in the earlier examples is plotted in Figure 9.13 for a constant supply voltage of 122.5 V. The output

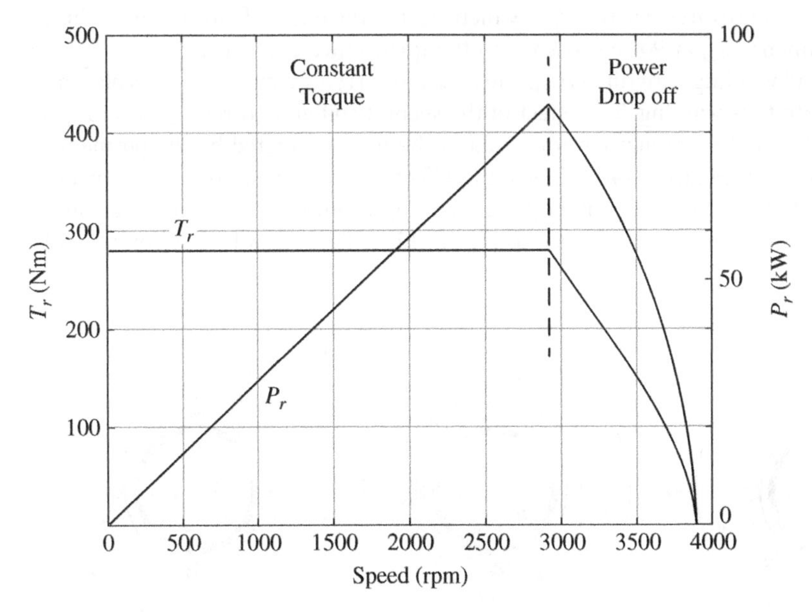

Figure 9.13 Plots of rotor torque and power versus speed for a constant supply voltage with torque limit.

torque is limited to the rated maximum value of 280 Nm, and then drops from the rated value of 280 Nm at the rated speed of 2728 rpm to zero at 3899 rpm, as just calculated.

The machine generates a full torque of 280 Nm over the speed range up until the full-power point of 80 kW at 2728 rpm. After 2728 rpm, the machine enters the power-drop-off mode as the back emf and synchronous reactance increase due to increasing current, and the resulting torque and power drop rapidly with speed.

In the next section, field weakening is introduced such that the machine can be run at full power at very high speeds.

9.2.5 High-Speed Operation of SPM Machine above Rated Speed

High-speed operation is common and desirable for all machines. We have investigated the field-weakening operation of the wound-field dc machine in Chapter 7 and of the induction machine in Chapter 8. In this section, the high-speed operation of the SPM machine in **field-weakening** mode is studied. Above the rated speed, a new current component must be introduced in order to weaken the field due to the permanent magnet.

Consider the earlier example of the single coil. As shown in Figure 9.14(a), the coil with turns a_{1+q} and a_{1-q} carries the quadrature current $I_{ph,q}$ and generates the electromagnetic torque $3kI_{ph,q}$ for a three-phase machine by interacting with the permanent magnet. From the left-hand rule, the interaction between the current and the magnet results in a clockwise torque on the conductor, but since the stator conductor is physically fixed, a counterclockwise torque acts on the rotor magnet to spin the rotor.

Next, we will place a virtual second winding, or the direct-current winding, at 90° to the first winding. The rationale for this second winding is more easily understood when the machine is analyzed from a *dq* perspective in Chapter 10. Simplistically, the second winding is carrying the direct current $I_{ph,d}$ which models the effect of introducing a direct current component $I_{ph,d}$ at 90° electrically to the quadrature current $I_{ph,q}$.

Thus, a second winding with turns a_{1+d} and a_{1-d} and carrying a current $I_{ph,d}$ is placed at 90° to the quadrature winding. The effect of the second coil and current $I_{ph,d}$ is to generate a magnetic field to strengthen (or weaken) the field generated by the permanent magnet, as shown in Figure 9.14(b). Thus, the effect of introducing the direct or field winding at 90° to the quadrature or torque-generating winding is to introduce a couple of new torque-generating components into the machine. These new torque components

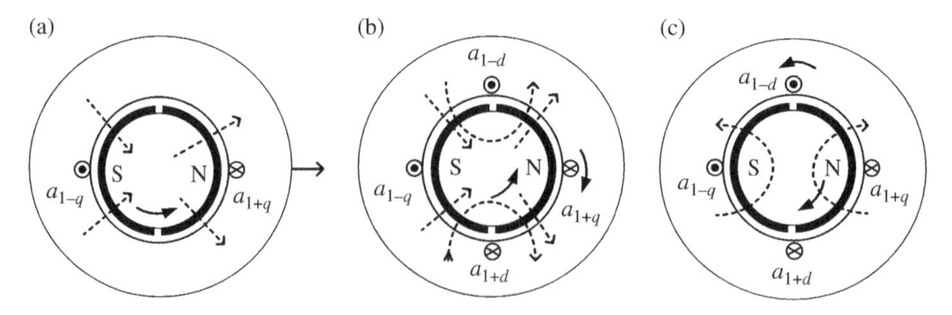

Figure 9.14 SPM torque generation due to (a) *q* winding only, (b) *q* winding due to *d* winding, and (c) *d* winding due to *q* winding.

cancel out for an SPM ac machine, but as we shall see in the Chapter 10, they do not cancel out for an IPM ac machine, resulting in a significant contribution to the overall machine torque.

First, consider the torque exerted on coil a_{1+q}-a_{1-q} due to current $I_{ph,d}$. From the right-hand rule, the magnetic field due to $I_{ph,d}$ can be seen as adding to the field from the permanent magnet, thereby generating additional machine torque when interacting with the current $I_{ph,q}$ in coil a_{1+q}–a_{1-q}. For a three-phase machine, this adds the term $+3^{p}/_{2}\, L_s\, I_{ph,d}\, I_{ph,q}$ to the usual torque component $3\, k\, I_{ph,q}$.

Second, consider the torque on the coil a_{1+d}–a_{1-d} due to current $I_{ph,q}$. Again, applying the left-hand rule, it can be seen that the torque created by coil a_{1+q}–$a_{1-q}$with current $I_{ph,q}$ is counterclockwise, in effect creating the term $-3^{p}/_{2}\, L_s\, I_{ph,d}\, I_{ph,q}$. Thus, there is no net torque in an SPM ac machine due to the interaction between the two currents.

Thus, the electromagnetic torque is given by

$$T_{em} = 3k\, I_{ph,q} + 3\frac{p}{2} L_s\, I_{ph,d}\, I_{ph,q} - 3\frac{p}{2} L_s\, I_{ph,d}\, I_{ph,q} \tag{9.42}$$

which results in the earlier torque expression:

$$T_{em} = 3\ k\, I_{ph,q} \tag{9.43}$$

A positive $I_{ph,d}$ strengthens the field, whereas a negative $I_{ph,d}$ weakens the field. Let

$$\bar{I}_{ph} = I_{ph,d} + jI_{ph,q} \tag{9.44}$$

Since

$$\bar{V}_{ph} = \bar{E}_{ph} + R_s\bar{I}_{ph} + j\omega_e L_s\, \bar{I}_{ph} \tag{9.45}$$

we get

$$\bar{V}_{ph} = jE_{ph,q} + R_s\left(I_{ph,d} + jI_{ph,q}\right) + j\omega_e L_s\left(I_{ph,d} + jI_{ph,q}\right) \tag{9.46}$$

$$\Rightarrow \bar{V}_{ph} = V_{ph,d} + jV_{ph,q} = +R_s I_{ph,d} - \omega_e L_s I_{ph,q} + j\left(E_{ph,q} + R_s I_{ph,q} + \omega_e L_s I_{ph,d}\right) \tag{9.47}$$

Thus,

$$V_{ph,q} = E_{ph,q} + R_s I_{ph,q} + \omega_e L_s I_{ph,d} \tag{9.48}$$

and

$$V_{ph,d} = +R_s I_{ph,d} - \omega_e L_s I_{ph,q} \tag{9.49}$$

The field-weakening operation of the SPM machine is now illustrated using the phasor diagram of Figure 9.15. The circular limit represents the maximum rms per-phase supply voltage V_{ph}. As we operate above the rated speed, the back emf increases such that the supply voltage is less that the back emf $E_{ph,q}$. However, by introducing a field weakening, we create a voltage drop across the synchronous inductance which cancels out a portion of the back emf, enabling the inverter to supply the required voltage and current.

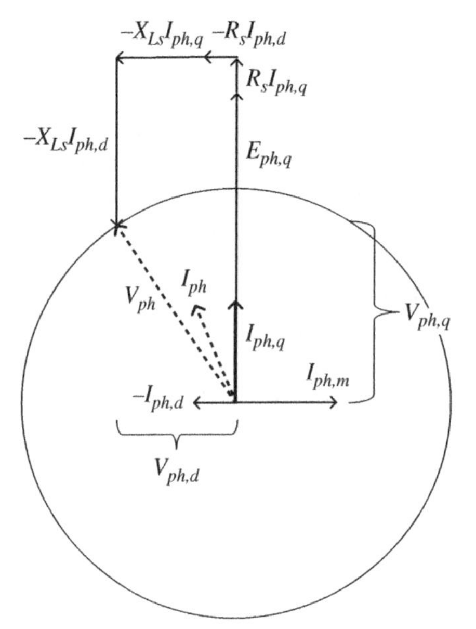

Figure 9.15 Phasor diagram for motoring during field weakening.

The **equivalent magnetizing current** due to the permanent magnet is $I_{ph,m}$, which is defined by the equation

$$E_{ph,q} = k\omega_r = \frac{p}{2} L_s I_{ph,m} \omega_r \tag{9.50}$$

Thus, the equivalent ac current source to represent the magnet is given by

$$I_{ph,m} = \frac{2k}{pL_s} \tag{9.51}$$

This equivalent magnetizing current is the reference phasor for the analysis in this chapter.

Operating the machine at high speeds above the rated speed results in a back emf $E_{ph,q}$ that is greater than the supply voltage V_{ph}. Introducing the negative field or direct-axis current effectively creates a back emf across the synchronous inductance, canceling out the excessive back emf. Thus, to size the field-weakening current $I_{ph,d}$ for the field-weakening mode we can use the following:

Below the rated speed, let

$$E_{ph,q} = k\,\omega_r \tag{9.52}$$

and

$$I_{ph,d} = 0 \tag{9.53}$$

Above the rated speed in field weakening, the field current is controlled such that the back emf at the rated speed equals the back emf at the operating speed:

$$E_{ph,q(rated)} = k\omega_{r(rated)} = k\omega_r + \frac{p}{2}\omega_r L_s I_{ph,d} \tag{9.54}$$

Thus, the required field current is given by

$$I_{ph,d} = \frac{2k}{pL_s}\left(\frac{\omega_{r(rated)}}{\omega_r} - 1\right) = I_{ph,m}\left(\frac{\omega_{r(rated)}}{\omega_r} - 1\right) \tag{9.55}$$

The torque–speed characteristic of the field-weakened SPM machine can be easily generated by modifying Equation (9.47) to get

$$V_{ph}{}^2 = \left(R_s I_{ph,d} - \omega_e L_s I_{ph,q}\right)^2 + \left(E_{ph,q} + R_s I_{ph,q} + \omega_e L_s I_{ph,d}\right)^2 \tag{9.56}$$

This equation can be rearranged to get a quadratic expression which can easily be solved for the current.

A simpler solution can be derived by ignoring the voltage drops across the per-phase resistances, in which case Equation (9.56) reduces to

$$V_{ph}{}^2 = \left(\omega_e L_s I_{ph,q}\right)^2 + \left(E_{ph,q} + \omega_e L_s I_{ph,d}\right)^2 \tag{9.57}$$

Thus, for a given supply voltage and speed, V_{ph}, $E_{ph,q}$, and $I_{ph,d}$ are all defined, and $I_{ph,q}$ can be determined from

$$I_{ph,q} = \frac{\sqrt{V_{ph}{}^2 - \left(E_{ph,q} + \omega_e L_s I_{ph,d}\right)^2}}{\omega_e L_s} \tag{9.58}$$

9.2.5.1 Example: Motoring using SPM Machine in Field Weakening

Using the machine of the earlier examples, determine the applied per-phase current and voltage, machine efficiency, and power factor for full-power operation at 3.75 times the rated speed with field weakening.

Simplify the analysis by ignoring the resistive elements when estimating the d-axis current and the supply voltages.

Ignore saturation.

Solution:

The rated conditions are +280 Nm, 80 kW, and 285.71 rad/s.

The new frequencies are

$$\omega_r = 3.75\,\omega_{r(rated)} = 3.75 \times 285.71\,\text{rad/s} = 1071.41\,\text{rad/s}$$

$$\omega_e = 3.75\,\omega_{e(rated)} = 3.75 \times 1142.8\,\text{rad/s} = 4285.5\,\text{rad/s}$$

In constant-power mode, the desired rotor torque is related to the rated torque as follows:

$$T_r = \frac{P_r}{\omega_r} = \frac{80000}{1071.41}\,\text{Nm} = 74.67\,\text{Nm}$$

The electromagnetic torque is

$$T_{em} = T_r + T_{nl} = 74.67\,\text{Nm} + 2\,\text{Nm} = 76.67\,\text{Nm}$$

The per-phase current is given by

$$I_{ph,q} = \frac{T_{em}}{3k} = \frac{76.67}{0.3 \times 3}\,\text{A} = 85.19\,\text{A}$$

The back emf is given by

$$E_{ph,q} = k\,\omega_r = 0.3 \times 1071.41\,\text{V} = 321.42\,\text{V}$$

The equivalent ac current source to represent the magnet is

$$I_{ph,m} = \frac{2k}{pL_s} = \frac{2 \times 0.3}{8 \times 0.2 \times 10^{-3}}\,\text{A} = 375\,\text{A}$$

The field current is given by

$$I_{ph,d} = I_{ph,m}\left(\frac{\omega_{r(rated)}}{\omega_r} - 1\right) = 375 \times \left(\frac{1}{3.75} - 1\right)\text{A} = -275\,\text{A}$$

The supply voltage components are

$$V_{ph,d} = +R_s I_{ph,d} - \omega_e L_s I_{ph,q} = 0.020 \times (-275)\,\text{V} - 4285.5 \times 0.2 \times 10^{-3} \times 85.19\,\text{V}$$

$$= -78.52\,\text{V}$$

$$V_{ph,q} = E_{ph,q} + R_s I_{ph,q} + \omega_e L_s I_{ph,d}$$

$$= 321.42\,\text{V} + 0.020 \times 85.19\,\text{V} + 4285.5 \times 0.2 \times 10^{-3} \times (-275)\,\text{V} = 87.41\,\text{V}$$

and

$$V_{ph} = \sqrt{V_{ph,d}^2 + V_{ph,q}^2} = \sqrt{(-78.52)^2 + 87.41^2}\,\text{V} = 117.5\,\text{V}$$

The supplied per-phase current is

$$I_{ph} = \sqrt{I_{ph,d}^2 + I_{ph,q}^2} = \sqrt{275^2 + 85.19^2}\,\text{A} = 287.9\,\text{A} \tag{9.59}$$

The machine useful output power is

$$P_r = T_r \omega_r = 80\,\text{kW} \tag{9.60}$$

The machine power loss is the sum of the ohmic and no-load losses:

$$P_{m\,(loss)} = T_{nl}\omega_r + 3R_s I_{ph}^2 = 2 \times 1071.41\,\text{W} + 3 \times 0.020 \times 287.9^2\,\text{W} = 7.116\,\text{kW} \tag{9.61}$$

Thus, the input power to the machine is

$$P = P_r + P_{m\,(loss)} = 80\,\text{kW} + 7.116\,\text{kW} = 87.116\,\text{kW} \tag{9.62}$$

The machine efficiency is

$$\eta = \frac{P_r}{P} = \frac{80}{87.116} \times 100\,\% = 91.83\,\% \tag{9.63}$$

The apparent power of the machine is given by

$$|\bar{S}| = 3\,V_{ph} I_{ph} = 3 \times 117.5 \times 287.9\,\text{VA} = 101.48\,\text{kVA} \tag{9.64}$$

The power factor, PF, is

$$PF = \frac{P}{|\bar{S}|} = \frac{87.116}{101.48} = 0.8585, \text{ lagging} \tag{9.65}$$

9.2.6 Machine Characteristics for Field-Weakened Operation

In this section, the SPM machine of Section 9.2.2.1 is modeled for use as a BEV traction machine. The torque, voltage, and current characteristics are shown in Figure 9.16. Figure 9.16(a) plots the maximum torque and power versus speed. Figure 9.16(b) shows

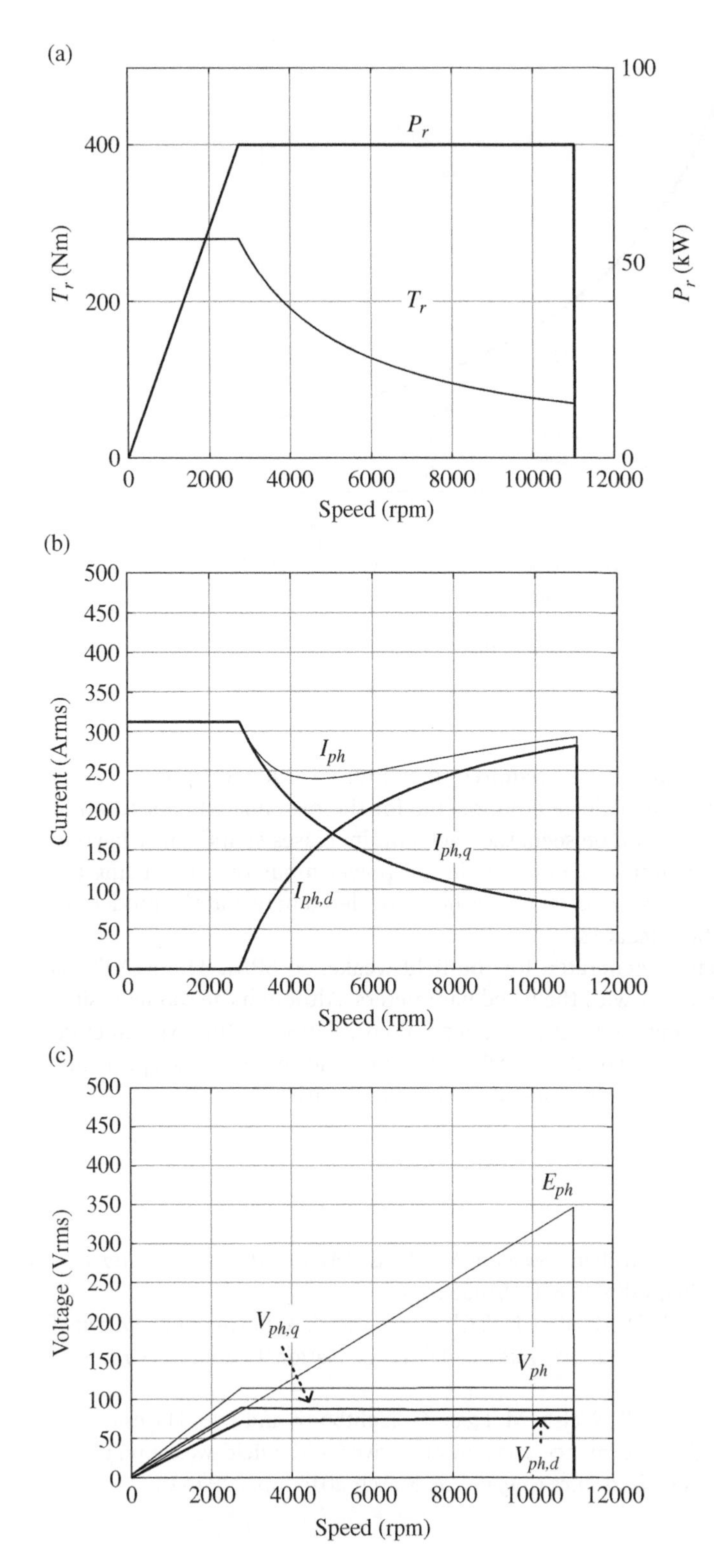

Figure 9.16 Characteristic plots for field-weakened SPM machine without saturation.

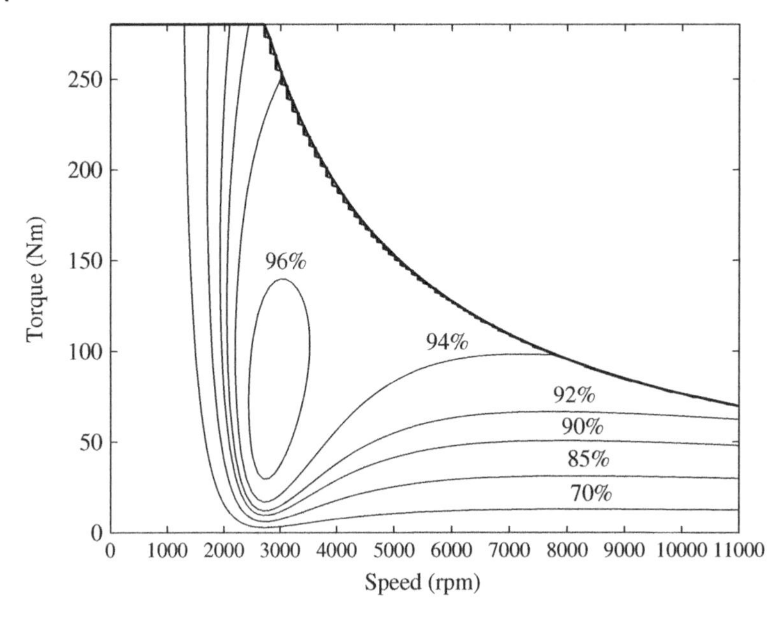

Figure 9.17 Efficiency map for field-weakened SPM machine.

the currents required to achieve the required maximum torque or power, while Figure 9.16(c) shows the various voltage components for the machine when saturation is omitted from the model. As can be seen, the back emf increases significantly beyond the rated speed, but the machine is maintained in full-power mode by introducing the direct-axis field-weakening current. Saturation is not considered here, but the model can be modified to consider the effects [3].

Finally, an efficiency map is generated for the field-weakened SPM. The MATLAB code is provided in the appendix with the listed parameters. Although the model is simple, the efficiency map of Figure 9.17 shows a general correlation to the experimental efficiency map of Chapter 6 Figure 6.13. As desired, the efficiency map shows a peak efficiency in the mid-range for torque and speed – where most driving occurs.

References

1 J. R. Hendershot Jr and T. J. E. Miller, *Design of Brushless Permanent-Magnet Motors*, Oxford Magna Physics, 1994, ISBN 978-1-881855.03.3.
2 R. H. Staunton, C. W. Ayers, L. D. Marlino, J. N. Chiasson, and T. A. Burress, *Evaluation of 2004 Toyota Prius Hybrid Electric Drive System*, Oak Ridge National Laboratory report, May 2006.
3 J. Bermingham, G. O'Donovan, R. Walsh, M. Egan. G. Lightbody, and J. G. Hayes, "Optimized control of high-performance servo-motor drives in the field-weakening region," *IEEE Applied Power Electronics Conference*, March 2016, pp. 2794–2800.

Further Reading

1 T. A. Lipo, *Introduction to AC Machine Design*, 3rd edition, University of Wisconsin, 2007, ISBN 0-9745470-2-6.
2 N. Mohan, *Advanced Electric Drives Analysis Control and Modelling using MATLAB/ Simulink®*, John Wiley & Sons, 2014.
3 P. Krause, O. Wasynczuk, S. D. Sudhoff, and S. Pekarek, *Analysis of Electrical Machinery and Drive Systems*, 3rd edition, Wiley-IEEE Press, 2013, ISBN: 978-1-118-02429-4.

Problems

9.1 An electric drive features a four-pole SPM machine with the following parameters: per-phase stator resistance $R_s = 10$ mΩ, machine constant $k = 0.2$ V/rad/s, synchronous inductance $L_s = 50$ μH, and no-load torque $T_{nl} = 3$ Nm. The machine is rated at 600 Nm and 310 kW.

Determine the applied per-phase voltage and current, machine efficiency, and power factor for the rated operating points.

[Ans. 124.7 V, 1005 A, 90.68%, 0.9092]

9.2 For the machine of Problem 9.1, determine the applied per-phase voltage and current, machine efficiency, and power factor for the rated operating points under full regeneration at rated speed.

[Ans. 106.6 V, 995 A, 89.92%, −0.8760]

9.3 For the machine of Problem 9.1, determine the applied per-phase voltage and current, machine efficiency, and power factor for a rotor torque of 100 Nm at half the rated speed.

[Ans. 53.6 V, 171.7 A, 93.97%, 0.9966]

9.4 For the machine of Problem 9.1, determine the applied per-phase voltage and current, machine efficiency, and power factor for the rated operating points if the machine constant and inductance drop by 30% due to saturation.

[Ans. 101 V, 1435.7 A, 83.02%, 0.858]

9.5 For the machine of Problem 9.1, determine the applied per-phase voltage and current, machine efficiency, and power factor for the rated power at twice the rated speed.

[Ans. 125 V, 1120 A, 88.38%, 0.835]

9.6 For the machine of Section 9.2.5.1, determine the applied per-phase voltage and current, machine efficiency, and power factor for motoring at 25% of the rated power at 1.5 times rated speed.

[Ans. 89.3 V, 136.2 A, 91.03%, 0.6021]

9.7 For the machine of Section 9.2.5.1, determine the applied per-phase voltage and current, machine efficiency, and power factor when regenerating 25% of the rated power at 1.5 times rated speed.

[Ans. 86 V, 134.5 A, 90.3%, −0.521]

MATLAB Code

```
% This program generates an efficiency map for a SPM based on
nominal BEV parameters
% Kevin Davis and John Hayes

close all;clear all; clc;

vmax       = 160/3.6;        % max speed in m/s
r          = 0.315;          % radius
ng         = 8.19;           % gear ratio
waxlemax   = vmax/r;                %max axle angular speed
wrmax      = waxlemax*ng;           %max rotor angular speed
Nrmax      = wrmax*60/(2*pi);       % Maximum speed in rpm

Trrated = 280;               % Rated torque (Nm)
Prrated = 80000;             % Rated power (W)
wrrated = Prrated/Trrated;   % Rated speed (rad/s)
Nrrated = wrrated/2/pi*60    % Rated speed (Rpm)
p = 8;                       % Number of poles

k      =  0.3;               % machine constant (Nm/A)
Rs     =  0.02;              % stator resistance (ohm)
Ls     =  0.2*10^-3;         % phase inductance (H)
Iphm   =  2*k/p/Ls           % Equivalent magnetizing current
Tnl    =  1;                 % no-load torque
%-----------------------------------------------------------------
%Make out a range of torque-speed values to be used for
efficiency map
tr = [0:1:Trrated]; % list of torque values in increments of 1Nm
Nr = [0:100:Nrmax]; % specifies range of speed values in increment
                     100rpm
%-----------------------------------------------------------------
[X,Y] =meshgrid(Nr,tr); % defines x and y axis of torque-speed plot
Pr     =X*pi/30.*Y;        % Calculating the motor output or rotor
                             power based on P=Tw
Pr(Pr>Prrated)=0;          % limiting power to 80kW

Iphd =-Iphm*(1-Nrrated./X); % field-weakening current above
                              rated speed
Iphd(Nr>Nrrated)=0;           % and zero below rated speed
```

```
Pin=Pr+3*Rs*((Y./k/3).^2)+(Tnl.*X.*(pi/30))+3*Rs*((Iphd).^2);
                    % motor input power
Eff=Pr./Pin;        % efficiency map based on the torque-speed
                      ranges given
Tlim=Prrated./(Nr.*(pi/30)); % setting a torque limited curve
                               based on max power and speed
Tlim(Tlim>Trrated)=Trrated;  % max torque value set
%------------------------------------------------------------

%Plotting outputs
[C,h]=contour(X,Y,Eff,[0.7 0.85 0.9 0.92, 0.94, 0.96],'k');%
using this command as I want to display values using the next
command
% no number in above command so only 10 contour levels displayed
clabel(C,h,'manual') ; % this displays the contour values
%l=colorbar; % adds a colorbar to the right of the plot
%set(l,'ylim',[0.7 0.96]); % limits range of colorbar to
values shown
hold on
plot(X,Tlim,'LineWidth',3,'color','k');
    xlabel('Speed (rpm)');
    ylabel('Torque (Nm)');
%xlabel('Speed (rpm)');ylabel('Torque (Nm)');zlabel
('Efficiency (%)');
%title ('BEV Motor Efficiency Map');
```

10

Interior-Permanent-Magnet AC Machine

The interior-permanent-magnet (IPM) machine is the machine of choice for many of the automotive manufacturers. The machine has similar magnet-torque-generating characteristics to its sibling, the surface-permanent-magnet (SPM) machine, but it also can generate significant reluctance torque, an attribute which increases the torque density of the machine. In-depth experimental characterizations of various IPM machines are presented in [1–5]. Note that Chapter 9 is essential reading.

10.1 Machine Structure and Torque Equations

Let us start with the operation of the salient-pole SPM ac machine as shown in Figure 10.1. As shown in Figure 10.1(a), the coil with turns a_{1+q} and a_{1-q} carries current $I_{ph,q}$ and generates electromagnetic torque $3\,k\,I_{ph,q}$ for a three-phase machine, by interacting with the permanent magnet. From the left-hand rule, the interaction between the current and the magnet produces a clockwise torque on the conductor, but since the conductor is stationary, there instead is a counterclockwise torque on the magnet. This torque component is termed the **synchronous permanent-magnet** or **magnet torque** T_{pm} and is expressed as

$$T_{pm} = 3\,k I_{ph,q} \tag{10.1}$$

in a three-phase machine.

Next, place the second coil with turns a_{1+d} and a_{1-d} and carrying a current $I_{ph,d}$ at 90° to the first winding. The effect of the second coil and current $I_{ph,d}$ is to generate a magnetic field to strengthen or weaken the field generated by the permanent magnet, as shown in Figure 10.1(b).

First, consider the torque exerted on coil $a_{1+q} - a_{1-q}$ carrying current $I_{ph,q}$ due to current $I_{ph,d}$. From the right-hand rule, the magnetic field due to $I_{ph,d}$ can be seen as adding to the field from the permanent magnet, thereby generating additional machine torque when interacting with the current $I_{ph,q}$ in coil $a_{1+q} - a_{1-q}$. For a three-phase machine, this adds the term $+3\frac{p}{2}L_{md}I_{ph,d}I_{ph,q}$ to the usual torque component $3\,k\,I_{ph,q}$, a similar term to that derived in Chapters 7 and 16 for the torque of a wound-field dc machine.

Electric Powertrain: Energy Systems, Power Electronics and Drives for Hybrid, Electric and Fuel Cell Vehicles,
First Edition. John G. Hayes and G. Abas Goodarzi.

Companion website: www.wiley.com/go/hayes/electricpowertrain

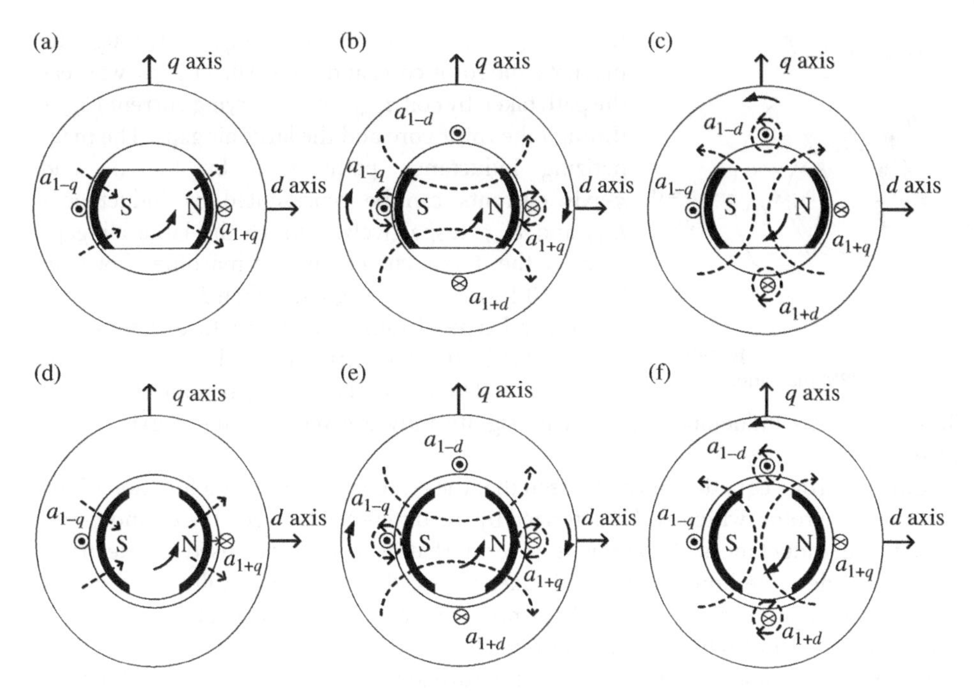

Figure 10.1 Two elementary salient-pole machines: (a–c) $L_{md} > L_{mq}$, (d–f) $L_{md} < L_{mq}$.

Simply put, the additional field results in a buildup of lines of flux which act to torque the rotor in a counterclockwise or positive direction. The inductance L_{md} is the magnetizing inductance along the direct axis.

Next, let us consider the torque on the coil $a_{1+d} - a_{1-d}$ carrying current $I_{ph,d}$ due to current $I_{ph,q}$. Again, applying the left-hand rule, it can be seen that the rotor torque created by coil $a_{1+q} - a_{1-q}$with current $I_{ph,q}$ is clockwise, in effect creating the term $-3\frac{p}{2}L_{mq}I_{ph,d}I_{ph,q}$. The inductance L_{mq} is the magnetizing inductance along the quadrature axis.

This resultant torque is termed the **synchronous reluctance** or **reluctance torque** T_{sr} and is given by

$$\begin{aligned} T_{sr} &= 3\frac{p}{2}L_{md}I_{ph,d}I_{ph,q} - 3\frac{p}{2}L_{mq}I_{ph,d}I_{ph,q} \\ &= 3\frac{p}{2}\left(L_{md} - L_{mq}\right)I_{ph,d}I_{ph,q} \end{aligned} \tag{10.2}$$

In an IPM machine, the electromagnetic torque is the sum of the magnet torque and the reluctance torque:

$$T_{em} = T_{pm} + T_{sr} \tag{10.3}$$

$$\Rightarrow T_{em} = 3kI_{ph,q} + 3\frac{p}{2}\left(L_{md} - L_{mq}\right)I_{ph,d}I_{ph,q} \tag{10.4}$$

where the currents are the per-phase rms quantities.

It is also clear from the salient-pole structure that the magnetic path seen by the currents from the d and q windings differ. The path taken by the flux generated by

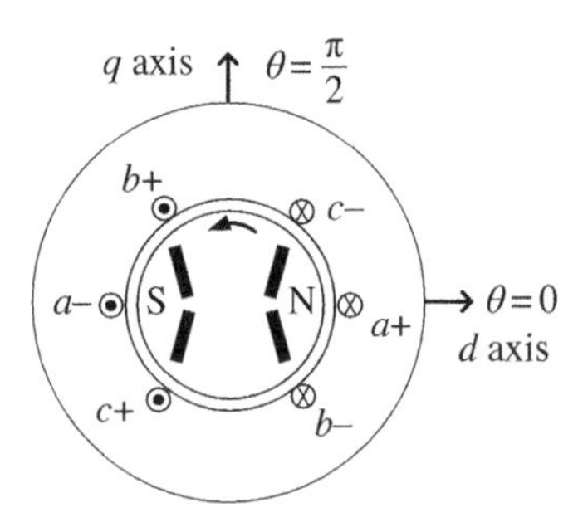

Figure 10.2 Elementary two-pole, three-phase IPM machine.

the coil $a_{1+d} - a_{1-d}$ carrying current $I_{ph,d}$ is through the magnets, the rotor core, and the small air gaps, whereas the path taken by coil $a_{1+q} - a_{1-q}$ carrying current $I_{ph,q}$ is through the rotor core and the large air gaps. The magnetizing inductance paths seen by the d- and q-axis currents can be represented by inductances L_{md} and L_{mq}, respectively. Due to the large air gaps seen by the q-axis current in the machine shown in Figure 10.1(a)–(c), L_{md} is greater than L_{mq}.

A second type of salient-pole structure is shown in Figure 10.1(d)–(f). This structure is closer to the automotive IPM. There is no large air gap along the q-axis. Thus, the q-axis inductance L_{mq} can be significantly greater than the d-axis inductance L_{md}.

From the torque equation, it can be seen that the reluctance torque is a function of the saliency of the rotor, which is characterized by the difference between the d- and q-axis magnetizing inductances. **Saliency** is a term that is commonly used in electrical machines to describe how the magnetic axes physically and magnetically stand out or apart from one another. There is little or no saliency in an SPM machine, but there is clearly a high level of saliency in an IPM machine.

The typical IPM machine has a similar structure to the elementary machine shown in Figure 10.2. As before, the magnetic field interacts with the q-axis current component to generate the magnet torque while the two current components interact to generate the reluctance torque.

The IPM has some key advantages over the SPM:

1) The issue of bonding the magnets to the rotor for the SPM is eliminated by embedding the magnet into the rotor. This enables the rotor to spin at higher speeds.
2) The concern of very high voltages at high speeds is mitigated by reducing the span of the magnet and the resulting magnet torque while compensating for this reduction with reluctance torque. This can also increase the overall torque density of the machine.

10.2 *d*- and *q*-Axis Inductances

The operation of the IPM is now investigated further. Various modes of interest are shown in Figure 10.3. The interaction between the magnets and the ***q*-axis** currents to generate the magnet torque is shown in Figure 10.3(a). The flux patterns generated by the ***d*-axis** current are shown in Figure 10.3(b). As can be seen, the magnets lie across a significant section of the flux path. The flux pattern for the q-axis current is shown in Figure 10.3(c). The flux paths for the q-axis current tend to go around the magnets and not through them, as is the case for the d-axis currents.

Let us now consider the flux patterns in a real machine, with the 2004 Toyota Prius traction motor being the example. The stator and rotor laminations for the 2004 Toyota Prius traction motor are shown in Figure 10.4.

The eight pairs of magnets fit into the eight V shapes on the rotor. The machine has eight magnets, resulting in four north-south pole pairs. In one electrical cycle, the rotor

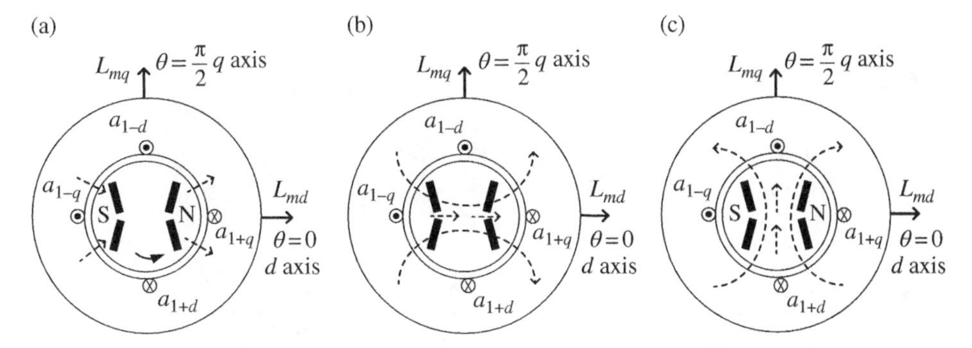

Figure 10.3 Elementary IPM machine showing flux patterns for (a) synchronous torque, (b) *d*-axis, and (c) *q*-axis.

Figure 10.4 2004 Toyota Prius traction motor stator and rotor iron laminations.

completes a quarter rotation mechanically. Thus, the machine is quarter symmetric. Finite element modeling can be used to model the flux patterns within the machine to determine the machine inductances and torque generation. There are many software packages available to model electric machines. The two-dimensional flux patterns shown in Figure 10.5 have been generated using the FEMM software package [6,7].

The flux patterns due to the d- and q-axis currents are shown in Figure 10.5(a) and (b), respectively. The flux generated by the d-axis current follows a pattern through the magnets. The flux generated by the q-axis current follows a pattern around the magnets and does not go through the magnets. In both cases, there is significant local saturation. Plots of the d- and q-axis inductances, generated using FEMM, are shown in Figure 10.6. As can be seen, the q-axis inductance experiences significant saturation as the current increases. The d-axis inductance also saturates but to a far lesser amount due to the low permeability of the magnets. Plots of the flux linkages, λ_{md} and λ_{mq}, are shown in Figure 10.6(b), and these curves trace the magnetization curves of their respective magnetic paths.

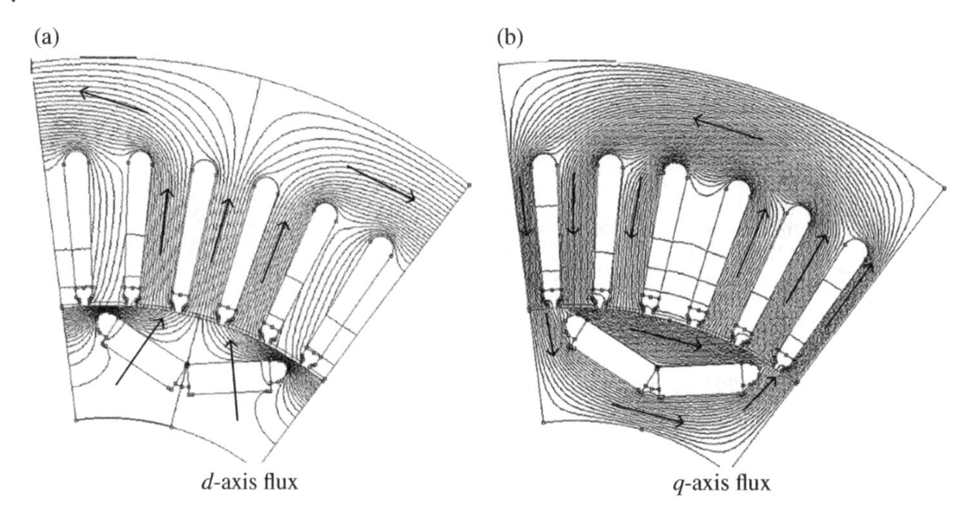

Figure 10.5 d- and q-axis flux patterns [7].

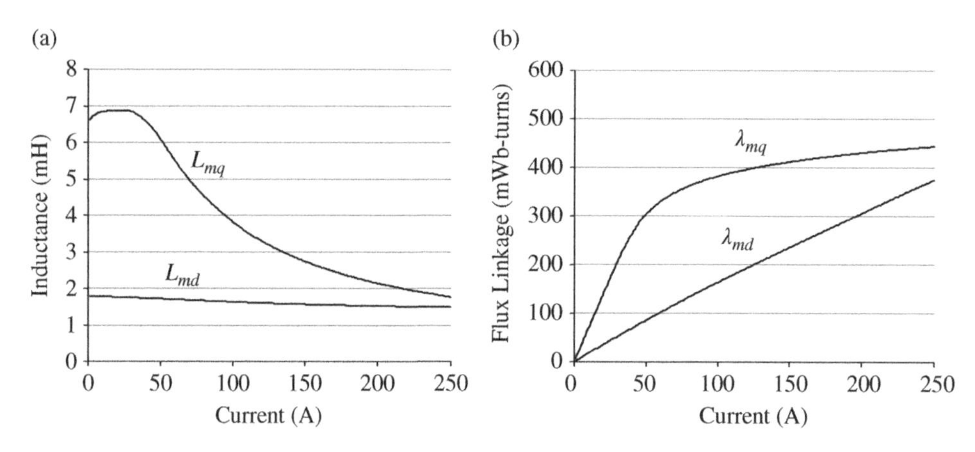

Figure 10.6 Simulated 2004 Toyota Prius motor d- and q-axis (a) inductances, and (b) flux linkages, generated using FEMM. [7].

While finite element analysis is required to characterize a machine, it is useful to run a simple calculation to provide an estimate of the inductances. In these calculations, we simply assume that the length of the pole is the full half circumference of the rotor, that the q-axis flux has to travel through the air gap, and that the d-axis flux has to travel through both the air gap and the magnets. The permeability of the iron is assumed to be infinite compared to that of the air. The permeability of the permanent magnet is close to air, as discussed in Chapter 16, Section 16.4.

The magnetizing inductance of the d-axis L_{md} and the magnetizing inductance of the q-axis L_{mq} are then approximated by

$$L_{md} \approx \frac{3}{2} \times \frac{\mu_0 N^2 \pi r l}{2(l_{pm} + l_g)} \tag{10.5}$$

and

$$L_{mq} \approx \frac{3}{2} \times \frac{\mu_0 N^2 \pi r l}{2 l_g} \tag{10.6}$$

where N is the number of stator turns per phase, l is the rotor length, r is the rotor radius, and l_{pm} and l_g are the thicknesses of the permanent magnet and the length of the air gap, respectively.

10.2.1 Example: Estimating the *d*-axis and *q*-axis Inductances for 2004 Toyota Prius Motor

Per Reference [2], the parameters for the machine are as follows: $p = 8$, $N = 18$, $l_{pm} = 6.5$ mm, $l_g = 0.73$ mm, $l = 8.36$ cm, and $r = 8.03$ cm. Estimate values for the d-axis and q-axis inductances.

Solution:

The two estimates of the inductances are as follows:

$$L_{md} \approx \frac{3}{2} \times \frac{\mu_0 N^2 \pi r l}{2\left(l_{pm} + l_g\right)} = \frac{3}{2} \times \frac{4\pi \times 10^{-7} \times 18^2 \times \pi \times 0.0803 \times 0.0836}{2 \times (6.5 + 0.73) \times 10^{-3}} = 0.9\text{mH}$$

$$L_{mq} \approx \frac{3}{2} \times \frac{\mu_0 N^2 \pi r l}{2 l_g} = \frac{3}{2} \times \frac{4\pi \times 10^{-7} \times 18^2 \times \pi \times 0.0803 \times 0.0836}{2 \times 0.73 \times 10^{-3}} = 9\text{mH}$$

These two values are in the general range of the inductances plotted in Figure 10.6.

10.3 IPM Machine Test

A dynamometer is used to load the drive machine with a load motor and to characterize machine operation with speed and power. The load motor acts as a brake on the motor being tested. The term **brake horsepower** refers to the testing of a drive motor on a **dynamometer** and braking it with a brake or load motor in order to determine the machine torque and power. An example of a dynamometer at Oak Ridge National Laboratory is presented in Figure 10.7.

Figure 10.7 Dynamometer setup for machine testing [5]. (Courtesy of Oak Ridge National Laboratory, US Dept. of Energy.)

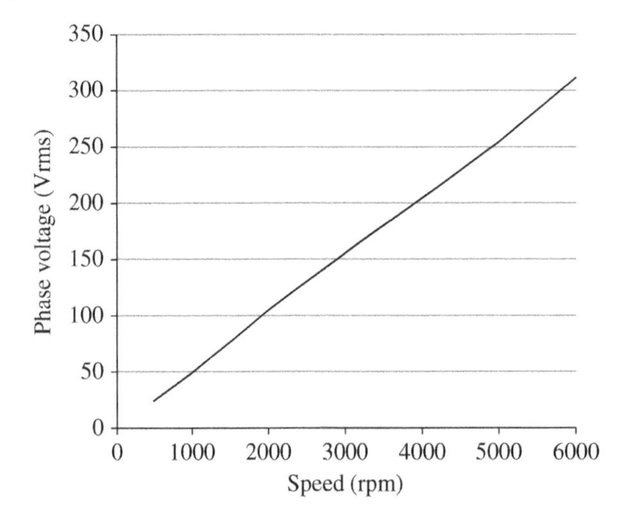

Figure 10.8 Phase rms voltage versus speed for 2004 Toyota Prius traction motor based on data from Reference [1].

10.3.1 No-Load Spin Test

The load motor can be used to spin up the drive motor and determine the relationship of back emf and speed [1]. The results for the 2004 Toyota Prius motor are shown in Figure 10.8. The machine generates a phase voltage of about 315 V_{rms} at 6000 rpm. The machine constant can be determined as follows:

$$k = \frac{E_{ph}}{\omega_r} = \frac{315}{2\pi \times \frac{6000}{60}} \frac{\text{V}}{\text{rads}^{-1}} = 0.50 \frac{\text{V}}{\text{rads}^{-1}} \tag{10.7}$$

10.3.2 DC Torque Test

A significant advantage of the SPM and IPM machines over the induction motor is that the synchronous machine can be excited by dc current at zero speed. The 2004 Toyota Prius motor torque is measured using a dc current excitation as shown in Figure 10.9. The dc current flows into phase a and splits in half, returning through phases b and c.

The machine is excited with the dc current, and the rotor is physically aligned at angle γ.Angle γ is termed the electrical angle and correlates to the physical angle multiplied by the number of pole pairs. In the dc test, the rotor is physically rotated a fixed amount of degrees for each test, in effect creating d- and q-axis components of current, and related magnet and reluctance torque components, as shown in Figure 10.10(a) for current and Figure 10.10(b) for torque.

It can easily be shown that the torque generated by the dc test is as follows:

$$T_r = T_{pm} + T_{sr} = 3k\frac{I_q}{\sqrt{2}} + 3\frac{p}{2}\left(L_{md} - L_{mq}\right)\frac{I_d}{\sqrt{2}}\frac{I_q}{\sqrt{2}} \tag{10.8}$$

where I_d and I_q are the direct and quadrature components of the dc input current. The input dc current correlates to the peak of the per-phase current for an ac excitation.

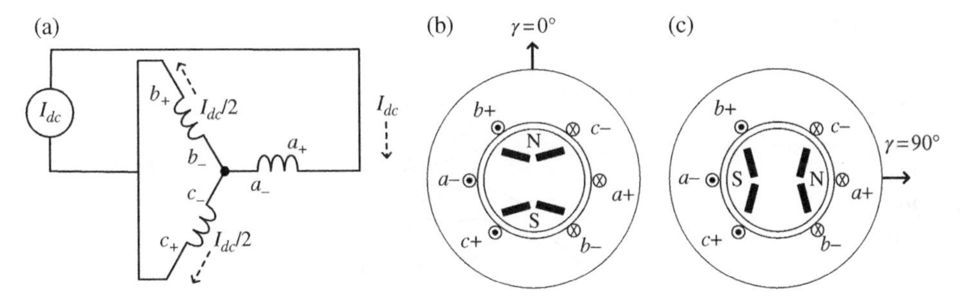

Figure 10.9 IPM torque versus dc test setup.

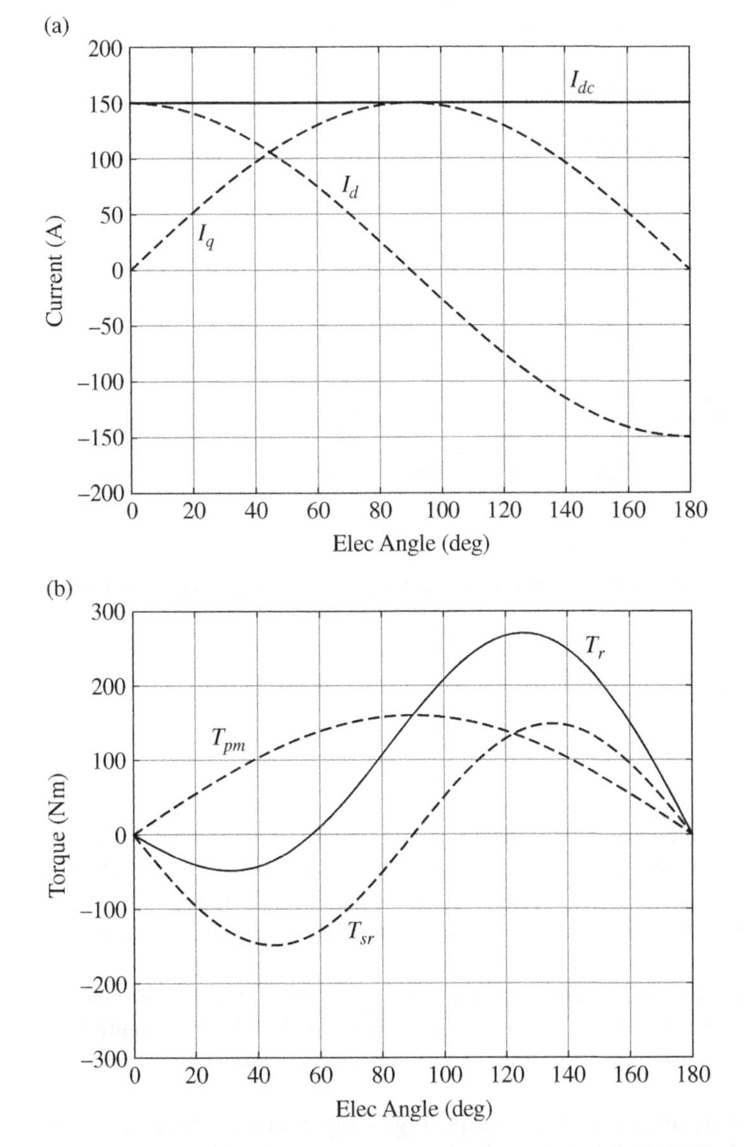

Figure 10.10 Calculated currents versus torque components for 2004 Toyota Prius motor at 150 A.

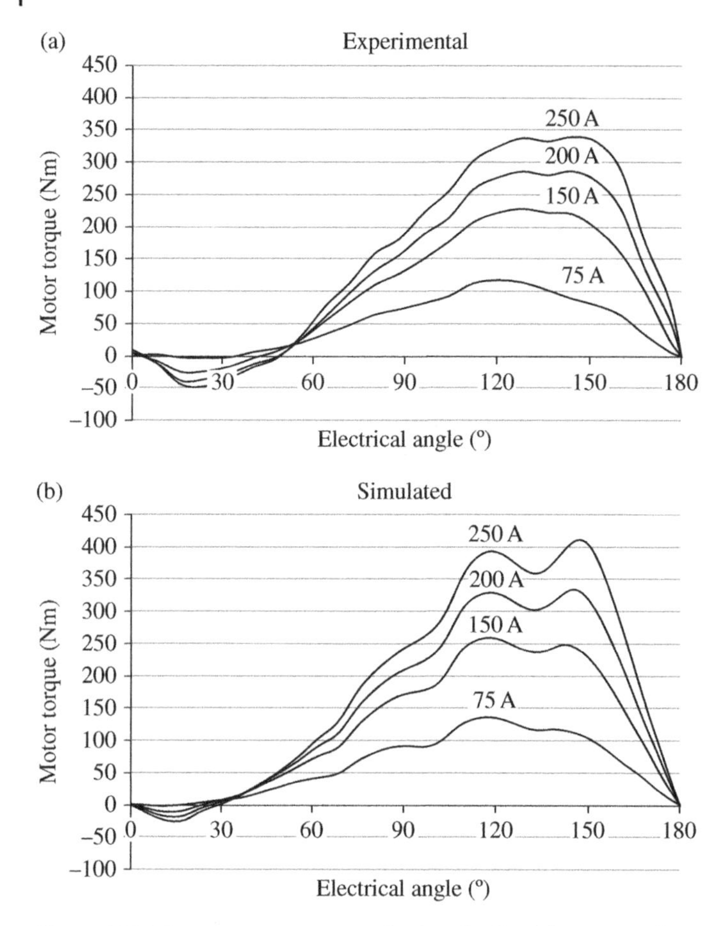

Figure 10.11 Rotor torque versus angle plots for fixed dc currents for (a) experimental [1] and (b) simulated [7].

The currents are given by

$$I_d = I_{dc} \cos \gamma \tag{10.9}$$

$$I_q = I_{dc} \sin \gamma \tag{10.10}$$

At $\gamma = 0°$, no torque is generated as I_q equals zero. At $\gamma = 90°$, only magnet torque is developed as no reluctance torque is developed because I_d equals zero. The reluctance torque peaks positively at $\gamma = 135°$. The overall torque developed by the machine tends to peak at about $\gamma = 120°$ if the two components of torque are approximately equal at their peaks.

Experimental results based on Reference [1] are shown in Figure 10.11(a). Simulated torque versus current characteristics are developed using FEMM and are presented in Figure 10.11(b). The experimental and the FEMM-simulated results show a good correlation.

Starting at $\gamma = 0°$, no magnet or reluctance torque is generated. Rotor torque T_r builds up as the rotor is shifted to $\gamma = 90°$. At this angle, the magnets are aligned with the stator

Table 10.1 DC test summary for 2004 Toyota Prius motor [1].

I_{dc} (A)	T_r (90°) (Nm)	k (Nm/A)	T_r (120°) (Nm)	T_{pm} (120°) (Nm)	I_q(120°) (A)	T_{sr} (120°) (Nm)	I_d(120°) (A)	$L_{md} - L_{mq}$ (mH)
75	75	0.47	118	65	65	53	−37.5	−3.6
150	136	0.43	223	118	130	105	−75	−1.8
200	162	0.38	277	140	173	137	−100	−1.3
250	191	0.36	324	165	217	159	−125	−1.0

currents such that maximum magnet torque is generated but no reluctance torque. Thus, at 90° electrical

$$T_{sr}(90°) = 0, \text{ and } T_{pm}(90°) = T_r(90°) = 3k\frac{I_{dc}}{\sqrt{2}} \tag{10.11}$$

Therefore, the machine constant k can be calculated at this test point:

$$k = \frac{\sqrt{2}\,T_r(90°)}{3I_{dc}} \tag{10.12}$$

The values are tabulated in Table 10.1. The k values of 0.47 to 0.36 Nm/A predicted by the torque tests correlate with the k value of 0.5 Nm/A calculated earlier using the back emf and speed. Clearly, the machine constant drops with current as the machine is saturating at higher currents.

The overall torque is at a maximum at $\gamma = 120°$ and is logged in Table 10.1. The contribution by the magnet torque remains high, and the magnet torque and quadrature current components are cos30° times the peak values at $\gamma = 90°$. The current and torque contributions are as follows:

$$I_q(120°) = I_{dc} \times \sin 120° \tag{10.13}$$

and

$$I_d(120°) = I_{dc} \times \cos 120° \tag{10.14}$$

$$T_{pm}(120°) = T_{pm}(90°) \times \cos 30° \tag{10.15}$$

$$T_{sr}(120°) = T_r(120°) - T_{pm}(120°) \tag{10.16}$$

Since

$$T_{sr} = 3\cdot\frac{p}{2}\cdot\left(L_{md} - L_{mq}\right)\cdot\frac{I_d}{\sqrt{2}}\cdot\frac{I_q}{\sqrt{2}} \tag{10.17}$$

we get an expression for the inductance difference at this operating condition:

$$\left(L_{md} - L_{mq}\right) = \frac{4T_{sr}}{3p\,I_d I_q} \tag{10.18}$$

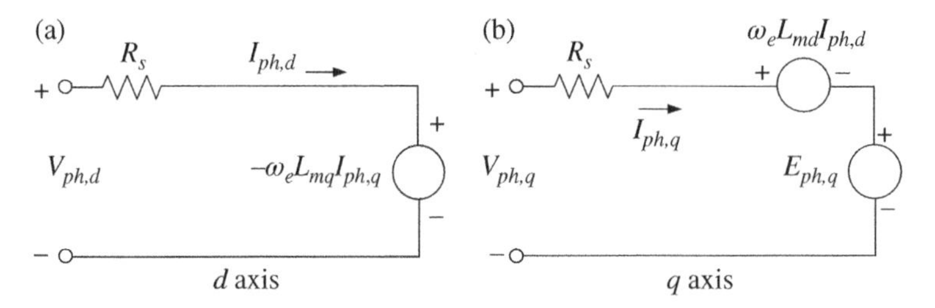

Figure 10.12 Per-phase direct and quadrature equivalent circuits of IPM ac machine.

10.4 Basic Theory and Low-Speed Operation

The equivalent circuit equation is similar to Equation (9.47) and is

$$\bar{V}_{ph} = V_{ph,d} + jV_{ph,q} = +R_s I_{ph,d} - \omega_e L_{mq} I_{ph,q} + j\left(E_{ph,q} + R_s I_{ph,q} + \omega_e L_{md} I_{ph,d}\right) \tag{10.19}$$

and so

$$V_{ph,q} = E_{ph,q} + R_s I_{ph,q} + \omega_e L_{md} I_{ph,d} \tag{10.20}$$

$$V_{ph,\mathrm{d}} = +R_s I_{ph,d} - \omega_e L_{mq} I_{ph,q} \tag{10.21}$$

The per-phase equivalent circuits of the IPM are as shown in Figure 10.12. The circuit is decomposed in order to separately show the direct and quadrature voltages as a function of their various components. Decoupled circuits are necessary as the direct and quadrature circuits feature different inductances.

The per-phase equivalent circuit is better understood by reviewing the phasor diagram of Figure 10.13. The phasor diagram is similar to that of the field-weakened SPM. The significant difference is that the d-axis current is utilized over the full speed range for the IPM and not just above the base speed as for the SPM. The significantly larger q-axis inductance can result in a large voltage drop, which may be the most significant component at high speeds for the IPM.

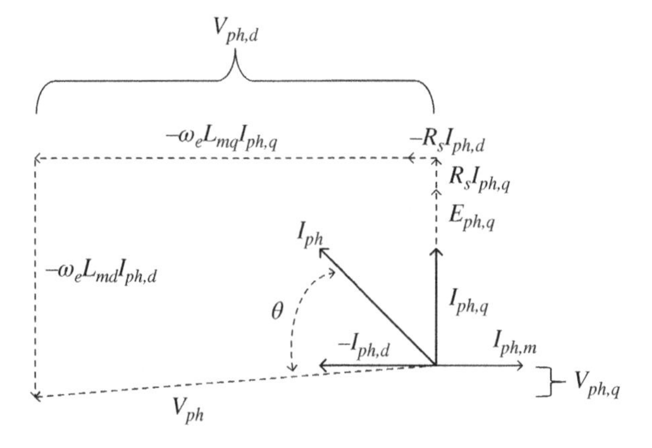

Figure 10.13 Phasor diagram for IPM ac machine in motoring mode with negative $I_{ph,d}$.

The input power can be expressed as

$$P = 3\left(V_{ph,q}I_{ph,q} + V_{ph,d}I_{ph,d}\right) \tag{10.22}$$

or

$$P = 3\left[\left(E_{ph,q} + R_s I_{ph,q} + \omega_e L_{md} I_{ph,d}\right) \times I_{ph,q} + \left(R_s I_{ph,d} - \omega_e L_{mq} I_{ph,q}\right) \times I_{ph,d}\right] \tag{10.23}$$

and rearranged in terms of the machine ohmic loss and electromechanical power:

$$\begin{aligned} P &= 3R_s\left(I_{ph,q}{}^2 + I_{ph,d}{}^2\right) + 3E_{ph,q}I_{ph,q} + 3\omega_e\left(L_{md} - L_{mq}\right)I_{ph,d}I_{ph,q} \\ &= \text{ohmic loss} + \text{electromechanical power} \end{aligned} \tag{10.24}$$

The electromagnetic power is given by

$$P_{em} = 3E_{ph,q}I_{ph,q} + 3\omega_e\left(L_{md} - L_{mq}\right)I_{ph,d}I_{ph,q} \tag{10.25}$$

10.4.1 Example: Motoring at Rated Condition

An eight-pole motor designed for use in a hybrid electric vehicle (HEV) is operating in forward-motoring mode and develops an output rotor torque of +324 Nm at 1500 rpm.

Use the parameters determined in the final row of Table 10.1.

Split the magnet and reluctance torques 165 Nm :159 Nm as per the final row.

Let the per-phase stator resistance R_s = 69 mΩ, and let the no-load torque T_{nl} equal zero (for simplicity).

Let the inductances be L_{md}= 1.6 mH and L_{mq}= 2.6 mH at this condition.

Determine the applied per-phase voltage and current, and machine efficiency and power factor for this operating point.

Solution:

The rotor speeds in hertz and rad/s are given as follows:

$$\omega_r = 2\pi\frac{N_r}{60} = 2\pi\frac{1500}{60}\text{rad/s} = 157.08\,\text{rad/s} \quad \text{and} \quad f_r = \frac{\omega_r}{2\pi} = \frac{157.08}{2\pi}\text{Hz} = 25\,\text{Hz}$$

The electrical input frequencies are

$$\omega_e = \frac{p}{2}\omega_r = \frac{8}{2} \times 157.08 = 628.3\,\text{rad/s} \quad \text{and} \quad f_e = \frac{\omega_e}{2\pi} = \frac{628.3}{2\pi}\text{Hz} = 100\,\text{Hz}$$

The electromagnetic torque is

$$T_{em} = T_r + T_{nl} = 324\,\text{Nm} + 0\,\text{Nm} = 324\,\text{Nm}$$

The magnet and reluctance torques are

$$T_{pm} = 165\,\text{Nm} \quad \text{and} \quad T_{sr} = 159\,\text{Nm}$$

The per-phase quadrature current is given by

$$I_{ph,q} = \frac{T_{pm}}{3k} = \frac{165}{3 \times 0.36}\text{A} = 152.8\,\text{A}$$

From Figure 10.6, the d-axis inductance is approximately 1.6 mH at this condition.

From Table 10.1 the inductance difference is 1.0 mH, which implies that the q-axis inductance is 2.6 mH.

The per-phase direct current is given by

$$I_{ph,d} = \frac{2T_{sr}}{3p(L_{md} - L_{mq})I_{ph,q}} = \frac{2 \times 159}{3 \times 8 \times (1.6 \times 10^{-3} - 2.6 \times 10^{-3}) \times 152.8} \mathrm{A} = -86.7\,\mathrm{A}$$

The supplied per-phase current is as follows:

$$|\bar{I}_{ph}| = \sqrt{I_{ph,d}^2 + I_{ph,q}^2} = \sqrt{(-86.7)^2 + 152.8^2}\,\mathrm{A} = 175.7\,\mathrm{A}$$

The back emf is given by

$$E_{ph,q} = k\omega_r = 0.36 \times 157.08\,\mathrm{V} = 56.55\,\mathrm{V}$$

The quadrature component of the supply voltage is

$$V_{ph,q} = E_{ph,q} + R_s I_{ph,q} + \omega_e L_{md} I_{ph,d} = 56.55\mathrm{V} + 0.069 \times 152.8\mathrm{V} + 628.3 \times 1.6 \times 10^{-3} \times (-86.7)\mathrm{V} = -20.1\,\mathrm{V}$$

The direct component of the supply voltage is

$$V_{ph,d} = R_s I_{ph,d} - \omega_e L_{mq} \cdot I_{ph,q} = 0.069 \times (-86.7)\mathrm{V} - 628.3 \times 2.6 \times 10^{-3} \times 152.8\,\mathrm{V} = -255.6\,\mathrm{V}$$

The supplied per-phase voltage is as follows:

$$|\bar{V}_{ph}| = \sqrt{V_{ph,d}^2 + V_{ph,q}^2} = \sqrt{(-20.1)^2 + (-255.6)^2}\,\mathrm{V} = 256.4\,\mathrm{V}$$

The machine useful output power is

$$P_r = T_r \omega_r = 324 \times 157.08\,\mathrm{W} = 50.89\,\mathrm{kW}$$

The machine power loss is the sum of the ohmic and no-load losses:

$$P_{m\,(loss)} = T_{nl}\omega_r + 3R_s I_{ph}^2 = 0 \times 157.08\,\mathrm{W} + 3 \times 0.069 \times 175.7^2\,\mathrm{W} = 6.39\,\mathrm{kW}$$

Thus, the input power to the machine is

$$P = P_r + P_{m\,(loss)} = 50.89\,\mathrm{kW} + 6.39\,\mathrm{kW} = 57.28\,\mathrm{kW}$$

The machine efficiency is given by

$$\eta_m = \frac{P_r}{P} = \frac{50.89}{57.28} \times 100\% = 88.84\%$$

The apparent power of the machine is given by

$$|\bar{S}| = 3V_{ph}I_{ph} = 3 \times 256.4 \times 175.7\,\mathrm{VA} = 135.1\,\mathrm{kVA}$$

The power factor PF is

$$PF = \frac{P}{|\bar{S}|} = \frac{57.28}{135.1} = 0.424, \text{lagging}$$

While the required machine torque has been generated in the example, the phase current is high, resulting in a low power factor. IPM machines are tested over a range of currents in order to determine the optimum operating condition for a given torque and speed.

Note that the per-phase values of current calculated in this example are equal to the dc currents presented in the final row of Table 10.1 divided by √2, as expected.

There are two principal operating modes resulting in optimal operation.

10.4.2 Maximum Torque per Ampere (MTPA)

In the **MTPA mode**, the IPM is controlled such that the maximum torque is generated for a given per-phase current.

10.4.3 Maximum Torque per Volt (MTPV) or Maximum Torque per Flux (MTPF)

In the **MTPV mode**, the IPM is controlled such that the maximum torque is generated for a given per-phase voltage. This is especially important when the machine is operating in constant power mode and close to the voltage limits.

10.5 High-Speed Operation of IPM Machine

The IPM machine can be operated at very high speeds – typically up to four or five times the rated speed. The operation of the machine above the rated speed is similar to operation below or at the rated speed as a significant d-axis current is required to both weaken the field of the magnet and provide reluctance torque.

At high speeds, the voltage is the limiting factor, and so it is usual to operate in MTPV mode. Again, the machine is characterized experimentally in order to generate the optimum supply voltages and currents.

10.5.1 Example: Motoring at High Speed using IPM Machine

The motor of the previous example is operating in forward-motoring mode and develops an output rotor torque of +81 Nm at 6000 rpm.

Determine the applied per-phase voltage and current, and machine efficiency and power factor for this operating point.

For simplicity, maintain k, L_{md}, and $I_{ph,d}$ at the values estimated for the condition in the previous example, and use $L_{md} - L_{mq} = -3.6$ mH, from the 75 A row of Table 10.1 to determine L_{mq}, as $I_{ph,q}$ is relatively low for this condition. Thus, $L_{mq} = 5.2$ mH. Let $T_{nl} = 0$ Nm.

Solution:

The rotor speed in hertz and rad/s is given as follows:

$$\omega_r = 2\pi\frac{N_r}{60} = 2\pi\frac{6000}{60}\text{rad/s} = 628.3\,\text{rad/s and } f_r = \frac{\omega_r}{2\pi} = \frac{628.3}{2\pi}\text{Hz} = 100\,\text{Hz}$$

The electrical input frequencies are

$$\omega_e = \frac{p}{2}\omega_r = \frac{8}{2} \times 628.3 = 2513\,\text{rad/s and } f_e = \frac{\omega_e}{2\pi} = \frac{2513}{2\pi}\text{Hz} = 400\,\text{Hz}$$

The electromagnetic torque is

$$T_{em} = T_r + T_{nl} = 81\,\text{Nm} + 0\,\text{Nm} = 81\,\text{Nm}$$

The per-phase direct current is as in the previous example:

$$I_{ph,d} = -86.7\text{A}$$

The per-phase quadrature current can be determined from Equation (10.4):

$$I_{ph,q} = \frac{T_r}{3k + 3\frac{p}{2}(L_{md} - L_{mq})I_{ph,d}}$$
$$= \frac{81}{3 \times 0.36 + 3 \times \frac{8}{2} \times (1.6 \times 10^{-3} - 5.2 \times 10^{-3}) \times (-86.7)}\text{A} = 16.8\text{A}$$

The supplied per-phase voltage is as follows:

$$|\bar{I}_{ph}| = \sqrt{I_{ph,d}^2 + I_{ph,q}^2} = \sqrt{(-86.7)^2 + 16.8^2}\text{A} = 88.3\text{A}$$

The back emf is given by

$$E_{ph,q} = k\omega_r = 0.36 \times 628.3\text{V} = 226.2\text{V}$$

The magnet and reluctance torques are

$$T_{pm} = 3kI_{ph,q} = 3 \times 0.36 \times 16.8\text{Nm} = 18.1\,\text{Nm}$$

$$T_{sr} = 3\frac{p}{2}(L_{md} - L_{mq})I_{ph,q}I_{ph,q} = 3 \times \frac{8}{2} \times (0.0016 - 0.0052) \times 16.8 \times (-86.7) = 62.9\text{Nm}$$

The magnet and reluctance torques are split in the ratio 22:78 for this MTPV condition.

The various voltage components are

$$V_{ph,q} = E_{ph,q} + R_s I_{ph,q} + \omega_e L_{md} I_{ph,d}$$
$$= 226.2\text{V} + 0.069 \times 16.8\text{V} + 2513 \times 1.6 \times 10^{-3} \times (-86.7)\text{V} = -121.2\text{V}$$

$$V_{ph,d} = R_s I_{ph,d} - \omega_e L_{mq} \cdot I_{ph,q} = 0.069 \times (-86.7)\text{V} - 2513 \times 5.2 \times 10^{-3} \times 16.8\text{V}$$
$$= -225.5\text{V}$$

$$|\bar{V}_{ph}| = \sqrt{V_{ph,d}^2 + V_{ph,q}^2} = \sqrt{(-225.5)^2 + (-121.2)^2}\text{V} = 256\text{V}$$

The machine power loss is

$$P_{m\,(loss)} = T_{nl}\omega_r + 3R_s I_{ph}^2 = 0 \times 628.3\text{W} + 3 \times 0.069 \times 88.3^2\text{W} = 1.61\text{kW}$$

Thus, the input power to the machine is

$$P = P_r + P_{m\,(loss)} = 81 \times 628.3\text{W} + 1.61\text{kW} = 50.89\text{kW} + 1.61\text{kW} = 52.5\text{kW}$$

The machine efficiency is given by

$$\eta_m = \frac{P_r}{P} = \frac{50.89}{52.5} \times 100\% = 96.9\%$$

The apparent power of the machine is given by

$$|\bar{S}| = 3\,V_{ph}I_{ph} = 3 \times 256 \times 88.3\,\text{VA} = 67.81\,\text{kVA}$$

The power factor is

$$PF = \frac{P}{|\bar{S}|} = \frac{52.5}{67.81} = 0.774, \text{lagging}$$

10.6 dq Modeling of Machines

We have analyzed several machines over the last several chapters. The three-phase ac machines can appear more complex from a control perspective compared to the simpler dc machines. In all these machines, we can see that the torque is directly related to the supply currents.

The control of an ac machine can be simplified by recognizing that the currents being supplied have a common frequency and result in a spinning magnetic field. We came across the concept of the space-vector current in Chapter 8, Section 8.1. The space-vector current is rotating at the synchronous frequency and comprises contributions from the three-phase currents. While the magnitudes of the phase currents are time varying, the **space-vector current** has a constant magnitude and is spinning at the synchronous frequency. In this section, the space-vector current is represented by its **direct** and **quadrature** current components, which are also allowed to spin at the synchronous frequency, as shown in Figure 10.14(a). Thus, the ac machine can be modeled as if it were a dc machine controlled by the current alone. As the axes spin, they can be referenced to the stationary phase a-axis, as shown in Figure 10.14(b).

For the SPM and IPM machines, the stator **direct d-axis** is aligned with the axis of the permanent magnet while the **quadrature q-axis** is aligned at 90° to the d-axis, similar to the earlier work on the per-phase models.

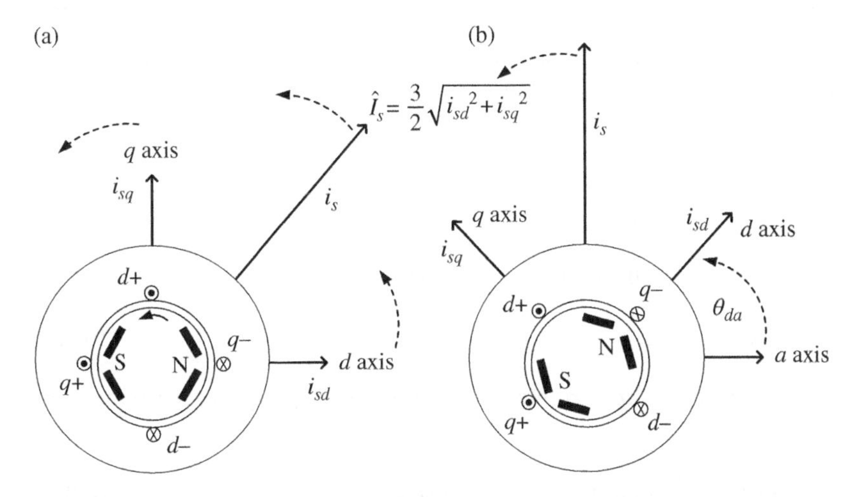

Figure 10.14 d- and q-axis representations.

We have a number of options for how we transform the space-vector current to the d- and q-axes currents. First, we consider the constant current transformation, and then the constant power transformation. Both are commonly used.

These transformations are often known as the **Clarke–Park transformations**. Robert Park (1902–1994) contributed to the theory in the 1920s. Edith Clarke (1883–1959) made some great contributions to power systems analysis, while also blazing a pioneering trail as a female engineer. She became the first female professor of electrical engineering in the United States in 1947.

10.6.1 Constant Current Transformation

The d- and q-axis current components can be related to the time-varying per-phase currents as follows. The earlier formula of Equation (8.1) for the space-vector current is rewritten as follows in exponential form:

$$\vec{i}_s^{\,a}(t) = i_a(t)e^{j0^\circ} + i_b(t)e^{j120^\circ} + i_c(t)e^{j240^\circ} \tag{10.26}$$

This expression for $\vec{i}_s(t)$ is with reference to the stationary a-axis. Let θ_{da} be the angle between the rotating d-axis and the stationary a-axis. Thus, the space-vector current $\vec{i}_s(t)$ with reference to the d-axis is as follows:

$$\vec{i}_s(t) = \vec{i}_s^{\,a}(t)e^{-j\theta_{da}} \tag{10.27}$$

Thus:

$$\vec{i}_s(t) = i_a(t)e^{-j\theta_{da}} + i_b(t)e^{-j(\theta_{da}-120^\circ)} + i_c(t)e^{-j(\theta_{da}-240^\circ)} \tag{10.28}$$

For this transformation, the peak space-vector current is 3/2 times the vector sum of the d- and q-axis currents [8].

Thus, the space-vector current with respect to the d-axis is simply

$$\vec{i}_s(t) = \frac{3}{2}\left(i_{sd} + ji_{sq}\right) \tag{10.29}$$

In the steady state, the stator d-axis current i_{sd} and stator q-axis current i_{sq} are equal to the peak of the earlier per-phase direct and quadrature currents:

$$i_{sd} = \sqrt{2}I_{ph,d} \tag{10.30}$$

$$i_{sq} = \sqrt{2}I_{ph,q} \tag{10.31}$$

The magnitude of the rotating space-vector current $\hat{I}_s$ is

$$\hat{I}_s = \frac{3}{2}\sqrt{i_{sd}^{\,2} + i_{sq}^{\,2}} = \frac{3}{2}\sqrt{2}I_{ph} \tag{10.32}$$

We can express the d and q currents in terms of their per-phase currents and vice versa:

$$i_{sd} + ji_{sq} = \frac{2}{3}\left(i_a(t)e^{-j\theta_{da}} + i_b(t)e^{-j(\theta_{da}-120^\circ)} + i_c(t)e^{-j(\theta_{da}-240^\circ)}\right) \tag{10.33}$$

In matrix form:

$$\begin{bmatrix} i_{sd} \\ i_{sq} \end{bmatrix} = \frac{2}{3}\begin{bmatrix} \cos(\theta_{da}) & \cos(\theta_{da}-120°) & \cos(\theta_{da}-240°) \\ -\sin(\theta_{da}) & -\sin(\theta_{da}-120°) & -\sin(\theta_{da}-240°) \end{bmatrix}\begin{bmatrix} i_a(t) \\ i_b(t) \\ i_c(t) \end{bmatrix} \tag{10.34}$$

The inverse form is given by

$$\begin{bmatrix} i_a(t) \\ i_b(t) \\ i_c(t) \end{bmatrix} = \begin{bmatrix} \cos(\theta_{da}) & -\sin(\theta_{da}) \\ \cos(\theta_{da}-120°) & -\sin(\theta_{da}-120°) \\ \cos(\theta_{da}-240°) & -\sin(\theta_{da}-240°) \end{bmatrix}\begin{bmatrix} i_{sd} \\ i_{sq} \end{bmatrix} \tag{10.35}$$

The earlier expression, Equation, (10.4) for torque in terms of the per-phase components was

$$T_{em} = 3kI_{ph,q} + 3\frac{p}{2}\left(L_{md} - L_{mq}\right)I_{ph,d}I_{ph,q}$$

and can be rewritten in terms of the d- and q-axis components as

$$T_{em} = 3\frac{k}{\sqrt{2}}i_{sq} + \frac{3p}{2\,2}\left(L_{md} - L_{mq}\right)i_{sd}i_{sq} \tag{10.36}$$

We know that

$$k = \frac{p}{2}\lambda \tag{10.37}$$

where λ is the flux linkage due to the permanent magnets. Again we note here that the flux linkage λ is treated in this textbook as an rms quantity for ac machines. It is very common for peak flux linkage to be used, in which case an adjustment by a factor of $\sqrt{2}$ is required.

Thus, the machine torque can be represented in the following commonly used form:

$$T_{em} = \frac{3}{2}\frac{p}{\sqrt{2}}\lambda i_{sq} + \frac{3}{2}\frac{p}{2}\left(L_{md} - L_{mq}\right)i_{sd}i_{sq} \tag{10.38}$$

which can be rearranged as follows:

$$T_{em} = \frac{3}{2}\frac{p}{2}\left(\sqrt{2}\lambda + L_{md}i_{sd}\right)i_{sq} - \frac{3}{2}\frac{p}{2}L_{mq}i_{sq}i_{sd} \tag{10.39}$$

Torque can then be expressed in terms of flux linkage and current as

$$T_{em} = \frac{3}{2}\frac{p}{2}\left(\lambda_{sd}i_{sq} - \lambda_{sq}i_{sd}\right) \tag{10.40}$$

where

$$\lambda_{sd} = \sqrt{2}\lambda + L_{md}i_{sd} \text{ and } \lambda_{sq} = L_{mq}i_{sq} \tag{10.41}$$

For the SPM ac machine, $L_{md} = L_{mq}$, and we simply get

$$T_{em} = \frac{3}{2}\frac{p}{\sqrt{2}}\lambda i_{sq} \tag{10.42}$$

In the steady state, the d and q voltages are related to the per-phase components by

$$v_{sd} = \sqrt{2} V_{ph,d} \tag{10.43}$$

$$v_{sq} = \sqrt{2} V_{ph,q} \tag{10.44}$$

Per Equation (10.22), the machine power in per-phase quantities can be expressed as

$$P = 3\left(V_{ph,q} I_{ph,q} + V_{ph,d} I_{ph,d}\right)$$

This power expression can be rewritten in terms of d and q components as follows, since the d and q components are equal to the peak of the related per-phase component:

$$P = \frac{3}{2}\left(v_{sq} i_{sq} + v_{sd} i_{sd}\right) \tag{10.45}$$

10.6.2 Constant Power Transformation

This transformation is also commonly used and is briefly presented here [9]. For this transformation, the peak space-vector current is $\sqrt{\frac{3}{2}}$ times the vector sum of the d- and q-axis currents:

$$\vec{i}_s(t) = \sqrt{\frac{3}{2}}\left(i_{sd} + j i_{sq}\right) \tag{10.46}$$

In the steady state, the d- and q-axis currents can be easily calculated on the basis of the earlier per-phase direct and quadrature currents:

$$i_{sd} = \sqrt{3} I_{ph,d} \tag{10.47}$$

$$i_{sq} = \sqrt{3} I_{ph,q} \tag{10.48}$$

The magnitude of the rotating space-vector current $\hat{I}_s$ is

$$\hat{I}_s = \sqrt{\frac{3}{2}}\sqrt{i_{sd}^2 + i_{sq}^2} = \sqrt{\frac{3}{2}}\sqrt{3} I_{ph} \tag{10.49}$$

Similarly, the d and q voltages are related to the per-phase components by

$$v_{sd} = \sqrt{3} V_{ph,d} \tag{10.50}$$

$$v_{sq} = \sqrt{3} V_{ph,q} \tag{10.51}$$

The torque is given by

$$T_{em} = \sqrt{3} k\, i_{sq} + \frac{p}{2}\left(L_{md} - L_{mq}\right) i_{sd} i_{sq} \tag{10.52}$$

or

$$T_{em} = \frac{p}{2}\left(\lambda_{sd} i_{sq} - \lambda_{sq} i_{sd}\right) \tag{10.53}$$

In this case, since the d and q components are equal to $\sqrt{3}$ times the rms of the related per-phase component, the power is given by

$$P = v_{sq} i_{sq} + v_{sd} i_{sd} \tag{10.54}$$

References

1 R. H. Staunton, C. W. Ayers, L. D. Marlino, J. N. Chiasson, and T. A. Burress, *Evaluation of 2004 Toyota Prius Hybrid Electric Drive System*, Oak Ridge National Laboratory report, May 2006.
2 T. A. Burress, C. L. Coomer, S. L. Campbell, L. E. Seiber, L. D. Marlino, R. H. Staunton, J. P. Cunningham, and H. T. Lin, *Evaluation of the 2007 Toyota Camry Hybrid Synergy Electric Drive System*, Oak Ridge National Laboratory report, 2008.
3 T. A. Burress, S. L. Campbell, C. L. Coomer, C. W. Ayers, A. A. Wereszczak, J. P. Cunningham, L. D. Marlino, L. E. Seiber, and H. T. Lin, *Evaluation of the 2010 Toyota Prius Hybrid Synergy Drive System*, Oak Ridge National Laboratory report, 2011.
4 J. S. Hsu, C. W. Ayers, C. L. Coomer, R. H. Wiles, S. L. Campbell, K. T. Lowe, and R. T. Michelhaugh, *Report on Toyota/Prius Motor Torque Capability, Torque Property, No-load Back Emf, and Mechanical Losses*, Oak Ridge National Laboratory report, 2004.
5 C. W. Ayers, J. S. Hsu, C. W. Miller, G. W. Ott, Jr., and C. B. Oland, *Evaluation of 2004 Toyota Prius Hybrid Electric Drive System Interim Report*, Oak Ridge National Laboratory report, 2004.
6 *Finite Element Method Magnetics*, a free software by David Meeker: www.femm.info.
7 K. Davis, Y. Lishchynskyy, O. Melon, and J. G. Hayes, *FEMM Modelling of an IPM* AC *Machine*, University College Cork, 2017 (files available on Wiley book web site).
8 D. W. Novotny and T. A. Lipo, *Vector Control and Dynamics of AC Drives*, Oxford Science Publications, 1996.
9 N. Mohan, *Advanced Electric Drives Analysis Control and Modelling using MATLAB/ Simulink®*, John Wiley & Sons, 2014.

Further Reading

1 T. Nakada, S. Ishikawa, and S. Oki, "Development of an electric motor for a newly developed electric vehicle," SAE Technical Paper 2014-01-1879.
2 D. G. Dorrell, A. M. Knight, and M. Popescu, "Performance improvement in high-performance brushless rare-earth magnet motors for hybrid vehicles by use of high flux-density steel," *IEEE Transactions on Magnetics*, 47 (10), pp. 3016–3019, October 2011.
3 S. T. Lee, T. A. Burress, and L. M. Tolbert, "Power-factor and torque calculation with consideration of cross saturation of the interior permanent magnet synchronous motor with brushless field excitation," *IEEE Electric Machines and Drives*, 2009.
4 G. Choi and T. M. Jahns, "Design of electric machines for electric vehicles based on driving schedules," *IEEE Electric Machines and Drives*, 2013.
5 M. Meyer and J. Bocker, "Optimum control for interior permanent magnet synchronous motors (IPMSM) in constant torque and flux weakening range," *Power Electronics and Motion Control Conference*, 2006.

6 D. Hu, Y. M. Alsmadi and L. Xu, "High fidelity nonlinear IPM modeling based on measured stator winding flux linkage," *IEEE Transactions on Industrial Applications*, 51 (4), pp. 3012–3019, July–August 2015.
7 K. M. Rahman and S. Hiti, "Identification of machine parameters of a synchronous machine," *IEEE Transactions on Industrial Applications*, 41, pp. 557–565, April 2005.
8 B. A. Welchko, T. M. Jahns, W. L. Soong, and J. M. Nagashima, "IPM synchronous machine drive response to symmetrical and asymmetrical short-circuit faults," *IEEE Transactions on Energy Conversion*, 18, pp. 291–298, June 2003.

Problems

10.1 The 2007 Toyota Camry eight-pole traction motor has been characterized as shown in Figure 10.15 and Table 10.2.

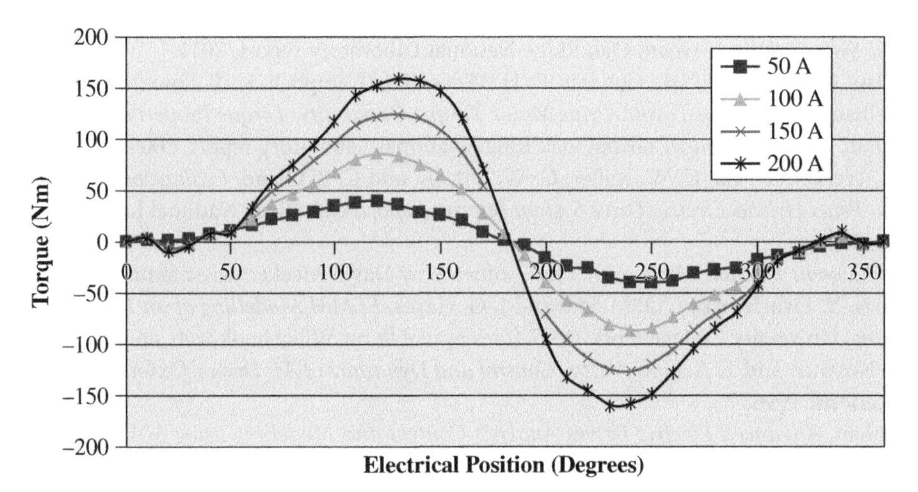

Figure 10.15 Torque versus dc current for 2007 Toyota Camry [2]. (Courtesy of Oak Ridge National Laboratory, US Dept. of Energy.)

Table 10.2 DC current test summary for 2007 Camry [2].

I_{dc} (A)	T_r (90°) (Nm)	k (Nm/A)	T_r (120°) (Nm)	T_{pm} (120°) (Nm)	I_q(120°) (A)	T_{sr} (120°) (Nm)	I_d(120°) (A)	$L_{md} - L_{mq}$ (mH)
50	30	**0.28**	45	**26**	**43**	**19**	**–25**	**-2.9**
100	55	**0.26**	85	**48**	**87**	**37**	**–50**	**–1.9**
150	76	**0.24**	122	**66**	**130**	**56**	**–75**	**–1.0**
200	95	**0.22**	160	**82**	**173**	**78**	**–100**	**-0.75**

Verify the machine parameters as shown in boldface in the table.

10.2 A HEV motor designed for use in the 2007 Toyota Camry is operating in forward-motoring mode and develops an output rotor torque of +160 Nm at a power of 60 kW.

Determine the applied per-phase voltage and current, and the machine efficiency and power factor for this operating point.

Split the magnet and reluctance torques in the ratio 50:50 for this operating point.

Use the parameters determined in the 200 A row of Table 10.2.

Let the per-phase stator resistance R_s = 22 mΩ and the no-load torque T_{nl} = 2 Nm.

Let the d-axis inductance L_{md}= 0.7 mH at this condition.

[Ans. 268.6 V, 142.9 A, 96.62%, 0.539]

10.3 Recalculate the answers in the previous problem, this time allowing for a 25:75 ratio for magnet:reluctance torque.

Use the parameters determined in the 100 A row of Table 10.2.

[Ans. 173.2 V, 148.7 A, 96.45%, 0.805]

10.4 The HEV motor of the previous problem is operating at high speed in forward-motoring mode at the rated power of 60 kW with an output torque of 80 Nm.

Split the magnet and reluctance torques 20:80 for this operating point.

Determine the applied per-phase voltage and current, and the machine efficiency and power factor for this operating point.

Use the parameters determined in the 50 A row of Table 10.2.

[Ans. 212.8 V, 98.6 A, 96.55%, 0.987]

10.5 The 2010 Toyota Prius 8-pole traction motor has been characterized as shown in Table 10.3.

Table 10.3 DC current test summary for 2010 Toyota Prius motor [3].

I_{dc} (A)	T_r (90°) (Nm)	k (Nm/A)	T_r (120°) (Nm)	T_{pm} (120°) (Nm)	I_q(120°) (A)	T_{sr} (120°) (Nm)	I_d(120°) (A)	$L_{md} - L_{mq}$ (mH)
50	32	**0.3**	54	**28**	**43**	**26**	**−25**	**−4.0**
100	65	**0.3**	105	**56**	**87**	**49**	**−50**	**−1.9**
150	80	**0.25**	145	**69**	**130**	**76**	**−75**	**−1.3**

Verify the machine parameters as shown in boldface in the table.

Determine the applied per-phase voltage and current, and the machine efficiency and power factor when the rotor develops +150 Nm at the rated power of 60 kW.

Split the magnet and reluctance torques 1:2 for this operating point.

Use the parameters determined in the 150 A row of Table 10.3.

Let per-phase stator resistance R_s = 77 mΩ and the no-load torque T_{nl} = 2 Nm. Let the d-axis inductance L_{md}= 1.0 mH at this condition.

[Ans. 260.8 V, 117.5 A, 93.77%, 0.696]

10.6 The HEV motor of the previous problem is operating in forward-motoring mode at a power of 60 kW with an output torque of 50 Nm.

Determine the applied per-phase voltage and current, and the machine efficiency and power factor for this operating point.

Use the parameters determined in the 50 A row of Table 10.3, and let the d-axis inductance L_{md} = 1.0 mH. Split the magnet and reluctance torques in the ratio 20:80.

[Ans. 284.2 V, 75.6 A, 94.16%, 0.988]

Assignments

10.1 Download FEMM [6], and model the IPM machine [7]. See the references for the various machine parameters.

Part 3

Power Electronics

11

DC-DC Converters

"It has today occurred to me that an amplifier using semiconductors rather than vacuum is in principle possible." William Shockley in 1939. Shockley, Brattain, and Bardeen, all of Bell Labs, were awarded the Noble Prize for Physics in 1956 for the earlier invention of the semiconductor transistor.

In this chapter, the reader is introduced to power electronic power converters. The commonly used buck (step-down) and boost (step-up) topologies are analyzed for the continuous, boundary, and discontinuous conduction modes (CCM, BCM, and DCM, respectively) of operation. Power semiconductors are briefly introduced, and the power losses of the insulated-gate bipolar transistor (IGBT) and diode are estimated. Topics such as the sizing of passive components and the benefits of interleaving are covered. Examples are reviewed based on HEV and FCEV dc-dc converters. The concepts of root-mean-squared (rms) and direct-current (dc) quantities are reviewed as they are key to learning about the various converters.

11.1 Introduction

Many of the advances in the evolving automotive powertrain are due to the advent of power electronics and power semiconductors. By utilizing power electronic converters, voltages of a given magnitude and frequency can be converted to voltages of virtually any magnitude and frequency. One of the major functions of power electronics, therefore, is to provide the electrical power conversion which can achieve maximum functionality from electromechanical devices, energy sources, and electric loads. A second factor driving power electronics is efficiency. Efficient power conversion typically results in reduced material, cooling, power demand, cost, and volume while increasing reliability for many electronic products such as laptop computers and smartphones.

Power electronic converters enable electrical power conversion by periodically switching an available power source in and out of a circuit. These converters use semiconductor devices to act as switches, and feature energy storage devices, such as inductors and

Electric Powertrain: Energy Systems, Power Electronics and Drives for Hybrid, Electric and Fuel Cell Vehicles, First Edition. John G. Hayes and G. Abas Goodarzi.

Companion website: www.wiley.com/go/hayes/electricpowertrain

capacitors, to store energy and filter the sharp-edged waveforms created by the fast switchings. Transformers are used for isolation and safety in addition to voltage and current conversions. These converters are known as **switch-mode power converters**. Significant electrical noise, known as **electromagnetic interference** (**EMI**), can be generated by the fast switching. EMI can affect the operation of other electronic and electrical equipment and is regulated by agencies around the world, such as the Federal Communications Commission (FCC) in the United States.

Electric power conversion can be described as falling into two categories, alternating current (ac) and direct current (dc). Thus, there are four possible types of electrical power conversion: ac-dc, dc-dc, dc-ac, and ac-ac.

Electronic circuits typically require dc voltage sources in order to function. Ac voltages are used for bulk power generation, transmission, and local distribution. Thus, an ac-dc converter is used to interface the ac grid to the dc battery.

Dc-dc converters are power electronics circuits which convert a dc voltage to a different dc voltage level. For example, the voltage of a fuel cell can drop significantly with increased power demand, and a dc-dc converter is required to step up the fuel cell voltage to a higher level for a more efficient powertrain.

The electric traction motor used in the modern electric powertrain requires ac voltages, rather than the dc of the battery. A **dc-ac inverter** is a power electronics circuit which converts dc to ac. Power inverters are the topic of Chapter 13.

Ac-dc rectifiers are power electronics circuits which convert ac to dc. Rectifiers are discussed in depth in Chapter 14.

Ac-ac power conversion is often required for electrical machines and loads at very high voltages, and is usually not considered for automotive applications.

The circuit diagram for the power stages of the 2010 Toyota Prius is shown in Figure 11.1. The vehicle uses a dc-dc converter to step up the battery voltage from about 200 V to a level between 200 V and 650 V in order to operate the powertrain most efficiently. Dc-ac inverters are used to interface the high-voltage dc link to the electric motor and generator. These types of dc-dc converters and dc-ac inverters enable a bidirectional flow of power. Thus, the dc-dc converter can take power from the battery when motoring and can recharge the battery when power is available from the generator or from the traction motor during regenerative braking.

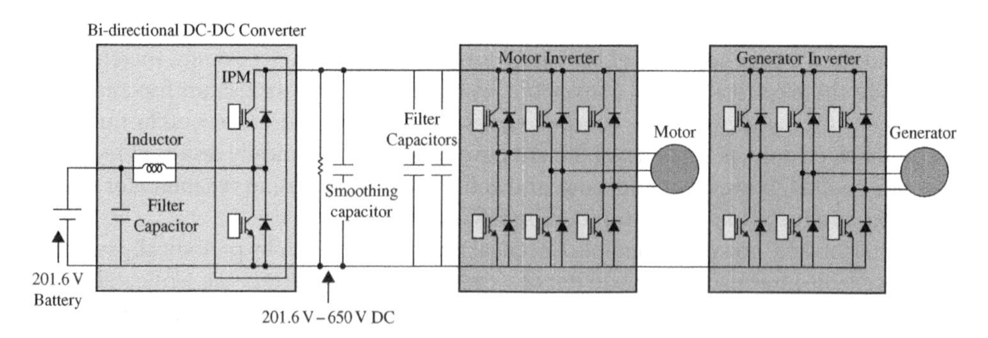

Figure 11.1 Power controller diagram for 2010 Toyota Prius [1]. (Courtesy of Oak Ridge National Laboratory, US Dept. of Energy.)

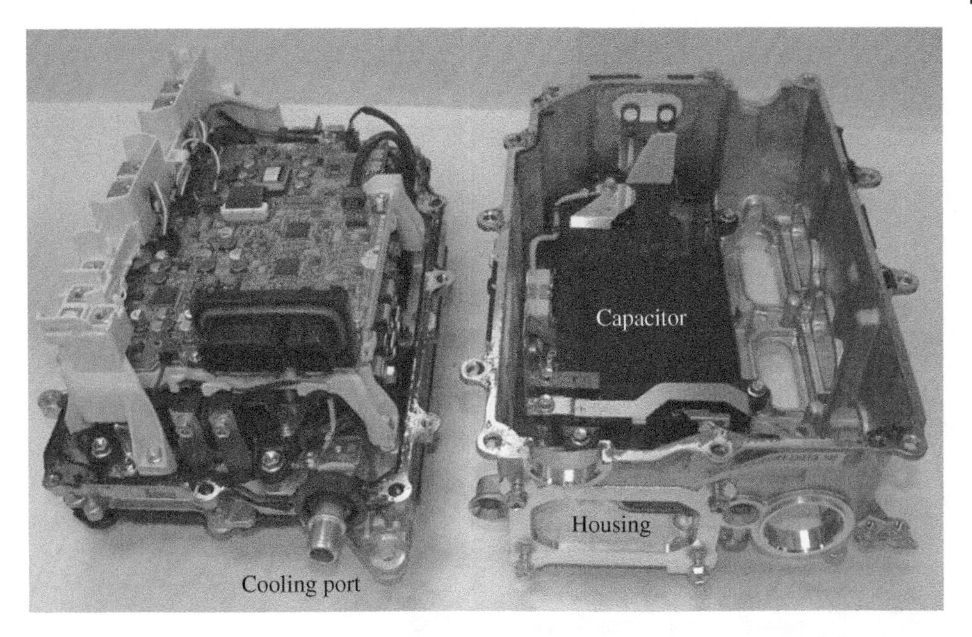

Figure 11.2 Views of the 2010 Toyota Prius power control unit [1]. (Courtesy of Oak Ridge National Laboratory, US Dept. of Energy.)

The power electronic converters and controllers for the 2010 Toyota Prius vehicle are integrated into a single box which is known as the power control unit (PCU). Views of the PCU are shown in Figure 11.2 and Figure 11.3 [1]. The PCU housing itself is made from aluminum and features an integrated heat sink. A cooling system is required as the power converters can generate significant power loss. The opened PCU is shown in Figure 11.2. The large capacitor assembly is noted, as is the cooling port for the liquid coolant.

Views of the power semiconductors for the dc-dc converter and the two dc-ac inverters are shown in Figure 11.3(a). The semiconductors are integrated into a single module which is mounted onto the PCU heat sink. A side view of the semiconductors is shown in Figure 11.3(b), where we can see the multiple thin bond wires from the connectors to the silicon dies and the thin silicon dies themselves.

The earliest generations of power converters were based on vacuum tubes, a technology which has almost been made obsolete with the invention of the semiconductor by William Shockley and others at Bell Labs in the early 1950s. The first power semiconductor device was a diode, which was later followed by the bipolar junction transistor (BJT) and the silicon-controlled rectifier (SCR), also known as the thyristor. Power diodes and thyristors continue to play a significant role in power conversion. The diode is ubiquitous, while the thyristor is commonly used for ac-dc and ac-ac power conversion.

Modern switch-mode power converters were first commercialized in the 1970s using the BJT as the main switching device. The metal-oxide-semiconductor field-effect transistor (MOSFET) took over in the late 1970s and is still the semiconductor switch of choice for many applications, especially for operating voltages less than 400 V.

(a)

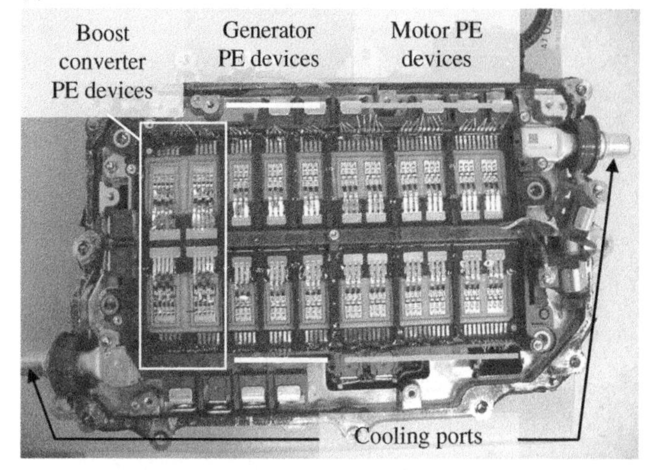

(b)

Figure 11.3 Views of boost and inverter silicon for the 2010 Toyota Prius [1]. (Courtesy of Oak Ridge National Laboratory, US Dept. of Energy.)

The MOSFET surpassed the BJT as it is more efficient and easier to control. The invention of the IGBT pushed the semiconductor to achieve dominance at higher voltages where greater efficiencies can be achieved [2]. The IGBT dominates for high-power automotive power converters due to its efficiency and short-circuit capability. Wide-band-gap semiconductor technologies are shifting from niche applications into the mainstream of power conversion – based on the efficiency advantages of gallium nitride and silicon carbide.

The IGBT is paired with the silicon diode in this chapter in order to estimate the semiconductor power loss, whereas the MOSFET is paired with the silicon carbide diode in Chapter 14 for the power-factor-correction boost stage. A discussion of silicon-carbide devices to compete with silicon is presented in Appendix III.

11.2 Power Conversion – Common and Basic Principles

Switch-mode technology has significant advantages over the competitive linear technology. Consider the simple example of an automotive power conversion going from a relatively high voltage of 14 V from the alternator on a vehicle to a lower voltage of 5 V in

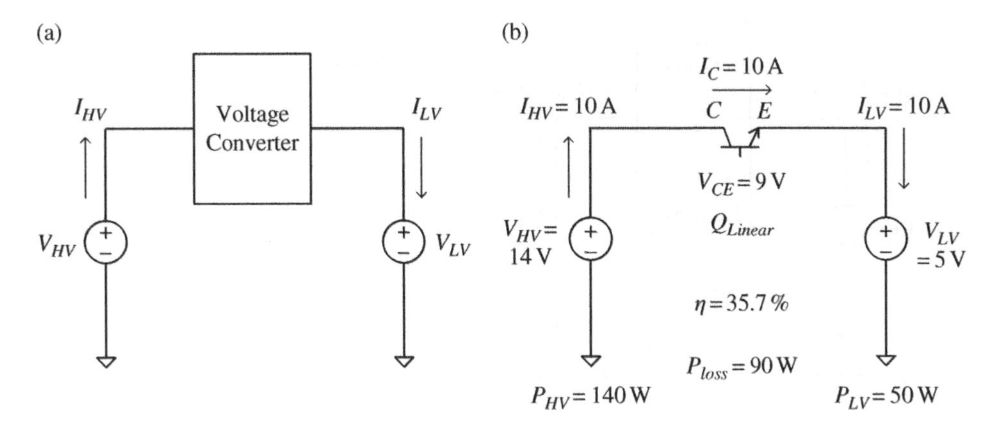

Figure 11.4 (a) Voltage conversion from 14 V to 5 V and (b) linear converter.

order to run vehicle loads. For consistency in dealing with the various voltages on the vehicle, the higher and lower voltage levels are represented by V_{HV} and V_{LV}, respectively, as shown in Figure 11.4(a). If a **linear power converter** is used to supply the low-voltage load from the high-voltage source, then the circuit will have a form similar to Figure 11.4 (b). For the linear power converter, a bipolar transistor, shown here as an NPN transistor, but typically a Darlington pair, is operated in the linear region, such that the desired output voltage V_{LV} is supplied to the load, and the voltage difference between the source and the load is dropped across the transistor. This method of converting electrical power has the advantages of simplicity, very high dynamic performance, fast response to any voltage or load changes, and the generation of little electrical noise.

The disadvantages of linear power converters are the large power loss, poor efficiency, and the resulting large physical size and mass required to dissipate the significant heat generated by the transistor's operation in the linear region. For the simple example shown, let us assume that the current to the load I_{LV} is 10 A; therefore, the load power consumption P_{LV} is 5 V times 10 A, which equals 50 W. In a linear converter, the current flowing from the source I_{HV} is also the transistor collector current I_C and is also the load current. Thus, the power pulled from the high-voltage source P_{HV} is 14 V times 10 A, which equals 140 W, and the difference between the source and load powers of 90 W is dissipated as heat from the linear transistor. The efficiency of the converter η is the ratio of load power to the source power and is given as follows:

$$\eta = \frac{P_{LV}}{P_{HV}} = \frac{50}{140} \times 100\% = 35.7\%$$

Switch-mode power converters operate by pulsing energy from the source to the load at a very high frequency, as shown in Figure 11.5. Similar to the linear power converter, the switch-mode power converter also uses a semiconductor device. However, in switch-mode converters, the switch is not operated in the linear region but is operated in a very efficient conduction mode, such that the switch is turned fully on with a minimal voltage drop across the device. A switch-mode converter providing the voltage conversion would be expected to have efficiencies in the mid-to-high 90s, in sharp contrast to the calculated efficiency of 35.7% for the preceding example.

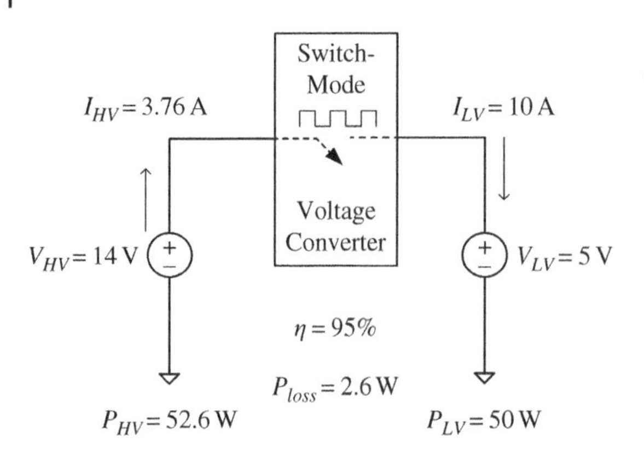

Figure 11.5 Switch-mode power converter.

Again, let us consider the voltage conversion from 14 V to 5 V at 50 W output. If the switch-mode converter efficiency is 95%, then the converter only pulls 3.76 A from the source, a significant reduction from the case of the linear converter. The power loss in the converter is only 2.6 W, resulting in a significant reduction in the size and weight of the switch-mode converter compared to the linear converter.

Although the switching power converter is more efficient and compact than the linear power converter, the switching power converter is more complex, has a lower bandwidth, and generates unwanted EMI.

11.2.1 The Basic Topologies

The next step is to realize a switch-mode power converter. There are three basic converters, all of which use a switch Q, a diode D, and an inductor L, in various configurations. These are the buck or step-down converter, the boost or step-up converter, and the buck-boost or step-down/step-up converter, as shown in Figure 11.6(a), (b), and (c), respectively.

The **buck converter** converts power from a high-voltage source to a lower voltage.

The **boost converter** converts power from a low-voltage source to a higher voltage.

The **buck-boost converter** converts power from an input voltage source to an output voltage and can step down or step up the output voltage.

The buck converter is initially analyzed, and the method is then applied to the boost converter. The buck-boost converter is briefly considered in the Appendix at the end of this chapter.

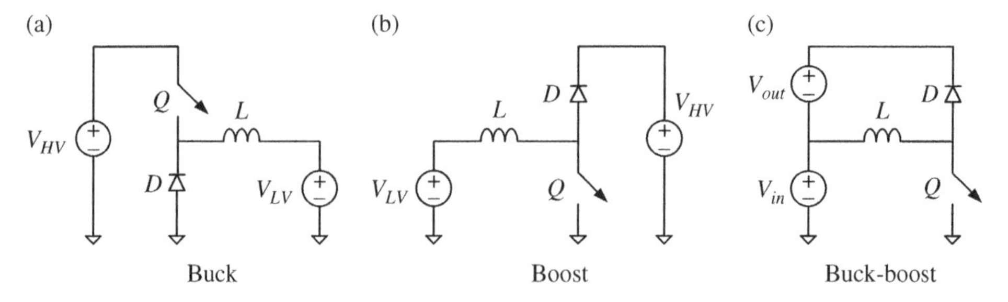

Figure 11.6 Buck, boost, and buck-boost converters.

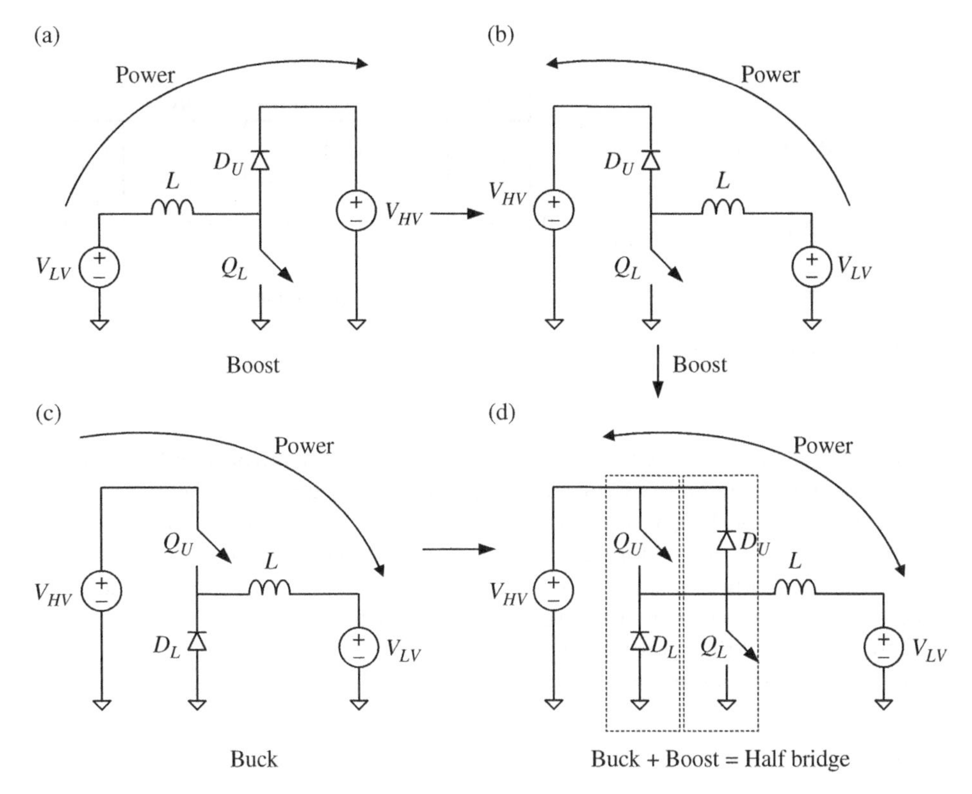

Figure 11.7 Half-bridge converter.

11.2.2 The Half-Bridge Buck-Boost Bidirectional Converter

The buck and boost converters can be integrated to create a bidirectional buck-boost converter. This converter is commonly used for hybrid electric vehicles as it enables discharge of the low-voltage battery to a higher voltage during motoring using a boost and charging of the low-voltage battery from the high-voltage link using a buck converter. This integration of a buck converter and a boost converter is often known as a **half–bridge converter**. The half-bridge can be explained as follows. The boost converter of Figure 11.7(a) can be redrawn as Figure 11.7(b). The switch and diode of the buck converter of Figure 11.7(c) are then integrated with the boost components of Figure 11.7(b) to create the buck-boost half-bridge of Figure 11.7(d). The subscripts U and L are used to designate the upper and lower switches and diodes.

11.3 The Buck or Step-Down Converter

The buck or step-down converter produces a lower average output voltage V_{LV} than the input voltage V_{HV}. The converter has three basic power components to perform the voltage conversion. First, a controlled power semiconductor switch, shown as Q in Figure 11.8, is pulsed on and off at a high frequency in order to transfer energy from

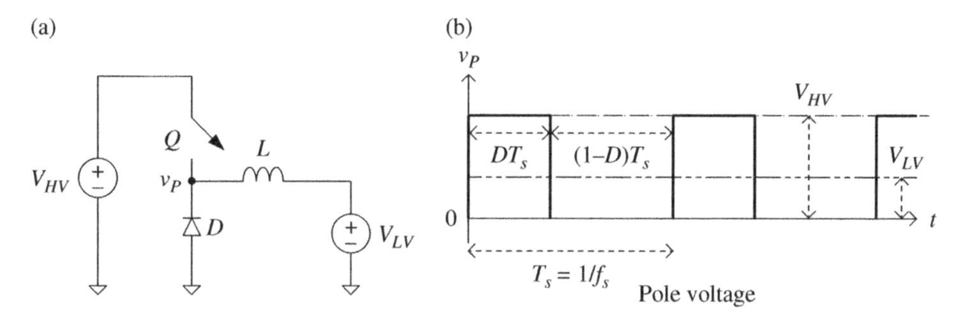

Figure 11.8 Buck or step-down (a) converter and (b) pole voltage.

the source V_{HV} to the load V_{LV}. The output from the switch is a controlled pulse train of a controlled switching frequency f_s, or switching period T_s, and a controlled conduction time within that period $D\ T_s$, where D is termed the **duty cycle** of the switch, or the portion of time during the period for which the switch is conducting. The output voltage is known as the pole output voltage v_p and is shown in Figure 11.8(a). The pole output voltage waveform $v_p(t)$ is as shown in Figure 11.8(b). As can be seen, the pole voltage is pulsing between 0 V and the source voltage V_{HV}.

Note that the switching frequency f_s is simply the inverse of the period T_s:

$$f_s = \frac{1}{T_s} \tag{11.1}$$

A dc or constant voltage is required at the load, and so a second component, the inductor L is required in order to remove the ac component of the pole voltage.

The inductor functions in the circuit by storing energy. The stored energy within the inductor cannot instantly flow from the source to the load. Current conduction must be maintained through the inductor when the switch Q is pulsed on and off. A third component, known as the **inverse diode**, is required as an uncontrolled switch in order to provide a conduction path for the inductor current when the switch Q is not conducting.

Note that D is used in this text both as a label to describe a diode and as a variable to describe the duty cycle. The terms are obviously not interchangeable, and the meaning should be obvious from the context.

In general, from a power perspective, dc-dc converters require additional filtering components. Typically, the on-vehicle voltage sources, such as the battery, fuel cell, or generator, are modeled to include the voltage source itself and the internal inductance of the voltage source and the required cabling. In addition, components such as the battery or fuel cell should be filtered such that they supply dc only, as the ac components generate electrical noise and additional heating, reducing lifetimes for these critical components. There are the three essential components of the buck converter plus the discrete filter capacitors on both voltage links and internal or added inductances between the dc links and the sources, as shown in Figure 11.9. Thus, the high-voltage source, shown as $V_{S(HV)}$, has the internal inductance L_{HV} and the dc link capacitance C_{HV}, while the low-voltage source, shown as $V_{S(LV)}$, has the internal inductance L_{LV} and the dc link capacitance C_{LV}.

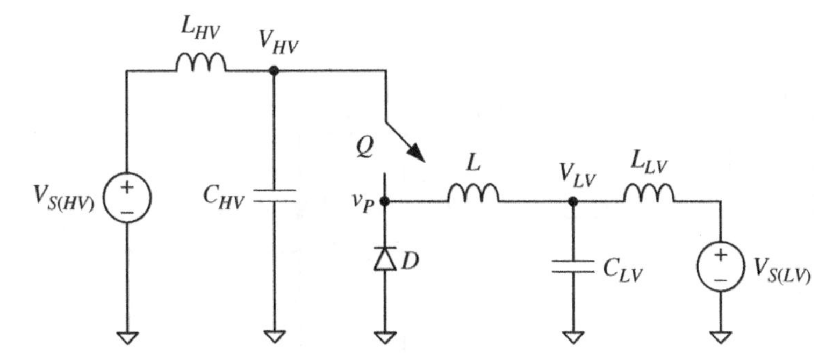

Figure 11.9 A buck converter with source and load filter components.

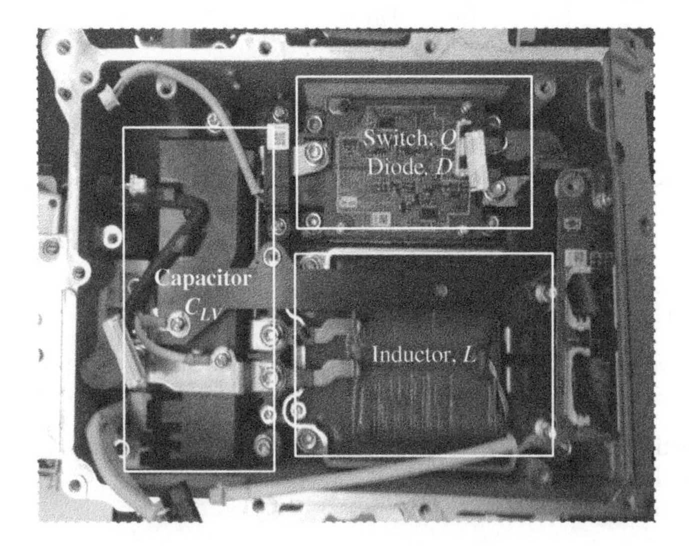

Figure 11.10 HEV bidirectional dc-dc converter.

The additional components can be of significant physical size and mass. An automotive dc-dc converter is shown in Figure 11.10. The approximate areas for the various components are outlined.

11.3.1 Analysis of Voltage Gain of Buck Converter in CCM

In CCM, the inductor of the buck converter is always conducting current. This mode of operation is relatively easy to analyze. First, the relationship between the voltage gain of the converter and the duty cycle must be determined.

Neglecting parasitics, in a buck converter, the dc or average output voltage equals the average of the pole voltage. If we ignore the switch and diode voltage drops, then the pole voltage is pulsing from the dc return or 0 V to the high-voltage dc link voltage V_{HV} as shown in Figure 11.8(b), and again in Figure 11.11(a). The pole is at V_{HV} for the time

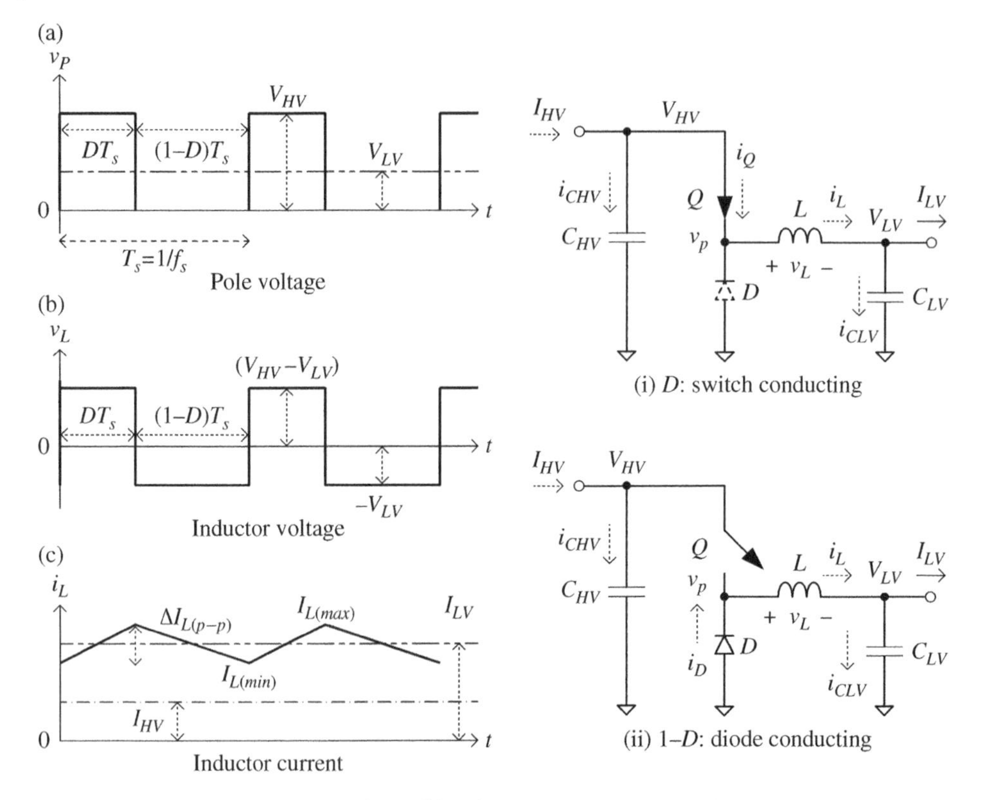

Figure 11.11 Buck CCM (a) pole voltage, (b) inductor voltage, and (c) inductor current.

period DT_s and at 0 V for the time period $(1-D)\ T_s$. The circuits are shown in Figure 11.11 (i) and (ii) for the two operating modes.

The dc output voltage is given by

$$V_{LV} = \frac{1}{T_s}[V_{HV}DT_s + 0 \times (1-D)T_s] = DV_{HV} \tag{11.2}$$

Thus, the duty cycle of a buck converter represents the voltage gain of the buck converter and is expressed as

$$D = \frac{V_{LV}}{V_{HV}} \tag{11.3}$$

It is noted here that the parasitic voltage drops of the various passive and active components are not considered. This is reasonable for high-voltage dc-dc converters, but the derivation of the duty cycle should be expanded to consider parasitics for low-voltage converters.

This relationship can also be shown in integral form as

$$V_{LV} = \frac{1}{T_s}\int_0^{T_s} v_P(t)dt = \frac{1}{T_s}\int_0^{DT_s} V_{HV}dt + \frac{1}{T_s}\int_{DT_s}^{T_s} 0\,dt = DV_{HV} \tag{11.4}$$

It is also productive to derive the voltage relationships by considering the inductor voltage and current. The inductor voltage and current are illustrated in Figure 11.11 (b) and (c), respectively. The inductor current increases linearly while the switch is closed during DT_s as a voltage of $(V_{HV}-V_{LV})$ appears across the inductor. When the switch opens and the diode is conducting during (1–D) T_s, the inductor current linearly decreases as there is a negative voltage of $-V_{LV}$ across the inductor.

Applying Faraday's law to determine the back emf across the inductor yields:

$$v_L(t) = v_p(t) - V_{LV} = L\frac{di_L(t)}{dt} \tag{11.5}$$

Rearranging the preceding equation in terms of the change in inductor current $i_L(t)$, we get

$$di_L(t) = \frac{1}{L}\left[v_p(t) - V_{LV}\right]dt. \tag{11.6}$$

Assuming that the inductor current is changing linearly in the steady-state operation, the change in inductor current from peak to peak can be described by $\Delta I_{L(p\text{-}p)}$.

During time DT_s:

$$\Delta I_{L(p-p)} = \frac{1}{L}(V_{HV} - V_{LV})DT_s = \frac{1}{f_s L}(V_{HV} - V_{LV})D \tag{11.7}$$

During time $(1-D)T_s$:

$$\Delta I_{L(p-p)} = \frac{V_{LV}(1-D)T_s}{L} = \frac{V_{LV}(1-D)}{f_s L} \tag{11.8}$$

Equating the two expressions for the change in current $\Delta I_{L(p\text{-}p)}$, we get

$$\Delta I_{L(p-p)} = \frac{1}{f_s L}(V_{HV} - V_{LV})D = \frac{V_{LV}(1-D)}{f_s L} \tag{11.9}$$

Rearranging the preceding equation again yields the earlier expression, Equation (11.3), for the voltage gain.

11.3.1.1 Analysis of Buck Converter in CCM

It is necessary to know the average, rms, minimum, and maximum currents in each component in order to design and specify the components of a buck converter.

The inductor current can be seen to have two current components: (1) a dc, or constant, or average component which flows to the load and (2) an ac or time-varying periodic component which is shunted from the load by the low-voltage capacitor due to the low ac impedance of the capacitor. The dc current is shown as I_{LV} in Figure 11.12(a), while the ac current is represented by the periodic triangular waveform oscillating between $I_{L(min)}$ and $I_{L(max)}$ for a total amplitude within the cycle of $\Delta I_{L(p\text{-}p)}$.

The inductor current is conducted by the switch as i_Q during D, as shown in Figure 11.12(b), and by the diode as i_D during (1-D), as shown in Figure 11.12(c).

The high-voltage and low-voltage capacitor currents can be determined by simply subtracting the respective dc currents from the switch and inductor currents, respectively, as shown in Figure 11.12(d) and (e).

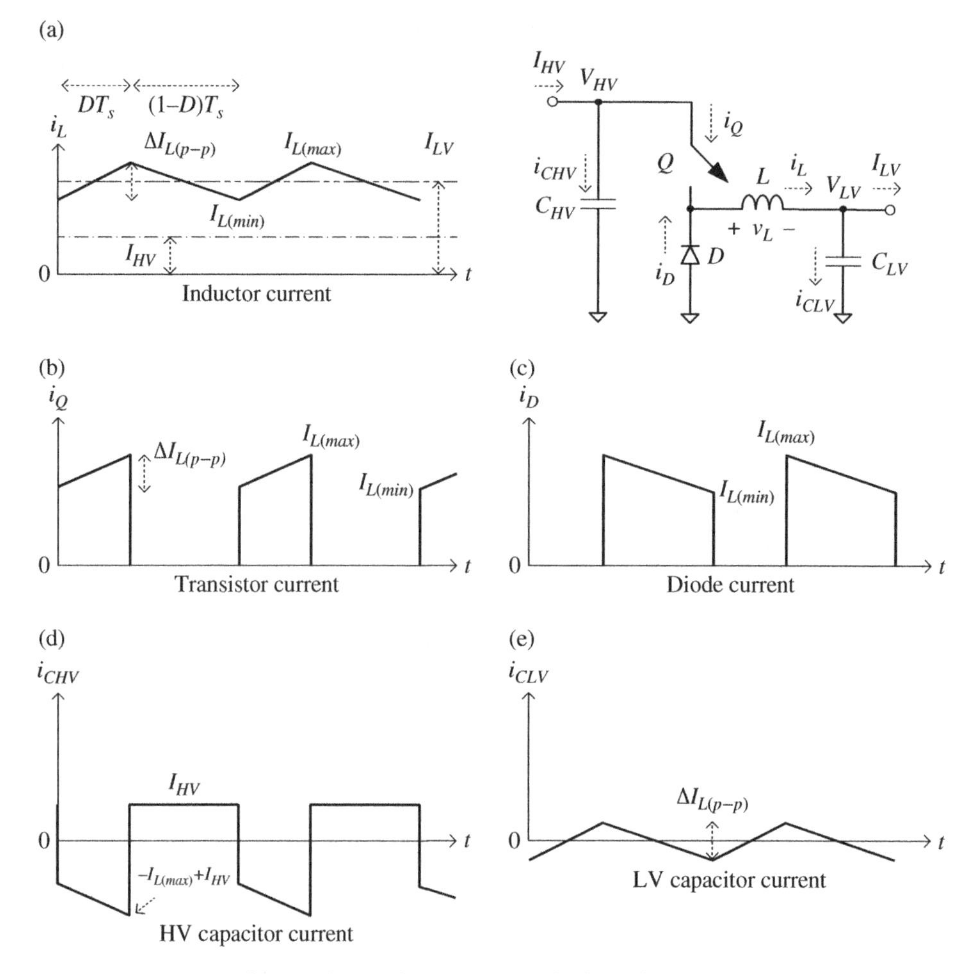

Figure 11.12 Currents for (a) inductor, (b) transistor, (c) diode, (d) high-voltage capacitor, and (e) low-voltage capacitor.

11.3.1.2 Determining Low-Voltage Capacitor RMS Current

The rms component of the low-voltage capacitor current $I_{CLV(rms)}$ is of initial interest as it is the rms of a triangular waveform. The rms value is given by

$$I_{CLV(rms)} = \sqrt{\frac{1}{T_s}\int_0^{T_s} i_{CLV}(t)^2 dt} \tag{11.10}$$

From an examination of the current waveform in Figure 11.12(e), it can easily be shown that the low-voltage capacitor current i_{CLV} has the following two straight line segments:

During time DT_s:

$$i_{CLV}(t) = \frac{\Delta I_{L(p-p)}}{DT_s}t - \frac{\Delta I_{L(p-p)}}{2} \tag{11.11}$$

During time $(1-D)T_s$:

$$i_{CLV}(t) = -\frac{\Delta I_{L(p-p)}}{(1-D)T_s}t + \frac{\Delta I_{L(p-p)}}{2}\frac{(1+D)}{(1-D)} \tag{11.12}$$

Substituting Equation (11.11) and Equation (11.12) into Equation (11.10) gives

$$I_{CLV(rms)} = \sqrt{\frac{1}{T_s}\int_0^{T_s} i_{CLV}(t)^2 dt}$$

$$= \sqrt{\frac{1}{T_s}\left[\int_0^{DT_s}\left(\frac{\Delta I_{L(p-p)}}{DT_s}t - \frac{\Delta I_{L(p-p)}}{2}\right)^2 dt + \int_{DT_s}^{T_s}\left(-\frac{\Delta I_{L(p-p)}}{(1-D)T_s}t + \frac{\Delta I_{L(p-p)}}{2}\frac{(1+D)}{(1-D)}\right)^2\right] dt} \tag{11.13}$$

Although this equation appears complex, it reduces to a simple expression for the rms value of a triangular waveform with no average component:

$$I_{CLV(rms)} = \frac{\Delta I_{L(p-p)}}{\sqrt{12}} \tag{11.14}$$

Knowing the dc load current and the capacitor rms current, it can similarly be derived that the rms value of the inductor $I_{L(rms)}$ is given by

$$I_{L(rms)} = \sqrt{I_{L(dc)}^2 + I_{CLV(rms)}^2} = \sqrt{I_{LV}^2 + \frac{\Delta I_{L(p-p)}^2}{12}} \tag{11.15}$$

where $I_{L(dc)} = I_{LV}$ is the dc current flowing through the inductor to the low-voltage output.

When designing or specifying the inductor, the maximum current is also of interest. The inductor peak current is simply given by

$$I_{L(max)} = I_{LV} + \frac{\Delta I_{L(p-p)}}{2} \tag{11.16}$$

The switch and diode currents can now be determined. The average currents of the switch and diode, designated $I_{Q(dc)}$ and $I_{D(dc)}$, respectively, are easily determined and are given by

$$I_{Q(dc)} = D I_{LV} \tag{11.17}$$

and

$$I_{D(dc)} = (1-D) I_{LV} \tag{11.18}$$

The rms currents of the switch and diode, designated $I_{Q(rms)}$ and $I_{D(rms)}$, respectively, can be determined by analysis and simply work out to be

$$I_{Q(rms)} = \sqrt{D} I_{L(rms)} \tag{11.19}$$

and

$$I_{D(rms)} = \sqrt{(1-D)} I_{L(rms)} \tag{11.20}$$

In addition to the dc and rms currents, the instantaneous turn-on and turn-off currents in the switch and diode are of interest:

$$I_{Q(off)} = I_{D(on)} = I_{L(max)} = I_{LV} + \frac{\Delta I_{L(p-p)}}{2} \tag{11.21}$$

and

$$I_{Q(on)} = I_{D(off)} = I_{L(min)} = I_{LV} - \frac{\Delta I_{L(p-p)}}{2} \tag{11.22}$$

where $I_{Q(on)}$ and $I_{Q(off)}$ and $I_{D(on)}$ and $I_{D(off)}$ are the turn-on and turn-off currents in the switch and diode, respectively, and $I_{L(min)}$ is the minimum current in the inductor for a given cycle.

The rms current in the high-voltage link capacitor $I_{CHV(rms)}$ can be determined by subtracting the dc current coming from the high-voltage source I_{HV} from the rms current in the switch $I_{Q(rms)}$:

$$I_{CHV(rms)} = \sqrt{I_{Q(rms)}^2 - I_{HV}^2} \tag{11.23}$$

Note that it is common to use the term **current-ripple ratio r_i**, which is defined as

$$r_i = \frac{\Delta I_{L(p-p)}}{I_{L(dc)}} \tag{11.24}$$

and is often expressed as a percentage.

11.3.1.3 Capacitor Voltages

The voltage ripple on the dc capacitors is also of interest. From Gauss's law of electricity, the capacitor charge Q is equal to the product of the capacitance C and the capacitor voltage V_C:

$$Q = CV_C \tag{11.25}$$

The capacitor voltage changes over the switching period as the capacitor is charged and discharged. The peak-to-peak voltage ripple on the capacitor $\Delta V_{C(p-p)}$ is directly proportional to the change in stored charge ΔQ as the capacitor charges and discharges. Thus:

$$\Delta V_{C(p-p)} = \frac{\Delta Q}{C} \tag{11.26}$$

As the various component currents have already been determined, it is easy to determine ΔQ as it is simply the change in charge due to the capacitor current.

The high-voltage and low-voltage capacitor voltages and currents are shown in Figures 11.13(a) and (b).

Referring to Figure 11.13(a), the peak-to-peak voltage ripple on the high-voltage capacitor $\Delta V_{CHV(p-p)}$ is determined by calculating the change in charge ΔQ_{CHV} on the basis of the negative or positive current. Using the current-time area product of the current and the diode conduction time gives the following answer:

$$\Delta V_{CHV(p-p)} = \frac{\Delta Q_{CHV}}{C_{HV}} = \frac{I_{HV}(1-D)T_s}{C_{HV}} = \frac{I_{HV}(1-D)}{f_s C_{HV}} \tag{11.27}$$

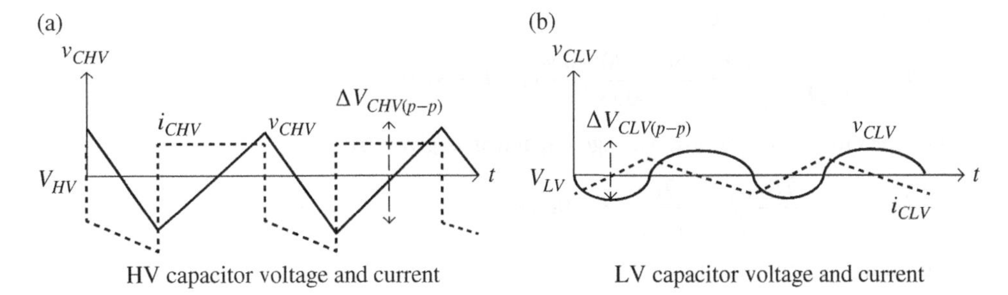

Figure 11.13 Capacitor voltages (solid) and currents (dashed).

Similarly, the peak-to-peak voltage ripple on the low-voltage capacitor $\Delta V_{CLV(p-p)}$ is estimated by determining the charge ΔQ_{CLV} on the basis of the negative or positive current. In this case, the current-time area product is based on a simple triangular current waveform, and the result is as follows:

$$\Delta V_{CLV(p-p)} = \frac{\Delta Q_{CLV}}{C_{LV}} = \frac{1}{2}\frac{\Delta I_{L(p-p)}}{2}\frac{T_s}{2}\frac{1}{C_{LV}} = \frac{\Delta I_{L(p-p)} T_s}{8 C_{LV}} = \frac{\Delta I_{L(p-p)}}{8 f_s C_{LV}} \tag{11.28}$$

At this point, all the current and voltages have been determined in order to analyze a given design, or design a converter based on a set of specifications, as shown in the next two examples.

We also use the term **voltage-ripple ratio r_v**, which is defined as

$$r_v = \frac{\Delta V_{C(p-p)}}{V} \tag{11.29}$$

and is often expressed as a percentage.

11.3.1.4 Example: Designing Buck Converter for CCM Operation

A hybrid electric vehicle requires a 20 kW bidirectional converter to generate a 500 V dc link voltage from the 200 V NiMH battery. The switching frequency is 10 kHz.

Determine the component parameters in order to have a 28% current-ripple ratio on the inductor and a 0.5% voltage-ripple ratio on the high- and low-voltage capacitors.

Determine the various component currents.

Assume ideal components, and ignore the power loss.

Solution:

The duty cycle is

$$D = \frac{V_{LV}}{V_{HV}} = \frac{200\,\text{V}}{500\,\text{V}} = 0.4$$

The average inductor current equals the output load current and is given by

$$I_{LV} = I_{L(dc)} = \frac{P}{V_{LV}} = \frac{20000}{200}\,\text{A} = 100\,\text{A}$$

The inductor ripple current is specified as

$$\Delta I_{L(p-p)} = r_i I_{L(dc)} = 0.28 \times 100\,\text{A} = 28\,\text{A}$$

The desired inductance is

$$L = \frac{(V_{HV} - V_{LV})D}{f_s \Delta I_{L(p-p)}} = \frac{(500-200)0.4}{10000 \times 28} \text{H} = 428.5\,\mu\text{H}$$

The rms current in the low-voltage capacitor is given by

$$I_{CLV(rms)} = \frac{\Delta I_{L(p-p)}}{\sqrt{12}} = \frac{28}{\sqrt{12}} \text{A} = 8.083\,\text{A}$$

The inductor rms current is

$$I_{L(rms)} = \sqrt{I_{L(dc)}{}^2 + I_{CLV(rms)}{}^2} = \sqrt{100^2 + 8.083^2} \text{A} = 100.3\,\text{A}$$

The inductor, switch, and diode maximum and minimum currents are

$$I_{L(\max)} = I_{Q(off)} = I_{D(on)} = I_{L(dc)} + \frac{\Delta I_{L(p-p)}}{2} = 114\,\text{A}$$

and

$$I_{L(min)} = I_{Q(on)} = I_{D(off)} = I_{L(dc)} - \frac{\Delta I_{L(p-p)}}{2} = 86\,\text{A}$$

The switch rms and average currents are

$$I_{Q(rms)} = \sqrt{D} I_{L(rms)} = 63.44\,\text{A}$$

and

$$I_{Q(dc)} = D I_{L(dc)} = 40\,\text{A}$$

The diode rms and average currents are

$$I_{D(rms)} = \sqrt{1-D} I_{L(rms)} = 77.69\,\text{A}$$

and

$$I_{D(dc)} = (1-D) I_{L(dc)} = 60\,\text{A}$$

The average high-voltage input current is

$$I_{HV} = \frac{P}{V_{HV}} = \frac{20000}{500} \text{A} = 40\,\text{A}$$

The rms current in the high-voltage link capacitor is

$$I_{CHV(rms)} = \sqrt{I_{Q(rms)}{}^2 - I_{HV}{}^2} = \sqrt{63.44^2 - 40^2} \text{A} = 49.24\,\text{A}$$

Finally, the capacitances can be determined by rearranging Equation (11.27) and Equation (11.28):

$$C_{HV} = \frac{I_{HV}(1-D)}{f_s \Delta V_{CHV(p-p)}} = \frac{I_{HV}(1-D)}{f_s r_v V_{HV}} = \frac{40 \times (1-0.4)}{10000 \times 0.005 \times 500} \text{F} = 960\,\mu\text{F}$$

$$C_{LV} = \frac{\Delta I_{L(p-p)}}{8 f_s \Delta V_{CLV(p-p)}} = \frac{\Delta I_{L(p-p)}}{8 f_s r_v V_{LV}} = \frac{28}{8 \times 10000 \times 0.005 \times 200} \text{F} = 350\,\mu\text{F}$$

11.3.2 BCM Operation of Buck Converter

As the load current and power drop in the buck converter for given input and output voltages, the same analysis can be applied for these part-load conditions until the minimum inductor current drops to zero. The CCM condition at which the current drops to zero is known as **BCM**. This is a common operating mode for many power converters, even at full load, as it has the significant advantage compared to CCM that the diode turns off at zero current, eliminating reverse-recovery losses in the silicon diode.

The inductor waveform is shown in Figure 11.14. The complete set of BCM voltage and current waveforms is shown in Figure 11.15.

As can be seen, the switch now turns on at zero current, and the diode turns off at zero current. Interestingly, there have been no changes in the inductor ripple current and the low-voltage capacitor ripple current as these currents are not a function of the load in BCM and CCM – assuming, of course, that the inductance value is constant with the load, which may not necessarily be correct as discussed in Chapter 16. The average and rms currents seen by the inductor, switch, diode, and input capacitor are a function of the load and can change significantly compared to the earlier CCM waveforms.

The converter enters BCM operation when the minimum inductor current equals 0 A. At this condition, the dc current going through the inductor, and into the load, is equal to one half the peak-to-peak inductor current. Thus, the load current at which the converter enters BCM is given by

$$I_{LV} = \frac{\Delta I_{L(p-p)}}{2} = \frac{V_{LV}(1-D)}{2f_s L} \tag{11.30}$$

Knowing the BCM level, the various component currents can be determined using the equations developed for the CCM analysis.

11.3.2.1 Example of Buck in BCM

Determine the power level at which the converter enters BCM for the converter voltages used in the previous example.

Solution:

As before, the duty cycle is 0.4, and the inductor ripple current remains at

$$\Delta I_{L(p-p)} = 28\,\text{A}$$

From Equation (11.30), the average inductor current in BCM is

$$I_{LV} = I_{L(dc)} = \frac{\Delta I_{L(p-p)}}{2} = 14\text{A}$$

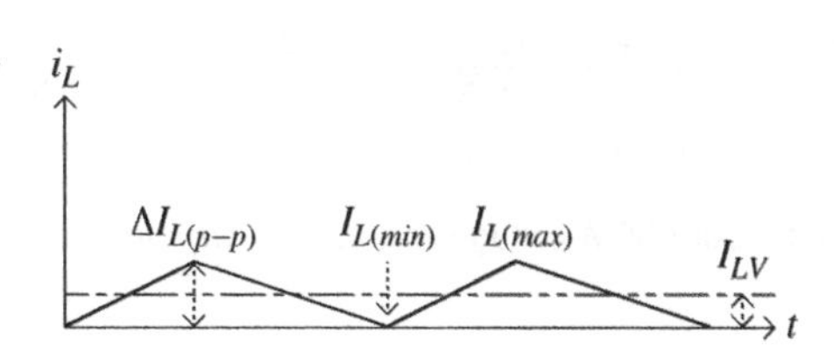

Figure 11.14 Buck BCM inductor current.

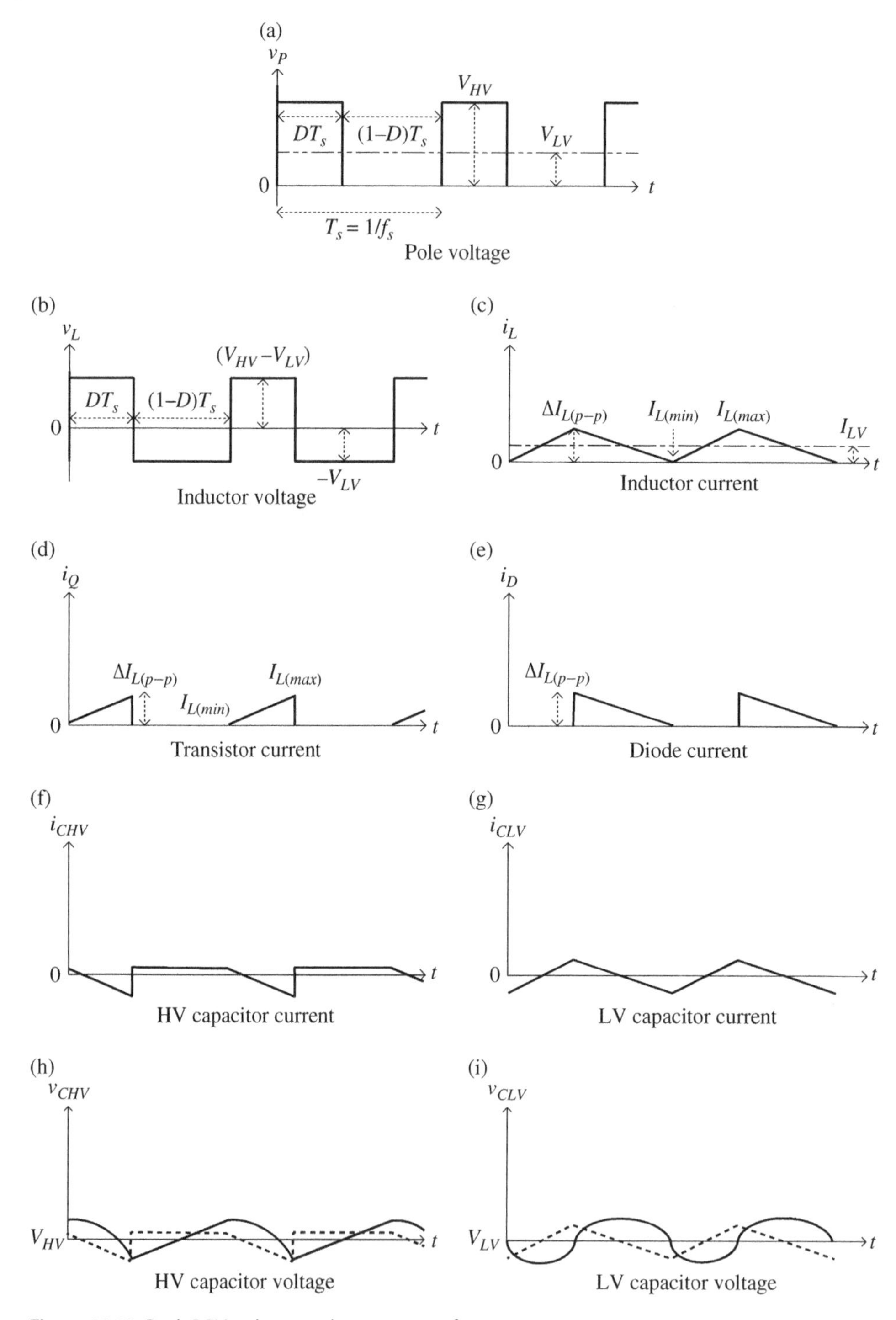

Figure 11.15 Buck BCM voltage and current waveforms.

The power at which this occurs is

$$P = V_{LV} I_{LV} = 200 \times 14\,\text{W} = 2.8\,\text{kW}$$

The low-voltage capacitor rms current remains

$$I_{CLV(rms)} = \frac{\Delta I_{L(p-p)}}{\sqrt{12}} = 8.083\,\text{A}$$

The inductor rms current reduces to

$$I_{L(rms)} = \sqrt{I_{LV}^2 + I_{CLV(rms)}^2} = \sqrt{14^2 + 8.083^2}\,\text{A} = 16.17\,\text{A}$$

The inductor, switch, and diode maximum and minimum currents are

$$I_{L(max)} = I_{Q(off)} = I_{D(on)} = I_{L(dc)} + \frac{\Delta I_{L(p-p)}}{2} = 28\,\text{A}$$

$$I_{L(min)} = I_{Q(on)} = I_{D(off)} = I_{L(dc)} - \frac{\Delta I_{L(p-p)}}{2} = 0\,\text{A}$$

The rms and average switch currents are

$$I_{Q(rms)} = \sqrt{D} I_{L(rms)} = \sqrt{0.4} \times 16.17\,\text{A} = 10.22\,\text{A}$$

and

$$I_{Q(dc)} = D I_{L(dc)} = 0.4 \times 14\,\text{A} = 5.6\,\text{A}$$

The rms and average diode currents are

$$I_{D(rms)} = \sqrt{1-D} I_{L(rms)} = \sqrt{(1-0.4)} \times 16.17\,\text{A} = 12.53\,\text{A}$$

and

$$I_{D(dc)} = (1-D) I_{L(dc)} = (1-0.4) \times 14\,\text{A} = 8.4\,\text{A}$$

The average high-voltage input current is

$$I_{HV} = \frac{P}{V_{HV}} = \frac{2800}{500}\,\text{A} = 5.6\,\text{A}$$

The rms current in the high-voltage link capacitor is

$$I_{CHV(rms)} = \sqrt{I_{Q(rms)}^2 - I_{HV}^2} = \sqrt{10.22^2 - 5.6^2}\,\text{A} = 8.55\,\text{A}$$

11.3.3 DCM Operation of Buck Converter

As the load current is reduced, the inductor current reduces in value and becomes discontinuous. The inductor current is no longer in CCM or BCM and is operating in DCM. This mode occurs commonly at light loads and has the advantages of zero-current turn-on and turn-off of the switch and diode, respectively, resulting in reasonably high efficiencies at light loads.

The pole voltage and inductor voltage and current waveforms for DCM operation are shown in Figure 11.16.

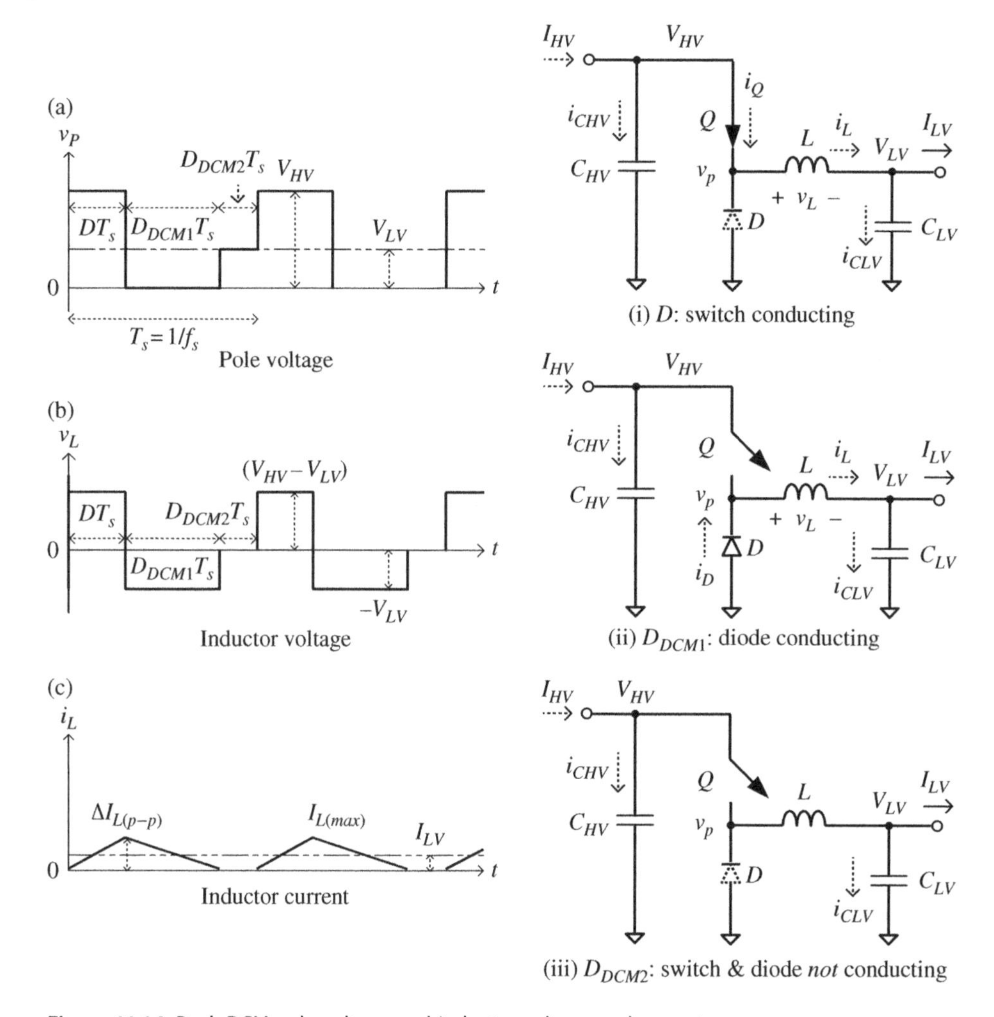

Figure 11.16 Buck DCM pole voltage and inductor voltage and current.

There are three time periods of interest within the switching period. During the first period DT_s, the switch is closed, and a positive voltage is applied across the inductor, causing the current to increase. During the time period $D_{DCM1}T_s$, the switch is open, and the diode is conducting, resulting in a negative voltage across the inductor and a decrease in the inductor current to zero. During the time period $D_{DCM2}T_s$, the inductor current is zero, and the load current is provided by the low-voltage capacitor. Note that during $D_{DCM2}T_s$ the inductor voltage and current are both zero, and the pole voltage equals the output voltage.

The rms and average values of the various waveforms are relatively straightforward to determine as the critical currents are discontinuous.

First, let us determine the relationship between the dc output current I_{LV} and the duty cycle during DCM. Two of the three time periods contribute to the average current.

$$
\begin{aligned}
I_{LV} = I_{L(dc)} &= \frac{1}{T_s}\int_0^{T_s} i_L(t)dt \\
&= \frac{1}{T_s}\left[\int_0^{DT_s} i_L(t)dt + \int_{DT_s}^{DT_s + D_{DCM1}T_s} i_L(t)dt + \int_{DT_s + D_{DCM1}T_s}^{T_s} i_L(t)dt\right] \\
&= \frac{1}{T_s}\left[\frac{1}{2}\Delta I_{L(p-p)}DT_s + \frac{1}{2}\Delta I_{L(p-p)}D_{DCM1}T_s + 0\times D_{DCM2}T_s\right] \\
&= \frac{\Delta I_{L(p-p)}}{2}(D + D_{DCM1})
\end{aligned}
\tag{11.31}
$$

The relationship between the duty cycle of the switch D and the duty cycle of the diode D_{DCM1} is easily determined. In the steady state, the inductor current increases by $\Delta I_{L(p-p)}$ during DT_s, and decreases by $\Delta I_{L(p-p)}$ during $D_{DCM1}T_s$.

Thus, assuming a linear current change, we get

$$\Delta I_{L(p-p)} = \frac{(V_{HV} - V_{LV})DT_s}{L} = \frac{V_{LV}D_{DCM1}T_s}{L} \tag{11.32}$$

This equation can be rearranged to show that

$$D_{DCM1} = \frac{(V_{HV} - V_{LV})}{V_{LV}}D \tag{11.33}$$

The three duty cycles sum to unity:

$$D + D_{DCM1} + D_{DCM2} = 1 \tag{11.34}$$

Thus, the third period of time is then easily determined and can be simplified to

$$
\begin{aligned}
D_{DCM2} = 1 - D - D_{DCM1} &= 1 - D - \frac{(V_{HV} - V_{LV})}{V_{LV}}D \\
&= 1 - \frac{V_{HV}}{V_{LV}}D
\end{aligned}
\tag{11.35}
$$

An expression for the duty cycle in terms of the dc input and output levels can be determined by rearranging Equation (11.31) to get

$$D + D_{DCM1} = \frac{2I_{LV}}{\Delta I_{L(p-p)}} \tag{11.36}$$

Substituting in Equation (11.32) and Equation (11.33) yields

$$D + \frac{(V_{HV} - V_{LV})}{V_{LV}}D = \frac{2LI_{LV}}{(V_{HV} - V_{LV})DT_s} \tag{11.37}$$

Rearranging the preceding equation gives an expression for D:

$$D = \sqrt{\frac{2V_{LV}}{V_{HV}(V_{HV} - V_{LV})}f_s LI_{LV}} \tag{11.38}$$

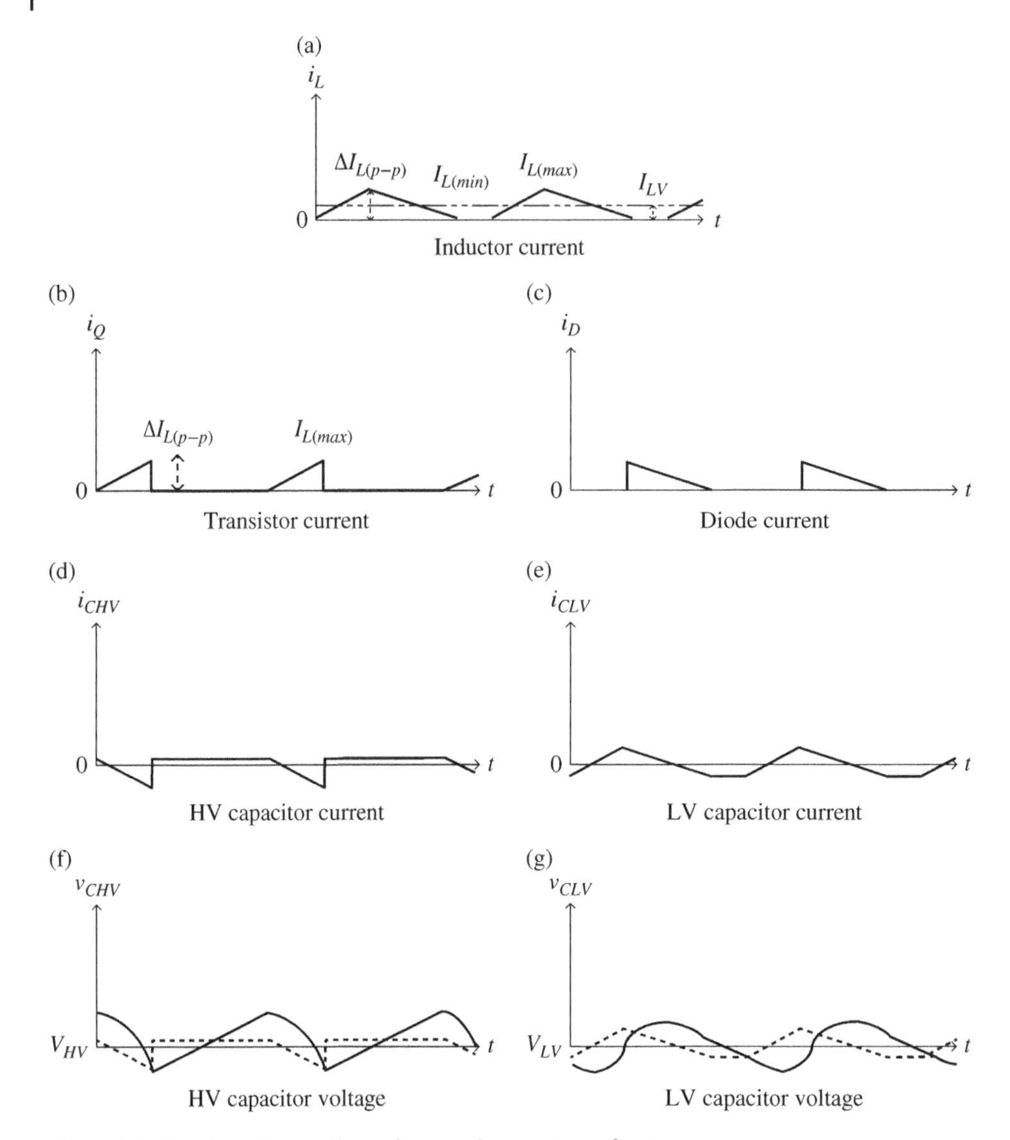

Figure 11.17 Buck DCM operation voltage and current waveforms.

Once the various duty cycles are known, the current values are easily be determined for a load condition.

The various converter waveforms in DCM are shown in Figure 11.17.

It is useful to note the following relationships between the peak, average, and rms currents for various waveforms. The various waveforms can be simply deconstructed in order to determine the rms and average components.

The switch current of Figure 11.17 is shown again in Figure 11.18.

The current in the switch is a ramp and is described by

$$i_Q(t) = \frac{\Delta I_{L(p-p)}}{DT_s}t \tag{11.39}$$

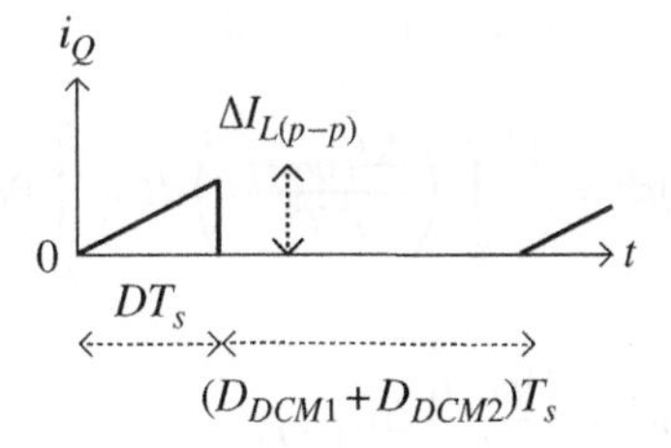

Figure 11.18 Switch current in DCM.

The rms value of the switch current is simply given by

$$\begin{aligned}
I_{Q(rms)} &= \sqrt{\frac{1}{T_s}\int_0^{T_s} i_Q(t)^2 dt} \\
&= \sqrt{\frac{1}{T_s}\int_0^{DT_s}\left(\frac{\Delta I_{L(p-p)}}{DT_s}t\right)^2 dt + \int_{DT_s}^{T_s} 0^2 dt} \\
&= \Delta I_{L(p-p)} \times \sqrt{\frac{1}{D^2 T_s^3}\left[\frac{t^3}{3}\right]_0^{DT_s}} \\
&= \sqrt{\frac{D}{3}}\Delta I_{L(p-p)}
\end{aligned} \tag{11.40}$$

The rms value of the diode current is similarly given by

$$I_{D(rms)} = \sqrt{\frac{1}{T_s}\int_0^{T_s} i_D(t)^2 dt} = \sqrt{\frac{D_{DCM1}}{3}}\Delta I_{L(p-p)} \tag{11.41}$$

The rms value of the inductor current is

$$I_{L(rms)} = \sqrt{\frac{1}{T_s}\int_0^{T_s} i_L(t)^2 dt} = \sqrt{\frac{(D+D_{DCM1})}{3}}\Delta I_{L(p-p)} \tag{11.42}$$

The rms value of the low-voltage capacitor current is

$$I_{CLV(rms)} = \sqrt{\frac{1}{T_s}\int_0^{T_s} i_{CLV}(t)^2 dt} = \sqrt{I_{L(rms)}^2 - I_{LV}^2} \tag{11.43}$$

The rms value of the high-voltage capacitor current is

$$I_{CHV(rms)} = \sqrt{\frac{1}{T_s}\int_0^{T_s} i_{CHV}(t)^2 dt} = \sqrt{I_{Q(rms)}^2 - I_{HV}^2} \tag{11.44}$$

Similarly, the average currents of the switch and diode are

$$I_{Q(dc)} = \frac{1}{T_s}\int_0^{T_s} i_Q(t)dt = \frac{1}{T_s}\int_0^{DT_s}\left(\frac{\Delta I_{L(p-p)}}{DT_s}t\right)dt + \int_{DT_s}^{T_s} 0dt = \frac{D}{2}\Delta I_{L(p-p)} \tag{11.45}$$

$$I_{D(dc)} = \frac{1}{T_s}\int_0^{T_s} i_D(t)dt = \frac{D_{DCM1}}{2}\Delta I_{L(p-p)} \tag{11.46}$$

11.3.3.1 Example: Buck Converter in DCM Operation

Knowing from the previous example that the converter operates in DCM below 2.8 kW for the given voltage conditions, determine the various converter currents when operating in DCM at 2 kW.

Solution:

The dc current in the low-voltage dc link is

$$I_{LV} = \frac{P}{V_{LV}} = \frac{2000}{200}\,\text{A} = 10\,\text{A}$$

The average high-voltage input current is

$$I_{HV} = \frac{P}{V_{HV}} = \frac{2000}{500}\,\text{A} = 4\,\text{A}$$

The duty cycle at this condition is

$$D = \sqrt{\frac{2V_{LV}}{V_{HV}(V_{HV} - V_{LV})} \times f_s L I_{LV}}$$

$$= \sqrt{\frac{2 \times 200}{500 \times (500 - 200)} \times 10000 \times 428.5 \times 10^{-6} \times 10} = 0.338$$

and

$$D_{DCM1} = \frac{(V_{HV} - V_{LV})}{V_{LV}}D = \frac{(500 - 200)}{200} \times 0.338 = 0.507$$

The inductor ripple current is

$$\Delta I_{L(p-p)} = \frac{(V_{HV} - V_{LV})D}{f_s L} = \frac{(500 - 200) \times 0.338}{10000 \times 428.5 \times 10^{-6}}\,\text{A} = 23.66\,\text{A}$$

The various rms currents are

$$I_{Q(rms)} = \sqrt{\frac{D}{3}}\Delta I_{L(p-p)} = \sqrt{\frac{0.338}{3}} \times 23.66 = 7.94\,\text{A}$$

$$I_{D(rms)} = \sqrt{\frac{D_{DCM1}}{3}}\Delta I_{L(p-p)} = \sqrt{\frac{0.507}{3}} \times 23.66\,\text{A} = 9.73\,\text{A}$$

$$I_{L(rms)} = \sqrt{\frac{(D + D_{DCM1})}{3}}\Delta I_{L(p-p)} = \sqrt{\frac{(0.338 + 0.507)}{3}} \times 23.66\text{A} = 12.56\text{A}$$

$$I_{CLV(rms)} = \sqrt{I_{L(rms)}{}^2 - I_{LV}{}^2} = \sqrt{12.56^2 - 10^2}\,\text{A} = 7.6\,\text{A}$$

and

$$I_{CHV(rms)} = \sqrt{I_{Q(rms)}{}^2 - I_{HV}{}^2} = \sqrt{7.94^2 - 4^2}\,\text{A} = 6.86\,\text{A}$$

The various dc currents are

$$I_{Q(dc)} = \frac{D}{2}\Delta I_{L(p-p)} = \frac{0.338}{2} \times 23.66\,\text{A} = 4\,\text{A}\,(= I_{HV}\,\text{as expected})$$

$$I_{D(dc)} = \frac{D_{DCM1}}{2}\Delta I_{L(p-p)} = \frac{0.509}{2} \times 23.66\,\text{A} = 6\,\text{A}$$

and

$$I_{L(dc)} = \frac{D + D_{DCM1}}{2}\Delta I_{L(p-p)} = \frac{0.338 + 0.509}{2} \times 23.66\,\text{A}$$
$$= 10\,\text{A}\left(= I_{Q(dc)} + I_{D(dc)}\ \text{as expected}\right)$$

11.4 The Boost or Step-up Converter

The **boost** or **step-up** converter produces a higher average output voltage V_{HV} than the input voltage V_{LV}. As with the buck converter, the boost converter has three basic power components to perform the voltage conversion. First, a controlled power semiconductor switch, shown as Q in Figure 11.19(a), is pulsed on and off at a high frequency in order to transfer energy from the source V_{LV} to the load V_{HV}. The output from the switch is a controlled pulse train of a controlled switching frequency f_s, or switching period T_s, and a controlled conduction time within that period DT_s, where D is termed the duty cycle or proportional on-time of the switch. The pole voltage waveform $v_p(t)$ is as shown in Figure 11.19(b). As can be seen, the pole voltage is pulsing between 0 V and V_{HV}. The input voltage V_{LV} is the average value of the pole voltage.

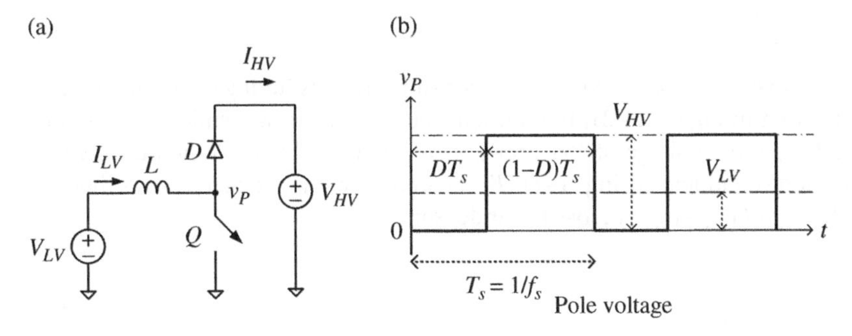

Figure 11.19 Boost or step-up converter.

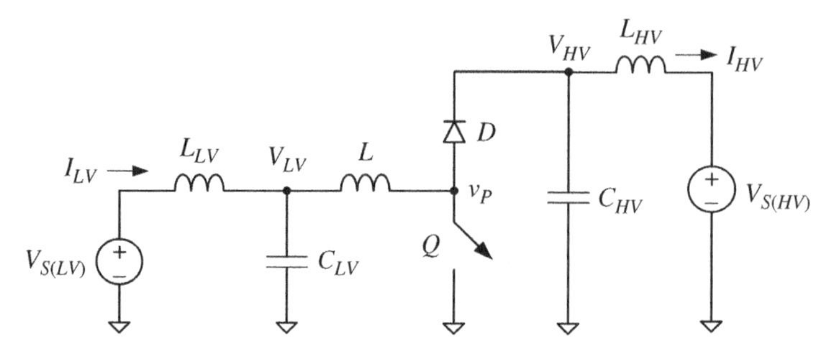

Figure 11.20 A boost converter with source and load filter components.

As with the buck converter, both voltage links feature internal or added inductances between the dc links and the sources, as shown in Figure 11.20. Thus, the high-voltage source, shown as $V_{S(HV)}$, has the internal inductance, L_{HV}, and the dc link capacitance C_{HV}, while the low-voltage source, shown as $V_{S(LV)}$, has the internal inductance L_{LV} and the dc link capacitance C_{LV}.

11.4.1 Analysis of Voltage Gain of Boost Converter in CCM

In CCM, the inductor of the boost converter always conducts current. First, the relationship between the voltage gain of the converter and the duty cycle is determined. Neglecting parasitics, in a boost converter, the dc or average input voltage equals the average of the pole voltage. Ignoring the switch and diode voltage drops, the pole voltage pulses from the dc return or 0 V to the high-voltage dc link voltage, V_{HV}. The pole is at V_{HV} for the time period $(1-D)\ T_s$ and at 0 V for the time period DT_s. Thus, the dc output voltage is given by

$$V_{LV} = \frac{1}{T_s}\int_0^{T_s} v_P(t)dt = \frac{1}{T_s}\int_0^{DT_s} 0dt + \frac{1}{T_s}\int_{DT_s}^{T_s} V_{HV}dt = (1-D)V_{HV} \tag{11.47}$$

Thus, the duty cycle of a boost converter is

$$D = 1 - \frac{V_{LV}}{V_{HV}} \tag{11.48}$$

The voltage gain of the converter can be expressed as

$$V_{HV}\Big/V_{LV} = \frac{1}{1-D} \tag{11.49}$$

It is also productive to derive the voltage relationships by considering the inductor current. As can be seen in Figure 11.21, the inductor current increases while the switch is closed during DT_s, and a voltage of V_{LV} appears across the inductor. When the switch opens and the diode conducts during $(1-D)\ T_s$, the inductor current decreases as there is a negative voltage of $(V_{LV}-V_{HV})$ across the inductor.

During time DT_s:

$$\Delta I_{L(p-p)} = \frac{V_{LV}}{L}DT_s \tag{11.50}$$

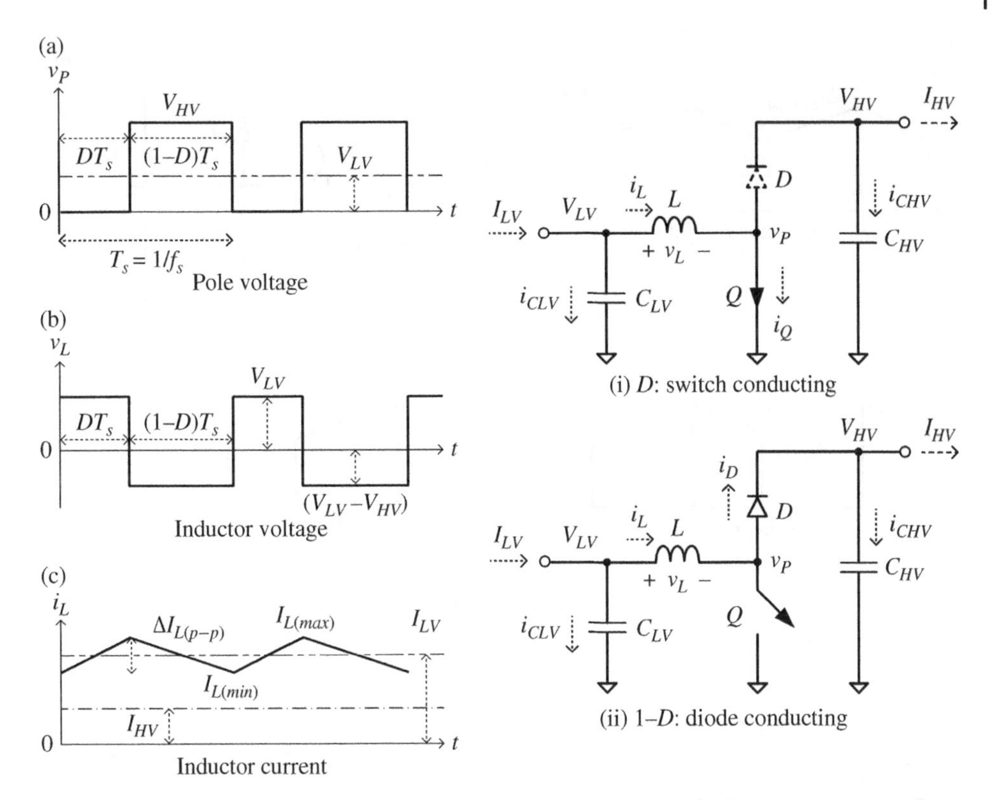

Figure 11.21 Boost CCM pole voltage and inductor voltage and current.

During time $(1-D)T_s$:

$$\Delta I_{L(p-p)} = \frac{-(V_{LV} - V_{HV})(1-D)T_s}{L} \tag{11.51}$$

Solving the two equations results in the earlier expression, Equation (11.48), for the voltage gain.

11.4.1.1 Analysis of Boost Converter in CCM

In order to design and specify the components of a boost converter, it is necessary to know the average, rms, minimum, and maximum currents in each component, the waveforms of which are shown in Figure 11.22. The inductor current $i_L(t)$ has the dc component, $I_{L(dc)} = I_{LV}$, flowing from the source and an ac component flowing into the low-voltage capacitor as $i_{CLV}(t)$, as shown in Figure 11.22(e).

As before for the CCM buck converter, the rms component of the capacitor triangular current is given by

$$I_{CLV(rms)} = \frac{\Delta I_{L(p-p)}}{\sqrt{12}} \tag{11.52}$$

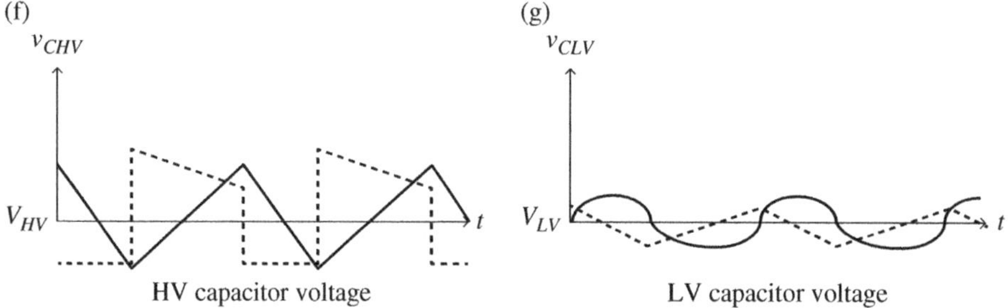

Figure 11.22 Boost CCM voltage and current waveforms.

Knowing the dc input current and the capacitor rms current, the rms value of the inductor $I_{L(rms)}$ is

$$I_{L(rms)} = \sqrt{I_{L(dc)}{}^2 + I_{CLV(rms)}{}^2} = \sqrt{I_{LV}{}^2 + \frac{\Delta I_{L(p-p)}{}^2}{12}} \tag{11.53}$$

The inductor peak current is simply given by

$$I_{L(max)} = I_{LV} + \frac{\Delta I_{L(p-p)}}{2} \tag{11.54}$$

The average currents of the switch and diode are given by

$$I_{Q(dc)} = D I_{LV} \tag{11.55}$$

$$I_{D(dc)} = (1-D) I_{LV} \tag{11.56}$$

The rms currents work out to be

$$I_{Q(rms)} = \sqrt{D} I_{L(rms)} \tag{11.57}$$

$$I_{D(rms)} = \sqrt{(1-D)} I_{L(rms)} \tag{11.58}$$

The instantaneous turn-on and turn-off currents in the switch and diode are

$$I_{Q(off)} = I_{D(on)} = I_{L(max)} = I_{LV} + \frac{\Delta I_{L(p-p)}}{2} \tag{11.59}$$

$$I_{Q(on)} = I_{D(off)} = I_{L(min)} = I_{LV} - \frac{\Delta I_{L(p-p)}}{2} \tag{11.60}$$

The rms current in the high-voltage link capacitor $I_{CHV(rms)}$ can be determined by subtracting the dc current I_{HV} from the rms current in the diode $I_{D(rms)}$:

$$I_{CHV(rms)} = \sqrt{I_{D(rms)}^2 - I_{HV}^2} \tag{11.61}$$

Referring to Figure 11.22(d) and (f), the peak-to-peak voltage ripple on the high-voltage capacitor $\Delta V_{CHV(p-p)}$ is determined by calculating the charge based on the negative or positive current. Using the current-time area product of the negative current and the diode conduction time gives the answer as follows:

$$\Delta V_{CHV(p-p)} = \frac{\Delta Q_{CHV}}{C_{HV}} = \frac{I_{HV} \cdot DT_s}{C_{HV}} \tag{11.62}$$

Similarly, the peak-to-peak voltage ripple on the high-voltage capacitor $\Delta V_{CLV(p-p)}$ is estimated by calculating the charge based on the negative or positive current. In this case, the current-time area product is the product of ½, ½ of the peak-to-peak inductor ripple current, and ½ of the period, because of the triangular waveform:

$$\Delta V_{CLV(p-p)} = \frac{\Delta Q_{CLV}}{C_{LV}} = \frac{1}{2}\frac{\Delta I_{L(p-p)}}{2}\frac{T_s}{2}\frac{1}{C_{LV}} = \frac{\Delta I_{L(p-p)} T_s}{8 C_{LV}} \tag{11.63}$$

At this point, all the currents and voltages for the CCM boost converter have been determined in order to analyze a given design or design a converter based on a set of specifications.

11.4.1.2 Example: Analyzing Boost for CCM Operation

The vehicle in the earlier buck example is now operating in **motoring** mode, and the bidirectional converter is required to act as a **boost** at full power.

Determine the currents in the various components. Assume ideal components.

Solution:

The duty cycle of the boost converter is

$$D = 1 - \frac{V_{LV}}{V_{HV}} = 1 - \frac{200\text{V}}{500\text{V}} = 0.6$$

The average inductor current equals the input current and is given by

$$I_{LV} = I_{L(dc)} = \frac{P}{V_{LV}} = \frac{20000}{200}\,\text{A} = 100\,\text{A}$$

The inductor ripple current is

$$\Delta I_{L(p-p)} = \frac{V_{LV} D}{f_s L} = \frac{200 \times 0.6}{10000 \times 428.5}\,\text{A} = 28\,\text{A}$$

The rms current in the low-voltage capacitor is given by

$$I_{CLV(rms)} = \frac{\Delta I_{L(p-p)}}{\sqrt{12}} = 8.083\,\text{A}$$

The inductor rms current is

$$I_{L(rms)} = \sqrt{I_{L(dc)}^{\ 2} + I_{CLV(rms)}^{\ 2}} = \sqrt{100^2 + 8.083^2}\,\text{A} = 100.3\,\text{A}$$

The inductor, switch, and diode maximum and minimum currents are

$$I_{L(max)} = I_{Q(off)} = I_{D(on)} = I_{L(dc)} + \frac{\Delta I_{L(p-p)}}{2} = 114\,\text{A}$$

$$I_{L(min)} = I_{Q(on)} = I_{D(off)} = I_{L(dc)} - \frac{\Delta I_{L(p-p)}}{2} = 86\,\text{A}$$

The switch rms and average currents are

$$I_{Q(rms)} = \sqrt{D} I_{L(rms)} = 77.69\text{A}$$

and

$$I_{Q(dc)} = D I_{L(dc)} = 60\,\text{A}$$

The diode rms and average currents are

$$I_{D(rms)} = \sqrt{1-D}\, I_{L(rms)} = 63.44\,\text{A}$$

and

$$I_{D(dc)} = (1-D) I_{L(dc)} = 40\,\text{A}$$

It can be seen that many of the currents are the same for the buck and boost examples and that the current values for the switch and diode are swapped for the same power and voltage levels.

11.4.2 BCM Operation of Boost Converter

As the load current and power drop in the boost converter for given input and output voltages, the converter enters **BCM,** when the CCM inductor current drops to zero. A number of boost converters operate in this mode as it can result in low semiconductor losses and compact inductors. Some designs of lower-power power-factor-correction boost converters operate in BCM in order to eliminate the reverse recovery of the silicon diode.

The earlier waveforms in Figure 11.22 have been modified to those of Figure 11.23 such that the converter is in BCM. As can be seen, the switch now turns on at zero current, and

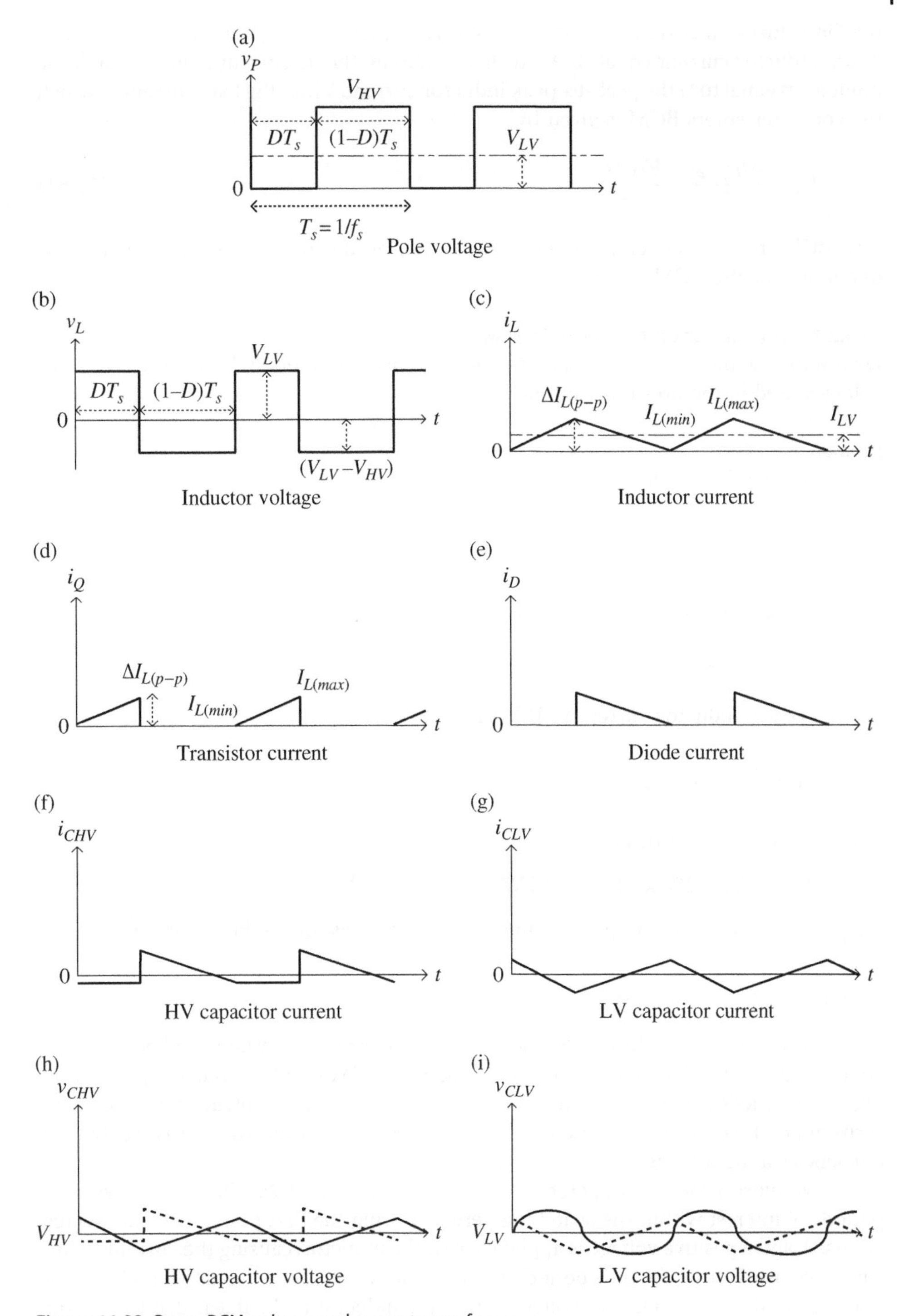

Figure 11.23 Boost BCM voltage and current waveforms.

the diode turns off at zero current. The converter enters BCM operation when the minimum inductor current equals 0 A. At this condition, the dc current going through the inductor is equal to ½ the peak-to-peak inductor current. Thus, the load current at which the converter enters BCM is given by

$$I_{LV} = \frac{\Delta I_{L(p-p)}}{2} = \frac{V_{LV} D}{2 f_s L} \tag{11.64}$$

In BCM, the various component currents can be determined using the equations developed for the CCM analysis.

11.4.2.1 Example: Boost Converter in BCM

Determine the power level at which the boost converter enters BCM for the converter voltages used in the previous example.

Solution:

As before, the duty cycle is

$$D = 1 - \frac{V_{LV}}{V_{HV}} = 1 - \frac{200}{500} = 0.6$$

The inductor ripple current remains

$$\Delta I_{L(p-p)} = 28\,\text{A}$$

The average inductor current in BCM is

$$I_{LV} = \frac{\Delta I_{L(p-p)}}{2} = 14\text{A}$$

The power at which this occurs

$$P = V_{LV} I_{LV} = 200 \times 14\text{W} = 2.8\text{kW}$$

Again, these current and power values are the same as for the buck converter.

11.4.3 DCM Operation of Boost Converter

As the load current is reduced, the inductor current reduces in value and becomes discontinuous. The inductor current is no longer in CCM or BCM, but is operating in **DCM**. This mode occurs commonly at light loads and has the advantages of zero-current turn-on and turn-off of the switch and diode respectively, resulting in reasonably high efficiencies at light loads.

The waveforms for DCM operation are shown in Figure 11.24. There are three time periods of interest within the switching period. During the first period DT_s, the switch is closed, and a positive voltage is applied across the inductor, causing the current to linearly increase. During the time period $D_{DCM1} T_s$, the switch is open, and the diode is conducting, resulting in a negative voltage across the inductor and a linear decrease in the inductor current to zero. During the time period $D_{DCM2} T_s$, the inductor current is zero, and the load current is provided by the high-voltage capacitor.

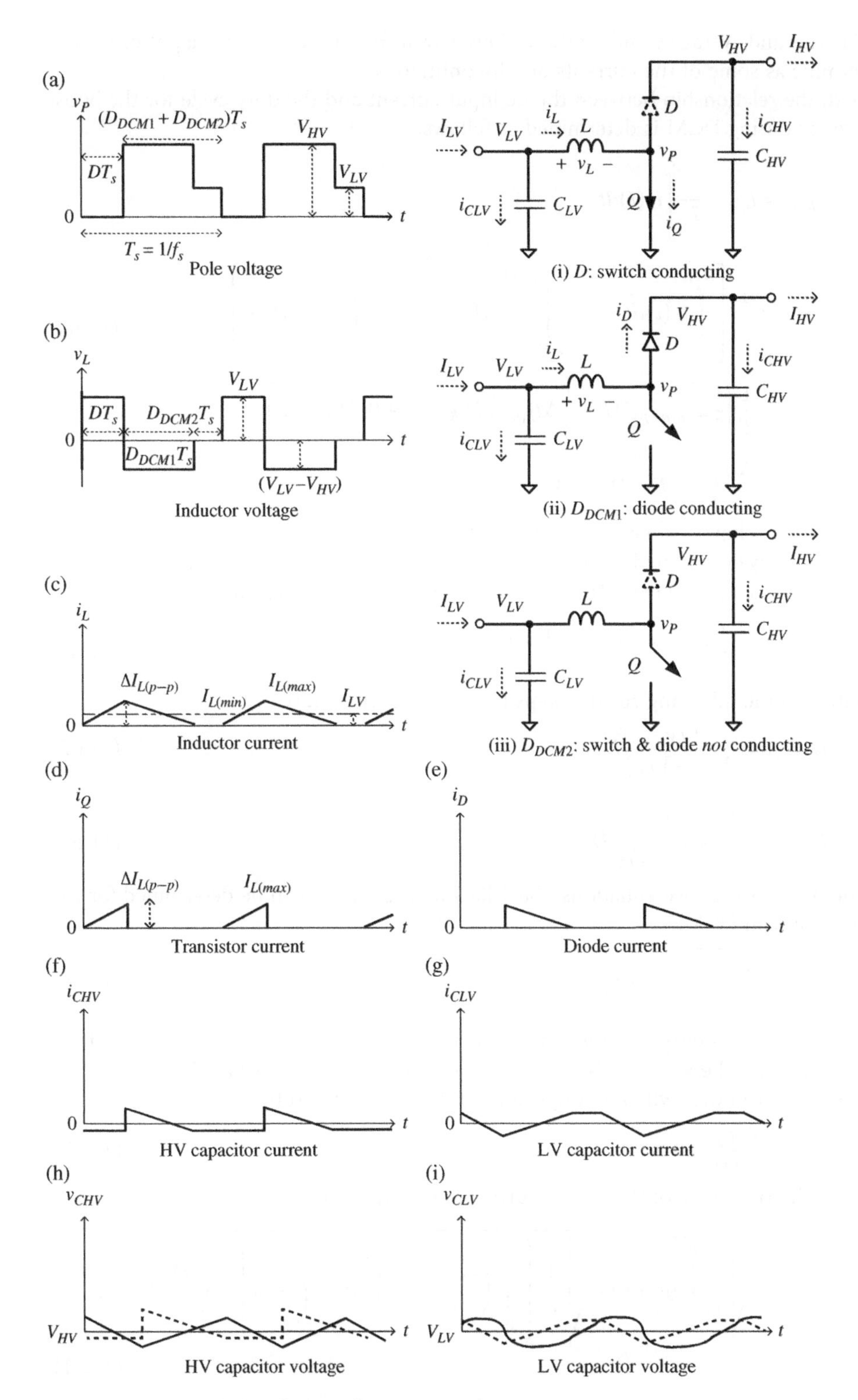

Figure 11.24 Boost DCM voltage and current waveforms.

The rms and average values of the various waveforms are relatively straightforward to determine as some of the currents are discontinuous.

First, the relationship between the dc input current and the duty cycle for the boost converter during DCM is determined as follows:

$$I_{L(dc)} = I_{LV} = \frac{1}{T_s}\int_0^{T_s} i_L(t)dt$$

$$= \frac{1}{T_s}\left[\int_0^{DT_s} i_L(t)dt + \int_{DT_s}^{DT_s + D_{DCM1}T_s} i_L(t)dt + \int_{DT_s + D_{DCM1}T_s}^{T_s} i_L(t)dt\right] \tag{11.65}$$

$$= \frac{1}{T}\left[\frac{1}{2}\Delta I_{L(p-p)}DT + \frac{1}{2}\Delta I_{L(p-p)}D_{DCM1}T + 0 \cdot D_{DCM2}T\right]$$

$$= \frac{\Delta I_{L(p-p)}}{2}(D + D_{DCM1})$$

The relationship between the duty cycle of the switch D and the duty cycle of the diode D_{DCM1} is easily determined as the inductor current increases by $\Delta I_{L(p-p)}$ during DT_s, and decreases by $\Delta I_{L(p-p)}$ during $D_{DCM1}T_s$, while in the steady state. Thus:

$$\Delta I_{L(p-p)} = \frac{V_{LV}DT_s}{L} = \frac{(V_{HV} - V_{LV})D_{DCM1}T_s}{L} \tag{11.66}$$

resulting in the following relationships between the three duty cycles:

$$D_{DCM1} = \frac{V_{LV}}{(V_{HV} - V_{LV})}D \tag{11.67}$$

and

$$D_{DCM2} = 1 - \frac{V_{HV}}{V_{HV} - V_{LV}}D \tag{11.68}$$

Combining the above equations, the following expression can be determined for the duty cycle D in DCM:

$$D = \sqrt{\frac{(V_{HV} - V_{LV})}{V_{HV}V_{LV}}2f_sLI_{LV}} \tag{11.69}$$

Once the various duty cycles are known, the current values are easily determined for a load condition. The various currents can be derived as in Section 11.3.3.

The current in the switch is simply a ramp and is described by

$$i_Q(t) = \frac{\Delta I_{L(p-p)}}{DT_s}t \tag{11.70}$$

Thus, the rms value of the switch current is simply given by

$$I_{Q(rms)} = \sqrt{\frac{1}{T_s}\int_0^{T_s} i_Q(t)^2dt} = \sqrt{\frac{1}{T_s}\left[\int_0^{DT_s}\left(\frac{\Delta I_{L(p-p)}}{DT_s}t\right)^2dt + \int_{DT_s}^{T_s}0^2dt\right]} = \sqrt{\frac{D}{3}}\Delta I_{L(p-p)} \tag{11.71}$$

The rms value of the diode current is similarly given by

$$I_{D(rms)} = \sqrt{\frac{1}{T_s}\int_0^{T_s} i_D(t)^2 dt} = \sqrt{\frac{D_{DCM1}}{3}}\Delta I_{L(p-p)} \tag{11.72}$$

The rms value of the inductor current is

$$I_{L(rms)} = \sqrt{\frac{1}{T_s}\int_0^{T_s} i_L(t)^2 dt} = \sqrt{\frac{(D + D_{DCM1})}{3}}\Delta I_{L(p-p)} \tag{11.73}$$

The rms value of the low-voltage capacitor current is

$$I_{CLV(rms)} = \sqrt{\frac{1}{T_s}\int_0^{T_s} i_{CLV}(t)^2 dt} = \sqrt{I_{L(rms)}{}^2 - I_{LV}{}^2} \tag{11.74}$$

The rms value of the high-voltage capacitor current is

$$I_{CHV(rms)} = \sqrt{\frac{1}{T_s}\int_0^{T_s} i_{CHV}(t)^2 dt} = \sqrt{I_{D(rms)}{}^2 - I_{HV}{}^2} \tag{11.75}$$

Similarly, the average currents of the switch, diode, and inductor are

$$I_{Q(dc)} = \frac{1}{T_s}\int_0^{T_s} i_Q(t) dt = \frac{1}{T_s}\left[\int_0^{DT_s}\left(\frac{\Delta I_{L(p-p)}}{DT_s}t\right)dt + \int_{DT_s}^{T_s} 0\,dt\right] = \frac{D}{2}\Delta I_{L(p-p)} \tag{11.76}$$

$$I_{D(dc)} = \frac{1}{T_s}\int_0^{T_s} i_D(t) dt = \frac{D_{DCM1}}{2}\Delta I_{L(p-p)} \tag{11.77}$$

$$I_{L(dc)} = \frac{1}{T_s}\int_0^{T_s} i_L(t) dt = \frac{D + D_{DCM1}}{2}\Delta I_{L(p-p)} = I_{LV} \tag{11.78}$$

11.4.3.1 Example: Boost Converter in DCM Operation

Knowing from the previous example that the converter enters DCM at 2.8 kW for the given voltage conditions, determine the duty cycle and the various converter currents when operating in DCM at 2 kW.

Solution:

The dc current in the low-voltage dc link is

$$I_{LV} = \frac{P}{V_{LV}} = \frac{2000}{200}\,\text{A} = 10\,\text{A}$$

The average high-voltage input current is

$$I_{HV} = \frac{P}{V_{HV}} = \frac{2000}{500} \text{A} = 4 \text{A}$$

The duty cycle at this condition is

$$D = \sqrt{\frac{(V_{HV} - V_{LV})}{V_{LV} V_{HV}} \cdot 2 f_s L I_{LV}} = \sqrt{\frac{(500 - 200)}{200 \times 500} \times 2 \times 10000 \times 428.5 \times 10^{-6} \times 10} = 0.507$$

and

$$D_{DCM1} = \frac{V_{LV}}{(V_{HV} - V_{LV})} D = \frac{200}{(500 - 200)} \times 0.507 = 0.338$$

It is left as an exercise for the reader to calculate the various currents and note that they are similar to the values of the earlier example on the buck converter operating in DCM.

11.5 Power Semiconductors

Semiconductor technology is a key enabler of modern power converters. Millions of transistors can be integrated onto a piece of silicon to act as a single power switch – or can be integrated to operate as a digital controller. Power semiconductors have hugely benefited from the advances at the signal level, and many variations on power semiconductors exist. New diamond-like semiconductor materials, such as gallium nitride (GaN) and silicon carbide (SiC), are competing with traditional silicon materials to enable high-efficiency power conversion. See Appendix III for a discussion of SiC devices.

The modern automotive powertrain power converter is based on the **IGBT**. The IGBT is closely related to its sibling, the **MOSFET**. In effect, the IGBT has one additional silicon layer compared to the MOSFET. The effect of the additional layer is to make the IGBT look like a hybrid between the MOSFET and the BJT. The IGBT is the preferred device for high-voltage converters and inverters due to its lower power loss and its short-circuit capability.

From a semiconductor perspective, the IGBT is a minority-carrier device, similar to the BJT, thyristor, and *pn* diode, while the MOSFET is a majority-carrier device, similar to the junction-field-effect transistor (JFET) and Schottky diode. A characteristic of minority-carrier devices is that the device can generate significant additional power loss when it is turned off. This phenomenon is manifested as the **tail current** in the IGBT and as **reverse recovery** in the power diode. A **minority-carrier device** is a semiconductor in which the mobile hole or electron, or minority carrier, flows through semiconductor regions which are very conductive to carriers of the opposite polarity, the majority carrier. A **majority-carrier device** is a semiconductor in which the mobile hole or electron, or majority carrier, flows through semiconductor regions which are very conductive to the majority carrier.

The symbol for the IGBT is as shown in Figure 11.25(a). The IGBT has three terminals: the gate (G) for control, and the collector (C) and emitter (E) for power. The IGBT must be paired with a separate discrete inverse diode in order to function in a typical power converter. The vertically diffused power MOSFET, with the symbol shown in Figure 11.25(b), actually comes with a free inverse diode as part of its basic construction.

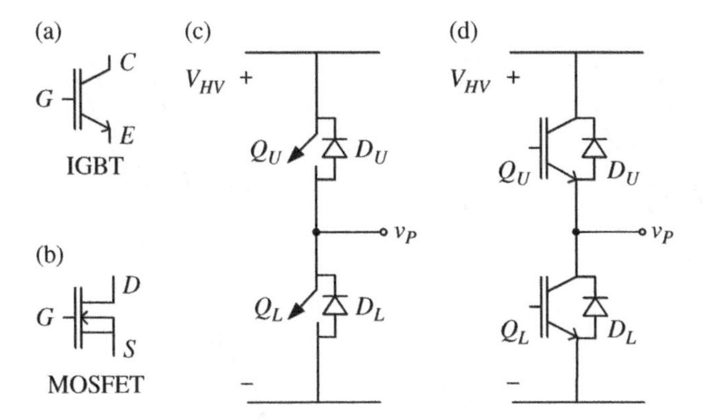

Figure 11.25 (a) IGBT, (b) MOSFET, and half-bridges with (c) generic switch and (d) IGBT switch.

The technical specifications for a power device are quite detailed. Typical specifications are the rated collector-emitter voltage and the continuous dc current. For example, a 600 V, 200 A IGBT can withstand a voltage of 600 V from collector to emitter when switched off, and can continuously conduct 200 A from collector to emitter when switched on. Of course, the IGBT generates a significant power loss which leads to device heating and temperature rise. For example, the hotspot of the IGBT at the semiconductor junction will be at its maximum temperature of 175 °C when the device is conducting 200 A, while the case of the module is at 80 °C. While 80 °C may appear to be a very high temperature, it is easily reached on a vehicle in the engine compartment. The device is switched on and off by applying gating signals across the gate and emitter. The maximum gate voltage that can be applied is quite low, at 20 V typically, in order to avoid over-voltaging the gate dielectric layer. It is important to keep the semiconductor junction hotspot relatively low as device reliability decreases and failure rates increase with rising temperature.

Two switches and two diodes are often arranged in a module as a half bridge, as shown in Figure 11.25(c) and (d). Such half-bridge modules can be used in high-power rectifiers, converters and inverters.

11.5.1 Power Semiconductor Power Loss

A power semiconductor device has two main power loss components. **Conduction loss** is the power loss due to the voltage drop within the device as it conducts current. **Switching loss** is the power loss within the device as the devices switch on and off. A third loss mechanism, known as **leakage loss**, results from the application of a voltage across a device and tends to be relatively small compared to the other two loss mechanisms. See References [3–5] for comprehensive discussions of semiconductor losses.

11.5.1.1 Conduction Losses of IGBT and Diode

The output characteristic of an IGBT is similar to that of a diode or BJT. The collector-emitter characteristic of a representative 600 V, 200 A IGBT is shown in Figure 11.26(a). While the device actually begins to conduct at the gate threshold voltage, typically about 4 to 5 V, the gate-emitter voltage must be increased significantly above the threshold, 15

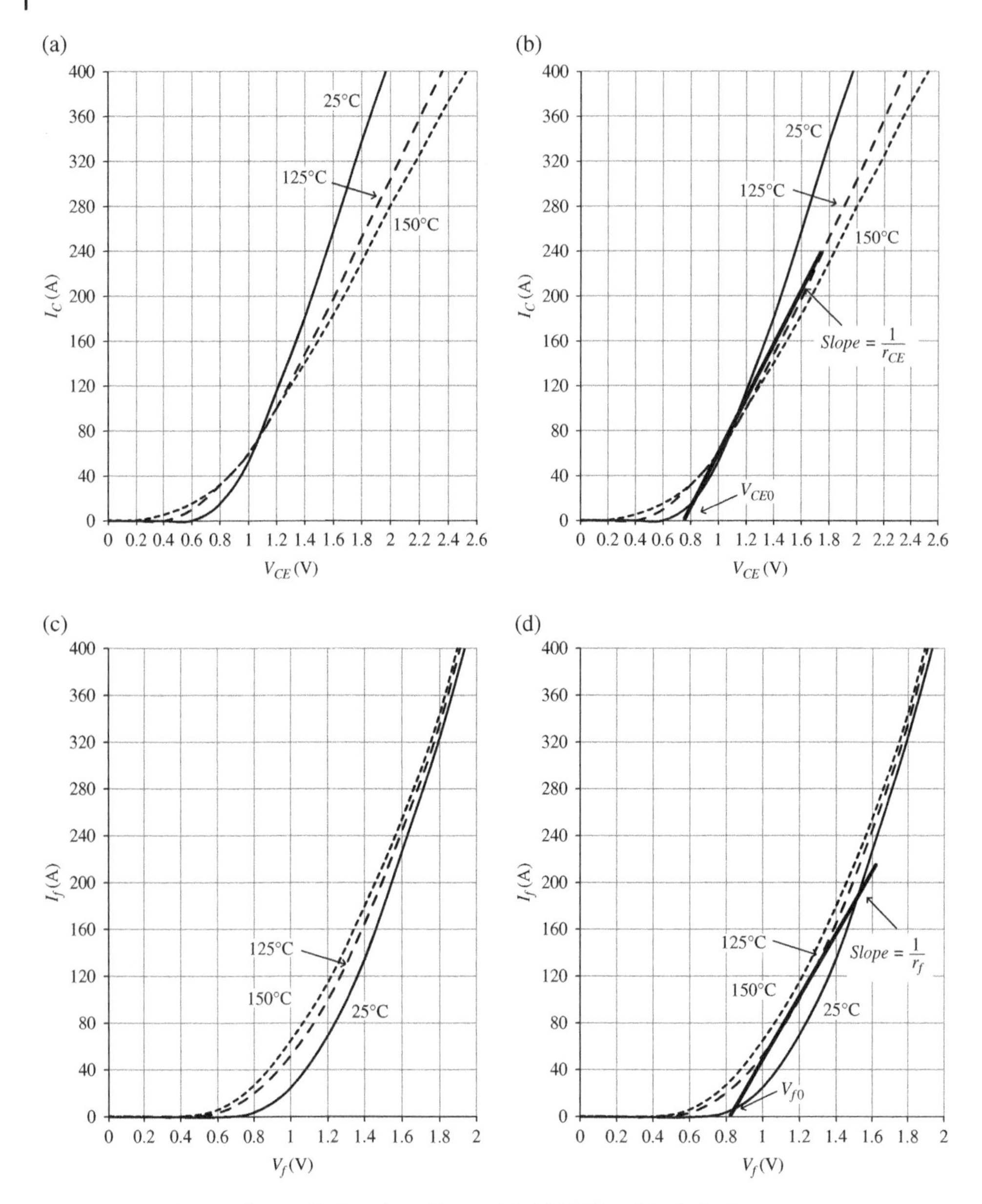

Figure 11.26 Output characteristics of 600 V, 200 A (a-b) IGBT and (c-d) diode.

V being a common level, in order to minimize the collector-emitter voltage drop and the associated conduction loss.

The output characteristic of the IGBT is the plot of the collector current i_C against the collector-emitter voltage v_{CE}, and can be simply modeled using a straight line approximation as follows:

$$v_{CE} = V_{CE0} + r_{CE} i_C \tag{11.79}$$

where V_{CE0} is the knee voltage and r_{CE} is the equivalent resistance, as shown in Figure 11.26(b). Note that the straight line should be drawn tangentially to the average current within the switch or diode while conducting. Thus, there can be variations in the parameters with the current level.

The conduction power loss of the IGBT $P_{Q(cond)}$ can then be simply modeled as

$$P_{Q(cond)} = V_{CE0}I_{Q(dc)} + r_{CE}I_{Q(rms)}^{2} \tag{11.80}$$

As the silicon power diode is also a minority-carrier device, the diode forward drop v_f, as a function of diode current i_f and power loss $P_{D(cond)}$ are similarly given by

$$v_f = V_{f0} + r_f i_f \tag{11.81}$$

and

$$P_{D(cond)} = V_{f0}I_{D(dc)} + r_f I_{D(rms)}^{2} \tag{11.82}$$

where V_{f0} is the knee voltage, and r_f is the equivalent resistance of the diode. A plot of the diode forward drop is presented in Figure 11.26(c). The diode characteristic can also be modeled by a straight line drawn tangentially to the current, as shown in Figure 11.26(d).

11.5.1.2 Example: Boost IGBT Conduction Losses

Determine the conduction losses for the IGBT and diode in the earlier CCM boost of Section 11.4.1.2 From Figure 11.26(b) and (d), the IGBT and diode have the following parameters: $V_{CE0} = 0.75$ V, $r_{CE} = 4.6$ mΩ, $V_{f0} = 0.85$ V, and $r_f = 3.6$ mΩ.

Solution:

The IGBT and diode conduction losses are simply

$$P_{Q(cond)} = V_{CE0}I_{Q(dc)} + r_{CE}I_{Q(rms)}^{2} = 0.75 \times 60\,\text{W} + 0.0046 \times 77.69^2\,\text{W} = 73\,\text{W}$$

and

$$P_{D(cond)} = V_{f0}I_{D(dc)} + r_f I_{D(rms)}^{2} = 0.85 \times 40\,\text{W} + 0.0036 \times 63.44^2\,\text{W} = 48\,\text{W}$$

11.5.1.3 Switching Losses of IGBT and Diode

Every instance that the current transitions from the switch to the diode, and vice versa, results in an energy loss and resulting power dissipation and temperature rise within the device. The IGBT and diode suffer from tail-current and reverse-recovery losses, respectively, as they turn off. The turn-on losses of the diode can be ignored, whereas the turn-on losses of the switch can be significant, especially as the diode reverse recovery at diode turn-off also affects the switch at turn-on.

These losses are often specified for the IGBT and diode. The energy losses for the representative module are shown in Figure 11.27 (a) and (b) for the IGBT and diode, respectively. The speeds of the turn-on and turn-off, and the resulting energy losses, are dependent on the gate voltages and the gate-drive resistor. Thus, these values are typically shown in the datasheets for the test. Energies E_{on} and E_{off} are the turn-on and turn-off energies of the IGBT, while E_{rec} is the turn-off energy of the diode due to reverse recovery. It is noteworthy that low-inductance connections are critical to power loss. Excessive parasitic inductances can cause increased power loss and voltage spikes.

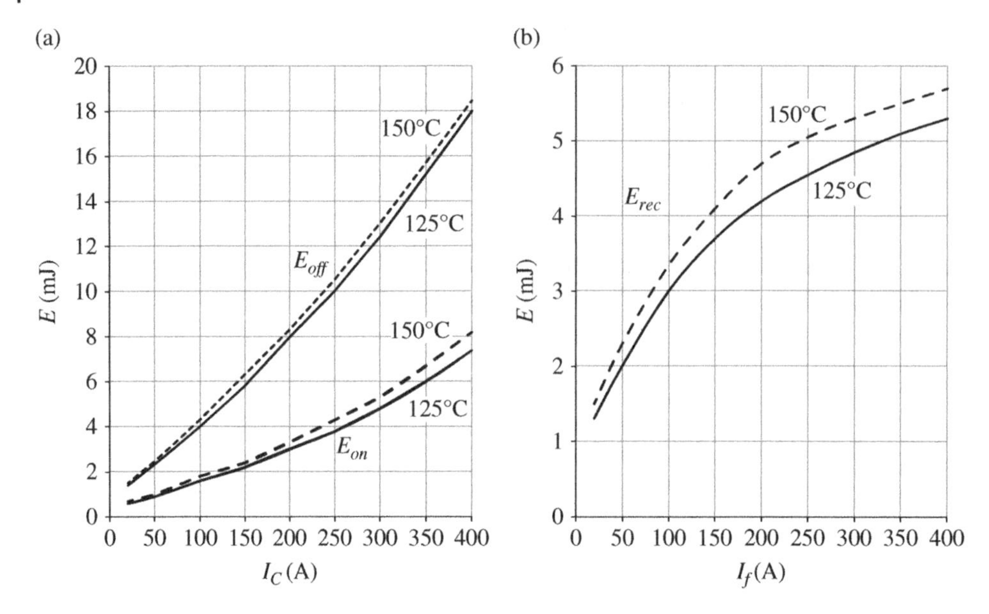

Figure 11.27 Switching losses for (a) IGBT and (b) diode at a test voltage of 300 V.

In order to determine the power loss, the various energy losses are simply added up and multiplied by the switching frequency. An additional caveat is that the test voltage may also need to be adjusted from that used in the test. For instance, the test voltage V_{test} in Figure 11.27 is 300 V. If we are testing at a different voltage, a reasonable assumption is to simply scale the energy loss linearly with voltage (or non-linearly as in [5]). Thus, the switching power loss $P_{Q(sw)}$ in the IGBT is equal to the sum of the voltage-adjusted turn-on and turn-off losses multiplied by the switching frequency:

$$P_{Q(sw)} = f_s\left(E_{on} + E_{off}\right)\frac{V_{HV}}{V_{test}} \tag{11.83}$$

The diode switching loss $P_{D(sw)}$ is similarly defined as

$$P_{D(sw)} = f_s E_{rec}\frac{V_{HV}}{V_{test}} \tag{11.84}$$

11.5.1.4 Example: Switching Losses of IGBT Module

Determine the switching losses in the IGBT and diode from the previous section. See Figure 11.27. Assume a junction temperature of 125 °C.

Solution:

The energy loss values are simply determined from the figures and substituted into the loss formulas. The earlier currents were

$$I_{L(max)} = I_{Q(off)} = I_{D(on)} = 114\,\text{A}$$

$$I_{L(min)} = I_{Q(on)} = I_{D(off)} = 86\,\text{A}$$

The approximate losses from the figures are as follows:

$$E_{on}\left(I_{Q(on)}\right) = E_{on}(86\,\text{A}) \approx 1.6\,\text{mJ}$$

$$E_{off}\left(I_{Q(off)}\right) = E_{off}(114\,\text{A}) \approx 4.7\,\text{mJ}$$

$$E_{rec}\left(I_{D(off)}\right) = E_{rec}(86\,\text{A}) \approx 2.8\,\text{mJ}$$

Thus, the power losses are as follows:

$$P_{Q(sw)} = f_s\left(E_{on} + E_{off}\right)\frac{V_{HV}}{V_{test}} = 10^4 \times (1.6 + 4.7) \times 10^{-3}\frac{500}{300}\,\text{W} = 105\,\text{W}$$

The diode switching loss is similarly defined:

$$P_{D(sw)} = f_s E_{rec}\frac{V_{HV}}{V_{test}} = 10^4 \times 2.8 \times 10^{-3} \times \frac{500}{300}\,\text{W} = 47\,\text{W}$$

11.5.2 Total Semiconductor Power Loss and Junction Temperature

The total semiconductor power loss is the sum of the switching and conduction power losses. Thus, the total IGBT power loss P_Q and the diode power loss P_D are as follows:

$$P_Q = P_{Q(cond)} + P_{Q(sw)} \tag{11.85}$$

and

$$P_D = P_{D(cond)} + P_{D(sw)} \tag{11.86}$$

Finally, a key semiconductor parameter is the hotspot temperature of the semiconductor junction. Excessive temperatures can result in semiconductor failure. Although the maximum temperature of the semiconductor is specified, the device is usually operated well below the maximum temperature in order to improve the reliability and lifetime of the semiconductor device.

The junction temperatures of the IGBT T_{JQ} and diode T_{JD} are given by

$$T_{JQ} = T_{HS} + R_{JQ-HS} \times P_Q \tag{11.87}$$

and

$$T_{JD} = T_{HS} + R_{JD-HS} \times P_D \tag{11.88}$$

where temperature T_{HS} is the temperature of the surface of the heat sink cooling the semiconductors, and $R_{JQ\text{-}HS}$ and $R_{JD\text{-}HS}$ are the thermal impedances of the IGBT and diode, respectively, in °C/W from the surface of the heat sink to the semiconductor junction. Note that the thermal impedance of the IGBT is typically less than that of the diode due to the larger die of the IGBT compared to the diode.

The above temperature relationship can be modeled by a simple electrical circuit as shown in Figure 11.28.

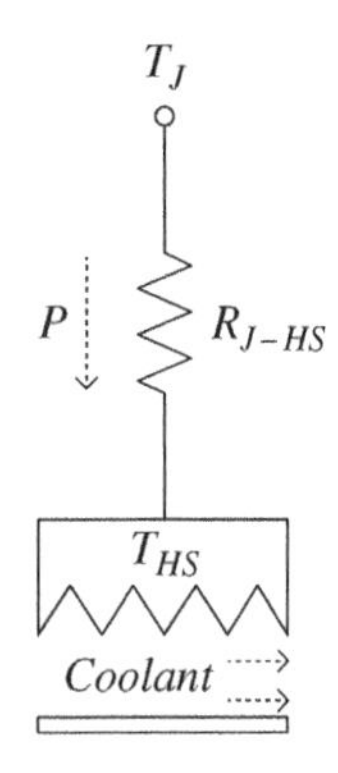

Figure 11.28 Thermal circuit model for heat flow from junction to heat sink.

11.5.2.1 Example: Total IGBT Module Loss and Die Temperatures

Determine the IGBT and diode power losses and their respective hotspot temperatures if the heat sink is maintained at 70 °C, and the thermal resistances of the IGBT and diode are 0.25 °C/W and 0.48 °C/W, respectively.

Solution:

The total losses are given by

$$P_Q = P_{Q(cond)} + P_{Q(sw)} = 73\,\text{W} + 105\,\text{W} = 178\,\text{W}$$

and

$$P_D = P_{D(cond)} + P_{D(sw)} = 48\,\text{W} + 47\,\text{W} = 95\,\text{W}$$

The junction temperatures are given by

$$T_{JQ} = T_{HS} + R_{JQ-HS} \times P_Q = 70°\text{C} + 0.25 \times 178°\text{C} = 114°\text{C}$$

and

$$T_{JD} = T_{HS} + R_{JD-HS} \times P_D = 70°\text{C} + 0.48 \times 95°\text{C} = 116°\text{C}$$

11.6 Passive Components for Power Converters

The inductors and capacitors can be a significant part of the cost and size of the dc-dc converter. The sizing of these components is covered in Chapter 16.

11.6.1 Example: Inductor Sizing

Determine the following inductor parameters for the buck converter of Example 11.3.1.4: area product, core area, and number of turns. Use the area product method, and assume that the core area equals the window area ($A_c = A_w$).

Let $k = 0.5$ be the copper fill factor, $J_{cu} = 6\text{A/mm}^2$ be the current density, and $B_{max} =$ 1.3 T be the maximum core flux density.

Solution:

From Chapter 16, Section 16.3.7, the area product AP is given by

$$AP = A_c A_w = \frac{L I_{L(rms)} I_{L(max)}}{k_{cu} J_{cu} B_{max}} = \frac{428.5 \times 10^{-6} \times 100.3 \times 114}{0.5 \times 6 \times 10^6 \times 1.3}\text{m}^4 = 1.256 \times 10^{-6}\text{m}^4 = 125.6\text{cm}^4$$

The area is given by

$$A_c = A_w = \sqrt{AP} = 1.12 \times 10^{-3}\,\text{m}^2 = 11.2\ \text{cm}^2$$

The number of turns is given by

$$N = \frac{L I_{L(\text{max})}}{B_{max} A_c} = \frac{428.5 \times 10^{-6} \times 114}{1.3 \times 1.12 \times 10^{-3}} = 34\ \text{turns}$$

11.6.2 Capacitor Sizing

The size of a capacitor can depend on many factors. An estimate of the physical size of a film capacitor can be made based on some simple calculations and assumptions, as presented in Chapter 16, Section 16.6.1

11.6.2.1 Example: Capacitor Sizing

Determine the length of foil and the foil volume of the 500 V, 960 μF capacitor required for the buck converter of Example 11.3.1.4 if the foil has the following specifications: dielectric strength $DS = 150$ V/μm, relative permittivity $\varepsilon_r = 2.2$, and foil width $w = 5$ cm. Allow for a voltage overshoot $V_{OS} = 100$ V on the capacitor.

Solution:

The thickness of the dielectric is

$$d = \frac{V_{HV} + V_{OS}}{DS} = \frac{500\text{V} + 100\text{V}}{150\text{V}/\mu\text{m}} = 4\mu\text{m}$$

As the width is specified, the length l is given by

$$l = \frac{Cd}{\varepsilon_r \varepsilon_0 w} = \frac{960 \times 10^{-6} \times 4 \times 10^{-6}}{2.2 \times 8.84 \times 10^{-12} \times 0.05}\text{m} = 3949\text{m}$$

The high-voltage capacitor foil volume V is

$$V = l \times w \times d = 3949 \times 0.05 \times 4 \times 10^{-6}\text{m}^3 = 0.632 \times 10^{-3}\text{m}^3 = 790\text{cm}^3$$

11.7 Interleaving

It can be understood from the previous examples that a significant mass and volume may be required for the passive components of the power converter. Thus, the size of a high-power system required for a fuel cell vehicle can take up a significant volume on the vehicle as the dc-dc converter interfacing the fuel cell is required to process all the fuel cell power. For many applications, the size of a power converter can be minimized by building multiple parallel stages and interleaving the operation of the stages. A two-phase interleaved boost converter is presented in Figure 11.29.

The main waveforms for the converter are shown in Figure 11.30. The turn-on of phase 2 is delayed by 180° with respect to phase 1 as shown in Figure 11.30(a) and (b). This results in the two-phase currents being out of phase as shown in Figure 11.30(c) and (d). The two-phase currents summed together gives the input current as shown in Figure 11.30(e).

It can be shown (and is left as an exercise for the student) that the peak-to-peak ripple of the input current $\Delta I_{int(p\text{-}p)}$ of an interleaved two-stage boost converter is as follows, the formula being dependent on whether the duty cycle is greater than or less than 0.5:

$$0 \le D \le 0.5 \quad \Delta I_{int(p-p)} = \frac{V_{LV}}{fL}\frac{(1-2D)}{(1-D)}D \qquad (11.89)$$

$$0.5 \le D \le 1.0 \quad \Delta I_{int(p-p)} = \frac{V_{LV}}{fL}(2D-1) \qquad (11.90)$$

We can appreciate interleaving by noting that the input ripple current is zero when the duty cycle is 0.5.

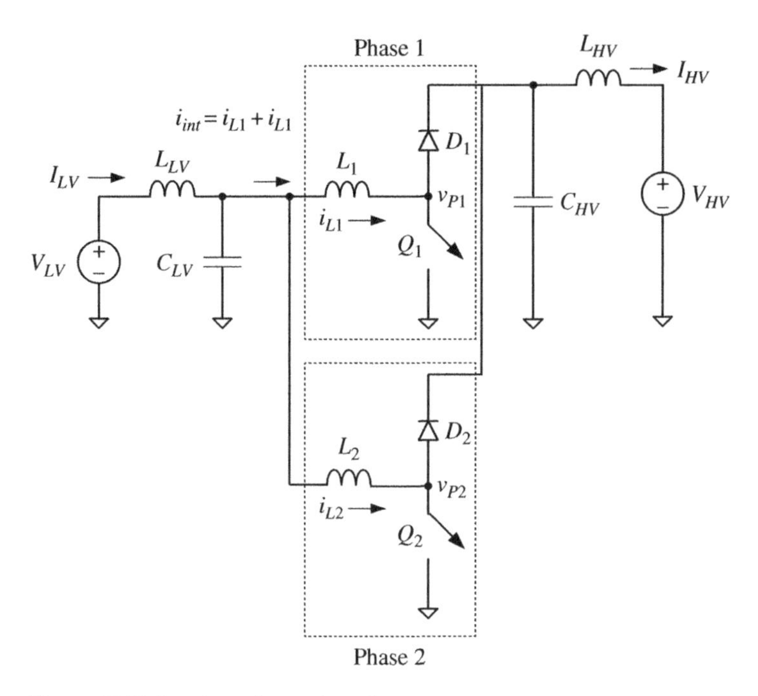

Figure 11.29 Interleaved two-phase boost converter.

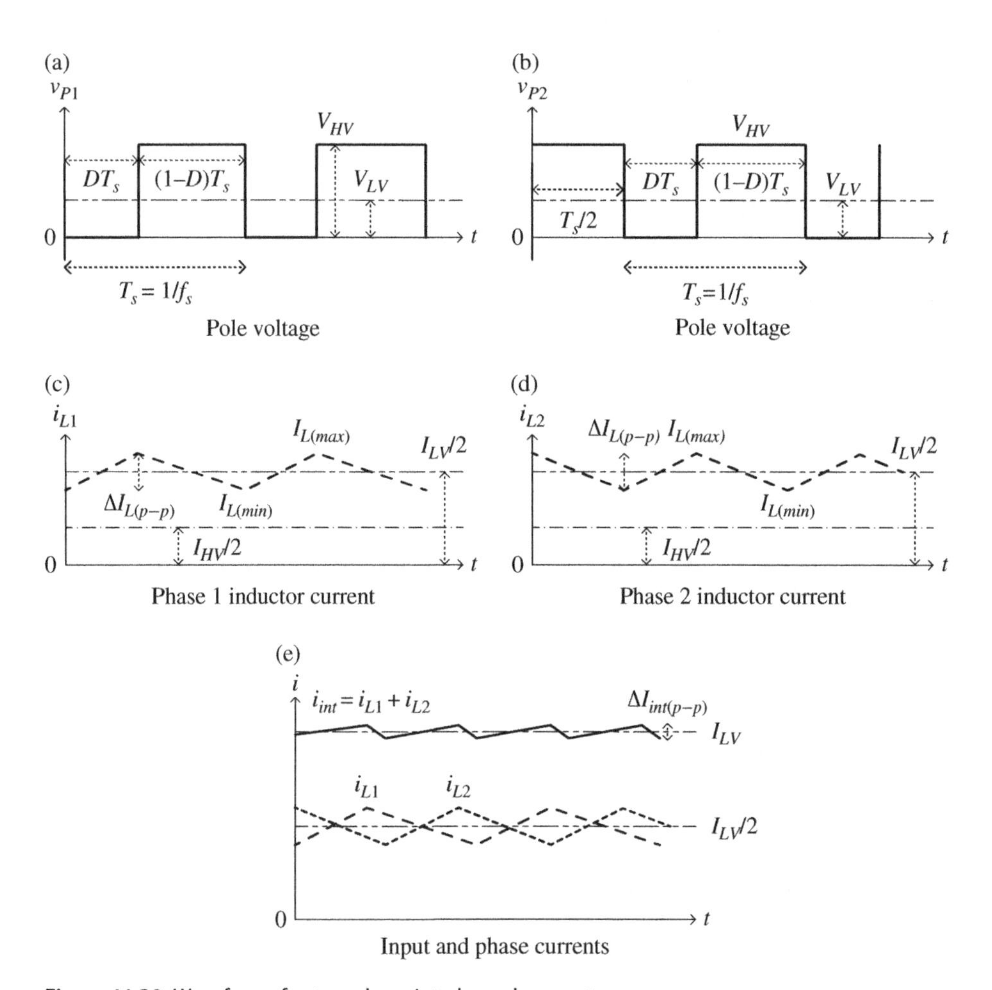

Figure 11.30 Waveforms for two-phase interleaved converter.

11.7.1 Example: Two-Phase Interleaved Boost Converter

A fuel cell vehicle features two interleaved 20 kW boost converters to provide a total output power of 40 kW. The converters are interleaved in order to reduce the ripple current coming from the source. The vehicle generates a 500 V dc link voltage when the fuel cell drops to 200 V at full power. Use the power stage of the previous boost example: $L = L_1 = L_2 = 428.5\ \mu H$.

i) Determine the peak-to-peak ripple on the input current.
ii) Determine the inductance and the area product which would be required if this converter was a single phase with the same low input ripple rather than a two-phase interleaved converter.

Solution:

i) As before, for a single-phase boost

$$D = 1 - \frac{V_{LV}}{V_{HV}} = 1 - \frac{200}{500} = 0.6$$

The per-phase current ripple remains

$$\Delta I_{L(p-p)} = \frac{V_{LV} D}{f_s L} = \frac{200 \times 0.6}{10000 \times 428.5} \text{A} = 28\,\text{A}$$

The input current ripple of the interleaved converter reduces to

$$\Delta I_{int(p-p)} = \frac{V_{LV}}{f_s L}(2D-1) = \frac{200}{10^4 \times 428.5 \times 10^{-6}}(2 \times 0.6 - 1) = 9.33\,\text{A} \tag{11.91}$$

Thus, the input current ripple at 9.33 A is one third of the per-phase current ripple, and the input capacitance can be reduced correspondingly. The area product for the two interleaved phases is simply twice that of the single-phase value of 125.6 cm^4, or 251.2 cm^4.

ii) For the second part of the question, the converter is a single stage which is designed with the same low input ripple current of 9.33 A as of the interleaved two-phase option. Thus, the phase inductance of the single-phase option has to increase by a factor of three compared to the two-phase design.

The inductor size increases to

$$L = 428.5\,\mu\text{H} \times \frac{28}{9.33} = 1286\,\mu\text{H}$$

The inductor dc current for a single phase is

$$I_{LV} = I_{L(dc)} = \frac{P}{V_{LV}} = \frac{40000}{200} \text{A} = 200\,\text{A}$$

The rms current in the low-voltage capacitor is given by

$$I_{CLV(rms)} = \frac{\Delta I_{L(p-p)}}{\sqrt{12}} = \frac{9.33}{\sqrt{12}} \text{A} = 2.69\,\text{A}$$

The inductor rms current is

$$I_{L(rms)} = \sqrt{I_{L(dc)}^2 + I_{CLV(rms)}^2} = \sqrt{200^2 + 2.69^2}\,\text{A} = 200.0\,\text{A}$$

The inductor maximum current is

$$I_{L(max)} = I_{L(dc)} + \frac{\Delta I_{L(p-p)}}{2} = 204.7\,\text{A}$$

The revised area product for a single-phase converter is

$$AP = \frac{LI_{L(rms)}I_{L(max)}}{k_{cu}J_{cu}B_{\max}} = \frac{1286 \times 10^{-6} \times 200 \times 204.7}{0.5 \times 6 \times 10^6 \times 1.3}\,\text{m}^4 = 1125\,\text{cm}^4$$

The inductor area product has increased to 1125 cm^4, which is over four times larger than the combined value of 251.2 cm^4 for the two inductances in the interleaved two-phase design.

References

1 T. A. Burress, S. L. Campbell, C. L. Coomer, C. W. Ayers, A. A. Wereszczak, J. P. Cunningham, L. D. Marlino, L. E. Seiber, and H. T. Lin, *Evaluation of the 2010 Toyota Prius Hybrid Electric Drive System*, Oak Ridge National Laboratory report, 2011.
2 D. Schneider, "How B. Jayant Baliga transformed power semiconductors," *IEEE Spectrum Magazine*, pp. 44–48, April 2014.
3 N. Mohan, T. M. Undeland, and W. P. Robbins, *Power Electronics Converters, Applications and Design*, 3rd edition, John Wiley & Sons, 2003.
4 D. Graovac and M. Purschel, "IGBT power losses calculation using the data sheet parameters," Infineon application note, 2009.
5 A. Wintrich, U. Nicolai, W. Tursky, and T. Reimann, *Application Manual Power Semiconductors*, SEMIKRON International Gmbh, 2015.

Further Reading

1 N. Mohan, *Power Electronics A First Course*, John Wiley & Sons, 2012.
2 R. W. Erickson, *Fundamentals of Power Electronics*, Kluwer Academic Publishers, 2000.
3 B. C. Barry, J. G. Hayes, and M. S. Ryłko, "CCM and DCM operation of the interleaved two-phase boost converter with discrete and coupled inductors," *IEEE Transactions on Power Electronics*, **30**, pp. 6551–6567, December 2015.

Problems

11.1 A hybrid electric vehicle uses a 30 kW bidirectional converter to generate a 650 V dc link voltage from the 288 V NiMH battery. The bidirectional converter has an inductance of 245 μH and switches at 10 kHz.

The vehicle is operating in generating mode, and the bidirectional converter is required to act as a buck at full power.

i) Calculate the rms currents in the low-voltage capacitor and in the inductor.
ii) Calculate the maximum, minimum, rms, and average currents in the IGBT and diode.
iii) Calculate the rms current in the high-voltage capacitor.
iv) Determine the low-voltage and high-voltage capacitor values if the peak-to-peak voltage ripple is 0.5%.

Ignore component losses.

[Ans. $I_{CLV(rms)} = 18.9$ A, $I_{L(rms)} = 105.87$ A, $I_{L(max)} = 136.9$ A, $I_{L(min)} = 71.44$ A, $I_{QU(rms)} = 70.47$ A, $I_{QU(dc)} = 46.15$ A, $I_{DL(rms)} = 79.0$ A, $I_{DL(dc)} = 58.01$ A, $I_{CHV(rms)} = 53.26$ A, $C_{HV} = 791$ μF, $C_{LV} = 568$ μF]

11.2 At what power does the converter in the previous problem reach BCM at a dc link voltage of 650 V? Recalculate the various currents.

[Ans. 9.43 kW, $I_{CLV(rms)} = 18.9$ A, $I_{L(rms)} = 37.8$ A, $I_{L(max)} = 65.46$ A, $I_{L(min)} = 0$ A, $I_{QU(rms)} = 25.16$ A, $I_{QU(dc)} = 14.5$ A, $I_{DL(rms)} = 28.21$ A, $I_{DL(dc)} = 18.23$ A, $I_{CHV(rms)} = 20.56$ A]

11.3 Assuming the above power converter is in DCM at 5 kW, recalculate the various currents.

[Ans. $I_{CLV(rms)} = 15.82$ A, $I_{L(rms)} = 23.49$ A, $I_{L(max)} = 47.68$ A, $I_{L(min)} = 0$ A, $I_{QU(rms)} = 15.64$ A, $I_{QU(dc)} = 7.69$ A, $I_{DL(rms)} = 17.53$ A, $I_{DL(dc)} = 9.67$ A, $I_{CHV(rms)} = 13.62$ A]

11.4 Assume the converter in the Problem 11.1 is in boost mode at 30 kW. Recalculate the various currents.

[Ans. $I_{CLV(rms)} = 18.9$ A, $I_{L(rms)} = 105.87$ A, $I_{L(max)} = 136.9$ A, $I_{L(min)} = 71.44$ A, $I_{QL(rms)} = 79.0$ A, $I_{QL(dc)} = 58.01$ A, $I_{DU(rms)} = 70.47$ A, $I_{DU(dc)} = 46.15$ A, $I_{CHV(rms)} = 53.26$ A]

11.5 Assuming the above boost power converter is in DCM at 5 kW, recalculate the various currents.

[Ans. $I_{CLV(rms)} = 15.82$ A, $I_{L(rms)} = 23.49$ A, $I_{L(max)} = 47.68$ A, $I_{L(min)} = 0$ A, $I_{QL(rms)} = 17.53$ A, $I_{QL(dc)} = 9.67$ A, $I_{DU(rms)} = 15.64$ A, $I_{DU(dc)} = 7.69$ A, $I_{CHV(rms)} = 13.62$ A]

11.6 For the converter of Problem 11.1:

i) Determine the typical power losses and junction temperatures due to conduction and switching in the IGBT and diode for this full-power condition when using a 1200 V, 300 A half-bridge module. Let $V_{CE0} = 0.75$ V and $r_{CE} = 4.4$ mΩ, and $V_{f0} = 0.7$ V and $r_f = 3.8$ mΩ. Use the loss curves of Figure 11.31. The heat sink is maintained at 80 °C, and the thermal resistances of the IGBT and diode are 0.125 °C/W and 0.202 °C/W, respectively.
ii) Determine the following inductor parameters: area product, core area, and number of turns. Use the area product method, and assume that the core area equals the window area ($A_c = A_w$), and that $k = 0.5$ is the copper fill factor, $J_{cu} = 6$A/mm^2 is the current density, and $B_{max} = 1.3$ T is the maximum core flux density.

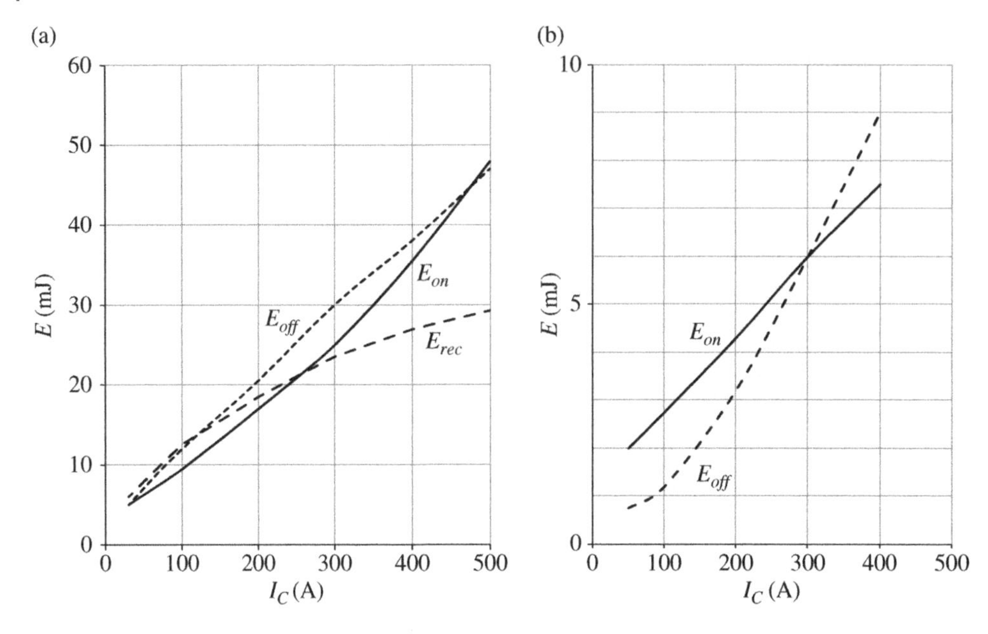

Figure 11.31 (a) Energy losses at 125 °C for 1200 V, 300 A IGBT at a test voltage of 600 V. (b) Energy losses at 150 °C for representative 1200 V, 300 A SiC module at a test voltage of 600 V.

iii) Determine the volume of the high-voltage capacitor, assuming a polypropylene film. Let the dielectric strength = 150 V/μm, relative permittivity = 2.2, and foil width = 5 cm. Allow for a 100 V design overshoot on the capacitor.

[Ans. $P_Q = 300.3$ W, $P_D = 172.6$ W, $T_{JQ} = 117.5$ °C, $T_{JD} = 114.9$ °C, $AP = 91$ cm^4, $A = 9.5$ cm^2, $N = 27$ turns, $V_{CHV} = 1017$ cm^3]

11.7 A fuel cell vehicle features two interleaved boost converters to provide a total output power of 72 kW. The converters are interleaved in order to reduce the ripple current coming from the source. The converters generate a 360 V dc link voltage when the fuel cell drops to 180 V at full power. Each converter has an inductance of 45 μH and switches at 16 kHz.

i) Determine the rms and peak currents in each phase inductor.
ii) Calculate the peak-to-peak ripple current from the fuel cell input capacitor.
iii) Determine the peak-to-peak current from the fuel cell input capacitor if the two phases were synchronized in phase rather than interleaved.

[Ans. $I_{L1(rms)} = I_{L2(rms)} = 203.2$ A, $I_{L1(max)} = I_{L2(max)} = 262.5$ A, $\Delta I_{int(p\text{-}p)} = 0$ A, $\Delta I_{in(p\text{-}p)} = 250$ A]

11.8 The half-bridge dc-dc converter of a HEV has an inductance of 428.5 μH and switches at 10 kHz when supplied by a 200 V NiMH battery.
The voltage on the dc link reduces proportionately with the motor power, and the dc link voltage drops to 350 V at 10 kW.

i) Calculate the switch rms and dc currents for this CCM generating/buck condition.

ii) Determine the typical power loss and junction temperatures at 125 °C due to conduction and switching in the IGBT for this partial-load condition when using the half-bridge module of Section 11.5.1.

Let $V_{CE0} = 0.65$ V and $r_{CE} = 5.9$ mΩ for the average current in the switch in this problem.

Use Figure 11.27 to estimate the switching loss for the module.

[Ans. $I_{Q(rms)} = 38.04$ A, $I_{Q(dc)} = 28.57$ A, $P_Q = 69$ W; $T_{JQ} = 97.3$ °C]

Assignments

11.1 The student is encouraged to experiment with circuit simulation software packages. The following simulation packages are among those which are commonly used.
   i) Simetrix, www.simetrix.co.uk
   ii) PSpice 9.1, available on various web sites.
   iii) Matlab/Simulink Simscape Power Systems, http://www.mathworks.com/products/simpower/
   iv) PSIM, www.powersimtech.com

11.2 Verify the waveforms and answers for the various examples and problems by simulation.

11.3 A basic laboratory experiment on a low-voltage, low-power buck converter entitled "Steady-state Operation of Buck Converter" is available on the companion website.

Appendix I

The simple Simulink schematic of Figure 11.32 simulates the buck CCM example. A resistive load is used rather than a battery to simplify the circuit. The resistor is set at 2 Ω, which is the equivalent resistive load of a 200 V battery pulling 100 A. Simulation waveforms are shown in Figure 11.33.

Appendix II: Buck-Boost Converter

The buck-boost converter is closely related to the buck and boost converters, and is the basis for the isolated flyback converter covered in Appendix II of Chapter 12. The relationship between voltage and the duty cycle can easily be determined for the buck-boost converter by applying the relationship already derived for the boost converter. The buck-boost converter is shown in Figure 11.34(a).

The voltage gain of the boost converter has been derived to be

$$\frac{V_{HV}}{V_{LV}} = \frac{1}{1-D}$$

The voltage gain for the buck-boost converter is easily derived if V_{in} is substituted for V_{LV} and $(V_{in} + V_{out})$ is substituted for V_{HV}:

$$\frac{V_{HV}}{V_{LV}} = \frac{V_{in} + V_{out}}{V_{in}} = \frac{1}{1-D} \qquad \text{(A11.II.1)}$$

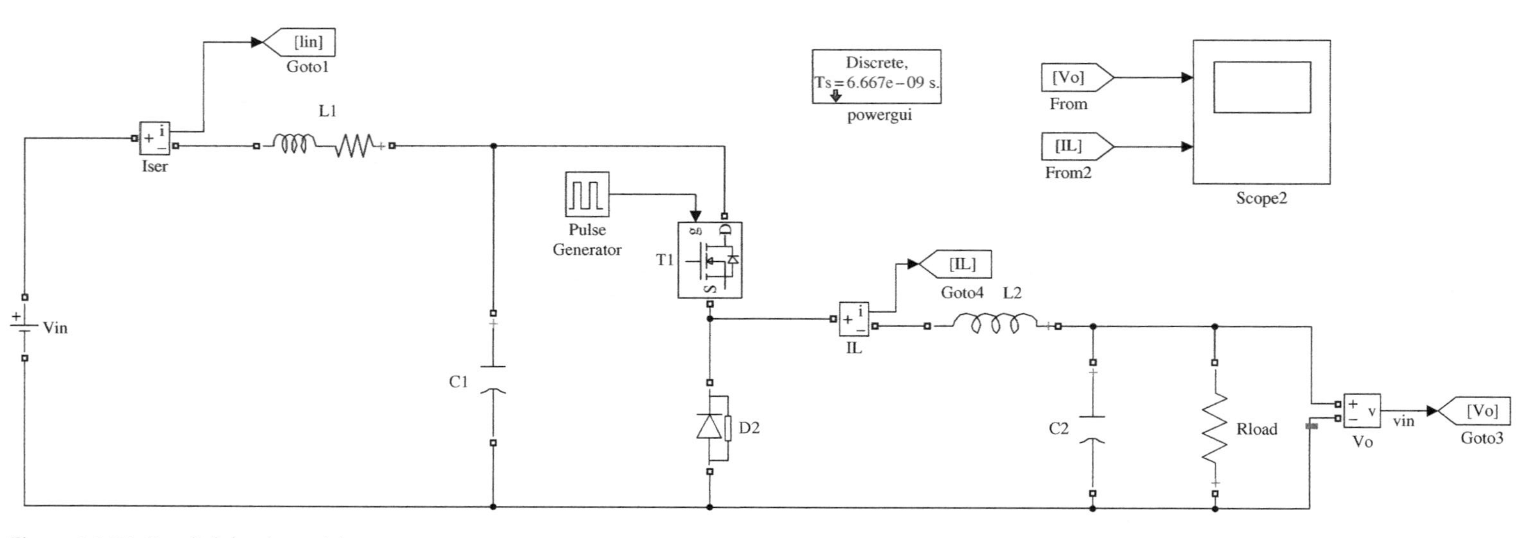

Figure 11.32 Simulink buck model.

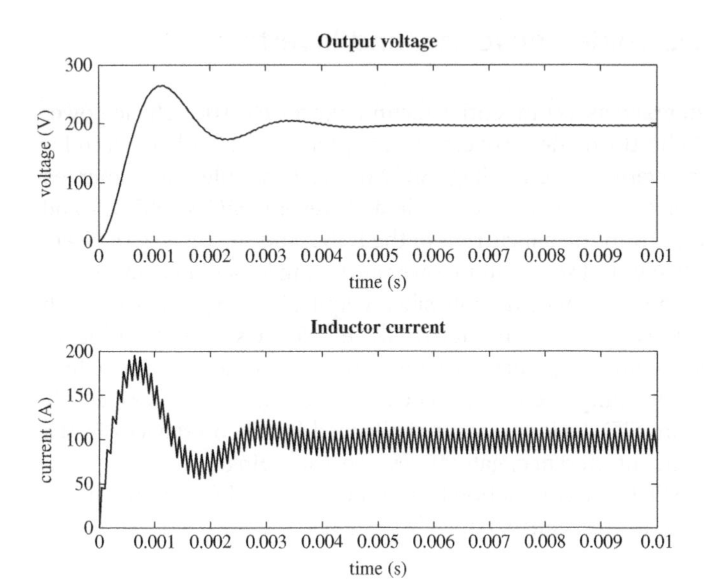

Figure 11.33 Simulation waveforms for a buck converter.

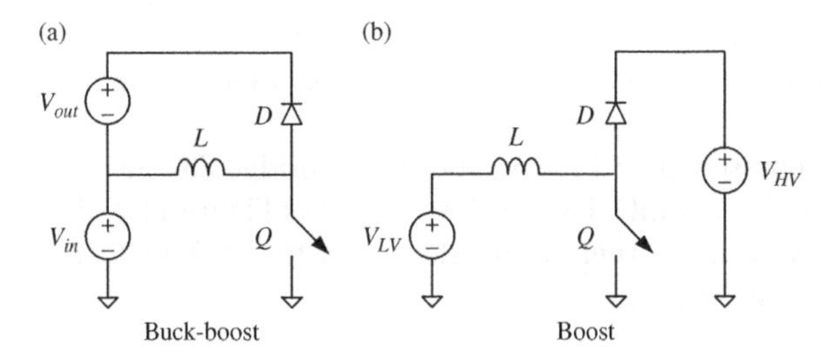

Figure 11.34 (a) Buck-boost and (b) boost converters.

Figure 11.35 Buck-boost converter.

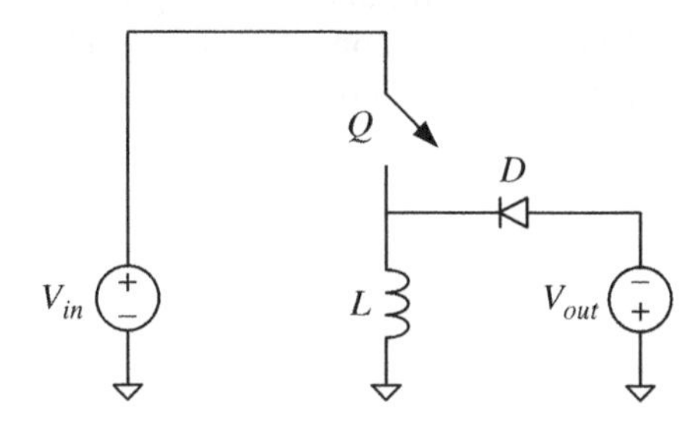

By rearranging the preceding equation, the duty cycle for the buck-boost converter is derived as

$$D = \frac{V_{out}}{V_{in} + V_{out}} \tag{A11.II.2}$$

The voltage gain can be expressed as

$$\frac{V_{out}}{V_{in}} = \frac{D}{1-D} \tag{A11.II.3}$$

The buck-boost converter is often sketched as shown in Figure 11.35.

Appendix III: Silicon Carbide Converters and Inverters

Silicon carbide (SiC) has emerged as a competitive technology for electric vehicle power semiconductors. The SiC Schottky diode is covered in Chapter 14 and has been used for many years for PFC boost converters. High-voltage SiC MOSFET modules have emerged for high-voltage, high-power dc-dc converters and dc-ac inverters. SiC MOSFETs and Schottky diodes do not suffer from the tail current of the IGBT and the reverse recovery of the silicon diode, respectively, and so, result in lower switching losses. The SiC MOSFET has substantially lower on-resistance than its silicon equivalent, especially for high voltage applications. Thus, these devices can improve the efficiencies, operate at higher temperatures and higher switching frequencies, require smaller volumes, and/or reduce the cooling requirements when compared to their silicon counterparts. This is especially so for high-voltage operation. SiC devices remain challenged by silicon devices in the areas of cost, electromagnetic interference, gate drives and packaging.

Similar to the silicon MOSFET, the conduction loss in the SiC MOSFET can be simply modelled by the drain-source on-resistance, typically designated $R_{DS(on)}$. The MOSFET conduction loss is given by

$$P_{Q(cond)} = R_{DS(on)} I_{Q(rms)}^2 \tag{A11.III.1}$$

Problems A11.III.1 below and 13.3 are presented for the reader to solve which are based on the comparable silicon problems already included in this textbook. For simplicity, diode switching losses are assumed negligible for the SiC Schottky diode. Compare your answers to those for silicon.

For both these problems, use a representative 1200 V, 300 A half-bridge SiC module with $R_{DS(on)}$ = 8.0 mΩ, V_{f0} =0.8 V and r_f = 4 mΩ. Use the loss curves of Figure 11.31(b). The heatsink is maintained at 80 °C, and the thermal resistances of the MOSFET and diode are 0.125 °C/W and 0.150 °C/W, respectively.

Problem A11.III.1

For the converter of Problem 11.6 (i): determine the typical power losses and junction temperatures due to conduction and switching in the SiC MOSFET and SiC Schottky diode for the full-power condition of Problem 11.1 when using the representative 1200 V, 300 A half-bridge SiC module.

[Ans. P_Q = 88.5 W, P_D = 71.4 W, T_{JQ} = 91.1 °C, T_{JD} = 90.7 °C]

12

Isolated DC-DC Converters

In this chapter, the reader is introduced to isolated dc-dc power converters. Isolated power converters enable safe and efficient power conversion and play key EV roles in the charging of batteries and the powering of auxiliary loads. This chapter introduces the isolated forward converter as a transformer-based buck converter and then introduces the full-bridge power converter as the main on-vehicle isolated power converter. Resonant converters play a significant role in power conversion, especially for inductive and wireless charging, and so the resonant LCLC full-bridge power converter is presented. Finally, the ubiquitous flyback converter, based on the buck-boost topology and widely used for lower power levels, is briefly introduced in Appendix II at the back of this chapter.

12.1 Introduction

In this chapter, isolated power converters are introduced and analyzed. These isolated converters are derived from the basic non-isolated power converter topologies discussed in Chapter 11.

12.1.1 Advantages of Isolated Power Converters

Transformer-isolated converters are required in applications for the following very fundamental reasons.

Safety

Electrical systems can present safety hazards to the consumer, and the risks are greater at higher voltages. In general, voltages below 40 V to 60 V are regarded as safe, but the levels can be lower depending on the conditions. Electric vehicles generally have high-voltage battery packs of about 400 V, and significant protections must be incorporated into the vehicle power architecture in order to ensure safe operation. The principal method to ensure a physical and electrical barrier between the high and low voltages is to use a transformer to provide galvanic isolation.

Efficiency

When significant voltage conversions are required, such as converting from a high-voltage battery at 400 V to the 12 V level required for on-vehicle accessory loads, it is

Electric Powertrain: Energy Systems, Power Electronics and Drives for Hybrid, Electric and Fuel Cell Vehicles, First Edition. John G. Hayes and G. Abas Goodarzi.

Companion website: www.wiley.com/go/hayes/electricpowertrain

necessary to use a transformer in order to optimally design and size a practical switch-mode power converter. The transformer-based circuits have reduced peak, mean, and rms currents, resulting in improved efficiency, better control, reduced voltage stresses, and reduced electromagnetic interference.

Multiple Outputs

There are many different low-voltage loads on the vehicle requiring different voltage levels such as the windshield defrost at 42 V and the accessory battery at 12 V. Similarly, there can be multiple voltage levels used in a smartphone or computer. The simplest way to provide multiple voltage outputs is to use a transformer with multiple secondary windings.

12.1.2 Power Converter Families

There are three main families of isolated power converters which are commonly used. These families are based on the buck converter, the buck-boost converter, and the resonant converter.

The common buck-derived converters are the forward converter, which is used for low to medium power, in the range 50 to 200 W, and the full bridge, which is used for high power, from about 200 W to many kilowatts. There are many variations on forward and full-bridge converters. For example, single-switch and two-switch forward converters can be used. The commonly used full-bridge converters are the hard-switched and the phase-shifted variations. Both types of full-bridge converters play a role on the vehicle. This chapter focuses on the basic hard-switched converter due to its relative simplicity. Other closely related variants are the push-pull and half-bridge converters. The forward converter is presented for teaching purposes in this chapter as it is based on the buck converter and links the buck and the full-bridge converters.

An example of an automotive full-bridge auxiliary power converter is shown in Figure 12.1, where the symbol Q marks the power switches, X marks the transformer, and L marks the output inductor. The high-voltage input terminals and the low-voltage output terminals are labeled HVIP and LVOP, respectively. The electromagnetic

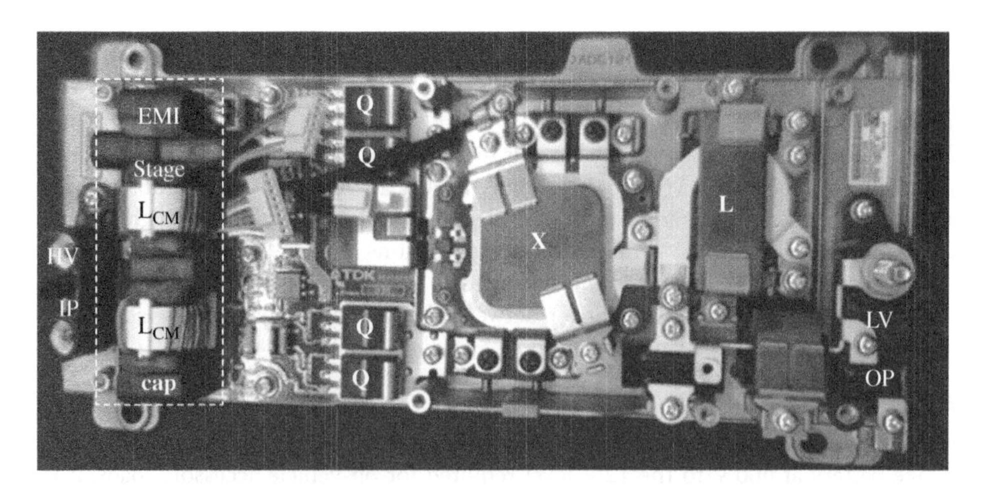

Figure 12.1 Automotive auxiliary power converter for a Ford vehicle.

interference (EMI) stage is appropriately marked and features two common-mode filter inductors (L_{CM}) and several capacitors (cap).

The transformer-isolated buck-boost converter is known as the **flyback**. The flyback converter is ubiquitous in the modern world and is typically the converter of choice for power converters at powers ranging from milliwatts to a couple of hundred watts. Again, there are many variations on flybacks, and the optimum topology can depend on the application. As the flyback converter is not used for high-power circuits, it is not discussed in the main chapter but it is briefly covered in Appendix II of this chapter.

Resonant converters have assumed a greater role in isolated power conversion over the last few decades. One of the first significant applications for resonant converters was as the power conversion stage for the inductive charging developed for the GM EV1 program in the mid 1990s [1–2]. The inductive power converters developed for EV charging at the time ranged from 1.5 kW to 130 kW. The resonant converter possesses particular attributes which make it very suitable for wireless power transfer. In the meantime, a class of resonant converters, known as the LLC converter, has been widely adapted for commercial and industrial power conversion [3].

12.2 The Forward Converter

The buck converter is redrawn in Figure 12.2(a) to show a resistive load R_O, and the labeling of the various components has been modified. The forward converter is derived from the buck converter, and the basic topology is as shown in Figure 12.2(b).

The output stage of the forward converter, with components R_O, L_O, C_O, and D_1, are identical to the redrawn buck.

The transformer is introduced to the circuit to create the forward converter. From a power and switch perspective, it does not matter whether the switch Q is placed between the positive terminal of the transformer and the input voltage supply or between the negative terminal and the dc return. However, placing the switch between the negative terminal and the dc return makes the control of the switch simpler to implement, and so the switch location is typically as in the diagram.

The introduction of the transformer requires an additional diode D_2 to ensure that the transformer only conducts when the primary-side switch Q conducts.

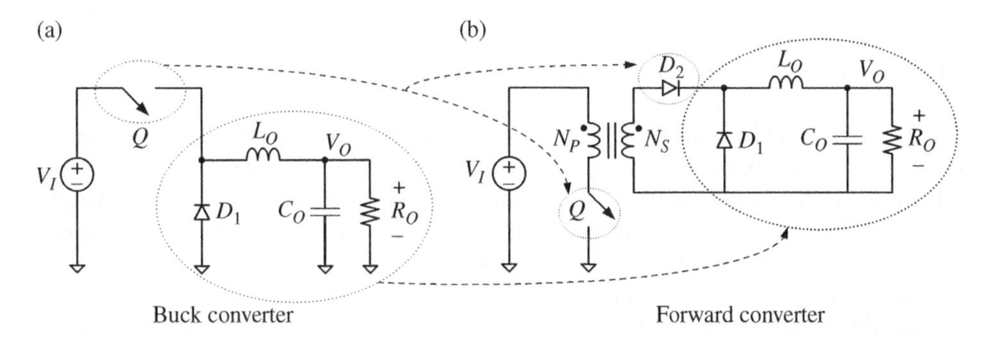

Figure 12.2 (a) Buck converter and (b) idealized forward converter.

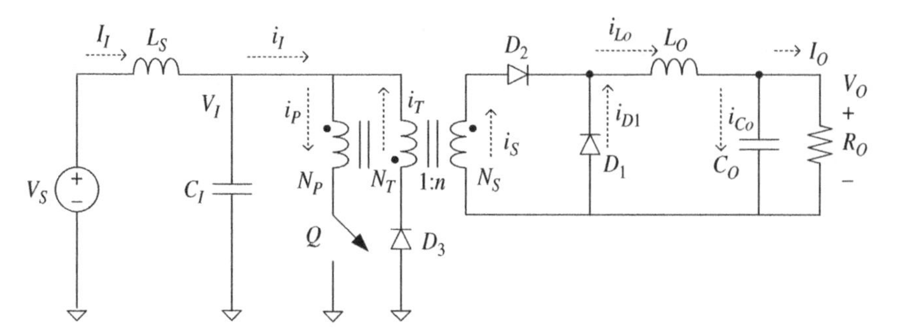

Figure 12.3 Forward converter.

A problem arises with the forward converter as shown in Figure 12.2(b). A transformer must be magnetized in order to operate as a transformer, as discussed in Chapter 16, Section 16.5. The magnetization is usually represented by a primary-side magnetizing inductance. However, the magnetizing inductance must be demagnetized each cycle when the primary-side switch is not conducting in order to avoid core saturation. The demagnetization in a forward converter is achieved by adding a third winding, often known as the tertiary winding N_T, and a diode D_3 so that the transformer demagnetizes when the switch stops conducting, as shown in Figure 12.3. The requirement to demagnetize the transformer puts a practical maximum limit on the duty cycle D of the primary-side switch. Typically, the tertiary winding has the same number of turns as the primary and is closely coupled to the primary winding in a bifilar winding. The maximum duty cycle is typically 0.5 for this condition of equal primary and tertiary turns.

The input source voltage V_S is also shown as having an internal inductance L_S. In this chapter, for simplicity, the source inductance and filter capacitance C_I are neglected, and the input is simply modeled as a voltage source V_I. The procedure of Chapter 11 can be used to determine the input capacitor current and voltage ripple.

It is worth investigating the operation of the forward converter in additional detail to illustrate the transformer action. There are three operating modes when operating in continuous conduction mode (CCM). When the switch Q is closed in the first mode, as shown in Figure 12.4(i), the magnetizing current i_m builds up in the magnetizing inductance L_m, and a reflected secondary current $i_s{}'$ flows by transformer action, induced by the secondary current i_s. No current flows in the tertiary winding when the switch is closed, and diode D_1 does not conduct in this mode as it is reverse biased.

In the second mode, as shown in Figure 12.4(ii), the switch Q is open, and the primary winding no longer conducts current. However, there is energy stored in the magnetizing inductance, which forces the tertiary magnetizing current i_T to flow in the tertiary winding by coupled-inductor transformer action, until the transformer is demagnetized. There is no secondary current due to the reverse-biased D_2, and the output current is supplied from the energy stored in L_O through D_1.

In the third mode, as shown in Figure 12.4(iii), there is no current in any of the transformer windings. The inductor current continues to flow through D_1.

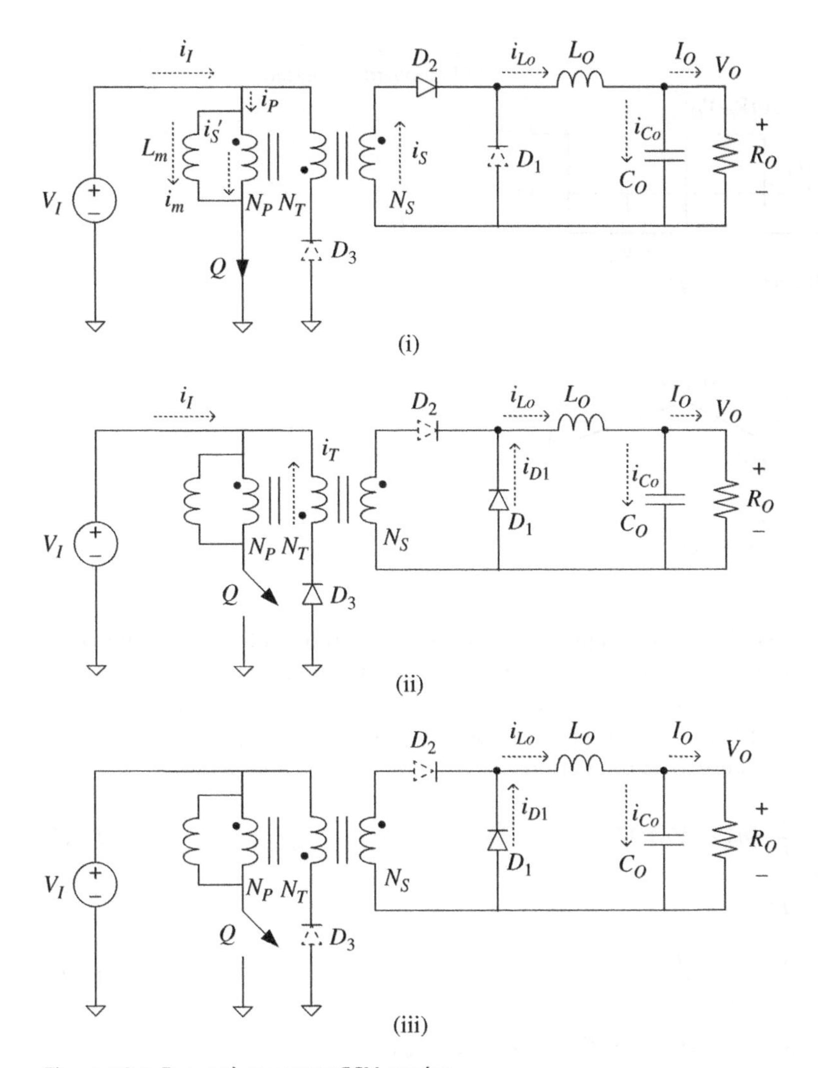

Figure 12.4 Forward converter CCM modes.

12.2.1 CCM Currents in Forward Converter

The duty cycle for the CCM buck converter, discussed in Chapter 11, is equal to the ratio of the output voltage to the input voltage. The presence of the transformer results in the duty cycle for the forward converter being the ratio of the output or secondary volts/turn to the input or primary volts/turn, as follows:

$$D = \frac{V_O / N_S}{V_I / N_P} = \frac{1}{n} \frac{V_O}{V_I} \tag{12.1}$$

where n is the **turns ratio** of the transformer N_S/N_P.

The output inductor waveforms for the forward converter are shown in Figure 12.5. The waveforms can be analyzed in a similar manner to the buck converter. As with

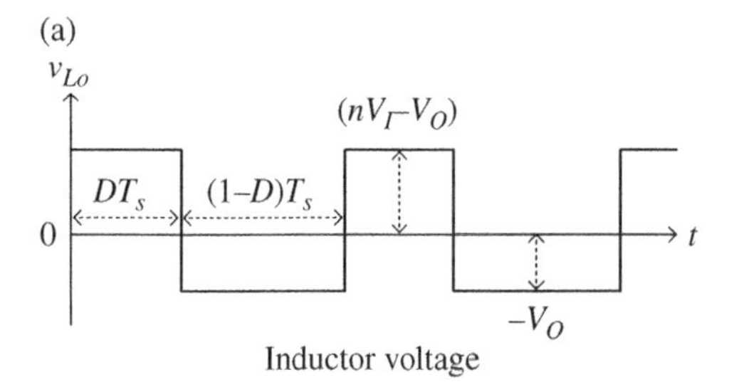

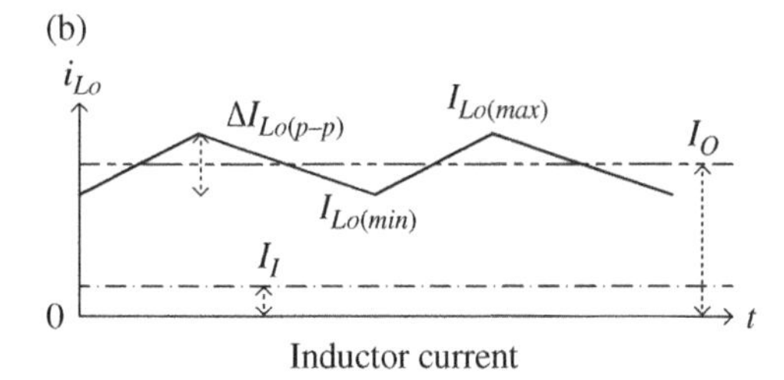

Figure 12.5 Output inductor waveforms for CCM forward converter.

the earlier converters, the change in peak-to-peak current can provide the basis for converter analysis. All device parasitics are ignored in order to simplify the analysis.

Assuming that the inductor current is changing linearly in steady-state operation, then the change in current from peak to peak can be described by $\Delta I_{Lo(p\text{-}p)}$.

During time DT_s, where T_s is the period:

$$\Delta I_{Lo(p-p)} = \frac{1}{L_O}(nV_I - V_O)DT_s \tag{12.2}$$

During time $(1\text{-}D)T_s$:

$$\Delta I_{Lo(p-p)} = \frac{V_O(1-D)T_s}{L_O} \tag{12.3}$$

Equating the two expressions for the current $\Delta I_{Lo(p\text{-}p)}$:

$$\Delta I_{Lo(p-p)} = \frac{1}{L_O}(nV_I - V_O)DT_s = \frac{V_O(1-D)T_s}{L_O} \tag{12.4}$$

Solving the two equations again yields the earlier expression for the relationship between the duty cycle and voltage gain in Equation (12.1).

The currents in the various components are illustrated in Figure 12.6 and are determined as follows. Diode D_2 conducts when the switch is conducting, and the current in D_2 is the secondary current i_s. The current changes from a minimum level $I_{s(min)}$ to a maximum level $I_{s(max)}$ over the cycle, and these values are given by

$$I_{Lo(min)} = I_{s(min)} = I_O - \frac{\Delta I_{Lo(p-p)}}{2} \tag{12.5}$$

$$I_{Lo(max)} = I_{s(max)} = I_O + \frac{\Delta I_{Lo(p-p)}}{2} \tag{12.6}$$

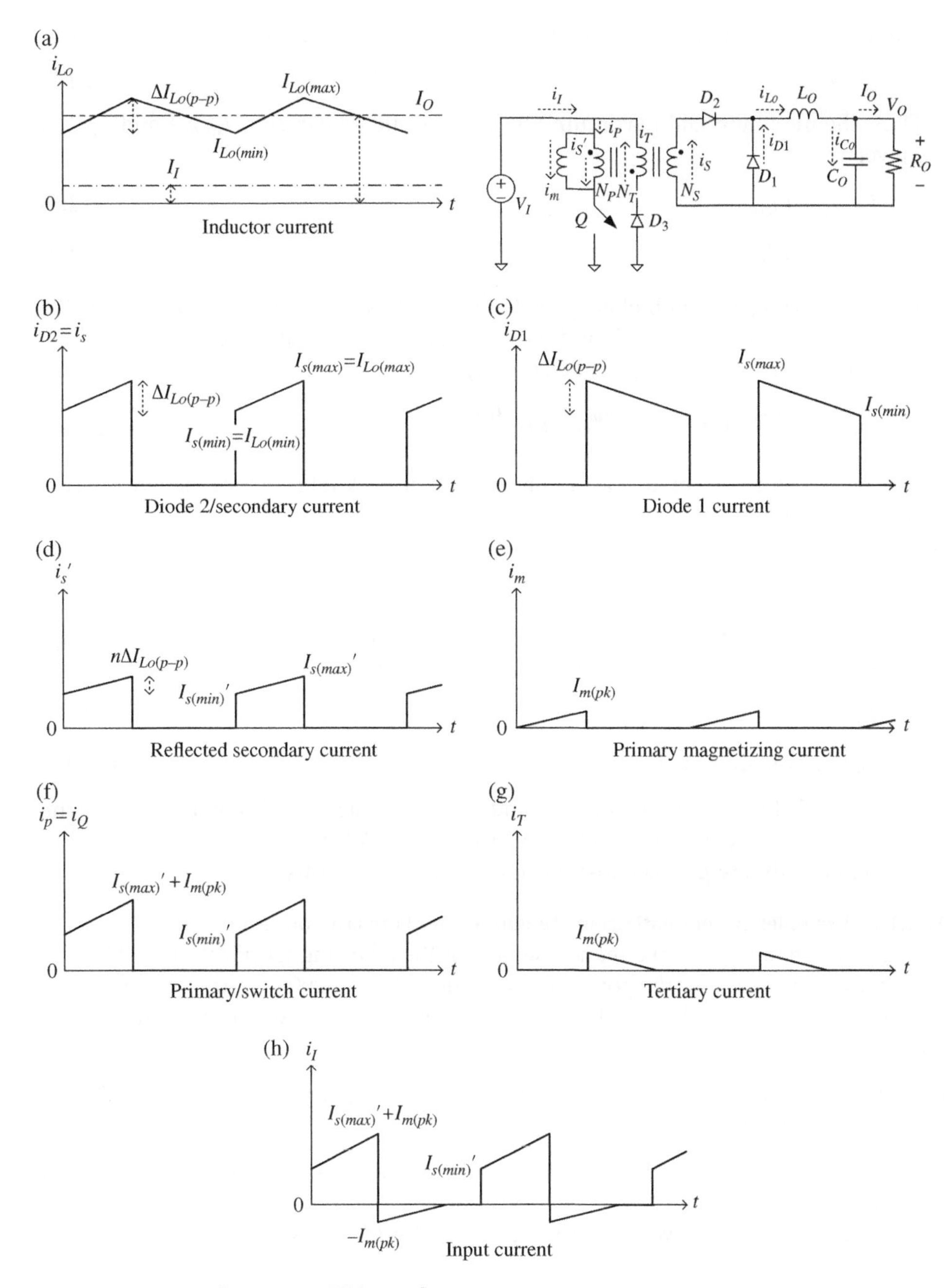

Figure 12.6 Forward converter CCM waveforms.

The secondary current is reflected to the primary using the transformer turns ratio:

$$i_s' = ni_s \tag{12.7}$$

Thus, the minimum and maximum reflected secondary currents, $I_{s(min)}'$ and $I_{s(max)}'$, are given by

$$I_{s(min)}' = nI_{s(min)} \tag{12.8}$$

$$I_{s(max)}' = nI_{s(max)} \tag{12.9}$$

The magnetizing current builds up while the switch is closed, as shown in Figure 12.4 (i). The switch conducts for DT_s, and the voltage across the inductance is V_I. Thus, the voltage across the magnetizing inductance v_{Lm} is

$$v_{Lm} = L_m \frac{di_m}{dt} \text{ or } V_I = L_m \frac{I_{m(pk)}}{DT_s} = f_s L_m \frac{I_{m(pk)}}{D} \tag{12.10}$$

The magnetizing current increases from zero to the peak value of the magnetizing current $I_{m(pk)}$, given by

$$I_{m(pk)} = \frac{DV_I}{f_s L_m} \tag{12.11}$$

The minimum and maximum primary and transistor currents are the sum of the magnetizing and reflected secondary currents and are given by

$$I_{p(min)} = I_{s(min)}' \tag{12.12}$$

$$I_{p(max)} = I_{s(max)}' + I_{m(pk)} \tag{12.13}$$

Knowing all these currents, the rms and average values for all components in the converter can be determined (See Appendix I). All the waveforms shown in Figure 12.6(b)–(h) are periodic discontinuous ramp waveforms.

12.2.1.1 Example: Current Ratings in Medium-Power Forward Converter

A single-switch forward converter is designed for the following specifications (which are typical practical values for the power level): input voltage of 200 V, output voltage of 12 V, output power of 180 W, switching frequency of 100 kHz, and a magnetizing inductance of 2 mH.

i) Size the output inductor for a peak-to-peak current ripple $r_i = 10\%$ at full load.
ii) Determine the various component currents.

Assume a unity turns ratio between the primary and tertiary windings and a maximum duty cycle of 50% at 200 V.

Ignore all losses, device voltage drops, and the effects of leakage.

Solution:

i) The secondary-primary turns ratio is determined by assuming that the converter operates at the maximum duty cycle of 50% at the minimum input voltage of 200 V. Thus, from Equation (12.1):

$$n = \frac{V_O}{DV_I} = \frac{12}{0.5 \times 200} = 0.12$$

The dc output current is the output power divided by the output voltage:

$$I_O = \frac{P_O}{V_O} = \frac{180}{12}\text{A} = 15\,\text{A}$$

The peak-to-peak output inductor ripple current is

$$\Delta I_{Lo(p-p)} = r_i I_O = 0.1 \times 15\,\text{A} = 1.5\,\text{A}$$

The output inductance is obtained by rearranging Equation (12.3) to get

$$L_O = \frac{V_O(1-D)}{f_s\,\Delta I_{Lo(p-p)}} = \frac{12 \times (1-0.5)}{100 \times 10^3 \times 1.5}\text{H} = 40\,\mu\text{H}$$

ii) The minimum and maximum secondary currents are given by

$$I_{s(min)} = I_{Lo(min)} = I_O - \frac{\Delta I_{Lo(p-p)}}{2} = 15\,\text{A} - \frac{1.5}{2}\text{A} = 14.25\,\text{A}$$

$$I_{s(max)} = I_{Lo(max)} = I_O + \frac{\Delta I_{Lo(p-p)}}{2} = 15\,\text{A} + \frac{1.5}{2}\,\text{A} = 15.75\,\text{A}$$

The rms current in the inductor can be determined using (11.15) as follows:

$$I_{Lo(rms)} = \sqrt{{I_O}^2 + \frac{{\Delta I_{Lo(p-p)}}^2}{12}} = \sqrt{15^2 + \frac{1.5^2}{12}}\text{A} = 15.01\,\text{A}$$

The rms and average currents in diode D_2 are

$$I_{D2(rms)} = I_{s(rms)} = \sqrt{\frac{D}{3}\left({I_{s(min)}}^2 + I_{s(min)}I_{s(max)} + {I_{s(max)}}^2\right)}$$

$$= \sqrt{\frac{0.5}{3}\left(14.25^2 + 14.25 \times 15.75 + 15.75^2\right)\text{A}} = 10.61\text{A}$$

and

$$I_{D2(dc)} = D\left(\frac{I_{s(min)} + I_{s(max)}}{2}\right) = 0.5\left(\frac{14.25 + 15.75}{2}\right)\text{A} = 7.5\,\text{A}$$

The rms and average currents in diode D_1 are

$$I_{D1(rms)} = \sqrt{\frac{(1-D)}{3}\left({I_{s(min)}}^2 + I_{s(min)}I_{s(max)} + {I_{s(max)}}^2\right)}$$

$$= \sqrt{\frac{(1-0.5)}{3}\left(14.25^2 + 14.25 \times 15.75 + 15.75^2\right)\text{A}} = 10.61\text{A}$$

and

$$I_{D1(dc)} = (1-D) \times \left(\frac{I_{s(min)} + I_{s(max)}}{2}\right) = (1-0.5) \times \left(\frac{14.25 + 15.75}{2}\right)\text{A} = 7.5\,\text{A}$$

The peak value of the magnetizing current is given by Equation (12.11):

$$I_{m(pk)} = \frac{DV_I}{f_s L_m} = \frac{0.5 \times 200}{100 \times 10^3 \times 2 \times 10^{-3}} \text{A} = 0.50 \text{ A}$$

Thus, the reflected secondary transistor currents are

$$I_{s(min)}{}' = nI_{s(min)} = 0.12 \times 14.25 \text{ A} = 1.71 \text{ A}$$

$$I_{s(max)}{}' = nI_{s(max)} = 0.12 \times 15.75 \text{ A} = 1.89 \text{ A}$$

The primary and transistor currents are the sum of the reflected secondary current and the magnetizing current:

$$I_{p(min)} = I_{s(min)}{}' = 1.71 \text{ A}$$

$$I_{p(max)} = I_{s(max)}{}' + I_{m(pk)} = 1.89 \text{ A} + 0.5 \text{ A} = 2.39 \text{ A}$$

The rms and average currents in switch Q are

$$I_{Q(rms)} = I_{p(rms)} = \sqrt{\frac{D}{3}\left(I_{p(min)}{}^2 + I_{p(min)} I_{p(max)} + I_{p(max)}{}^2\right)}$$

$$= \sqrt{\frac{0.5}{3}\left(1.71^2 + 1.71 \times 2.39 + 2.39^2\right)} \text{A} = 1.456 \text{ A}$$

and

$$I_{Q(dc)} = D \times \left(\frac{I_{p(min)} + I_{p(max)}}{2}\right) = 0.5 \times \left(\frac{1.71 + 2.39}{2}\right) \text{A} = 1.025 \text{ A}$$

Finally, the rms and average currents in the tertiary winding are

$$I_{T(rms)} = \sqrt{\frac{D}{3}} I_{m(pk)} = \sqrt{\frac{0.5}{3}} \times 0.5\text{A} = 0.204 \text{ A}$$

and

$$I_{T(dc)} = D \times \left(\frac{0 + I_{m(pk)}}{2}\right) = 0.5 \times \left(\frac{0 + 0.5}{2}\right) \text{A} = 0.125 \text{ A}$$

where a turns ratio of unity is assumed between the primary and tertiary windings, causing the tertiary windings to demagnetize over the same time interval DT_s that the magnetizing current took to build up when the switch was closed.

12.2.2 CCM Voltages in Forward Converter

Significant constraining factors for a single-switch forward converter are the relatively high voltages across the various semiconductor devices. The voltages across the various components are shown in Figure 12.7(a)–(d). The circuits of Figure 12.4 are repeated as Figure 12.7(i)–(iii).

As shown in Figure 12.7(a), the voltage across the output inductor v_L is the difference between the reflected input voltage and the output voltage.

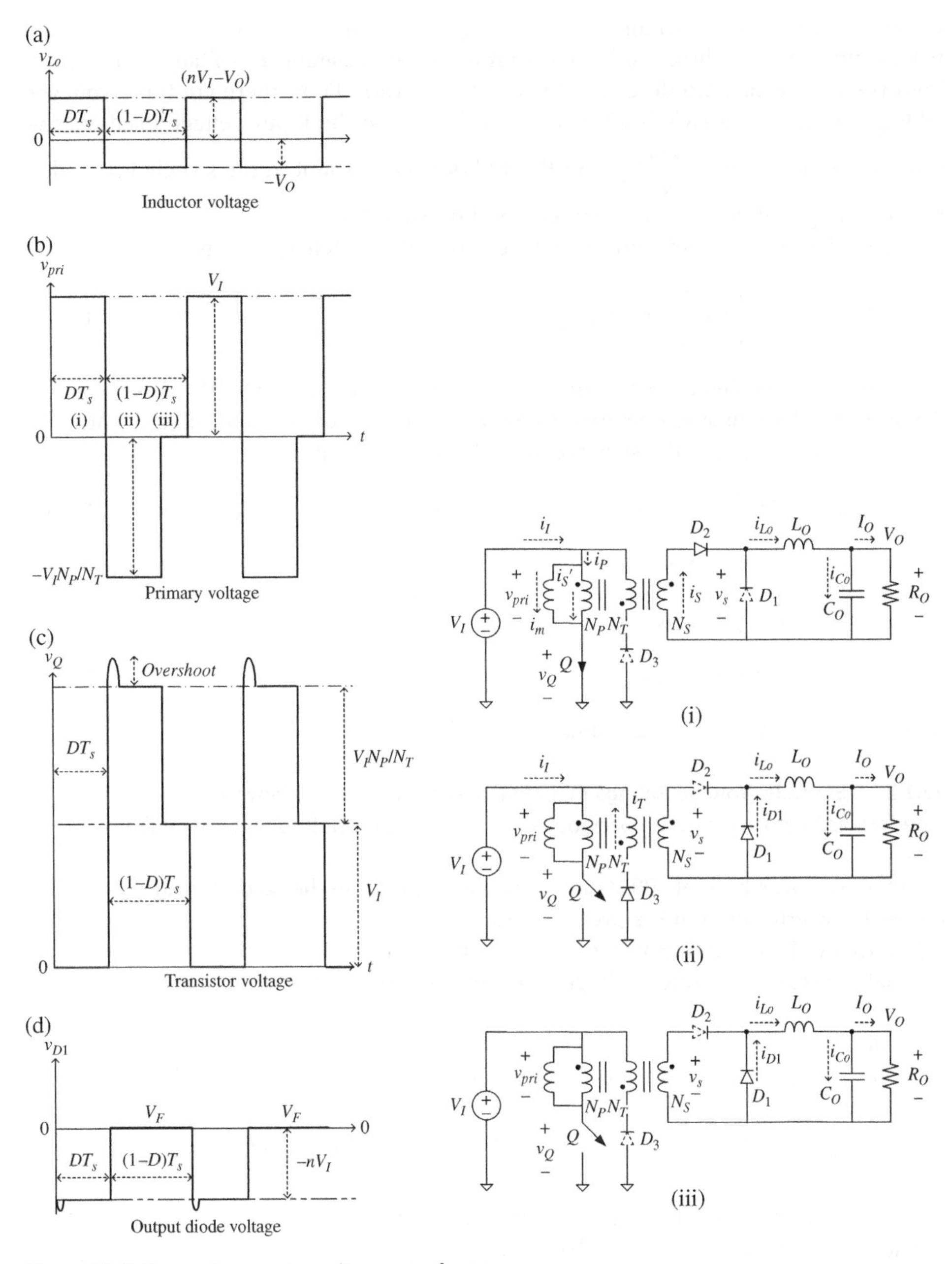

Figure 12.7 Forward converter voltage waveforms.

As shown in Figure 12.7(b), the voltage across the transformer primary v_{pri} equals the input voltage when the switch is closed, as seen from Figure 12.7(i). The voltage across the switch is shown in Figure 12.7(c). The voltage is very low when the switch is conducting current, but the voltage increases to $V_{Q(max)}$ when the switch stops conducting, and the tertiary winding conducts. The primary winding voltage is the reflected tertiary voltage when the tertiary winding is conducting current and being demagnetized, as seen from Figure 12.7(ii). These modes of operation result in significant voltage drops across both the transformer winding and the semiconductors. Thus, there are three voltages adding up across the switch when it is off. These are the input voltage itself V_I, the reflected tertiary voltage $\frac{N_P}{N_T}V_1$, and the additional overshoot in the switch $V_{Q(OS)}$ due to the leakage inductances of the circuit and transformer.

In equation form, the maximum voltage across the switch $V_{Q(max)}$ is

$$V_{Q(max)} = V_I\left(1 + \frac{N_P}{N_T}\right) + V_{Q(OS)} \tag{12.14}$$

It is necessary to derate semiconductor devices in order to operate the devices safely. A reasonable derating is to operate the device at up to 80% of the rated voltage. Thus, the rated voltage $V_{Q(rated)}$ of the switch can be determined from

$$V_{Q(max)} = 80\% \times V_{Q(rated)} \text{ or } V_{Q(rated)} = 1.25 \times V_{Q(max)} \tag{12.15}$$

The other semiconductor components also experience relatively high voltages. The voltage across the output diode D_1 is shown in Figure 12.7(d). The maximum voltage across the diode $V_{D1(max)}$ is

$$V_{D1(max)} = nV_I + V_{D1(OS)} \tag{12.16}$$

where $V_{D1(OS)}$ is the overshoot voltage across D_1.

12.2.2.1 Example: Voltage Ratings in a Medium-Power Forward Converter

The forward converter in the previous example has an input voltage range from 200 to 400 V.

Assume an overshoot of 100 V on the switch and 10 V on the secondary diodes for the forward converter used in the previous example.

Determine the maximum voltages seen by the devices.

What are reasonable rated voltages for these semiconductor components?

Solution:

The maximum voltage across the switch occurs at the maximum input voltage of 400 V:

$$V_{Q(max)} = V_I\left(1 + \frac{N_P}{N_T}\right) + V_{Q(OS)} = 400\,\text{V}\left(1 + \frac{1}{1}\right) + 100\,\text{V} = 900\,\text{V}$$

where a primary-tertiary turns ratio of unity is assumed.

The minimum rated voltage of the device is

$$V_{Q(rated)} \geq 1.25 \times V_{Q(max)} = 1.25 \times 900\,\text{V} = 1125\,\text{V}$$

Thus, a rated voltage of 1200 V, when rounded up to the nearest 100 V, would be appropriate for the switch when derated.

The secondary diode experiences a maximum voltage of

$$V_{D1(max)} = nV_I + V_{D1(OS)} = 0.12 \times 400\,\text{V} + 10\,\text{V} = 58\,\text{V}$$

The minimum rated voltage of the diode is

$$V_{D(rated)} \geq 1.25 \times V_{D(max)} = 1.25 \times 58\,\text{V} = 72.5\,\text{V}$$

Thus, a rated voltage of 80 V, when rounded to the nearest 10 V, would be appropriate for the diode when derated.

12.2.3 Sizing the Transformer

As discussed in Chapter 16, Section 16.5.3, the area product (AP) for a transformer fed by a duty cycle controlled voltage source supplying a transformer is

$$AP = \frac{DV_I\left(I_{p(rms)} + \frac{N_T}{N_P}I_{T(rms)} + \frac{N_S}{N_P}I_{s(rms)}\right)}{2f_s B_{max} k_{cu} J_{cu}} \tag{12.17}$$

where B_{max} is the maximum flux density, k_{cu} is the copper fill factor, and J_{cu} is the current density.

12.2.3.1 Example: AP of a Forward Converter Transformer

A transformer has the following parameters: copper fill factor $k_{cu} = 0.5$, current density $J_{cu} = 6\text{A/mm}^2$, and maximum core flux density $B_{max} = 20$ mT. The maximum core flux density is significantly lower than the value used in the examples in Chapter 11 as the high-frequency transformer is likely to be limited in size by its core loss.

Determine the AP for the transformer in the previous example.

Solution:

The AP is given by

$$AP = \frac{DV_I\left(I_{p(rms)} + \frac{N_T}{N_P}I_{T(rms)} + \frac{N_S}{N_P}I_{s(rms)}\right)}{2f_s B_{max} k_{cu} J_{cu}}$$
$$= \frac{0.5 \times 200 \times (1.456 + 1 \times 0.204 + 0.12 \times 10.61)}{2 \times 100 \times 10^3 \times 20 \times 10^{-3} \times 0.5 \times 6 \times 10^6}\text{m}^4 = 2.444\,\text{cm}^4 \tag{12.18}$$

12.3 The Full-Bridge Converter

The isolated full-bridge converter is the preferred topology for medium to high power, and is often the topology used on vehicles for charging the high-voltage battery and powering the low-voltage auxiliary loads. The relative voltage and current stresses of the full-bridge converter are lower than those of the forward converter, at the expense of additional components and complexity. The advantages are as follows: the full bridge has twice the power for the same current ratings, half the voltage on the switches, and twice the ripple frequency. On the downside, the full bridge has more semiconductors, and requires isolated gate drives with the danger of shoot-through.

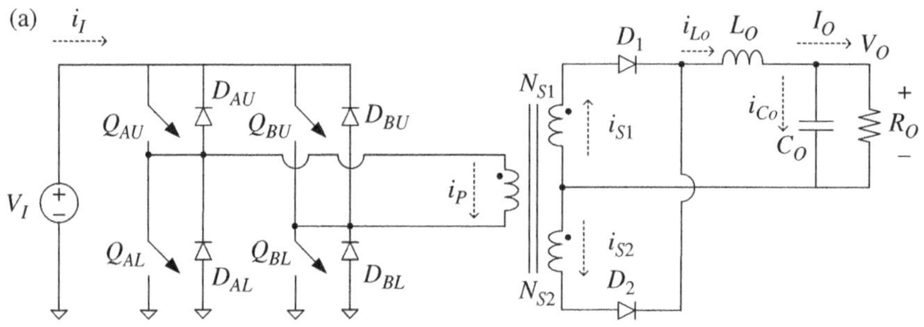

Full-bridge converter with center-tapped transformer and rectifier

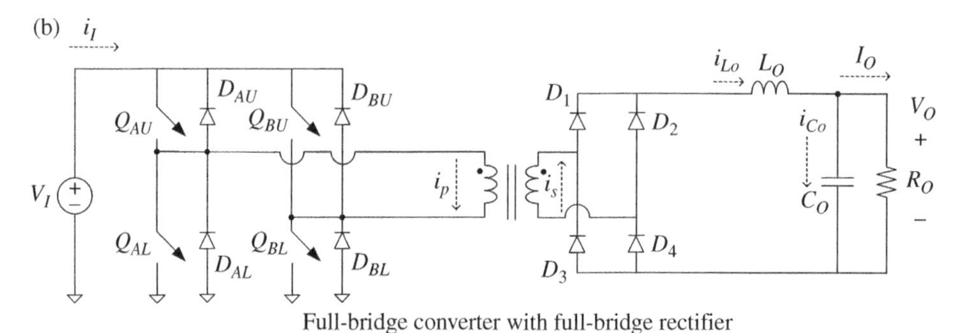

Full-bridge converter with full-bridge rectifier

Figure 12.8 Full-bridge converters.

The basic circuit topologies are shown in Figure 12.8. The full bridge features four semiconductor switches, each partnered with an inverse diode. The full bridge has two poles, *A* and *B*, and upper (*U*) and lower (*L*) switches and diodes. The full bridge is switched such that an ac square wave voltage appears across the primary of the transformer. For low-voltage applications, the transformer has two secondary windings which are center-tapped to connect to the output, as shown in Figure 12.8(a). This ensures that the output current only conducts through one diode at a time – resulting in improved efficiency. For high-voltage output applications, the relative effect of a diode drop on efficiency is not as significant, and a standard transformer with a single secondary can be connected to a full-bridge diode rectifier, as shown in Figure 12.8(b). The center-tapped rectifier, often featuring a low-conduction-drop Schottky diode, is suitable for auxiliary power converters outputting 12 V, whereas the full-bridge rectifier is suitable for high-voltage battery chargers with outputs of hundreds of volts. Note that a synchronous rectifier is often used for low-voltage outputs. A **synchronous rectifier** uses a controlled switch such as a MOSFET to replace the output diode. A low-voltage MOSFET can have a significantly lower conduction drop than a silicon or Schottky diode.

For both types of converters, the inductor current is filtered by an output capacitor such that dc waveforms are output from the converter. The load is represented generically by a resistor but could be a battery pack or the various auxiliary loads on a vehicle.

There are a number of full-bridge control schemes which can be used depending on the application. In this chapter, the basic hard-switched full bridge is considered. The phase-shifted full bridge is commonly used but is more complex [4].

12.3.1 Operation of Hard-Switched Full-Bridge Converter

There are four principal operating modes, labeled I to IV in Figure 12.9, Figure 12.10, and Figure 12.11. The switches are switched on and off according to the patterns shown in Figure 12.9. The pole A upper and lower switches, Q_{AU} and Q_{AL}, alternate on and off, and both have a 50% duty cycle. The duty cycles of the pole B upper and lower switches, Q_{BU} and Q_{BL}, are controlled to regulate the output voltage.

Modes I and II occur during the first half of the switching period. In mode I, gating signals are sent to switches Q_{AU} and Q_{BL}, as shown in Figure 12.9(a) and (c). This results in a positive voltage across the primary of the transformer, as shown in Figure 12.9(e). When Q_{BL} is turned off in mode II, diode D_{BU} becomes forward biased and maintains the primary magnetizing current flow. It is noteworthy that the magnetizing current in the transformer, as shown in Figure 12.9(f), ramps up and down when a net voltage is applied across the primary, but the magnetizing current remains constant when there is no applied voltage – as is expected from Faraday's law. Modes III and IV occur during the second half of the switching period. In mode III, gating signals are sent to switches Q_{AL} and Q_{BU}, as shown in Figure 12.9(b) and (d). This results in a negative voltage across the primary of the transformer. When Q_{BU} is turned off in mode IV, diode D_{BL} becomes forward biased and maintains the primary current flow. The switch and diode conductions in the various modes are noted in Table 12.1.

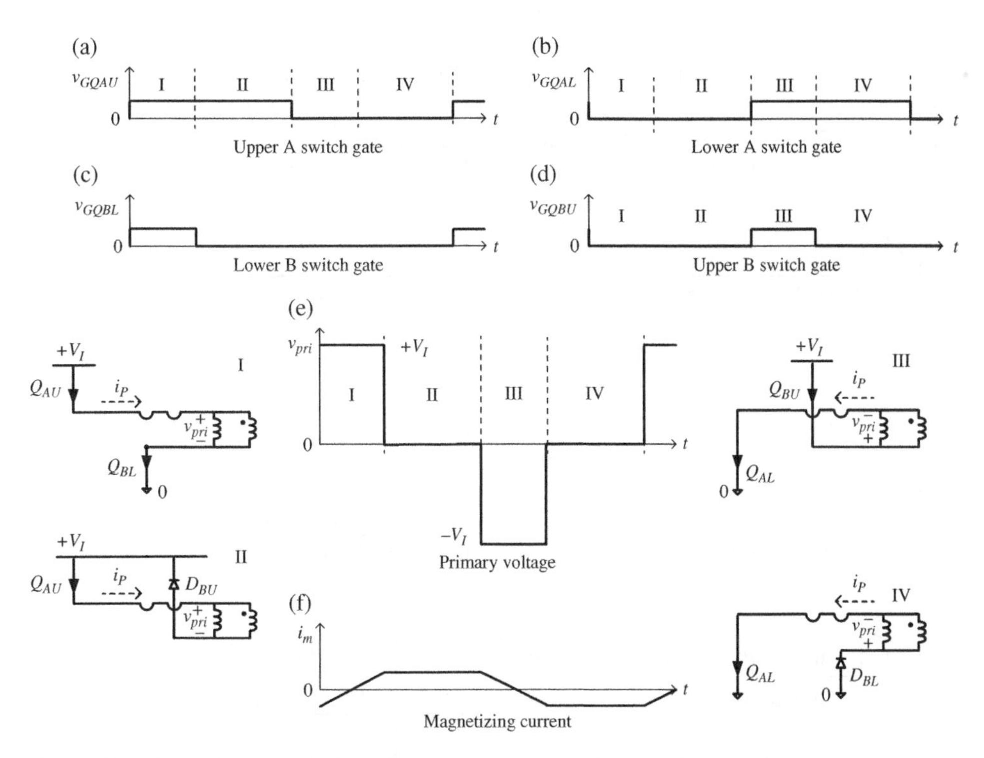

Figure 12.9 Gate voltages, switch configurations, primary voltage, and magnetizing current waveforms.

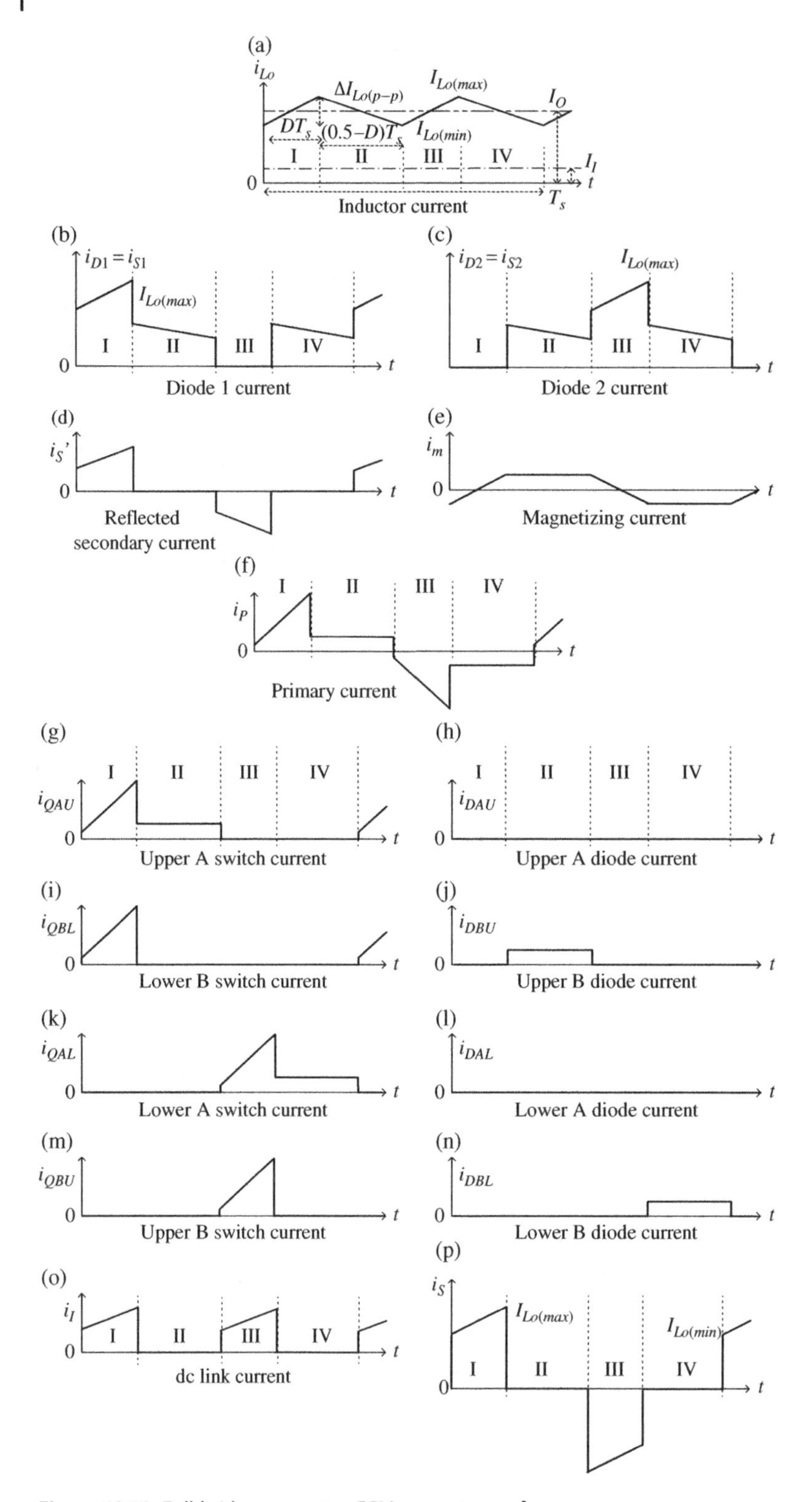

Figure 12.10 Full-bridge converter CCM current waveforms.

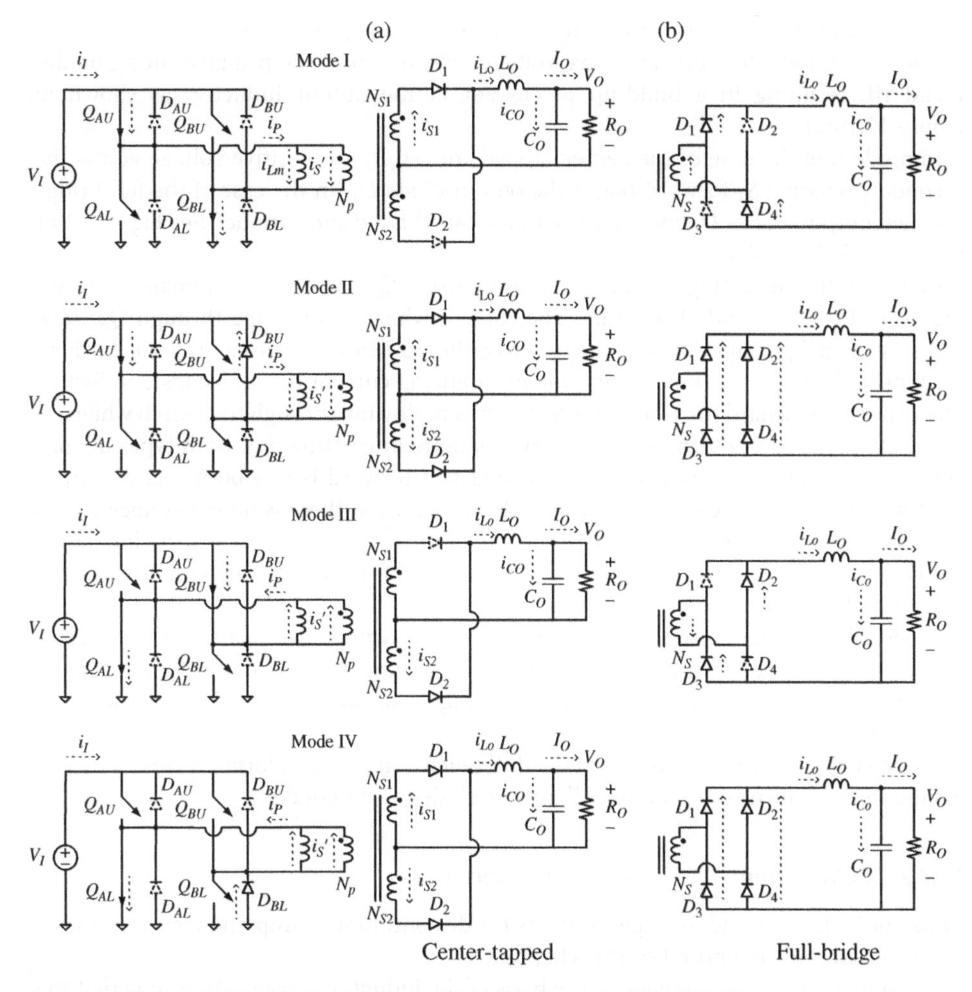

Figure 12.11 Principal modes for hard-switched full-bridge converter with (a) center tap and (b) full-bridge rectifier.

Table 12.1 Switch and diode conduction.

Mode	A-leg	B-leg
I	Q_{AU}	Q_{BL}
II		D_{BU}
III	Q_{AL}	Q_{BU}
IV		D_{BL}

The complete set of current waveforms is shown in Figure 12.10.

The output inductor sees a positive voltage reflected from the primary during modes I and III, resulting in a build-up of current in the output inductor, as shown in Figure 12.10(a).

In mode I, in the case of the center-tapped converter, the positive voltage across the secondary winding S_1 forward-biases the output diode D_1. In the case of the full-bridge rectifier, output diodes D_1 and D_4 are both forward-biased and conduct. See Figure 12.11 (a) for mode I circuits.

In mode II, the gating signal is removed from Q_{BL} while Q_{AU} remains on. See Figure 12.11(b) for mode II circuits. The current which was flowing through Q_{BL} now flows through D_{BU}, and the net voltage across the transformer primary is zero. As there is no net voltage across the primary, the magnetizing current remains constant. Switch Q_{BL} experiences a hard turn-off as there is a current flowing through the switch while it is gated off. The action on the secondary side is quite interesting. For both types of converters, the inductor current remains flowing and forward-biases both sets of output diodes such that equal currents flow in each winding, and there is no net voltage across the transformer secondary or primary. The inductor current decreases as the inductor's stored energy decreases due to the current flow to the load. The inductor back emf is correspondingly negative.

Mode III is similar to mode I, but the other two switches are activated in order to demagnetize the transformer and to forward-bias output diode D_2 for the center-tapped converter and diodes D_2 and D_3 for the full bridge. The switch from Q_{BU} experiences a hard turn-off.

Mode IV is similar to mode II with the difference that the transformer primary is connected to the return of the input. All output diodes again conduct.

12.3.2 CCM Currents in Full-Bridge Converter

Please note the rms and average formulas for discontinuous ramp and step waveforms presented in the Appendix I of this chapter.

The duty cycle can be derived on the basis of the inductor current. Assuming that the inductor current is changing linearly in the steady-state operation, the change in current from peak to peak can be described by $\Delta I_{Lo(p\text{-}p)}$. It is important to note that the switching period covers the durations of the four modes, as shown in Figure 12.10(a).

The current increases during time DT_s:

$$\Delta I_{Lo(p-p)} = \frac{1}{L_O}(nV_I - V_O)DT_s \tag{12.19}$$

and decreases during time $(0.5\text{-}D)T_s$:

$$\Delta I_{Lo(p-p)} = \frac{V_O(0.5-D)T_s}{L_O} \tag{12.20}$$

Equating the two expressions for the current $\Delta I_{Lo(p\text{-}p)}$:

$$\Delta I_{Lo(p-p)} = \frac{1}{L_O}(nV_I - V_O)DT_s = \frac{V_O(0.5-D)T_s}{L_O} \tag{12.21}$$

Solving the two equations results in the following expression for the relationship between duty cycle and voltage gain:

$$D = \frac{1}{2n}\frac{V_O}{V_I} \tag{12.22}$$

where n is the turns ratio of the transformer, $n = N_{S1}/N_P = N_{S2}/N_P$ for the center-tapped transformer, and $n = N_S/N_P$ for the full-bridge rectifier.

The current waveforms for the full-bridge converter are shown in Figure 12.10. The waveforms can be analyzed in a similar manner to the forward converter. Similar to the earlier converter, the change in peak-to-peak current can provide the basis for converter analysis. As noted in Equation (12.21), the change in the inductor peak-to-peak current is easily determined.

The inductor current changes from a minimum to a maximum over the half cycle, and these values are given by

$$I_{Lo(min)} = I_O - \frac{\Delta I_{Lo(p-p)}}{2} \tag{12.23}$$

$$I_{Lo(max)} = I_O + \frac{\Delta I_{Lo(p-p)}}{2} \tag{12.24}$$

The rms current in the inductor is

$$I_{Lo(rms)} = \sqrt{I_O^2 + \frac{\Delta I_{Lo(p-p)}^2}{12}} \tag{12.25}$$

The minimum and maximum secondary currents are given by

$$I_{s(min)} = I_{Lo(min)} \tag{12.26}$$

$$I_{s(max)} = I_{Lo(max)} \tag{12.27}$$

Let D_O represent the output diodes for either rectifier type.

The output diode current has three conduction segments within the period, modes I, II, and IV or modes II, III, and IV. The related rms value of the diode current $I_{Do(rms)}$ is given:

$$I_{Do(rms)} = \sqrt{\frac{D}{3}\left(I_{s(min)}^2 + I_{s(min)}I_{s(max)} + I_{s(max)}^2\right) + 2\times\frac{(0.5-D)}{3}\left(\left(\frac{I_{s(min)}}{2}\right)^2 + \frac{I_{s(min)}}{2}\times\frac{I_{s(max)}}{2} + \left(\frac{I_{s(max)}}{2}\right)^2\right)} \tag{12.28}$$

which simplifies to

$$I_{Do(rms)} = \sqrt{\left(\frac{1+2D}{12}\right)\left(I_{s(min)}^2 + I_{s(min)}I_{s(max)} + I_{s(max)}^2\right)} \tag{12.29}$$

The average value of the diode current $I_{Do(dc)}$ is

$$I_{Do(dc)} = D\left(\frac{I_{s(min)} + I_{s(max)}}{2}\right) + (1-2D)\left(\frac{I_{s(min)} + I_{s(max)}}{4}\right) = \frac{I_{s(min)} + I_{s(max)}}{4} \tag{12.30}$$

an expression which makes sense as the average current in each diode is equal to half the average inductor current.

Note that in the center-tapped converter, the secondary rms current in one winding is the same as the output diode current:

$$I_{s(rms)} = I_{Do(rms)} \tag{12.31}$$

However, in the full-bridge rectifier, the secondary current only flows during modes I and III, as shown by the waveform in Figure 12.10(p).

Thus, for the full-bridge rectifier:

$$I_{s(rms)} = \sqrt{\frac{2D}{3}\left(I_{s(min)}{}^2 + I_{s(min)}I_{s(max)} + I_{s(max)}{}^2\right)} \tag{12.32}$$

The minimum and maximum reflected secondary currents are given by

$$I_{s(min)}{}' = nI_{s(min)} \tag{12.33}$$

$$I_{s(max)}{}' = nI_{s(max)} \tag{12.34}$$

The peak value of the magnetizing current is half the peak-to-peak value and is given by

$$I_{m(pk)} = \frac{DV_I}{2f_s L_m} \tag{12.35}$$

Thus, the minimum and maximum primary currents are given by

$$I_{p(min)} = I_{s(min)}{}' - I_{m(pk)} \tag{12.36}$$

$$I_{p(max)} = I_{s(max)}{}' + I_{m(pk)} \tag{12.37}$$

The rms of the primary current waveform considers the four conduction modes, I–IV:

$$I_{p(rms)} = \sqrt{2 \times \frac{D}{3}\left(I_{p(min)}{}^2 + I_{p(min)}I_{p(max)} + I_{p(max)}{}^2\right) + (1-2D)I_{m(pk)}{}^2} \tag{12.38}$$

The rms currents in the A-leg switches are due to conduction modes I and II:

$$I_{QAU(rms)} = I_{QAL(rms)} = \sqrt{\frac{D}{3}\left(I_{p(min)}{}^2 + I_{p(min)}I_{p(max)} + I_{p(max)}{}^2\right) + (0.5-D)I_{m(pk)}{}^2} \tag{12.39}$$

The rms currents in the B-leg switches are due to conduction mode I only:

$$I_{QBU(rms)} = I_{QBL(rms)} = \sqrt{\frac{D}{3}\left(I_{p(min)}{}^2 + I_{p(min)}I_{p(max)} + I_{p(max)}{}^2\right)} \tag{12.40}$$

The average currents in the A-leg switches are

$$I_{QAU(dc)} = I_{QAL(dc)} = D\left(\frac{I_{p(min)} + I_{p(max)}}{2}\right) + (0.5-D)I_{m(pk)} \tag{12.41}$$

The average currents in the B-leg switches are

$$I_{QBU(dc)} = I_{QBL(dc)} = D\left(\frac{I_{p(min)} + I_{p(max)}}{2}\right) \tag{12.42}$$

The rms currents in the B-leg diodes are due to conduction modes II or IV:

$$I_{DBU(rms)} = I_{DBL(rms)} = \sqrt{(0.5-D)}I_{m(pk)} \tag{12.43}$$

The average currents in the B-leg diodes are

$$I_{DBU(dc)} = I_{DBL(dc)} = (0.5-D)I_{m(pk)} \tag{12.44}$$

For this CCM mode, the rms and average currents in the A-leg diodes are both zero.

12.3.2.1 Example: Current Ratings in a Medium-Power Full-Bridge Converter

A full-bridge converter with a high-voltage full-bridge rectifier is designed to the following specifications: input voltage of 380 V, output voltage of 400 V, output power of 6 kW, switching frequency of 100 kHz, and a magnetizing inductance of 0.5 mH.

i) Size the output inductor to ensure a peak-to-peak current ripple r_i = 10% at full load.
ii) Determine the various component currents.
iii) Determine the AP for the transformer with the following parameters: copper fill factor $k_{cu} = 0.5$, current density $J_{cu} = 6\text{A/mm}^2$, and maximum core flux density B_{max} = 20 mT.

Assume a maximum duty cycle of 0.45 at 400 V output.
Ignore all losses, device voltage drops, and the effects of leakage.

Solution:

i) The turns ratio is

$$n = \frac{V_O}{2DV_I} = \frac{400}{2 \times 0.45 \times 380} = 1.17$$

The dc output current is the output power divided by the output voltage:

$$I_O = \frac{P_O}{V_O} = \frac{6000}{400}\text{A} = 15\,\text{A}$$

The peak-to-peak output inductor ripple current is

$$\Delta I_{Lo(p-p)} = r_i I_O = 0.1 \times 15\,\text{A} = 1.5\,\text{A}$$

The output inductance is given by rearranging Equation (12.21):

$$L_O = \frac{V_O(0.5-D)}{f_s \Delta I_{Lo(p-p)}} = \frac{400 \times (0.5-0.45)}{100 \times 10^3 \times 1.5}\,\mu\text{H} = 133.3\,\mu\text{H}$$

ii) The inductor currents are

$$I_{Lo(min)} = I_O - \frac{\Delta I_{Lo(p-p)}}{2} = 15\text{A} - \frac{1.5}{2}\text{A} = 14.25\text{A}$$

$$I_{Lo(max)} = I_O + \frac{\Delta I_{Lo(p-p)}}{2} = 15\text{A} + \frac{1.5}{2}\text{ A} = 15.75\text{A}$$

$$I_{Lo(rms)} = \sqrt{I_O{}^2 + \frac{\Delta I_{Lo(p-p)}{}^2}{12}} = \sqrt{15 + \frac{1.5^2}{12}}\text{A} = 15.01\text{ A}$$

The minimum and maximum secondary currents are given by

$$I_{s(min)} = I_{Lo(min)} = 14.25\text{A}$$

$$I_{s(max)} = I_{Lo(max)} = 15.75\text{A}$$

For the full-bridge rectifier, the secondary current is given by

$$I_{s(rms)} = \sqrt{\frac{2D}{3}\left(I_{s(min)}{}^2 + I_{s(min)}I_{s(max)} + I_{s(max)}{}^2\right)}$$

$$= \sqrt{\left(\frac{2 \times 0.45}{3}\right)\left(14.25^2 + 14.25 \times 15.75 + 15.75^2\right)}\text{ A} = 14.24\text{ A}$$

The output diode currents are

$$I_{Do(rms)} = \sqrt{\left(\frac{1+2D}{12}\right)\left(I_{s(min)}{}^2 + I_{s(min)}I_{s(max)} + I_{s(max)}{}^2\right)}$$

$$= \sqrt{\left(\frac{1 + 2 \times 0.45}{12}\right)\left(14.25^2 + 14.25 \times 15.75 + 15.75^2\right)}\text{ A} = 10.34\text{ A}$$

$$I_{Do(dc)} = \frac{I_{s(min)} + I_{s(max)}}{4} = \frac{14.25 + 15.75}{4}\text{A} = 7.5\text{A}$$

The minimum and maximum reflected secondary currents are given by

$$I_{s(min)}{}' = nI_{s(min)} = 1.17 \times 14.25\text{ A} = 16.67\text{ A}$$

$$I_{s(max)}{}' = nI_{s(\max)} = 1.17 \times 15.75\text{ A} = 18.43\text{ A}$$

The peak value of the magnetizing current is

$$I_{m(pk)} = \frac{DV_I}{2f_sL_m} = \frac{0.45 \times 380}{2 \times 100 \times 10^3 \times 500 \times 10^{-6}}\text{ A} = 1.71\text{ A}$$

Thus, the minimum and maximum primary currents are given by

$$I_{p(min)} = I_{s(min)}{}' - I_{m(pk)} = 16.67\text{ A} - 1.71\text{ A} = 14.96\text{ A}$$

$$I_{p(max)} = I_{s(max)}{}' + I_{m(pk)} = 18.43\text{A} + 1.71\text{A} = 20.14\text{ A}$$

The rms of the primary current is

$$I_{p(rms)} = \sqrt{2 \times \frac{D}{3}\left(I_{p(min)}{}^2 + I_{p(min)}I_{p(max)} + I_{p(max)}{}^2\right) + (1-2D)I_{m(pk)}{}^2}$$

$$= \sqrt{2 \times \frac{0.45}{3}\left(14.96^2 + 14.96 \times 20.14 + 20.14^2\right) + (1-2\times 0.45) \times 1.71^2}\,\text{A} = 16.72\,\text{A}$$

The currents in the A-leg switches are

$$I_{QAU(rms)} = I_{QAL(rms)} = \sqrt{\frac{D}{3}\left(I_{p(min)}{}^2 + I_{p(min)}I_{p(max)} + I_{p(max)}{}^2\right) + (0.5-D)I_{m(pk)}{}^2}$$

$$= \sqrt{\frac{0.45}{3}\left(14.96^2 + 14.96 \times 20.14 + 20.14^2\right) + (0.5-0.45) \times 1.71^2}\,\text{A} = 11.82\,\text{A}$$

$$I_{QAU(dc)} = I_{QAL(dc)} = D\left(\frac{I_{p(min)} + I_{p(max)}}{2}\right) + (0.5-D)I_{m(pk)}$$

$$= 0.45 \times \left(\frac{14.96 + 20.14}{2}\right)\text{A} + (0.5-0.45) \times 1.71\,\text{A} = 7.98\,\text{A}$$

The currents in the B-leg switches are

$$I_{QBU(rms)} = I_{QBL(rms)} = \sqrt{\frac{D}{3}\left(I_{p(min)}{}^2 + I_{p(min)}I_{p(max)} + I_{p(max)}{}^2\right)}$$

$$= \sqrt{\frac{0.45}{3}\left(14.96^2 + 14.96 \times 20.14 + 20.14^2\right)}\,\text{A} = 11.82\,\text{A}$$

$$I_{QBU(dc)} = I_{QBL(dc)} = D\left(\frac{I_{p(min)} + I_{p(max)}}{2}\right)$$

$$= 0.45 \times \left(\frac{14.96 + 20.14}{2}\right)\text{A} = 7.90\text{A}$$

The currents in the B-leg diodes are

$$I_{DBU(rms)} = I_{DBL(rms)} = \sqrt{(0.5-D)}I_{m(pk)} = \sqrt{(0.5-0.45)} \times 1.71\text{A} = 0.38A$$

$$I_{DBU(dc)} = I_{DBL(dc)} = (0.5-D)I_{m(pk)} = (0.5-0.45) \times 1.71\,\text{A} = 0.09\,\text{A}$$

We can check this calculation by noting the following: the average input current should equal the sum of the average currents in the A- and B-leg switches minus the B-leg diode average current. The input current is given by

$$I_I = \frac{P_O}{V_I} = \frac{6000}{380}\text{A} = 15.79\,\text{A}$$

Alternatively,

$$I_I = I_{QAU(dc)} + I_{QBU(dc)} - I_{DBU(dc)} = 7.98\,\text{A} + 7.90\,\text{A} - 0.09\,\text{A} = 15.79\,\text{A}$$

which confirms the value.

iii) The transformer AP is given by

$$AP = \frac{DV\left(I_{p(rms)} + nI_{s(rms)}\right)}{2fB_{max}k_{cu}J_{cu}} = \frac{0.45 \times 380(16.72 + 1.17 \times 14.24)}{2 \times 100 \times 10^3 \times 20 \times 10^{-3} \times 0.5 \times 6 \times 10^6} \text{m}^4 = 47.55 \text{ cm}^4$$

for the full-bridge rectifier converter

12.3.3 CCM Voltages in the Full-Bridge Converter

A significant constraining factor for a single-switch forward is the relatively high voltage across the semiconductor device. The primary-side switches and diodes in a full-bridge converter are simply limited to the input voltage plus any overshoot.

In equation form, the maximum voltage across the switch $V_{Q(max)}$ is

$$V_{Q(max)} = V_I + V_{Q(OS)}$$

where $V_{Q(OS)}$ is the overshoot. Again, it is necessary to derate the semiconductor devices in order to operate the device safely. A reasonable derating is to operate the device at up to 80% of the rated voltage.

The voltage across the output diode should factor in the reflected primary voltage. The maximum voltage across the secondary diode $V_{D(max)}$ is

$$V_{D(max)} = nV_I + V_{D(OS)}$$

where $V_{D(OS)}$ is the diode overshoot.

12.3.3.1 Example: Voltage Ratings in a Full-Bridge Converter

The full-bridge converter in the previous example has an input voltage of 380 V.

i) Determine the maximum voltages seen by the devices. Assume an overshoot of 100 V on the switch and 100 V on the secondary diodes.
ii) What are reasonable rated voltages for these semiconductor components?

Solution:

i) The maximum voltage across the switch is

$$V_{Q(max)} = V_I + V_{Q(OS)} = 380 \text{ V} + 100 \text{ V} = 480 \text{ V}$$

The secondary diodes will see a maximum voltage of

$$V_{D(max)} = nV_I + V_{D(OS)} = 1.17 \times 380 \text{ V} + 100 \text{ V} = 545 \text{ V}$$

ii) The minimum rated voltage of the switch is

$$V_{Q(rated)} \geq 1.25 \times V_{Q(max)} = 1.25 \times 480 \text{ V} = 600 \text{ V}$$

A rated voltage of 600 V would be appropriate for the primary devices.

The minimum rated voltage of the diode is

$$V_{D(rated)} \geq 1.25 \times V_{D(max)} = 1.25 \times 545 \text{ V} = 681 \text{ V}$$

Thus, a rated voltage of 700 V, rounded up to the nearest 100 V, would be appropriate for the diode when derated.

12.4 Resonant Power Conversion

Resonant and soft-switched power converters are commonly used for power conversion. Soft switching is desirable in a power converter as the switching losses of the semiconductors can be significantly reduced, albeit at the expense of circuit complexity and conduction losses. Resonant converters form a class of switch-mode power converters which enable soft switching by using passive elements such as inductors and capacitors to generate quasi-sinusoidal currents and voltages within the power circuitry. The quasi-sinusoidal waveforms typically have a lower high-frequency content, with reduced harmonics, than the hard-switched quasi-linear waveforms which we have observed in the forward and full-bridge converters investigated in this chapter so far.

Resonant power converters have played a significant role in the development of inductive and wireless charging. There are a number of basic reasons for the special suitability of resonant converters:

i) The transformers by their nature are loosely coupled as significant displacements can occur between the primary and secondary windings, resulting in very high leakage inductances and a very low magnetizing inductance [5]. In addition, these inductances can exhibit significant variation during operation due to the alignment of the secondary winding on the vehicle and the off-vehicle primary winding. The long cables result in additional leakage inductance.
ii) Operation of the power circuitry should be at as high a frequency as possible in order to minimize the size and mass of the passive components, especially the loosely coupled transformer.
iii) Circuit waveforms should be as sinusoidal as possible in order to minimize high-frequency effects which can cause significant heating and excessive EMI. EMI can be a particularly difficult area to address and is a significant consideration when selecting a circuit topology for wireless charging.

A hard-switched converter requires a low-leakage transformer in order to function. A resonant converter with a series-resonant LC circuit can handle the high leakage and low magnetizing inductances. A capacitor can be added in parallel with the low magnetizing inductance to cancel the effects of the low inductance and provide a resonant boost for the converter. The parallel capacitor increases the resonant currents, which has the benefit of enabling soft switching over the entire operating region. While the resonant currents increase, the high gain transformer and capacitor boost help to reduce cable and primary currents and the associated copper losses.

12.4.1 LCLC Series-Parallel Resonant Converter

The series-parallel LCLC resonant converter was the technology of choice for the inductive charging system developed in the 1990s (see also Chapter 14). A simplified circuit diagram is as shown in Figure 12.12.

The full-bridge converter is a variation on the earlier hard-switched topology. The resonant full bridge has the additional discrete capacitors, C_{AU}, C_{AL}, C_{BU}, and C_{BL}, which are added across the switches and are necessary to enable a zero-voltage turn-off of the switch.

The series-resonant tank components are L_{os} and C_s. These components are shown in the diagram as two series sets. For simplicity, it is assumed that the cable inductance,

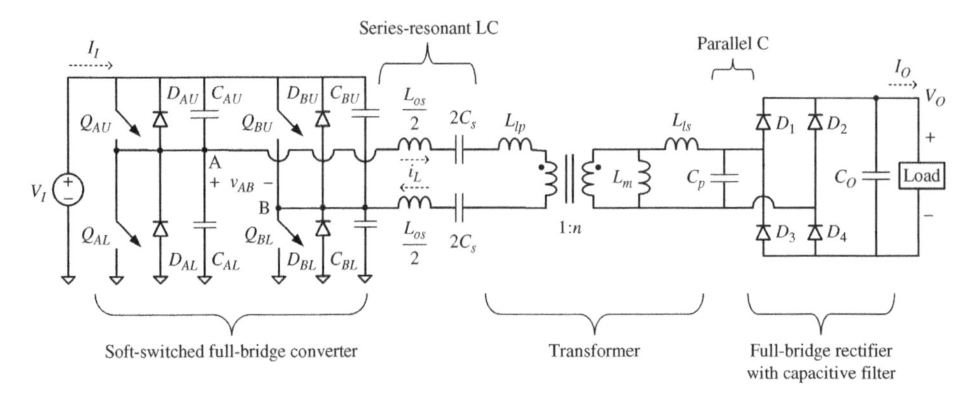

Figure 12.12 Inductive charging resonant topology.

which can be relatively significant, is integral to the series-resonant tank inductances. (Splitting the series tank as shown provides equal impedance on each pole output and aids in reducing common-mode EMI.)

The transformer is presented with explicit primary and secondary leakage inductances, L_{lp} and L_{ls}, magnetizing inductance L_m, and a transformer turns ratio n.

A discrete capacitor C_p is placed on the transformer secondary to create a resonance with the magnetizing inductance of the transformer.

The output rectifier is also a full bridge with diodes D_1–D_4, and the output is filtered using a capacitive filter.

12.4.2 Desirable Converter Characteristics for Inductive Charging

The following is a list of the general desired converter characteristics for inductive charging [6]:

1) *Unity transformer turns ratio.*
   The ac utility voltage is nominally 230 V for a single-phase input. The dc link voltage is typically in the 380 V to 400 V range from a boost-regulated PFC stage. The required battery charging voltage is generally in the 200 to 450 V range. Thus, a transformer turns ratio of unity is desirable to minimize primary- and secondary-side voltage and current stresses.
2) *Buck/boost voltage gain and current-source capability.*
   Based on Note (1) above, buck/boost operation is required when using a transformer ratio of unity in order to regulate over the full input and output voltage range. A desirable feature of the converter is that it should operate as a controlled current source.
3) *Capacitive output filter.*
   The vehicle inlet should have a capacitive rather than an inductive output filter stage to reduce the on-vehicle cost and weight, and also desensitize the converter to the operating frequency and power level.
4) *Monotonic power transfer curve over a wide load range.*
   Wide load operation is required for the likely battery voltage range. The output power should be monotonic with the controlling variable over the load range.

5) *Throttling capability down to no load.*
   All battery technologies require a wide charging current range, varying from a rated power charge to a trickle charge for battery equalization, especially at high voltages.
6) *High-frequency operation.*
   High-frequency operation is required to minimize vehicle inlet weight and cost and the general size of the passive components.
7) *Full-load operation at minimum frequency for variable frequency control.*
   Optimizing transformer operation over the load range suggests that full-load operation takes place at low frequencies.
8) *Narrow frequency range.*
   The charger should operate over as narrow a frequency range as possible for two principal reasons. First, the passive components can be optimized for a given range, and second, operation into the AM radio band is avoided as electromagnetic emissions in this region must be even more tightly controlled.
9) *Soft switching.*
   Soft switching of the converter stage is required to minimize semiconductor switching loss for high-frequency operation. Zero-voltage-switching of a power MOSFET with its slow integral diode can result in high efficiency and cost advantages. Similarly, soft recovery of the output rectifiers results in reduced power loss and a reduction in EMI.
10) *High Efficiency.*
    The power transfer must be highly efficient to minimize heat loss and to maximize the overall fuel economy of the electric vehicle charging and driving cycle.
11) *Secondary dv/dt control.*
    Relatively slow *dv/dt*'s on the cable and secondary result in reduced high-frequency harmonics and less parasitic ringing in the cable and secondary, minimizing electromagnetic emissions.
12) *CCM quasi-sinusoidal current waveforms.*
    Quasi-sinusoidal current waveforms over the load range result in reduced high-frequency harmonics and less parasitic ringing in the cable and secondary, minimizing electromagnetic emissions.

12.4.2.1 Basic Converter Operation

The full bridge operates as a square wave voltage generator v_{AB}, as shown in Figure 12.13 (a). A quasi-sinusoidal current i_{ser} flows from the full bridge into the series components. The current has the general form as shown in Figure 12.13(a).

The conduction of the various full-bridge devices are shown in Figure 12.13(b) and (c) for the switches and diodes, respectively. As can be seen, the switches turn on at zero current while the diodes turn off at zero current – avoiding diode reverse-recovery loss. The switches turn off and the diodes turn on at a finite current. However, the switches do not experience a turn-off loss as the additional discrete capacitors absorb the inductor current and enable a low-loss turn-off of the switch.

The sequence for the zero-voltage switching (ZVS) turn-off is as shown in Figure 12.14. Initially, the upper switch Q_U is conducting, as shown in Figure 12.14(a). The switch is rapidly turned off, and the inductor current rapidly transitions to the two capacitors, thus

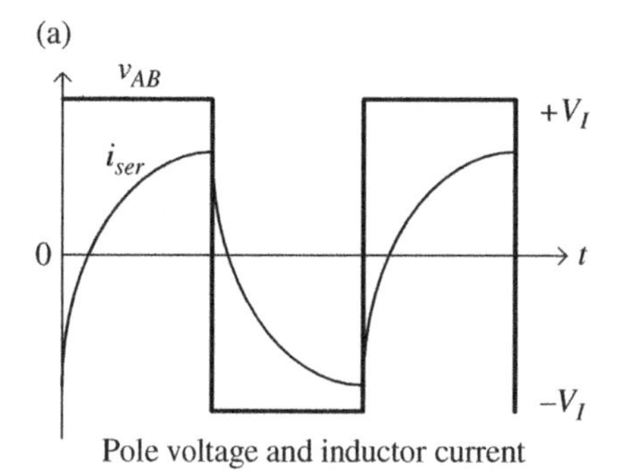

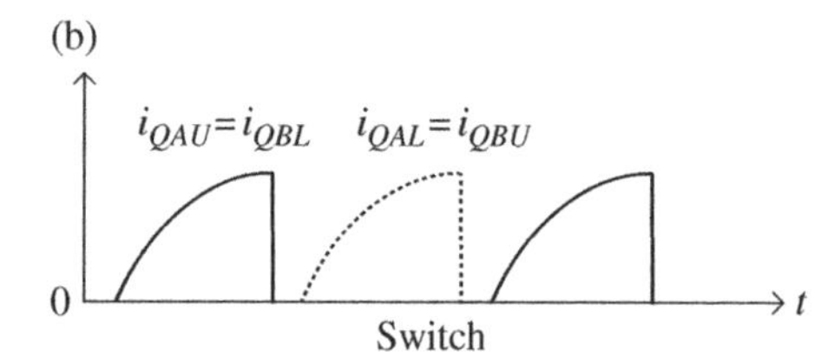

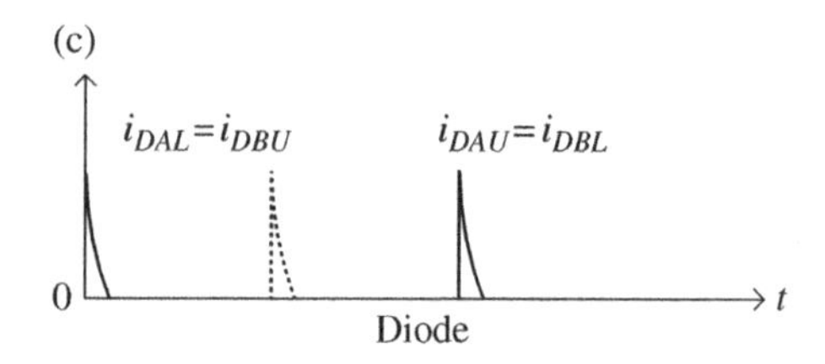

Figure 12.13 Pole voltage and inductor, switch, and diode currents.

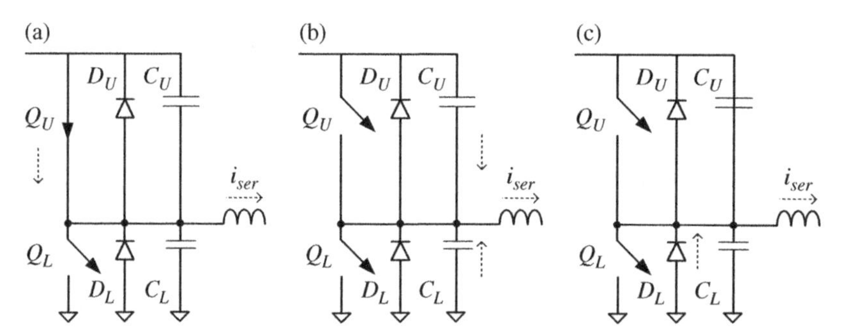

Figure 12.14 ZVS turn-off of the switch.

maintaining the switch voltage low while it turns off. The capacitors C_U and C_L charge and discharge, respectively, with time until the lower diode D_L is forward-biased and conducts the inductor current.

The inductor current flows into the cable and into the primary of the coupling transformer. A portion of the current charges and discharges the parallel capacitor, with the balance of the current flowing into the output filter capacitor when the diodes conduct.

The resonant converter of Figure 12.12 can be simplified to the more basic diagram of Figure 12.15, where the inverting full bridge is represented by a simple square wave. The

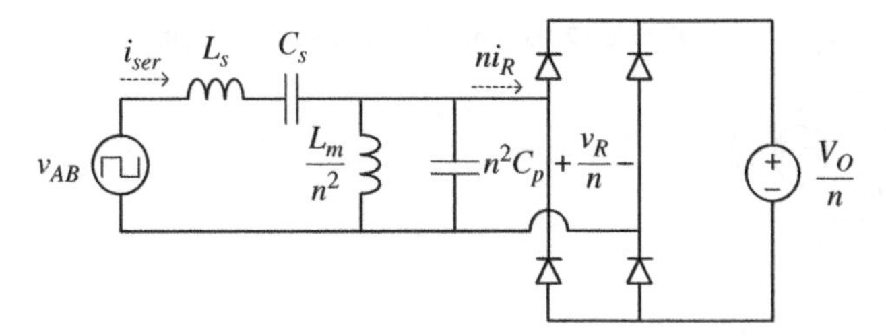

Figure 12.15 Simplified equivalent circuit.

secondary components are shown reflected to the primary. Variables i_R and v_R represent the current into and the voltage across the rectifier.

In this circuit, the equivalent series tank inductance L_s is given by

$$L_s = L_{os} + L_{lp} + \frac{L_{ls}}{n^2} \tag{12.45}$$

12.4.2.2 Design Considerations

There are a number of design considerations in selecting the various components.

The natural frequency of the series-resonant tank is given by

$$f_{ser} = \frac{1}{2\pi\sqrt{L_s C_s}} \tag{12.46}$$

The natural frequency of the parallel resonant tank is given by

$$f_{par} = \frac{1}{2\pi\sqrt{L_m C_p}} \tag{12.47}$$

The following two considerations apply:

i) The natural frequency of the parallel resonant tank should be less than the minimum switching frequency $f_{s(min)}$ in order to ensure a voltage gain for the power converter.
ii) The natural frequency of the equivalent series-resonant tank should be less than the minimum switching frequency and less than the natural frequency of the parallel resonant tank in order to ensure a zero-current turn-on of the switch.

Thus:

$$f_{ser} < f_{par} < f_{s(min)} \tag{12.48}$$

12.4.3 Fundamental-Mode Analysis and Current-Source Operation

Analysis of the LCLC converter can be quite complex [6, 7]. A simplified analysis, known as fundamental-mode analysis, can be used to explore the operation of the circuit at the switching or fundamental frequency [7]. The analysis is based on phasors rather than on time-domain analysis. In terms of phasors, let the impedances of the series and parallel tanks be

$$\mathbf{Z}_{ser} = j\omega L_s + \frac{1}{j\omega C_s} = j\left(\omega L_s - \frac{1}{\omega C_s}\right) \tag{12.49}$$

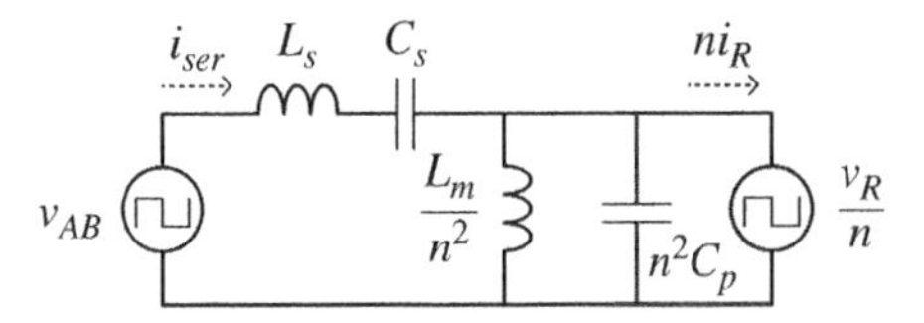

Figure 12.16 Simplified equivalent circuit.

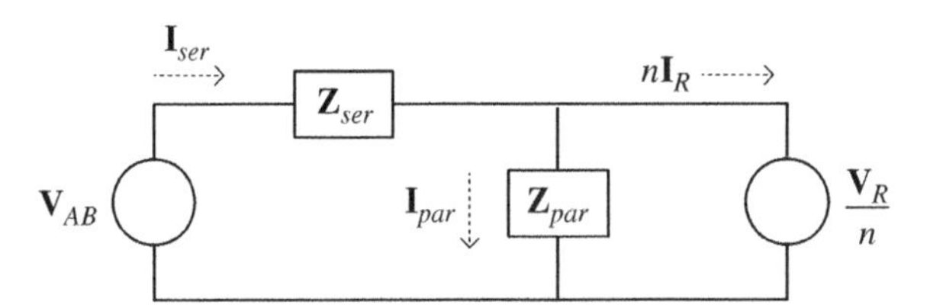

Figure 12.17 Equivalent circuit at fundamental frequency.

and

$$\mathbf{Z}_{par} = \frac{1}{\frac{1}{j\omega L_m / n^2} + \frac{1}{1/j\omega n^2 C_p}} = \frac{1}{j\omega n^2 C_p + \frac{n^2}{j\omega L_m}} = \frac{1}{jn^2\left(\omega C_p - \frac{1}{\omega L_m}\right)} \tag{12.50}$$

where ω is the switching frequency, and j is the imaginary operator.

A key characteristic of the series-parallel resonant converter is current-source operation. Current-source operation can be explained using the analysis in this section. First, Figure 12.15 is redrawn to represent the rectifier and output as an equivalent voltage source v_R as shown in Figure 12.16.

The circuit can then be redrawn as the simple circuit shown in Figure 12.17. The use of boldface emphasizes that the variable is a phasor with a magnitude and phase.

From Kirchhoff's voltage law:

$$\mathbf{V}_{AB} = \mathbf{Z}_{ser}\mathbf{I}_{ser} + \frac{\mathbf{V}_R}{n} \tag{12.51}$$

which can be expressed in terms of the parallel and rectifier currents as

$$\mathbf{V}_{AB} = \mathbf{Z}_{ser}\left(n\mathbf{I}_R + \mathbf{I}_{par}\right) + \frac{\mathbf{V}_R}{n} \tag{12.52}$$

since

$$\mathbf{I}_{ser} = n\mathbf{I}_R + \mathbf{I}_{par} \tag{12.53}$$

But

$$\mathbf{I}_{par} = \frac{\mathbf{V}_R}{n\mathbf{Z}_{par}} \tag{12.54}$$

Substituting Equation (12.54) into Equation (12.52) gives

$$\begin{aligned}\mathbf{V}_{AB} &= \mathbf{Z}_{ser}n\mathbf{I}_R + \mathbf{Z}_{ser}\frac{\mathbf{V}_R}{n\mathbf{Z}_{par}} + \frac{\mathbf{V}_R}{n} \\ &= n\mathbf{Z}_{ser}\mathbf{I}_R + \frac{\mathbf{V}_R}{n}\left(1 + \frac{\mathbf{Z}_{ser}}{\mathbf{Z}_{par}}\right)\end{aligned} \tag{12.55}$$

At the current-source frequency ω_{cs}, the following condition holds:

$$\frac{\mathbf{Z}_{ser}}{\mathbf{Z}_{par}} = -1 \tag{12.56}$$

At this frequency, the rectifier current is only a function of the input voltage and the series tank impedance:

$$\mathbf{I}_R = \frac{\mathbf{V}_{AB}}{n\mathbf{Z}_{ser}} \tag{12.57}$$

and is independent of the magnitude of the output voltage. Thus, the circuit acts as a current source.

The current-source frequency is determined by solving

$$\frac{\mathbf{Z}_{ser}}{\mathbf{Z}_{par}} = \frac{j\left(\omega_{cs}L_s - \dfrac{1}{\omega_{cs}C_s}\right)}{\dfrac{1}{jn^2\left(\omega_{cs}C_p - \dfrac{1}{\omega_{cs}L_m}\right)}} = -1 \tag{12.58}$$

which is the solution of a quadratic equation.

However, we now take a shortcut and estimate an approximate current-source frequency by assuming that the series inductance and parallel capacitance are the significant components at the current-source frequency. Thus:

$$\mathbf{Z}_{ser} \approx j\omega_{cs}L_s \text{ and } \mathbf{Z}_{par} \approx \frac{1}{j\omega_{cs}n^2C_p} \tag{12.59}$$

which results in the following simple solution for the current-source frequency:

$$\omega_{cs} \approx \frac{1}{n\sqrt{L_sC_p}} \tag{12.60}$$

and

$$f_{cs} \approx \frac{1}{2\pi n\sqrt{L_sC_p}} \tag{12.61}$$

We can estimate the output current at the current-source frequency by making some further simplifying assumptions.

First, we assume that the full-bridge inverter voltage v_{AB} can be represented by the rms value of the fundamental frequency of the square wave of magnitude V_I, as shown in Figure 12.18(a).

Using Fourier analysis, it can be shown that the relationship between the rms value of the fundamental $V_{AB(rms)}$ and the square wave of magnitude V_I is given by

$$V_{AB(rms)} = \frac{2\sqrt{2}}{\pi}V_I \tag{12.62}$$

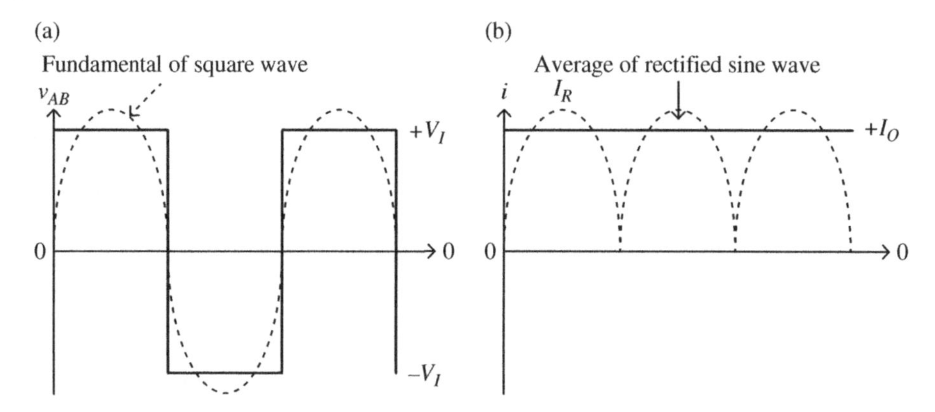

Figure 12.18 Simplified inverter and rectifier waveforms.

Second, we assume that the value of the output current I_O is the average of a rectified sine wave of rms value $I_{R(rms)}$, as shown in Figure 12.18(b), and is simply related by

$$I_O = \frac{2\sqrt{2}}{\pi} I_{R(rms)} \tag{12.63}$$

Substituting Equation (12.62) and Equation (12.63) into Equation (12.57) enables us to derive a relationship between the dc input voltage and the dc output current at the current-source frequency. Therefore:

$$I_{R(rms)} = \frac{V_{AB(rms)}}{nZ_{ser}(\omega_{cs})} = \frac{\frac{2\sqrt{2}}{\pi} V_I}{nZ_{ser}(\omega_{cs})} = \frac{\pi}{2\sqrt{2}} I_O \tag{12.64}$$

Thus:

$$I_O = \frac{8}{\pi^2} \frac{V_I}{nZ_{ser}(\omega_{cs})} \tag{12.65}$$

Note that the magnitude of the series impedance at a particular frequency is given by

$$Z_{ser}(\omega) = \left| \omega L_s - \frac{1}{\omega C_s} \right|$$

12.4.3.1 Example

The following specifications are representative values for the SAE J1773 inductive charging standard: $L_m = 45\ \mu H$, $C_p = 40\ nF$, $L_{os} = 17\ \mu H$, $L_{lp} = L_{ls} = 1\ \mu H$, $C_s = 0.33\ \mu F$, and $n = 1$.

Determine the series, parallel, and current-source frequencies.

Determine the approximate output current and output power at the current-source frequency when $V_I = 380$ V and $V_O = 400$ V.

Solution:

The equivalent series inductance is

$$L_s = L_{os} + L_{lp} + \frac{L_{ls}}{n^2} = 17\,\mu\text{H} + 1\,\mu\text{H} + \frac{1}{1^2} 1\,\mu\text{H} = 19\,\mu\text{H}$$

$$f_{ser} = \frac{1}{2\pi\sqrt{L_s C_s}} = \frac{1}{2\pi\sqrt{19 \times 10^{-6} \times 0.33 \times 10^{-6}}}\,\text{Hz} = 63.6\,\text{kHz}$$

$$f_{par} = \frac{1}{2\pi\sqrt{L_m C_p}} = \frac{1}{2\pi\sqrt{45 \times 10^{-6} \times 40 \times 10^{-9}}}\,\text{Hz} = 118.6\,\text{kHz}$$

$$f_{cs} \approx \frac{1}{2\pi n\sqrt{L_s C_p}} = \frac{1}{2\pi\sqrt{19 \times 10^{-6} \times 40 \times 10^{-9}}}\,\text{Hz} = 182.6\,\text{kHz}$$

The magnitude of the series impedance at the current-source frequency is given by

$$Z_{ser}(\omega_{cs}) = \left|\omega_{cs} L_s - \frac{1}{\omega_{cs} C_s}\right| = \left|2\pi \times 182.6 \times 10^3 \times 19 \times 10^{-6} - \frac{1}{2\pi \times 182.6 \times 10^3 \times 0.33 \times 10^{-6}}\right|\Omega$$

$$= 19.14\,\Omega$$

Thus, the dc output current is

$$I_O = \frac{8}{\pi^2}\frac{nV_I}{Z_{ser}(\omega_{cs})} = \frac{8}{\pi^2}\frac{1 \times 380}{19.14}\,\text{A} = 16.09\,\text{A}$$

and the output power is

$$P_O = V_O I_O = 400 \times 16.09\,\text{W} = 6.436\,\text{kW}$$

12.4.4 Simulation

The analytical modeling of the LCLC topology can be complex [1–7]. The converter can be modeled using any circuit simulator. A simplified Simulink circuit model is as shown in Figure 12.19.

The circuits waveforms for the input and rectifier voltages and the series tank current are presented in Figure 12.20.

The circuit is simulated for a dc link voltage of 380 V and two output voltages of 200 V and 400 V. The results of the simulation versus frequency are presented in Figure 12.21 for (a) 200 V and (b) 400 V. The output current, and the resulting power, can be controlled by increasing or decreasing the switching frequency. The series inductor current remains high even when the output current is low. Maintaining the series tank current high is necessary in order to ensure ZVS of the switches.

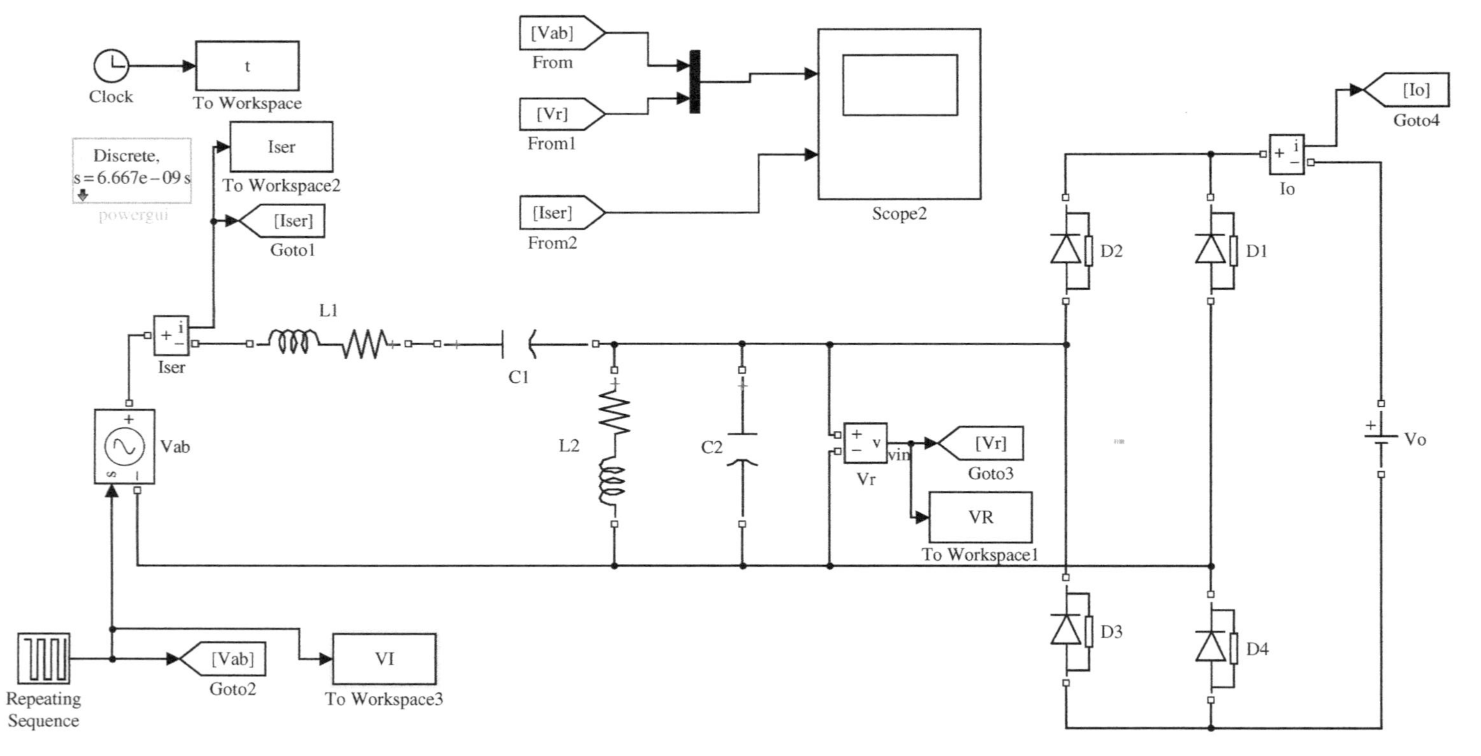

Figure 12.19 Simple Simulink LCLC circuit model

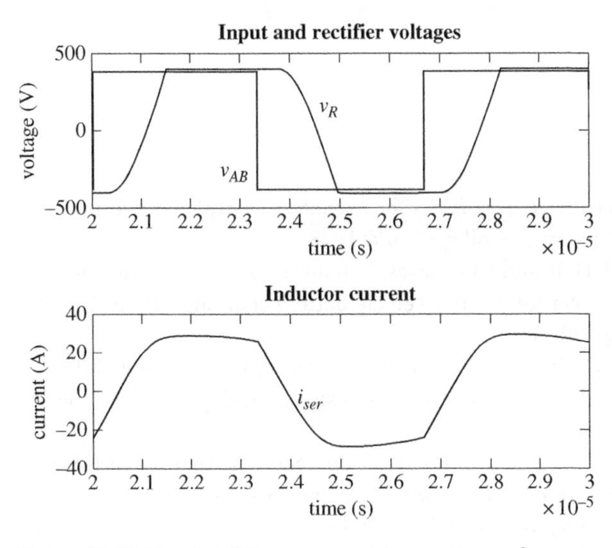

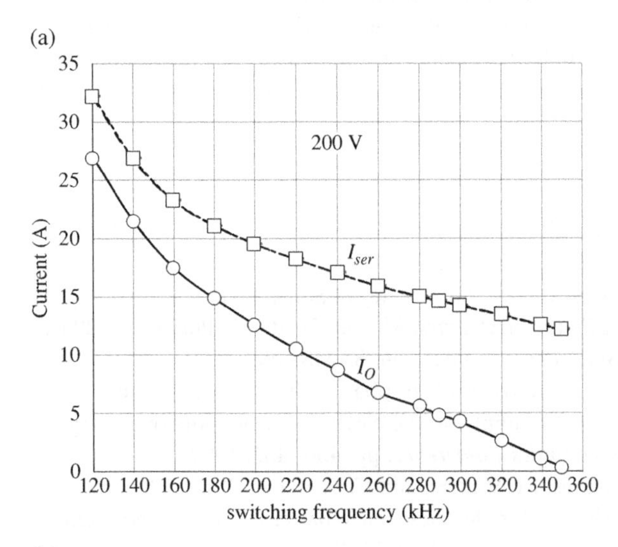

Figure 12.20 Simple LCLC circuit model current waveforms.

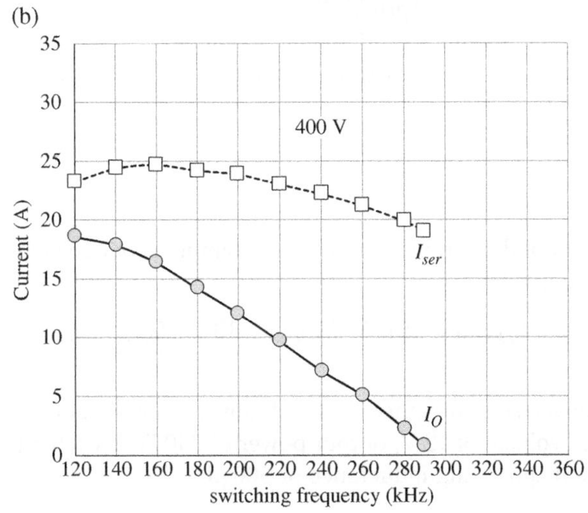

Figure 12.21 Rms series inductor current and dc output current at output voltages of (a) 200 V and (b) 400 V.

References

1 J. G. Hayes, *Resonant Power Conversion Topologies for Inductive Charging of Electric Vehicle Batteries*, PhD Thesis, University College Cork, 1998.
2 R. Severns, E. Yeow, G. Woody, J. Hall, and J. G. Hayes, "An ultra-compact transformer for a 100 W to 120 kW inductive coupler for electric vehicle battery charging," *IEEE Applied Power Electronics Conference*, pp. 32–38, 1996.
3 J. M. Leisten, "LLC design for UCC29950," Texas Instruments application note, 2015.
4 TI staff, "Phase-shifted full bridge dc/dc power converter design guide," Texas Instruments application note, 2014.
5 J. G. Hayes, N. O'Donovan, and M. G. Egan, "Inductance characterization of high-leakage transformers," *IEEE Applied Power Electronics Conference*, pp. 1150–1156, 2003.
6 J. G. Hayes, M. G. Egan, J. M. D. Murphy, S. E. Schulz, and J. T. Hall, "Wide load resonant converter supplying the SAE J-1773 electric vehicle inductive charging interface," *IEEE Transactions on Industry Applications*, pp. 884–895, July–August 1999.
7 J. G. Hayes and M. G. Egan, "Rectifier-compensated fundamental mode approximation analysis of the series-parallel LCLC family of resonant converters with capacitive output filter and voltage-source load," *IEEE Power Electronics Specialists Conference*, pp. 1030–1036, 1999.

Further Reading

1 N. Mohan, T. M. Undeland, and W. P. Robbins, *Power Electronics Converters, Applications and Design*, 3rd edition, John Wiley & Sons, 2003.
2 N. Mohan, *Power Electronics: A First Course*, John Wiley & Sons, 2012.
3 R. W. Erickson, *Fundamentals of Power Electronics*, Kluwer Academic Publishers, 2000.
4 R. E. Tarter, *Principles of Solid-State Power Conversion*, SAMS, 1985.
5 J. G. Hayes and M. G. Egan, "A comparative study of phase-shift, frequency, and hybrid control of the series resonant converter supplying the electric vehicle inductive charging interface," *IEEE Applied Power Electronics Conference*, pp. 450–457, 1999.
6 M. G. Egan, D. O'Sullivan, J. G. Hayes, M. Willers, and C. P. Henze, "Power-factor-corrected single-stage inductive charger for electric-vehicle batteries," *IEEE Transactions on Industrial Electronics*, 54 (2), pp. 1217–1226, April 2007.
7 R. Radys, J. Hall, J. Hayes, and G. Skutt, "Optimizing AC and DC winding losses in ultra-compact high-frequency power transformers," *IEEE Applied Power Electronics Conference*, pp. 1188–1195, 1999.

Problems

12.1 Determine the rms component of the various currents for Example 12.2.1.1 when the input voltage increases to 400 V.

[Ans. $I_{Lo(rms)} = 15.01$ A, $I_{s(rms)} = I_{D2(rms)} = 7.51$ A, $I_{D1(rms)} = 13$ A, $I_{p(rms)} = I_{Q(rms)}$ $=1.031$ A, $I_{T(rms)} = 0.144$ A]

12.2 A single-switch forward converter is designed to the following specifications: input voltage of 42 V, output voltage of 12 V, output power of 240 W, switching frequency of 200 kHz, and a magnetizing inductance of 500 μH.

i) Size the output inductor for a peak-to-peak current ripple of 10% at full load.
ii) Determine the various component currents at full load.
iii) Determine reasonable rated voltages, rounded up to the nearest 10 V, for the switch and output diodes if the input voltage can go as high as 60 V. Allow for overshoots of 10 V on both the switch and diodes.
iv) Determine the AP for the transformer. Let copper fill factor $k = 0.5$, current density $J_{cu} = 6\text{A/mm}^2$, and maximum core flux density $B_{max} = 20$ mT.

Assume a turns ratio of unity between the primary and tertiary windings and a maximum duty cycle of 50% at 42 V.

Ignore all losses, device voltage drops, and the effects of leakage.

[Ans. $I_{o(dc)} = 20$ A $I_{Lo(rms)} = 20.01$ A, $I_{s(rms)} = I_{D2(rms)} = 14.15$ A, $I_{D2(dc)} = 10$ A, $I_{D1(rms)} = 14.15$ A, $I_{D1(dc)} = 10$ A, $I_{p(rms)} = I_{Q(rms)} = 8.16$ A, $I_{T(rms)} = 0.086$ A, 170 V, 60 V, AP = 1.42 cm^4]

12.3 Determine the various primary-side currents for the full-bridge converter of Example 12.3.2.1 when the battery voltage decreases to 200 V.

[Ans: $I_{p(rms)} = 23.58$ A, $I_{QA(rms)} = 16.68$ A, $I_{QA(dc)} = 8.13$; A, $I_{QB(rms)} = 16.67$ A, $I_{QB(dc)} = 7.89$ A, $I_{DB(dc)} = 0.24$ A]

12.4 Recalculate the various currents for the full-bridge Example 12.3.2.1, but simplify the analysis by assuming that the magnetizing current is negligible and can be ignored.

[Ans: $I_{o(dc)} = 15$ A, $I_{Lo(rms)} = 15.01$ A, $I_{Do(rms)} = 10.34$ A, $I_{Do(dc)} = 7.5$ A, $I_{p(rms)} = 16.65$ A, $I_{QA(rms)} = 11.77$ A, $I_{QA(dc)} = 7.89$ A, $I_{QB(rms)} = 11.77$ A, $I_{QB(dc)} = 7.89$ A, $I_{DB(dc)} = 0$ A]

12.5 A full-bridge converter with a center-tapped rectifier is designed to the following specifications: input voltage of 380 V, output voltage of 15 V, output power of 2.4 kW, switching frequency of 100 kHz, and a magnetizing inductance of 1 mH.

i) Size the output inductor for a peak-to-peak current ripple of 10% at full load.
ii) Determine the primary and secondary rms currents at full load.
iii) Determine the AP for the transformer. Let the copper fill factor $k = 0.5$, current density $J_{cu} = 6\text{A/mm}^2$, and maximum core flux density $B_{max} = 20$ mT.

Assume a maximum duty cycle of 0.45 at 15 V output.

Ignore all losses, device voltage drops, and the effects of leakage.

Note that, as there are two identical secondaries, the AP is given by

$$AP = \frac{DV_I\left(I_p + nI_S \times 2\right)}{2f_s B_{max} k_{cu} J_{cu}}$$

[Ans. $L_o = 0.4688$ μH, $I_{s(rms)} = 110.3$ A, $I_{p(rms)} = 6.7$ A, AP = 23.3 cm^4]

12.6 Determine the currents for the preceding problem when the output voltage decreases to 12 V.

[Ans: $I_{s(rms)} = 131.3$ A, $I_{p(rms)} = 7.49$ A]

12.7 For the LCLC converter of Section 12.4.3.1, determine the approximate output current and output power at the current-source frequency when the input voltage is 200 V and the output voltage is 400 V.

[Ans. 8.47 A, 3.39 kW]

Assignments

12.1 Verify the waveforms and answers for the various examples and problems using a circuit simulator of your choice.

Appendix I: RMS and Average Values of Ramp and Step Waveforms

It can easily be shown that the rms and average values of a periodic discontinuous ramp waveform are given by

$$I_{rms}(\text{ramp}) = \sqrt{\frac{D_{ramp}}{3}\left(I_{min}{}^2 + I_{min}I_{max} + I_{max}{}^2\right)} \tag{A12-I.1}$$

and

$$I_{dc} = D_{ramp}\left(\frac{I_{min} + I_{max}}{2}\right) \tag{A12-I.2}$$

where I_{min} and I_{max} are the minimum and maximum currents, respectively, and D_{ramp} is the duty cycle of the ramp. These equations can be applied to all the current waveforms in Figure 12.6(b)–(g).

Similarly, the rms and average values of a periodic discontinuous step waveform are given by

$$I_{rms}(\text{step}) = \sqrt{D_{step}}I_{step} \tag{A12-I.3}$$

and

$$I_{dc}(\text{step}) = D_{step}I_{step} \tag{A12-I.4}$$

where I_{step} is the amplitude of the step current, and D_{step} is the duty cycle of the step.

If the waveform features a combination of discontinuous periodic ramps, as in Figure 12.10(b) and (c), then the rms and average of the waveforms are

$$I_{rms}(\text{ramp}_1 + \text{ramp}_2) = \sqrt{\frac{D_{ramp1}}{3}\left(I_{min1}{}^2 + I_{min1}I_{max1} + I_{max1}{}^2\right) + \frac{D_{ramp2}}{3}\left(I_{min2}{}^2 + I_{min2}I_{max2} + I_{max2}{}^2\right)} \tag{A12-I.5}$$

and

$$I_{dc}(\text{ramp}_1 + \text{ramp}_2) = D_{ramp1}\left(\frac{I_{min1} + I_{max1}}{2}\right) + D_{ramp2}\left(\frac{I_{min2} + I_{max2}}{2}\right) \tag{A12-I.6}$$

If the waveform features a combination of a discontinuous periodic ramp and a step, as in Figure 12.10(f), (g), and (k), then the rms and average of the waveforms are

$$I_{rms}(\text{ramp} + \text{step}) = \sqrt{\frac{D_{ramp}}{3}\left(I_{min}^2 + I_{min}I_{max} + I_{max}^2\right) + D_{step}I_{step}^2} \qquad \text{(A12-I.7)}$$

and

$$I_{dc}(\text{ramp} + \text{step}) = D_{ramp}\left(\frac{I_{min} + I_{max}}{2}\right) + D_{step}I_{step} \qquad \text{(A12-I.8)}$$

Appendix II: Flyback Converter

The transformer-coupled buck-boost converter is known as the flyback converter and is the commonly used switch-mode power converter for low-power applications. A simplified flyback converter is shown in Figure 12.22. The transformer acts to store energy in the primary winding and then release the energy to the secondary. Note that when the primary is energized, there is no secondary current, and vice versa. The inductive operation can be viewed as representing coupled-inductor action rather than transformer action.

The CCM duty cycle for the buck-boost converter is given by

$$D = \frac{V_{out}}{nV_{in} + V_{out}} \qquad \text{(A12-II.1)}$$

and the voltage gain is given by

$$\frac{V_{out}}{V_{in}} = n\frac{D}{1-D} \qquad \text{(A12-II.2)}$$

Figure 12.22 Flyback converter.

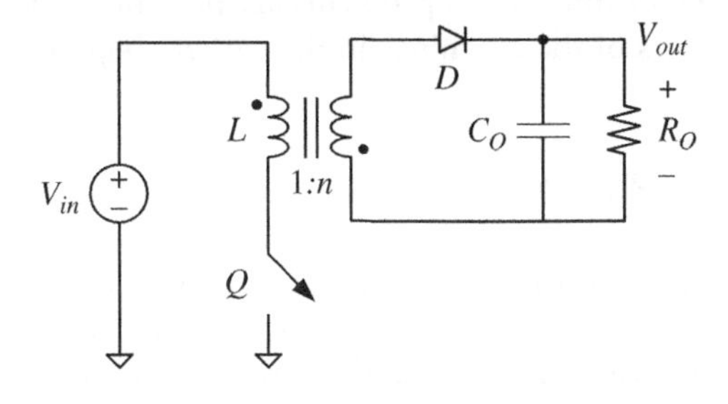

13

Traction Drives and Three-Phase Inverters

In this chapter, three-phase inverters are introduced and analyzed. The function of the inverter is to convert the dc voltage and current provided by the dc link, battery, or fuel cell to the ac voltages and currents required to power the ac machines. As discussed in the earlier chapters on the ac induction and permanent-magnet machines, the ac waveforms supplied by the inverter to the machine require variable voltages, currents, and frequencies in order to optimally drive the machines.

The chapter initially investigates the basic sinusoidal pulse-width-modulated (SPWM) inverter. The voltages and currents supplied by the SPWM inverter are sinusoidal at the fundamental frequency of operation. A third harmonic of the fundamental frequency can be added to the control waveforms to increase the voltage achievable from an inverter and decrease the current for a given power level. Modern power inverters are digitally modulated and controlled using a technique known as space-vector modulation (SVM), with results similar to the third harmonic addition.

Equations are derived for the rms and dc values of the currents in the inverter switches and diodes. The conduction and switching losses of the switches and diodes are then calculated, allowing the inverter efficiency and the semiconductor hot-spot temperatures to be estimated. Examples and problems are presented based on the voltages, currents, and power factors of the machines in the earlier chapters.

13.1 Three-Phase Inverters

The power stage of the three-phase inverter comprises six switches (Q_{UA}, Q_{UB}, Q_{UC}, Q_{LA}, Q_{LB}, and Q_{LC}) paired with six diodes (D_{UA}, D_{UB}, D_{UC}, D_{LA}, D_{LB}, and D_{LC}), as shown in Figure 13.1. Subscripts U and L stand for "upper" and "lower," respectively. Subscripts A, B, and C and a, b, and c represent each of the three phases.

Views of the 2010 Toyota Prius boost and inverter silicon switches were shown earlier in Chapter 11, Figure 11.3. For this vehicle, the power semiconductors are all packaged together into one module. Each motor pole features two semiconductor dies in parallel for each IGBT and diode, as shown in Figure 11.3, while each generator pole features one die for each IGBT and diode. The bond wires connecting the silicon die to the main connector bus bars are shown in the figure.

Electric Powertrain: Energy Systems, Power Electronics and Drives for Hybrid, Electric and Fuel Cell Vehicles, First Edition. John G. Hayes and G. Abas Goodarzi.

Companion website: www.wiley.com/go/hayes/electricpowertrain

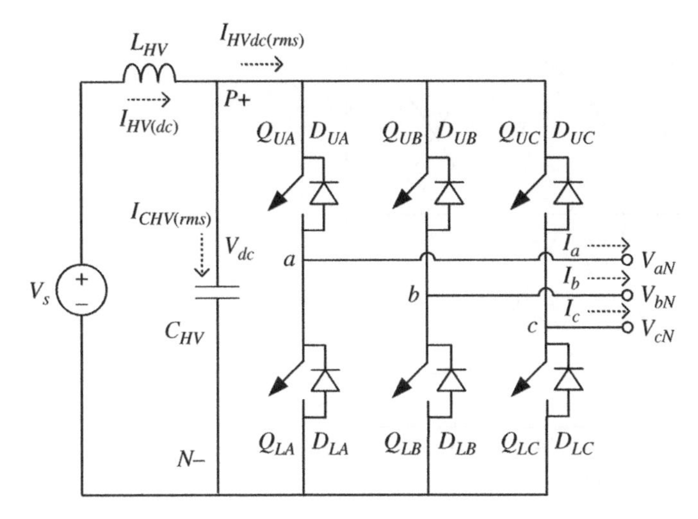

Figure 13.1 Three-phase inverter.

The machine is supplied with sine-wave voltages which can have significant harmonic content and distortion. Simple sinusoidal waveforms without distortion are shown in Figure 13.2. The amplitude of the voltage is limited by the available dc voltage of the battery pack, fuel cell, or power converter.

In the modern drive, the machine is driven with a quasi-sinusoidal voltage waveform output from a current-controlled voltage-source inverter, as shown in Figure 13.3(a).

A **pole** is a term often used to refer to a single leg of an inverter. The output of the inverter pole can range from 0 V to the dc link voltage V_{dc}. The desired sinusoidal voltage can be created within these limits, as shown in Figure 13.2(a). The output voltage from the pole has a dc bias of half the dc link voltage, or $V_{dc}/2$. The three pole voltages, v_{aN}, v_{bN}, and v_{cN}, are with respect to the negative dc link (*N*). When the three pole voltages are applied across the three-phase motor, the net effect is that the dc bias of the pole voltages cancel out, and the line or phase voltages supplied to the machine are ac only without a dc bias, as shown in Figure 13.2(b). The three-phase voltages, v_{an}, v_{bn}, and v_{cn}, are with respect to the machine neutral (*n*).

The equivalent three-phase circuit is shown in Figure 13.3(b), where the motor is represented by the three balanced loads of impedance *Z*.

13.2 Modulation Schemes

There are four basic modulation schemes for supplying ac voltages from the inverter. These are (1) sinusoidal modulation; (2) SVM, a commonly used advanced digital method; (3) sinusoidal modulation with a third harmonic addition, with similar results to SVM; and (4) overmodulation, with the limiting waveform being a square wave rather than a sinusoid.

We will briefly note the maximum ac voltages which the inverter can output with these schemes.

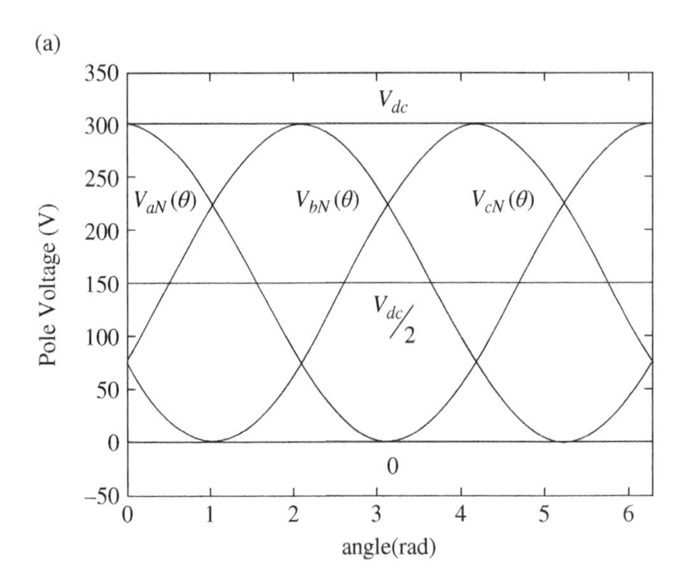

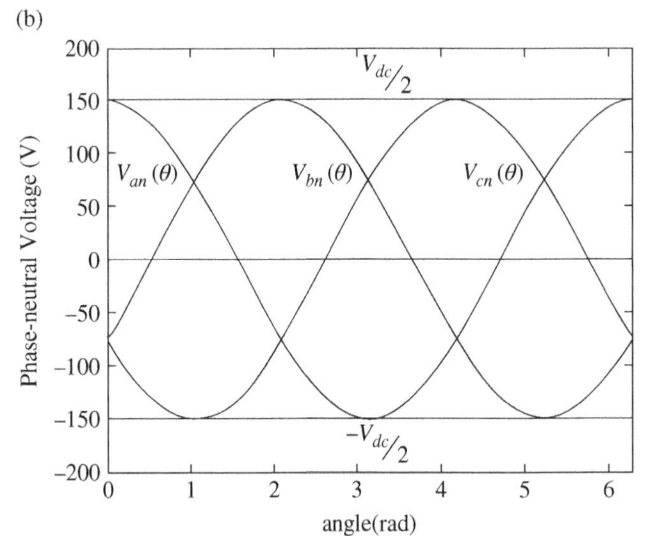

Figure 13.2 Sinusoidal (a) pole and (b) phase voltages.

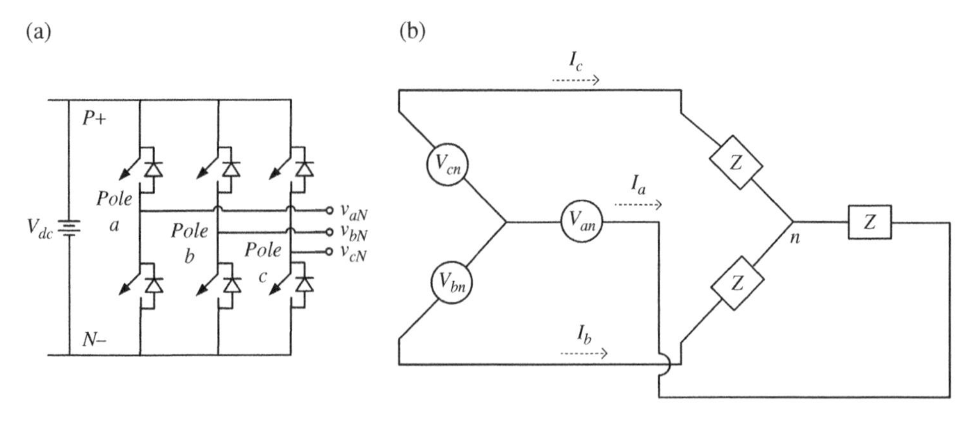

Figure 13.3 (a) PWM inverter and (a) ac three-phase circuit.

13.2.1 Sinusoidal Modulation

The pole output voltage is sinusoidal with a dc level equal to half the dc link voltage, as shown in Figure 13.2(a). At its minimum and maximum values, the pole voltage is oscillating sinusoidally between the dc negative at 0 V and the dc positive at $+V_{dc}$. Thus, the pole voltages for poles a, b, and c are

$$v_{aN}(\theta) = \frac{V_{dc}}{2} + \sqrt{2}V_{ph}\cos(\theta) \tag{13.1}$$

$$v_{bN}(\theta) = \frac{V_{dc}}{2} + \sqrt{2}V_{ph}\cos\left(\theta - \frac{2}{3}\pi\right) \tag{13.2}$$

$$v_{cN}(\theta) = \frac{V_{dc}}{2} + \sqrt{2}V_{ph}\cos\left(\theta - \frac{4}{3}\pi\right) \tag{13.3}$$

where V_{ph} is the rms value of the phase voltage and angle $\theta = \omega t$.

In a balanced three-phase system, the neutral of the machine is electrically at half the dc link voltage. Thus, the neutral voltage is

$$V_{nN} = \frac{V_{dc}}{2} \tag{13.4}$$

The phase-neutral voltage, as shown in Figure 13.2(b), is the difference between the pole voltage and the neutral:

$$v_{an}(\theta) = v_{aN}(\theta) - V_{nN} = \sqrt{2}V_{ph}\cos(\theta) \tag{13.5}$$

$$v_{bn}(\theta) = v_{bN}(\theta) - V_{nN} = \sqrt{2}V_{ph}\cos\left(\theta - \frac{2}{3}\pi\right) \tag{13.6}$$

$$v_{cn}(\theta) = v_{cN}(\theta) - V_{nN} = \sqrt{2}V_{ph}\cos\left(\theta - \frac{4}{3}\pi\right) \tag{13.7}$$

The peak value of the phase-to-neutral voltage equals $\sqrt{2}\,V_{ph}$ and is thus limited to half of the dc link voltage:

$$\sqrt{2}V_{ph} = \frac{V_{dc}}{2} \tag{13.8}$$

The maximum rms value of the phase-to-neutral voltage V_{ph}, when using basic sinusoidal modulation, is given by

$$V_{ph}(\text{maximum rms}) = \frac{V_{dc}}{2\sqrt{2}} = 0.354\,V_{dc} \tag{13.9}$$

The line-line voltages can be determined using basic phasor analysis. Taking the phase a voltage as the reference, the three line-line voltages, v_{ab}, v_{bc}, and v_{ca}, can be determined by multiplying the phase voltage by $\sqrt{3}$ and phase-shifting to get the resultant:

$$v_{ab}(\theta) = \sqrt{6}V_{ph}\cos\left(\theta + \frac{\pi}{6}\right) \tag{13.10}$$

$$v_{bc}(\theta) = \sqrt{6}V_{ph}\cos\left(\theta - \frac{\pi}{2}\right) \tag{13.11}$$

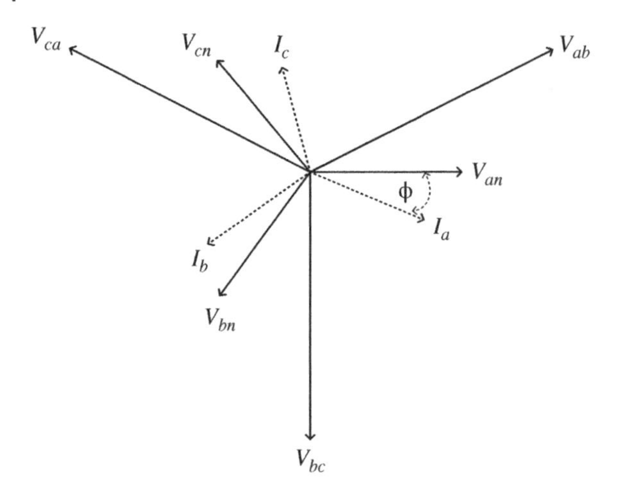

Figure 13.4 Phasor diagram showing phase and line voltages.

$$v_{ca}(\theta) = \sqrt{6}V_{ph}\cos\left(\theta - \frac{5\pi}{6}\right) \tag{13.12}$$

The various phase and line voltages can be represented in polar form as follows:

$$\bar{V}_{an} = V_{ph}\angle 0^\circ,\ \bar{V}_{bn} = V_{ph}\angle -120^\circ,\ \bar{V}_{cn} = V_{ph}\angle -240^\circ \tag{13.13}$$

$$\bar{V}_{ab} = \sqrt{3}V_{ph}\angle 30^\circ,\ \bar{V}_{bc} = \sqrt{3}V_{ph}\angle -90^\circ,\ \bar{V}_{ca} = \sqrt{3}V_{ph}\angle -210^\circ \tag{13.14}$$

A phasor diagram for a balanced three-phase system is shown in Figure 13.4. In a balanced three-phase system, the various phase and line voltages and currents are equal in magnitude and are displaced by 120° with respect to each other.

13.2.2 Sinusoidal Modulation with Third Harmonic Addition

The maximum output voltage from the inverter can be increased by adding a third harmonic while still outputting a sinusoidal phase voltage. SVM is a widely used digital modulation technique which achieves similar voltage levels. Adding the third harmonic to the pole voltage increases the maximum output phase voltage by 1/cos 30° = 1/0.866 = 1.1547 times the maximum value using regular sinusoidal modulation. Figure 13.5(a) shows the pole voltages and the third harmonic v_{3H} which has been added to each.

The pole voltage is given by

$$v_{aN}(\theta) = \frac{V_{dc}}{2} + \frac{1}{\cos 30^\circ}\frac{V_{dc}}{2}\cos(\theta) - \left(\frac{1}{\cos 30^\circ} - 1\right)\frac{V_{dc}}{2}\cos(3\theta) \tag{13.15}$$

The peak value of the pole voltage remains V_{dc}. It can be seen from Figure 13.5(a) that the addition of the third harmonic has effectively increased the rms value of the waveform while maintaining the same maximum and minimum dc levels compared to Figure 13.2(a).

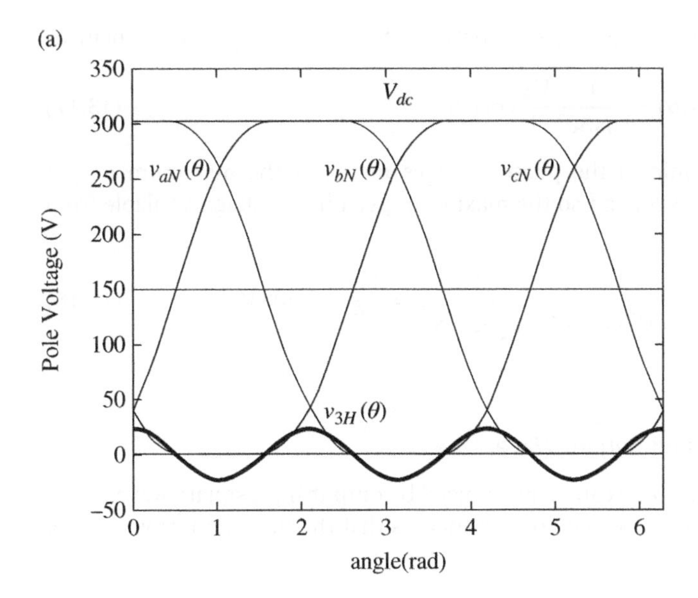

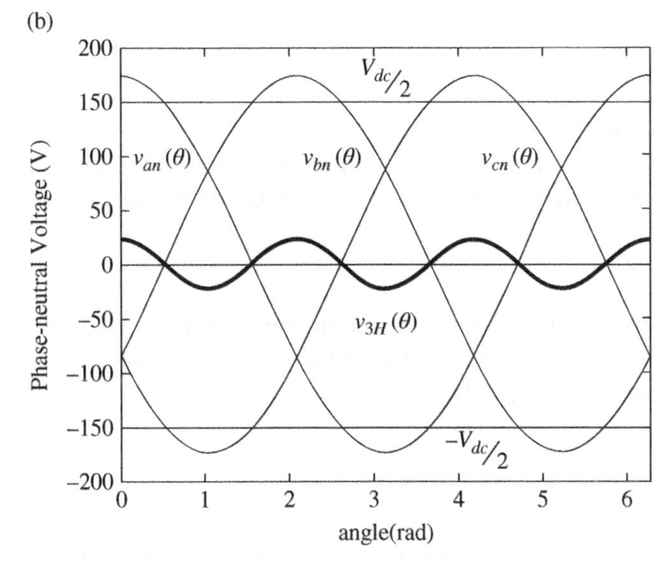

Figure 13.5 Sinusoidal phase voltages with third harmonic addition.

As the same third harmonic is added to each pole, the neutral also carries the same third harmonic. Thus, the neutral is at the level of half the dc link minus the third harmonic component:

$$v_{nN}(\theta) = \frac{V_{dc}}{2} - \left(\frac{1}{\cos 30^\circ} - 1\right)\frac{V_{dc}}{2}\cos(3\theta) \tag{13.16}$$

The phase-neutral voltage is the difference between the pole voltage and the neutral:

$$v_{an}(\theta) = v_{aN}(\theta) - v_{nN}(\theta) = \frac{1}{\cos 30^\circ}\frac{V_{dc}}{2}\cos(\theta) \tag{13.17}$$

Adding the third harmonic to the phase voltages increases the maximum output voltage by 1/cos 30°, or 15.47%, and so the maximum per-phase voltage available from the inverter is

$$V_{ph}(\text{maximum rms}) = \frac{V_{dc}}{2\sqrt{2}\cos 30^\circ} = \frac{V_{dc}}{2\sqrt{2}\times\frac{\sqrt{3}}{2}} = \frac{V_{dc}}{\sqrt{6}} = 0.408\,V_{dc} \tag{13.18}$$

13.2.3 Overmodulation and Square Wave

The phase voltage to the machine can be maximized by outputting a square wave rather than a sinusoid. It can be shown using Fourier analysis that the maximum rms value of the fundamental component of the square wave is given by

$$V_{ph}(\text{maximum rms}) = \frac{\sqrt{2}}{\pi}V_{dc} = 0.45\,V_{dc} \tag{13.19}$$

13.2.3.1 Example: AC Voltages Available from DC Link

What are the maximum rms phase voltages available from the inverter to drive the traction motor of an EV if the battery voltage is at 300 V? Consider the various modulation schemes.

Solution:

The dc link voltage V_{dc} is 300 V. On the basis of this section, the maximum rms phase voltages available from the inverter are 106.2 V, 122.4 V, and 135 V for sinusoidal, space-vector, and square-wave modulations, respectively.

13.3 Sinusoidal Modulation

The modulation and associated duty cycles of a three-phase inverter can be derived based on the earlier derivations of the duty cycles for the buck and boost converters in Chapter 11. Each pole of the three-phase inverter is essentially a half-bridge converter comprising buck and boost converters which enable bidirectional current and power flow to and from the dc link, as shown in Figure 13.6.

We know from the earlier analysis in Chapter 11 that the pole voltage is determined by the duty cycle of the upper switch in the case of buck operation, or by the upper diode in the case of boost operation. Let $d(\theta)$ be the duty cycle of the upper switch or upper diode. Thus, the duty cycle is given by

$$d(\theta) = \frac{v_{pole}(\theta)}{V_{dc}} \tag{13.20}$$

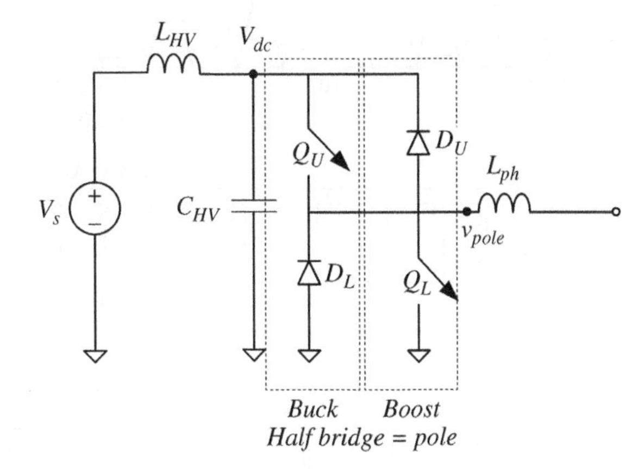

Figure 13.6 Single pole of the three-phase inverter.

Note that $d(\theta)$ varies with time, and so d is used rather than D.
Substitute this in the expression for the desired pole voltage, Equation (13.1), to get

$$d(\theta) = \frac{v_{pole}(\theta)}{V_{dc}} = \frac{1}{V_{dc}}\left\{\frac{V_{dc}}{2} + \sqrt{2}V_{ph}\cos(\theta)\right\} = \frac{1}{2} + \frac{\sqrt{2}V_{ph}}{V_{dc}}\cos(\theta) \tag{13.21}$$

Thus, the three duty cycles for poles a, b, and c are

$$d_a(\theta) = \frac{1}{2} + \frac{\sqrt{2}V_{ph}}{V_{dc}}\cos(\theta) \tag{13.22}$$

$$d_b(\theta) = \frac{1}{2} + \frac{\sqrt{2}V_{ph}}{V_{dc}}\cos\left(\theta - \frac{2}{3}\pi\right) \tag{13.23}$$

$$d_c(\theta) = \frac{1}{2} + \frac{\sqrt{2}V_{ph}}{V_{dc}}\cos\left(\theta - \frac{4}{3}\pi\right) \tag{13.24}$$

The various duty cycles can be determined by intersecting the references with a triangular waveform as shown in Figure 13.7. The Simulink simulation circuit is shown at the end of the chapter. The three duty cycles are as shown in the figure.

The resulting voltages feature a stream of pulses, as shown in Figure 13.8. The top waveforms are two of the three pole voltages, while the third waveform is the pole-pole or line-line voltage. A low-frequency averaging of these voltage waveforms synthesizes a sinusoidal voltage at the fundamental frequency. The resulting currents flowing into the three-phase load or machine are sinusoidal with a varying PWM ripple, as shown in the bottom plots of Figure 13.8.

13.3.1 Modulation Index *m*

The modulation index m is a commonly used term, and is defined as the ratio of the peak phase voltage to half the dc voltage:

$$m = \frac{\sqrt{2}V_{ph}}{\frac{V_{dc}}{2}} = \frac{2\sqrt{2}V_{ph}}{V_{dc}} \tag{13.25}$$

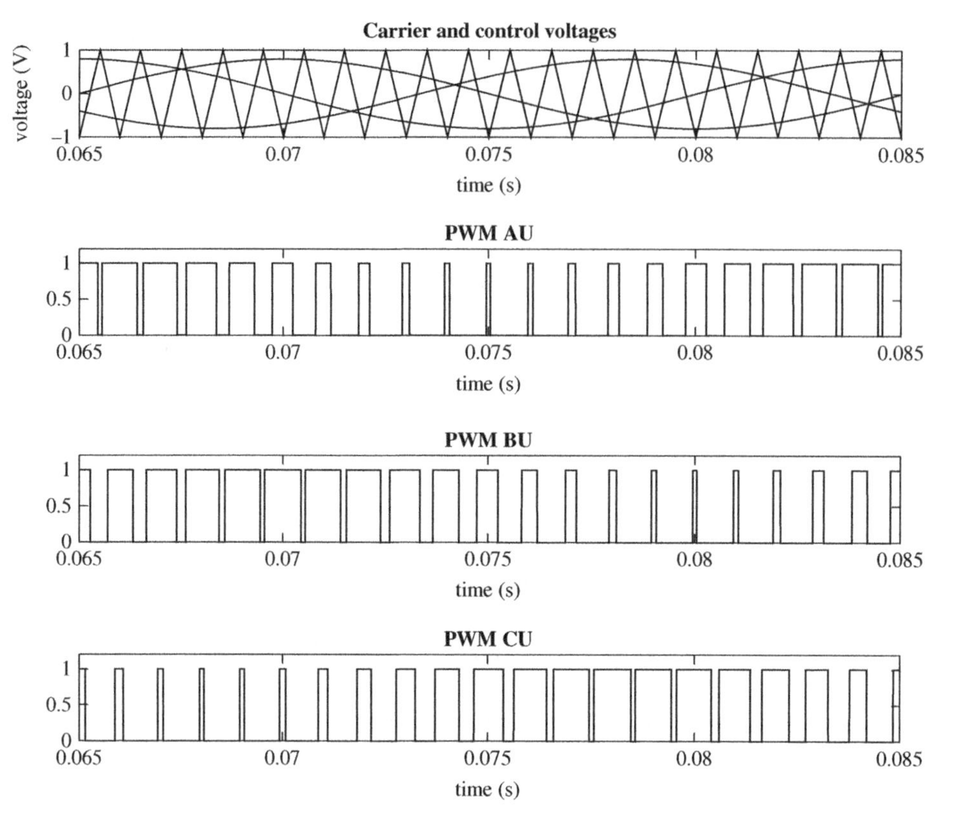

Figure 13.7 Three-phase sinusoidal PWM patterns.

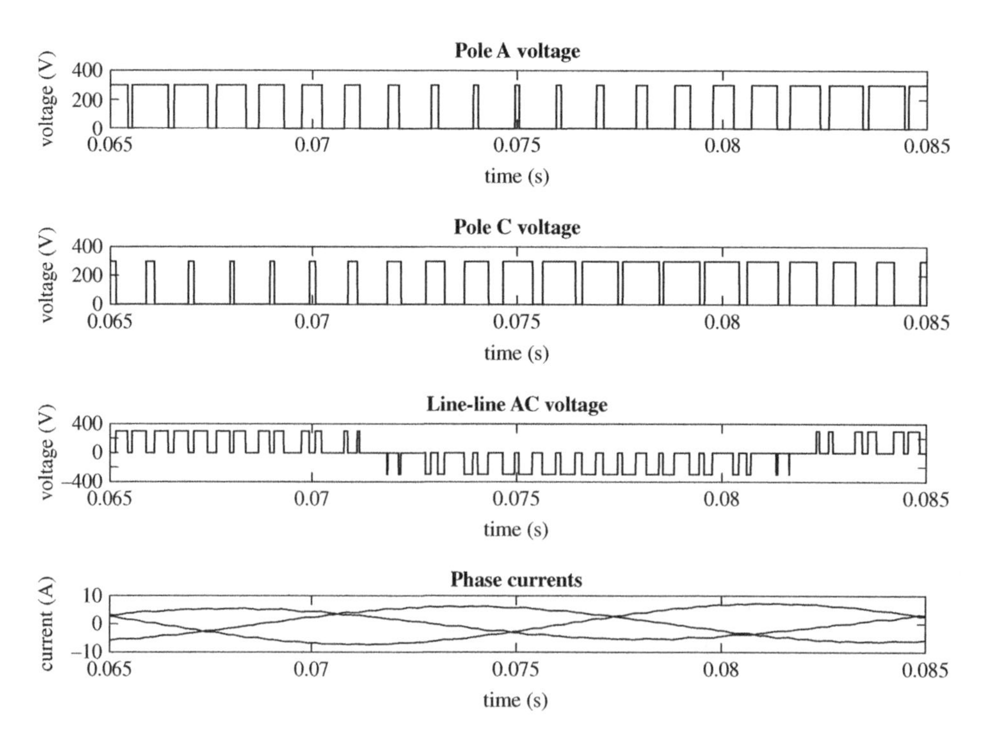

Figure 13.8 Three-phase sinusoidal PWM patterns.

The index m is less than 1 for sinusoidal modulation and greater than 1 for overmodulation.

13.3.2 Inverter Currents

The pole output current is fed into the phase winding of the star-connected electrical machine.

Let the per-phase voltage be the reference phasor with the following time-domain equation:

$$v_{ph}(\theta) = \sqrt{2}V_{ph}\cos(\theta) \tag{13.26}$$

Let the phase current be represented by

$$i_{ph}(\theta) = \sqrt{2}I_{ph}\cos(\theta - \phi) \tag{13.27}$$

where ϕ is the phase angle between the current and the voltage. The per-phase waveforms are shown in Figure 13.9. The phase current lags the phase voltage by ϕ. After the current turns positive at $\left(-\frac{\pi}{2} + \phi\right)$, the switch and diode combination Q_U and D_L conducts current for the next half cycle until $\left(\frac{\pi}{2} + \phi\right)$. After $\left(\frac{\pi}{2} + \phi\right)$, the switch and diode combination Q_L and D_U conducts again, and so on.

Once the per-phase voltage, current and phase angle (or power factor) are known, the average and rms current components in the various switches, diodes, and capacitors can be calculated and used to determine the various component stresses and power losses.

The average and rms currents in the switch and diode are easily derived as presented in the next two sections.

13.3.3 Switch, Diode, and Input Average Currents

The low-frequency time-averaged current in the switch $i_Q(\theta)$ is given by

$$i_Q(\theta) = d(\theta) \times i_{ph}(\theta) \tag{13.28}$$

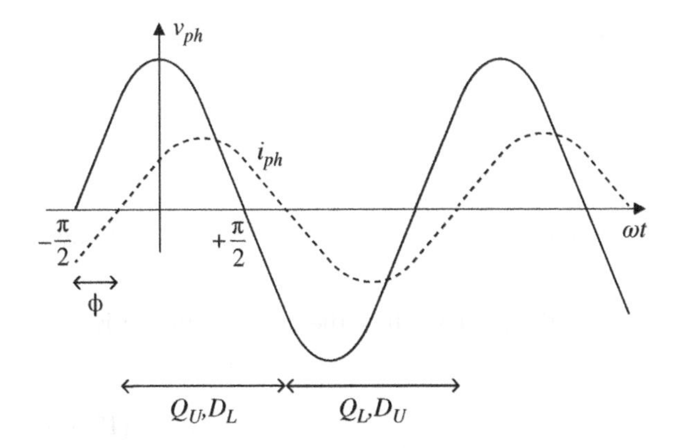

Figure 13.9 Per-phase voltage and current.

Substituting in Equation (13.21) and Equation (13.27) yields

$$i_Q(\theta) = \left(\frac{1}{2} + \frac{\sqrt{2}V_{ph}}{V_{dc}}\cos\theta\right) \times \sqrt{2}I_{ph}\cos(\theta - \phi) \tag{13.29}$$

or

$$i_Q(\theta) = \frac{I_{ph}}{\sqrt{2}}\left(\cos(\theta - \phi) + \frac{2\sqrt{2}V_{ph}}{V_{dc}}\cos\theta\cos(\theta - \phi)\right) \tag{13.30}$$

The average current in the switch over the cycle is given by

$$I_{Q(dc)} = \frac{1}{2\pi}\int_{-\frac{\pi}{2}+\phi}^{\frac{\pi}{2}+\phi} i_Q(\theta)d\theta \tag{13.31}$$

Note that the switch conducts for half the cycle.
Substituting in Equation (13.30) yields

$$I_{Q(dc)} = \frac{1}{2\pi} \times \frac{I_{ph}}{\sqrt{2}}\int_{-\frac{\pi}{2}+\phi}^{\frac{\pi}{2}+\phi} \left(\cos(\theta - \phi) + \frac{2\sqrt{2}V_{ph}}{V_{dc}}\cos\theta\cos(\theta - \phi)\right)d\theta \tag{13.32}$$

Since

$$\int_{-\frac{\pi}{2}+\phi}^{\frac{\pi}{2}+\phi} \cos(\theta - \phi)d\theta = 2 \text{ and } \int_{-\frac{\pi}{2}+\phi}^{\frac{\pi}{2}+\phi} \cos\theta\cos(\theta - \phi)d\theta = \frac{\pi}{2}\cos\phi \tag{13.33}$$

we get

$$I_{Q(dc)} = \frac{1}{2\pi} \times \frac{I_{ph}}{\sqrt{2}}\left(2 + \frac{2\sqrt{2}V_{ph}}{V_{dc}}\frac{\pi}{2}\cos\phi\right) = I_{ph}\left(\frac{1}{\sqrt{2}\pi} + \frac{V_{ph}}{2V_{dc}}\cos\phi\right) \tag{13.34}$$

Similarly, the low-frequency time-averaged current in the diode is given by

$$i_D(\theta) = (1 - d(\theta)) \times i_{ph}(\theta) \tag{13.35}$$

which expands to

$$i_D(\theta) = \left(\frac{1}{2} - \frac{\sqrt{2}V_{ph}}{V_{dc}}\cos\theta\right) \times \sqrt{2}I_{ph}\cos(\theta - \phi) \tag{13.36}$$

The diode conducts the current for half the period. Thus, the average current is

$$I_{D(dc)} = \frac{1}{2\pi}\int_{-\frac{\pi}{2}+\phi}^{\frac{\pi}{2}+\phi} i_D(\theta)d\theta \tag{13.37}$$

which simplifies to

$$I_{D(dc)} = \frac{1}{2\pi} \times \frac{I_{ph}}{\sqrt{2}} \left(2 - \frac{2\sqrt{2}V_{ph}}{V_{dc}} \frac{\pi}{2} \cos\phi \right) = I_{ph} \left(\frac{1}{\sqrt{2}\pi} - \frac{V_{ph}}{2V_{dc}} \cos\phi \right) \tag{13.38}$$

The dc input current from the source equals the dc link average current and is given by

$$I_{HV(dc)} = 3\frac{V_{ph}}{V_{dc}} I_{ph} \cos\phi \tag{13.39}$$

13.3.4 Switch, Diode, DC Link, and Input Capacitor RMS Currents

The time-averaged rms current in the switch is given by

$$I_{Q(rms)} = \sqrt{\frac{1}{2\pi} \int_{-\frac{\pi}{2}+\phi}^{\frac{\pi}{2}+\phi} d(\theta) \left(i_{ph}(\theta) \right)^2 d\theta} \tag{13.40}$$

Substituting in Equation (13.21) and Equation (13.27) yields

$$I_{Q(rms)} = \sqrt{\frac{1}{2\pi} \int_{-\frac{\pi}{2}+\phi}^{\frac{\pi}{2}+\phi} \left(\frac{1}{2} + \frac{\sqrt{2}V_{ph}}{V_{dc}} \cos\theta \right) \times \left(\sqrt{2} I_{ph} \cos(\theta - \phi) \right)^2 d\theta} \tag{13.41}$$

which simplifies to

$$I_{Q(rms)} = I_{ph} \sqrt{\frac{1}{\pi} \int_{-\frac{\pi}{2}+\phi}^{\frac{\pi}{2}+\phi} \left(\frac{1}{2} \cos^2(\theta - \phi) + \frac{\sqrt{2}V_{ph}}{V_{dc}} \cos^2(\theta - \phi) \cos\theta \right) d\theta} \tag{13.42}$$

Given that

$$\int_{-\frac{\pi}{2}+\phi}^{\frac{\pi}{2}+\phi} \cos^2(\theta - \phi) d\theta = \frac{\pi}{2} \quad \text{and} \quad \int_{-\frac{\pi}{2}+\phi}^{\frac{\pi}{2}+\phi} \cos^2(\theta - \phi) \cos\theta d\theta = \frac{4}{3} \cos\phi \tag{13.43}$$

the rms current in the switch is given by

$$I_{Q(rms)} = I_{ph} \sqrt{\frac{1}{\pi} \left(\frac{1}{2} \frac{\pi}{2} + \frac{\sqrt{2}V_{ph}}{V_{dc}} \frac{4}{3} \cos\phi \right)} = I_{ph} \sqrt{\frac{1}{4} + \frac{4\sqrt{2}V_{ph}}{3\pi V_{dc}} \cos\phi} \tag{13.44}$$

Similarly, the rms current in the diode is given by

$$I_{D(rms)} = I_{ph} \sqrt{\frac{1}{4} - \frac{4\sqrt{2}V_{ph}}{3\pi V_{dc}} \cos\phi} \tag{13.45}$$

The rms currents in the dc link $I_{HVdc(rms)}$ and the high-voltage input capacitor $I_{CHV(rms)}$ can similarly be derived, albeit with additional effort [1,2] and are given by

$$I_{HVdc(rms)} = I_{ph}\sqrt{\frac{\sqrt{6}}{\pi}\frac{V_{ph}}{V_{dc}}(1+4\cos^2\phi)} \tag{13.46}$$

and

$$I_{CHV(rms)} = I_{ph}\sqrt{\frac{\sqrt{6}}{\pi}\frac{V_{ph}}{V_{dc}} + \left(\frac{4\sqrt{6}}{\pi}\frac{V_{ph}}{V_{dc}} - 9\left(\frac{V_{ph}}{V_{dc}}\right)^2\right)\cos^2\phi} \tag{13.47}$$

13.3.5 Example: Inverter Currents

Determine the inverter currents for an 80 kW traction drive for the induction motor of Chapter 8, Example 8.3.4.1 supplied by a dc link voltage of 300 V. The per-phase quantities are 104.21 V and 301.5 A at a lagging PF of 0.9087.

Solution:
The various dc currents are

$$I_{Q(dc)} = I_{ph}\left(\frac{1}{\sqrt{2}\pi} + \frac{V_{ph}}{2V_{dc}}\cos\phi\right) = 301.5\left(\frac{1}{\sqrt{2}\pi} + \frac{104.21}{2\times 300}\times 0.9087\right)\text{A} = 115.45\,\text{A}$$

$$I_{D(dc)} = I_{ph}\left(\frac{1}{\sqrt{2}\pi} - \frac{V_{ph}}{2V_{dc}}\cos\phi\right) = 301.5\left(\frac{1}{\sqrt{2}\pi} - \frac{104.21}{2\times 300}\times 0.9087\right)\text{A} = 20.28\,\text{A}$$

$$I_{HV(dc)} = 3\frac{V_{ph}}{V_{dc}}I_{ph}\cos\phi = 3\frac{104.21}{300}301.5\times 0.9087\,\text{A} = 285.51\,\text{A}$$

The various rms currents are

$$I_{Q(rms)} = I_{ph}\sqrt{\frac{1}{4} + \frac{4\sqrt{2}}{3\pi}\frac{V_{ph}}{V_{dc}}\cos\phi} = 301.5\sqrt{\frac{1}{4} + \frac{4\sqrt{2}}{3\pi}\frac{104.21}{300}\times 0.9087}\,\text{A} = 199.87\,\text{A}$$

$$I_{D(rms)} = I_{ph}\sqrt{\frac{1}{4} - \frac{4\sqrt{2}}{3\pi}\frac{V_{ph}}{V_{dc}}\cos\phi} = 301.5\sqrt{\frac{1}{4} - \frac{4\sqrt{2}}{3\pi}\frac{104.21}{300}\times 0.9087}\,\text{A} = 74.19\,\text{A}$$

$$I_{HVdc(rms)} = I_{ph}\sqrt{\frac{\sqrt{6}}{\pi}\frac{V_{ph}}{V_{dc}}(1+4\cos^2\phi)} = 301.5\sqrt{\frac{\sqrt{6}}{\pi}\frac{104.21}{300}(1+4\times 0.9087^2)}\,\text{A}$$

$$= 325.48\,\text{A}$$

$$I_{CHV(rms)} = I_{ph}\sqrt{\frac{\sqrt{6}}{\pi}\frac{V_{ph}}{V_{dc}} + \left(\frac{4\sqrt{6}}{\pi}\frac{V_{ph}}{V_{dc}} - 9\left(\frac{V_{ph}}{V_{dc}}\right)^2\right)\cos^2\phi}$$

$$= 301.5\sqrt{\frac{\sqrt{6}}{\pi}\frac{104.21}{300} + \left(\frac{4\sqrt{6}}{\pi}\frac{104.21}{300} - 9\left(\frac{104.21}{300}\right)^2\right)\times 0.9087^2}\,\text{A}$$

$$= 156.28\,\text{A}$$

13.4 Inverter Power Loss

As previous discussed in Chapter 11 Section 11.5, there are two principal loss mechanisms in the power semiconductors: conduction and switching. The silicon IGBT and diode are discussed in this section. See Chapter 11, Appendix III and Problem 13.3 for a related discussion of silicon carbide devices.

13.4.1 Conduction Loss of IGBT and Diode

The conduction power loss of the IGBT switch $P_{Q(cond)}$ can be modeled as

$$P_{Q(\mathrm{cond})} = V_{CE0} I_{Q(dc)} + r_{CE} I_{Q(rms)}^2 \tag{13.48}$$

where V_{CE0} is the knee voltage, and r_{CE} is the equivalent resistance.

The conduction power loss of the diode $P_{D(cond)}$ can be modeled as

$$P_{D(\mathrm{cond})} = V_{f0} I_{D(dc)} + r_f I_{D(rms)}^2 \tag{13.49}$$

where V_{f0} is the knee voltage, and r_f is the equivalent resistance.

13.4.2 Switching Loss of IGBT Module

The switching loss of the IGBT switch $P_{Q(sw)}$ is given by

$$P_{Q(sw)} = \frac{f_s}{2}\left(E_{on} + E_{off}\right)\frac{V_{dc}}{V_{test}} \tag{13.50}$$

where the energies E_{on} and E_{off} are the turn-on and turn-off energies of the IGBT at the average switching current seen by the device over the half cycle, f_s is the switching frequency, and V_{test} is the voltage level used for the device testing.

For a sinusoidal excitation, the average switching current experienced by the switch and diode is given by

$$I_{Q(sw,avg)} = I_{D(sw,avg)} = \frac{2\sqrt{2}}{\pi} I_{ph} \tag{13.51}$$

the derivation of which is left as an exercise for the reader.

The diode switching loss $P_{D(sw)}$ is similarly defined as

$$P_{D(sw)} = \frac{f_s}{2} E_{rec} \frac{V_{dc}}{V_{test}} \tag{13.52}$$

where E_{rec} is the turn-off energy of the diode due to reverse recovery.

The total loss for the inverter is six times the conduction and switching losses for any IGBT and diode pair.

13.4.2.1 Example: Power Losses of Power Semiconductor Module

The inverter of the previous example is switching at 10 kHz. The characteristics of the 600 V, 600 A IGBT and diode half-bridge module are shown in Figure 13.10. The output characteristics of the IGBT and diode are shown in Figure 13.10(a) and (b), respectively.

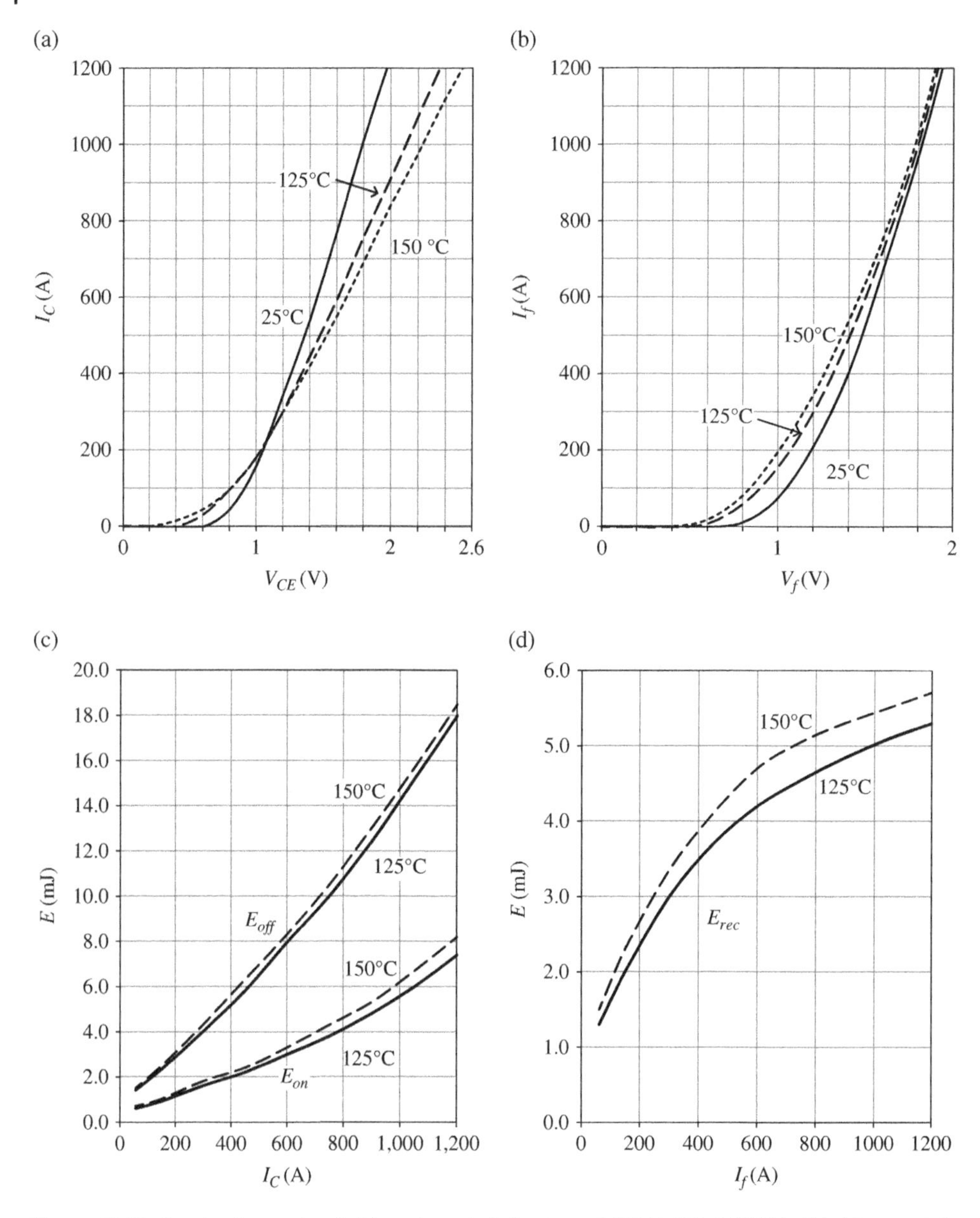

Figure 13.10 Conduction and switching characteristic plots of 600 V, 600 A IGBT half-bridge module (300 V test voltage).

Let the conduction losses be modeled by the following parameters: V_{CE0} =0.85 V, r_{CE} = 1.25 mΩ, V_{f0} =0.8 V, and r_f = 1.3 mΩ. The switch and diode switching energies at 125°C are plotted in Figure 13.10(c) and (d), respectively. The switching energies are measured at a test voltage of 300 V.

Determine the conduction and switching losses in the IGBT and diode from the previous example.

Solution:

The conduction loss of the switch is

$$P_{Q(\text{cond})} = V_{CE0}I_{Q(dc)} + r_{CE}{I_{Q(rms)}}^2 = 0.85 \times 115.45\,\text{W} + 0.00125 \times 199.87^2\,\text{W}$$
$$= 148.1\,\text{W}$$

The conduction loss of the diode is

$$P_{D(cond)} = V_{f0}I_{D(dc)} + r_f{I_{D(rms)}}^2 = 0.8 \times 20.28\,\text{W} + 0.0013 \times 74.19^2\,\text{W} = 23.4\,\text{W}$$

The average switching current is given by

$$I_{Q(sw,avg)} = \frac{2\sqrt{2}}{\pi}I_{ph} = \frac{2\sqrt{2}}{\pi}301.5\text{A} = 271.4\text{A}$$

The approximate losses from the figures are as follows:

$$E_{on}\left(I_{Q(sw,avg)}\right) = E_{on}(271.4\text{A}) \approx 1.5\,\text{mJ}$$
$$E_{off}\left(I_{Q(sw,avg)}\right) = E_{off}(271.4\text{A}) \approx 3.8\,\text{mJ}$$
$$E_{rec}\left(I_{Q(sw,avg)}\right) = E_{rec}(271.4\text{A}) \approx 2.8\,\text{mJ}$$

Thus, the power losses are

$$P_{Q(sw)} = \frac{f_s}{2}\left(E_{on} + E_{off}\right)\frac{V_{dc}}{V_{test}} = \frac{10^4}{2} \times (1.5 + 3.8) \times 10^{-3}\frac{300}{300}\,\text{W} = 26.5\,\text{W}$$

The diode switching loss is similarly defined.

$$P_{D(sw)} = \frac{f_s}{2}E_{rec}\frac{V_{dc}}{V_{test}} = \frac{10^4}{2} \times 2.8 \times 10^{-3} \times \frac{300}{300}\,\text{W} = 14\,\text{W}$$

13.4.3 Total Semiconductor Power Loss and Junction Temperature

The total semiconductor power loss is the sum of the switching and conduction power losses. Thus, the total IGBT loss P_Q and the total diode loss P_D are

$$P_Q = P_{Q(cond)} + P_{Q(sw)} \tag{13.53}$$

and

$$P_D = P_{D(cond)} + P_{D(sw)} \tag{13.54}$$

The junction temperatures of the IGBT T_{JQ} and diode T_{JD} are given by

$$T_{JQ} = T_{HS} + R_{JQ-HS} \times P_Q \tag{13.55}$$

and

$$T_{JD} = T_{HS} + R_{JD-HS} \times P_D \tag{13.56}$$

where the temperature T_{HS} is the temperature of the surface of the heat sink cooling the semiconductors, and $R_{JQ\text{-}HS}$ and $R_{JD\text{-}HS}$ are the thermal impedances of the IGBT and diode, respectively.

13.4.3.1 Example: Total IGBT Module Loss and Die Temperatures

Determine the IGBT and diode power losses and their respective hotspot temperatures if the heat sink is maintained at 80°C, and the thermal resistances of the IGBT and diode are 0.08°C/W and 0.16°C/W, respectively.

Solution:

The total losses are given by

$$P_Q = P_{Q(cond)} + P_{Q(sw)} = 148.1\,\text{W} + 26.5\,\text{W} = 174.6\,\text{W}$$

and

$$P_D = P_{D(cond)} + P_{D(sw)} = 23.4\,\text{W} + 14\,\text{W} = 37.4\,\text{W}$$

The hotspot temperatures are given by

$$T_{JQ} = T_{HS} + R_{JQ-HS} \times P_Q = 80^\circ\text{C} + 0.08 \times 174.6^\circ\text{C} = 94^\circ\text{C}$$

and

$$T_{JD} = T_{HS} + R_{JD-HS} \times P_D = 80^\circ\text{C} + 0.16 \times 37.4^\circ\text{C} = 86^\circ\text{C}$$

The inverter loss is

$$P_{loss} = 6 \times (P_Q + P_D) = 6 \times (174.6 + 37.4)\,\text{W} = 1272\,\text{W}$$

The inverter efficiency is

$$\eta_{inv} = \frac{P_o}{P_o + P_{loss}} \times 100\% = \frac{3V_{ph}I_{ph}\cos\phi}{3V_{ph}I_{ph}\cos\phi + P_{loss}} \times 100\%$$

$$= \frac{85652}{85652 + 1272} \times 100\% = 98.5\%$$

13.4.4 Example: Regenerative Currents

Determine the inverter currents and the component power loss for the 80 kW traction drive in the previous example when operating in regenerative mode. Assume similar per-phase quantities of 104.21 V and 301.5 A, but at a PF of −0.9087.

Solution:

The various dc currents are

$$I_{Q(dc)} = I_{ph}\left(\frac{1}{\sqrt{2}\pi} + \frac{V_{ph}}{2V_{dc}}\cos\phi\right) = 301.5\left(\frac{1}{\sqrt{2}\pi} + \frac{104.21}{2 \times 300} \times (-0.9087)\right)\text{A} = 20.28\,\text{A}$$

$$I_{D(dc)} = I_{ph}\left(\frac{1}{\sqrt{2}\pi} - \frac{V_{ph}}{2V_{dc}}\cos\phi\right) = 301.5\left(\frac{1}{\sqrt{2}\pi} - \frac{104.21}{2 \times 300} \times (-0.9087)\right)\text{A} = 115.45\,\text{A}$$

$$I_{HV(dc)} = 3\frac{V_{ph}}{V_{dc}}I_{ph}\cos\phi = 3\frac{104.21}{300}301.5 \times (-0.9087)\,\text{A} = -285.51\,\text{A}$$

The various rms currents are

$$I_{Q(rms)} = I_{ph}\sqrt{\frac{1}{4} + \frac{4\sqrt{2}}{3\pi}\frac{V_{ph}}{V_{dc}}\cos\phi} = 301.5\sqrt{\frac{1}{4} + \frac{4\sqrt{2}}{3\pi}\frac{104.21}{300} \times (-0.9087)}\,\text{A} = 74.19\,\text{A}$$

$$I_{D(rms)} = I_{ph}\sqrt{\frac{1}{4} - \frac{4\sqrt{2}}{3\pi}\frac{V_{ph}}{V_{dc}}\cos\phi}$$

$$= 301.5\sqrt{\frac{1}{4} - \frac{4\sqrt{2}}{3\pi}\frac{104.21}{300} \times (-0.9087)}\,\text{A}$$

$$= 199.87\,\text{A}$$

$$I_{HVdc(rms)} = I_{ph}\sqrt{\frac{\sqrt{6}}{\pi}\frac{V_{ph}}{V_{dc}}(1 + 4\cos^2\phi)}$$

$$= 301.5\sqrt{\frac{\sqrt{6}}{\pi}\frac{104.21}{300}(1 + 4 \times (-0.9087)^2)}\,\text{A}$$

$$= 325.48\,\text{A}$$

$$I_{CHV(rms)} = I_{ph}\sqrt{\frac{\sqrt{6}}{\pi}\frac{V_{ph}}{V_{dc}} + \left(\frac{4\sqrt{6}}{\pi}\frac{V_{ph}}{V_{dc}} - 9\left(\frac{V_{ph}}{V_{dc}}\right)^2\right)\cos^2\phi}$$

$$= 301.5\sqrt{\frac{\sqrt{6}}{\pi}\frac{104.21}{300} + \left(\frac{4\sqrt{6}}{\pi}\frac{104.21}{300} - 9\left(\frac{104.21}{300}\right)^2\right) \times (-0.9087)^2}\,\text{A}$$

$$= 156.28\,\text{A}$$

The currents in the diode and switch have swapped compared to the motoring example. Thus, when operating in regenerative mode, the diode losses increase, and the switch losses decrease compared to motoring mode.

References

1 J. W. Kolar and S. D. Round, "Analytical calculation of the RMS current stress on the DC-link capacitor of voltage-PWM converter systems," *IEE Electric Power Applications*, **153** (4), pp. 535–543, July 2006.

2 F. Renken, "Analytical calculation of the dc-link capacitor current for pulsed three-phase inverters," EPE PEMC 2004.

Further Reading

1 N. Mohan, *Advanced Electric Drives Analysis Control and Modelling using MATLAB/Simulink®*, John Wiley & Sons, 2014.

Problems

13.1 The electric drive example of Chapter 9 features an eight-pole SPM machine. The dc link voltage is 300 V, and the switching frequency is 10 kHz. The heat sink is at 80°C, and the thermal resistances of the IGBT and diode are 0.08°C/W and 0.16°C/W.

See Figure 13.10.
For simplicity, assume a junction temperature of 125°C, and let V_{CE0} =0.85 V, r_{CE} = 1.25 mΩ, V_{f0} =0.8 V, and r_f = 1.3 mΩ for the first three cases.

Determine the various dc and rms inverter currents, switch and diode losses, junction temperatures, and inverter efficiencies for the following conditions:

i) Motoring at the rated condition: V_{ph} = 116.6 V, I_{ph} = 313.33 A, PF = 0.7889;
ii) Generating at the rated condition: V_{ph} = 106.35 V, I_{ph} = 308.89 A, PF = −0.7479;
iii) Motoring at rated power at 3.75 times rated speed: V_{ph} = 117.5 V, I_{ph} = 287.9 A, PF = 0.8585;
iv) Motoring at 20 kW at half the rated speed: V_{ph} = 45.39 V, I_{ph} = 80 A, power factor = 0.9798. (Assume V_{CE0} =0.5 V, r_{CE} = 3.8 mΩ, V_{f0} =0.6 V, and r_f = 3.5 mΩ for this case.)

[Ans. (i) $I_{Q(dc)}$ = 118.56 A, $I_{D(dc)}$ = 22.49 A, $I_{Q(rms)}$ = 206.43 A, $I_{D(rms)}$ = 80.47 A, P_Q = 180.5 W, P_D = 40 W, T_{JQ} = 94.4 °C, T_{JD} = 86.4 °C, η= 98.5 %; ($E_{on}\approx$ 1.5 mJ, $E_{off}\approx$ 3.8 mJ, $E_{rec}\approx$ 2.8 mJ @ 282 A)

(ii) $I_{Q(dc)}$ = 28.58 A, $I_{D(dc)}$ = 110.47 A, $I_{Q(rms)}$ = 93.11 A, $I_{D(rms)}$ = 197.58 A, P_Q = 61.5 W, P_D = 153 W, T_{JQ} = 84.9 °C, T_{JD} = 104.5 °C, η= 98.3%; ($E_{on}\approx$ 1.5 mJ, $E_{off}\approx$ 3.8 mJ, $E_{rec}\approx$ 2.8 mJ @ 278 A)
(iii) $I_{Q(dc)}$ = 113.2 A, $I_{D(dc)}$ = 16.4 A, $I_{Q(rms)}$ = 193.52 A, $I_{D(rms)}$ = 63.19 A, P_Q = 169.5 W, P_D = 32 W, T_{JQ} = 93.6 °C, T_{JD} = 85.1 °C, η= 98.6 %; ($E_{on}\approx$ 1.5 mJ, $E_{off}\approx$ 3.8 mJ, $E_{rec}\approx$ 2.8 mJ @ 259 A)
(iv) $I_{Q(dc)}$ = 23.94 A, $I_{D(dc)}$ = 12.08 A, $I_{Q(rms)}$ = 46.58 A, $I_{D(rms)}$ = 32.1 A, P_Q = 31.2 W, P_D = 17.4 W, T_{JQ} = 82.5 °C, T_{JD} = 82.8 °C, η= 97.4 %, ($E_{on}\approx$ 0.7 mJ, $E_{off}\approx$ 1.5 mJ, $E_{rec}\approx$ 1.3 mJ @ 41 A)]

13.2 Determine the various dc and rms inverter currents, switch and diode losses, junction temperatures, and the inverter efficiency for the following condition: an induction machine motoring at rated power and twice the rated speed as in Chapter 8, Example 8.4.2.1. Use the conduction and switching parameters of Examples 13.4.2.1 and 13.4.3.1.

[Ans. (i) $I_{Q(dc)}$ = 116.11 A, $I_{D(dc)}$ = 19.65 A, $I_{Q(rms)}$ = 200.50 A, $I_{D(rms)}$ = 72.66 A, P_Q = 175.4 W, P_D = 36.6 W, T_{JQ} = 94.0 °C, T_{JD} = 85.9 °C, η= 98.6 %]

13.3 The 60 kW interior-permanent motor of Problem 10.5 has the following per-phase rated conditions of 260.8 V and 117.5 A at a power factor of 0.696. The dc link voltage is 700 V and the switching frequency is 10 kHz. Determine the various switch and diode currents and power losses, and the inverter efficiency for the rated condition when using (i) the representative 1200 V, 300 A half-bridge SiC module of Chapter 11, Appendix III and Figure 11.31(b), and (ii) the 1200 V, 300 A half-bridge silicon module of Problem 11.6 and Figure 11.31(a).

[Ans. $I_{Q(dc)}$ = 41.46 A, $I_{D(dc)}$ = 11.21 A, $I_{Q(rms)}$ = 74.84 A, $I_{D(rms)}$ = 36.09 A, $I_{Q(sw,avg)}$ = 105.79 A, (i) P_Q = 68.1 W, P_D = 14.2 W, η = 99.23%; ($E_{on} \approx$ 2.8 mJ, $E_{off} \approx$ 1.2 mJ, $E_{rec} \approx$ 0 mJ), (ii) P_Q = 187.2 W, P_D = 88.6 W, η = 97.48%; ($E_{on} \approx$ 10 mJ, $E_{off} \approx$ 12.5 mJ, $E_{rec} \approx$ 13 mJ)]

Assignments

13.1 The student is encouraged to experiment with a circuit simulation software package. The Simulink circuit of Figure 13.11 is used for the inverter simulation in this chapter.

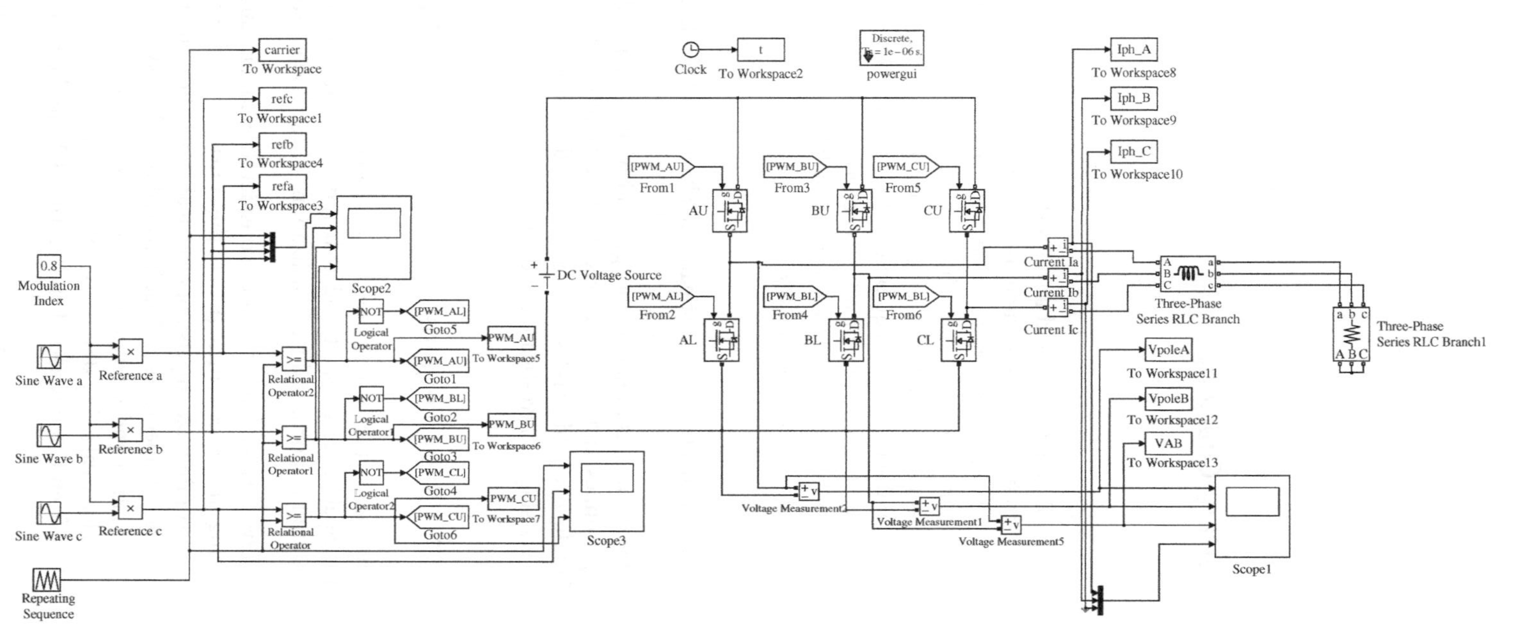

Figure 13.11 Simulink inverter circuit.

14

Battery Charging

> "Power can be, and at no distant date will be, transmitted without wires, for all commercial uses, such as the lighting of homes and the driving of aeroplanes. I have discovered the essential principles, and it only remains to develop them commercially. When this is done, you will be able to go anywhere in the world — to the mountain top overlooking your farm, to the arctic, or to the desert — and set up a little equipment that will give you heat to cook with, and light to read by. This equipment will be carried in a satchel not as big as the ordinary suit case. In years to come wireless lights will be as common on the farms as ordinary electric lights are nowadays in our cities."
>
> *Nikola Tesla (1856–1943)*

In this chapter, we investigate the charging of electric vehicles from the electric grid. Battery charging connects the vehicle to the electric grid, and many factors must be considered, such as available voltages and wiring, standardization, safety, communication, ergonomics, and more. Various charging architectures and charging standards are used. Conductive and wireless standards are discussed. The boost power-factor correction power stage is investigated in detail.

14.1 Basic Requirements for Charging System

Many important issues must be considered when selecting the charging system. The principal issues are safety, reliability, user-friendliness, power levels and charging times, communications, and standardization. These issues are briefly discussed as follows.

Safety: This is the most serious consideration for any automotive manufacturer introducing an electric vehicle (EV) to the consumer marketplace. The battery charger system must minimize the risk of electrical shock, fire, and injury to the end user for a wide range of operating and fault conditions. The system must provide various levels of insulation and safety checks in order to ensure a safe system. There are a number of electrical safety standards which are used around the world. The principal standards are from Underwriters Laboratories (UL) in the United States and VDE in Germany.

Electric Powertrain: Energy Systems, Power Electronics and Drives for Hybrid, Electric and Fuel Cell Vehicles, First Edition. John G. Hayes and G. Abas Goodarzi.

Companion website: www.wiley.com/go/hayes/electricpowertrain

It is necessary to obtain safety approvals from these types of agencies in order to sell products in many countries.

Reliability: The automotive environment is very harsh. The same performance is expected whether a car is driven in the dry heat of the Arizona desert, the freezing cold of Minnesota, or the humid conditions of Florida. The car is exposed to significant shock and vibration in addition to corrosive solvents, salt, water, and mud. The charger for the EV must have a long service life with daily operation. The electrical connector must be designed to withstand over 10,000 insertions and withdrawals in these harsh conditions and still remain safe for the consumer for the many fault scenarios. For example, the plug and cable should be able to withstand the vehicle weight in case the vehicle accidentally drives over the assembly.

User-friendliness: A consumer product such as a car requires that significant attention be paid to customer expectations and ergonomics. The present method of fueling a vehicle with an internal combustion engine is simple and straightforward. EV charging must also be simple and should pose minimal challenge to EV users or to young children whom a parent may send to "plug in the car." Greater emphasis has to be placed on the ergonomics of refueling an EV compared to a gasoline-fueled car. The basic reason for this is that the EV may require daily charging, whereas the gasoline-powered car may only be fueled once a week or less. Such ergonomic considerations as single-handed operation and intuitive insertion and withdrawal processes help the user.

Power levels and charging times: The charging time for an EV can range from tens of minutes, if high-power charging is used, to many hours, if low-power chargers are used. Thus, for EVs to gain widespread acceptance, the charging power levels should be maximized in order to reduce the charging times. However, in practice the power levels can be limited by the household electrical wiring, electrical grid impacts, battery chemistries and degradation due to high charging levels, and, of course, size and cost. Given the importance of optimizing society's use of energy, it is important to charge efficiently. However, as, discussed in Chapter 3, battery charging efficiency can drop with higher power levels.

Communication: At a basic level, the plug and cable assembly must not only transmit power, but must also provide a communications path between the charger and the vehicle in order to ensure a safe and optimized power flow. Communication has taken on a greater role in society in the twenty-first century as smartphones and the Internet are part of the overall communications and control interfaces. Simple messages relating to availability, maximum power output, charging time, and problem or fault reporting are also critical communications.

Standardization: Market acceptance of a new product can be accelerated by creating a product standard. A market standard can reduce the cost by ensuring a larger market with access to more charging points and ease of communication. An advantage of a commonly agreed standard is that the automotive companies do not have to compete on the charging but can focus on the vehicle.

Compliance: Many regulations already exist to ensure safe and reliable operation of electrical equipment. The equipment must also comply with standards to limit electrical noise emissions, commonly known as electromagnetic interference (EMI), an area of government regulations in order to protect other electrical devices, for example, cardiac pacemakers, and to limit susceptibility and increase immunity to unwanted emissions from other equipment or from events, such as lightning strikes. Charging systems must comply with all these regulations.

14.2 Charger Architectures

Before the various charging topics are investigated in more depth, it is useful to consider the basic electrical power conversions required to charge a battery.

A basic block diagram of the charging power flow is shown in Figure 14.1. A wonder of the modern world is the electrical grid. The electrical grid provides alternating current (ac) voltages and currents at an electrical frequency. Voltages and frequencies vary around the globe and will be discussed in the next section. The ac supply frequency is low and is typically 50 Hz or 60 Hz. A battery requires direct current (dc) electricity, and so the first stage of the power conversion is to rectify and filter the ac waveform from the grid to dc. However, this first-stage conversion to dc cannot be supplied directly to the battery, as a transformer is typically required to order to provide electrical safety to the user. Thus, the dc is chopped to create a stream of very high-frequency ac (hfac) waveforms. Frequencies of tens or hundreds of kilohertz are common to minimize transformer size. A transformer is used to provide a physical barrier in the electrical path between the ac grid and the battery and to minimize failure modes which could result in life-threatening mishaps. The high-frequency ac is rectified and filtered to create dc to supply the battery.

There are a number of options for charging the vehicle, and there are a variety of charging technologies available for EVs. Choices and decisions must be made by the various manufacturers, infrastructure providers, and the consumers with respect to the following areas: conductive or wireless/inductive charging, high-power or rapid charging, on-board versus off-board chargers, and integral charging.

Conductive charging is the common approach to charging a vehicle. Conductive charging simply means that the vehicle is electrically connected to the off-board powering system by a conductive plug and socket assembly, similar to the operation of common household electrical appliances. Generally, vehicles feature a low-to-medium power on-board charger, with ac being supplied from the electrical grid to the vehicle. Figure 14.1 is modified to that shown in Figure 14.2 to include all the charging components on the vehicle.

While low-to-medium power chargers are expected to be the commonly used charging approach, an enabler for EVs is the capability to rapidly charge using a high-power charger. High-power chargers can be very large physically and are designed as stationary off-board devices to be operated in a similar manner to a gasoline pump at a filling station. In this case, the charger is off-board, and dc is supplied on-board. The conductive dc block diagram is shown in Figure 14.3.

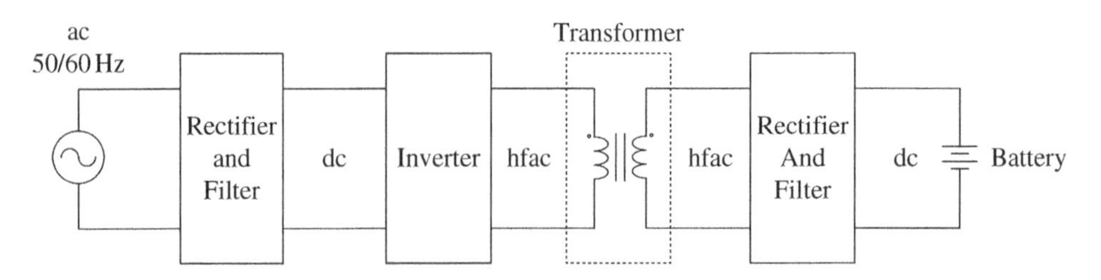

Figure 14.1 Basic power block diagram for battery charging.

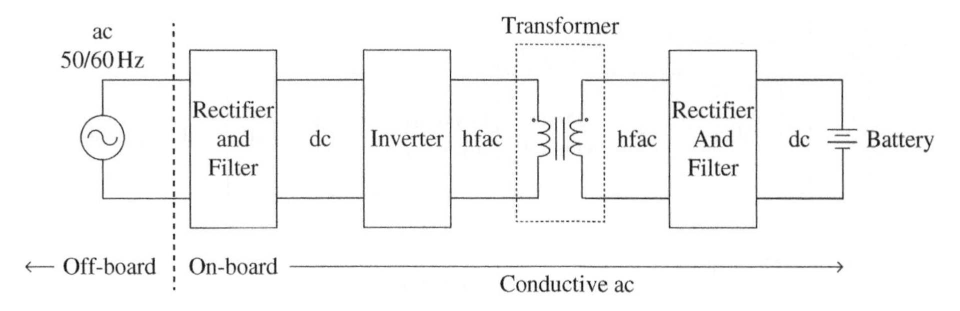

Figure 14.2 Conductive ac charging power block diagram.

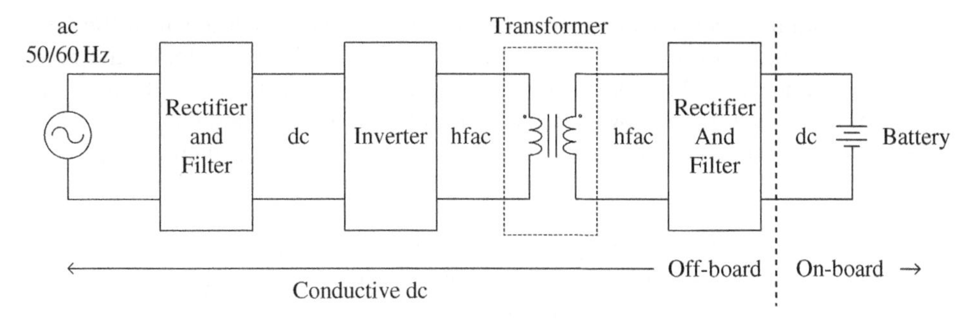

Figure 14.3 Conductive dc charging power block diagram.

Wireless or inductive charging does not connect the vehicle to the electrical grid by a conductive coupling using copper wires. Instead, the magic of transformer coupling is used to couple power from the grid to the vehicle without conductive contacts. Such an approach can result in safety enhancements and consumer ease of use, but also comes with significant engineering challenges. For wireless charging, the partition between the on-board and off-board components of the charging system is within the transformer itself, as shown in Figure 14.4.

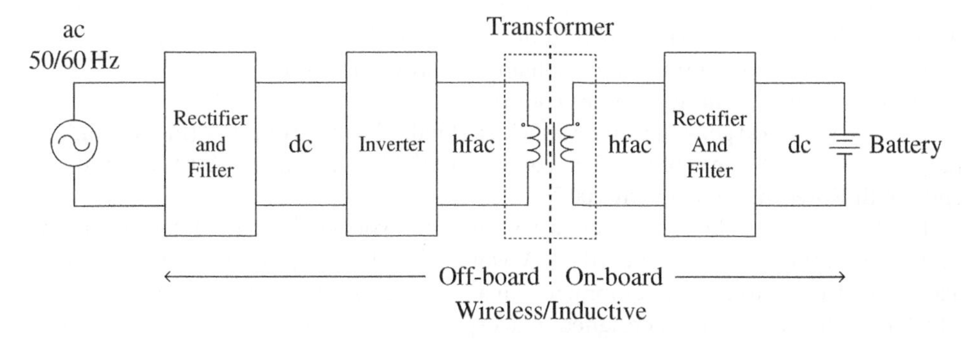

Figure 14.4 Wireless/inductive ac charging power block diagram.

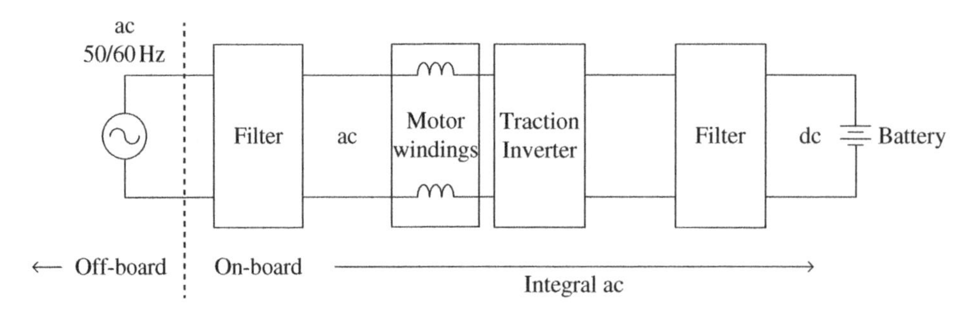

Figure 14.5 Integral ac charging power block diagram.

Integral charging was originally employed for the prototype predecessor of the GM EV1, known as the GM Impact, and the technology has since been developed and used by such companies as AC Propulsion and Renault. As shown in Figure 14.5, the technology reconfigures the traction power electronics and machine to re-employ the components to also perform the power conversions required for charging. Such an approach reduces the overall parts count on the vehicle while requiring additional measures on electrical safety and isolation.

A significant advantage of the integral charging approach is that the power flow can be made bidirectional. Thus, the vehicle can supply power back to the local grid if required – such operation is known as **vehicle to grid (V2G)**.

Microgrid operation is also possible. A **microgrid** is a term for a local power grid which can be connected to the main grid, or can be disconnected from the main grid for independent operation if required.

14.3 Grid Voltages, Frequencies, and Wiring

Electrical equipment, especially mobile devices, are designed to operate globally. There are many configurations of wiring and voltage, power, and frequency around the world. This section briefly considers the principal configurations.

First, the world is divided in terms of the electrical frequency of operation. Most countries, with a handful of exceptions, operate at either 50 Hz or 60 Hz. Japan is the notable exception as it has two separate grids, one with 50 Hz to the east and one with 60 Hz to the west. The simplest broad classifications would be that the world operates at 50 Hz except for North and Central America, many countries in South America, South Korea, Saudi Arabia, and a handful of other countries.

Similarly, the world generally operates with a standard voltage of 230 V (in the nominal range of 220–240 V), except for North and Central America, where 120 V is standard, and South Korea and Japan with 100 V.

Note that there can be significant variations in the voltage during regular operation. Equipment working off the standard 230 V would likely be designed to operate off a voltage range of 180 V to 270 V. Equipment can also be designed for universal operation, in which case the input stage is designed to accept a voltage range of 80 V to 270 V. Note that the voltages just discussed are all root-mean-squared (rms) values.

The basic residential wiring system provides a phase voltage of 230 V, 50 Hz between the line and neutral, as shown in Figure 14.6(a). The **neutral** is typically grounded by physically connecting the neutral to a copper grounding rod which is driven into the ground at a location close to the residence. The three wires of line, neutral, and ground are hard-wired into the charging assembly.

The single-phase connection is typically provided by a three-phase transformer. Commercial premises often have a three-phase connection in order to power electric motors, fans, compressors, and so on. The typical three-phase configuration is shown in Figure 14.6(b). Three phase can be an option for many charger power levels, in which case the five wires of the three lines, neutral, and ground are supplied to the charging equipment. Simply multiply the phase voltage by $\sqrt{3}$ in order to get the appropriate line voltage. Thus, 400 V is the line voltage when the phase-to-neutral voltage is 230 V.

It is common in the 60 Hz regions for 100/120 V to be available. Again, basic household wiring provides a single-phase connection, as shown by Figure 14.7(a). As the 120 V system outputs a relatively low power, it is usual to have a higher voltage available. It is common to have the grounded midpoint single-phase wiring system of Figure 14.7(b), from which two 120 V outputs are available. Note that these outputs are 180° out of phase with each other, which means that the sum of the two outputs is 240 V, the high-voltage output.

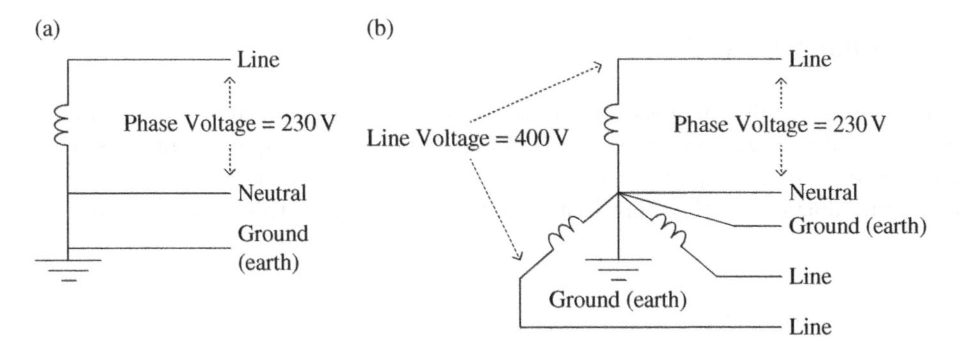

Figure 14.6 Basic 230 V wiring system.

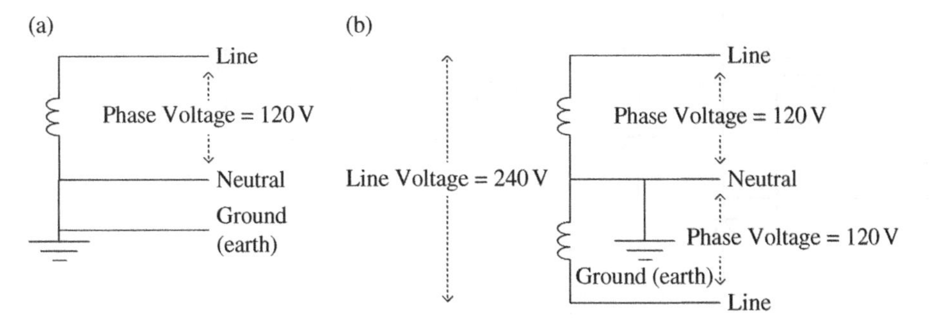

Figure 14.7 Basic 100/120 V wiring system.

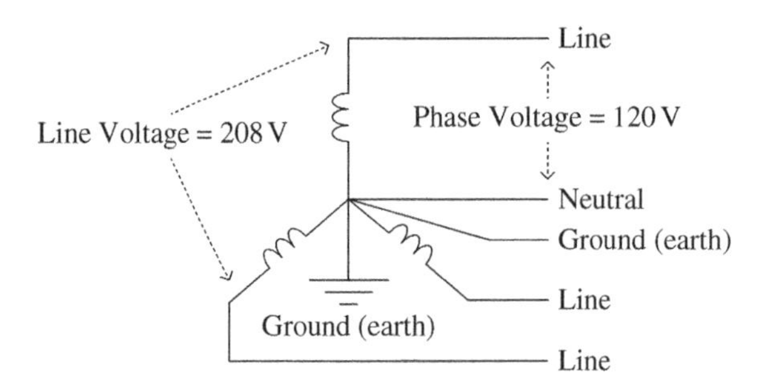

Figure 14.8 Three-phase 208 V system.

The 120 V output, shown in Figure 14.7(a), may itself be the output of a three-phase transformer with a nominal 208 V line voltage, as shown in Figure 14.8.

The 240 V winding of Figure 14.7(b) can be the output of a single-phase winding of a three-phase star or delta transformer.

The voltages mentioned so far in this section are the typical voltages considered for household and commercial premises. Higher voltages are likely to be used for high-power charging. These again are three phase, and the wiring configurations can be in star or delta and may feature auxiliary windings.

14.4 Charger Functions

EV chargers are similar in operation but have some key differences compared to battery chargers used in other applications, such as mobile phones and laptop computers. The basic functions of a low-power battery charger are shown in Figure 14.9. First, we will review the functions of the more basic charger and then turn to the functions of the

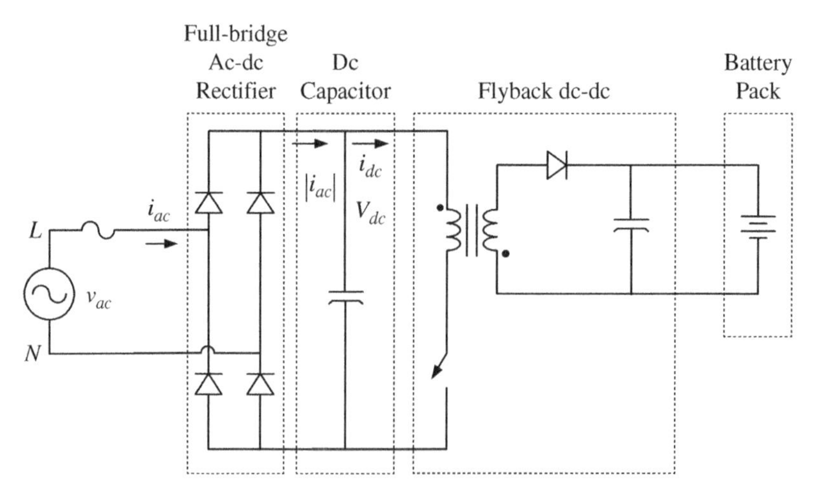

Figure 14.9 Low-power charger.

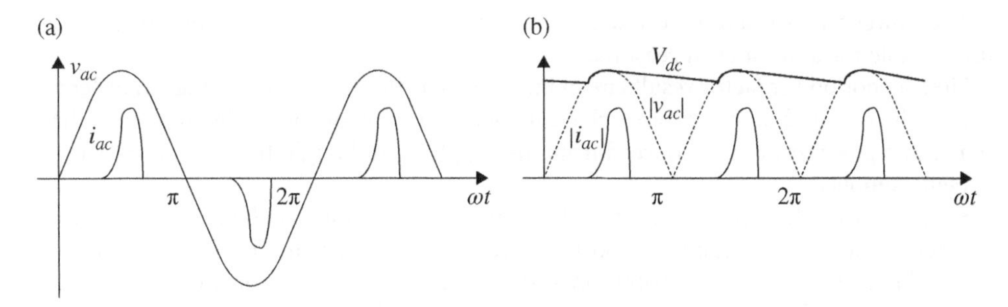

Figure 14.10 (a) Input and (b) rectifier-capacitor waveforms.

automotive charger. The related voltage and current waveforms are shown in Figure 14.10. The basic charger power blocks are as follows.

Ac-dc rectifier: The function of the ac-dc rectifier is to rectify, or make positive, the input ac voltage v_{ac} and current i_{ac} when they are in the negative half cycle. Thus, the output of the diode rectifier is always positive, as can be seen in the waveforms of the rectifier output voltage $|v_{ac}|$ and current $|i_{ac}|$.

Dc capacitor: The dc capacitor is charged to the peak ac voltage when the rectified voltage $|v_{ac}|$ exceeds the capacitor voltage. This only happens during a portion of the cycle, and there is a surge of current from the ac input through the diodes and into the capacitor. Thus, the current waveform has a sharp pulsed waveform.

Dc-dc converter: The dc-dc converter converts the high voltage on the dc capacitor to a safe lower voltage for input to the laptop or mobile phone for use in charging. The simplest and most cost-effective dc-dc is the switch-mode flyback converter, which switches at a high frequency and has the transformer isolation that is essential for safety.

At this point it is useful to identify some very commonly used power terms.

14.4.1 Real Power, Apparent Power, and Power Factor

The **apparent power** S is the product of the rms voltage V and the rms current I. This product is also known as the volt-ampere product. In equation form:

$$S = VI \tag{14.1}$$

The apparent power has the units of volt-amperes (symbol VA).

The **real power** P is the true power delivered to an electrical circuit. The real power is the power measured by a wattmeter and has the units of watts (symbol W).

In a power circuit, the apparent power can vary significantly from the real power due to the distortion introduced by the power-stage components, such as the diode-capacitive rectifier above, or by the load itself.

The **power factor** PF is the ratio of the real power to the apparent power:

$$PF = \frac{P}{VI} = \frac{P}{S} \tag{14.2}$$

The power factor is a dimensionless quantity and has no units. A low power factor is undesirable for a number of reasons.

First, a poor power factor results in an increased current for a given voltage in order to supply the required power. From a charging perspective, a low power factor means that maximum power cannot be sourced from the supply even though the maximum current is being supplied.

Second, the diode-capacitor front-end, described earlier, results in high-frequency harmonics of the fundamental 50 or 60 Hz waveform. These harmonics create increased power losses through the distribution system, increasing conductor and transformer temperatures and reducing the efficiency and system reliability.

Third, the current distortion results in voltage distortion of the supply voltage, which affects the supply and the other loads fed from the supply.

Fourth, commercial and industrial customers of the power utilities can be financially penalized for demanding currents with a low power factor.

Global standards have been developed to govern power quality and harmonic distortion. IEC 61000 contains commonly referenced standards addressing harmonics, EMI, and other grid-related issues.

Thus, simple diode–rectifier–capacitor front-ends are only permitted in low-power applications. At levels above hundreds of watts, the simple capacitive filter is buffered with another switch-mode power converter, known as the **boost converter**. The boost converter serves to maintain the input current waveform identical with the input voltage waveform and so eliminates any harmonic distortion and improves the power factor to unity. The power-factor-corrected boost converter is shown in Figure 14.11, and the waveforms are presented in Figure 14.12. A basic requirement for the boost converter is that the output dc voltage must be greater than the peak of the input ac voltage. Note that the power converter has been changed from the simple flyback to the full-bridge (see Chapter 12) for higher power.

A more detailed overview of the EV battery charging system is shown in Figure 14.13. This charging system is representative of the on-board conductive systems. The circuit has a number of different functions as follows:

- **RCCB**: The **residual current circuit breaker** (RCCB) detects an imbalance in the line and neutral currents, usually between about 5 to 20 mA, and triggers a circuit breaker to take the charger off-line to prevent fatalities. This circuitry is also known as a **ground-fault circuit interrupter** (**GFCI**).
- **EMI filter**: Switching power electronics can generate significant radiated and conducted noise, known as electromagnetic interference. A high-current filter with common-mode and differential-mode stages is required to meet legal emission standards in the United States (FCC) and the EU (VDE). FCC Part 15b is commonly referenced in the United States, while the VDE B standard is commonly referenced in Europe.
- **Rectifier**: A simple diode bridge rectifies the 50/60 Hz ac waveform.
- **Boost PFC**: A boost converter, typically switching at tens or hundreds of kHz, chops up the low-frequency rectified power and boosts it to a voltage level of about 400 V_{dc}, a value higher than the peak ac value.
- **Dc link**: An electrolytic capacitor is usually used for bulk storage to filter the 50/60 Hz component.

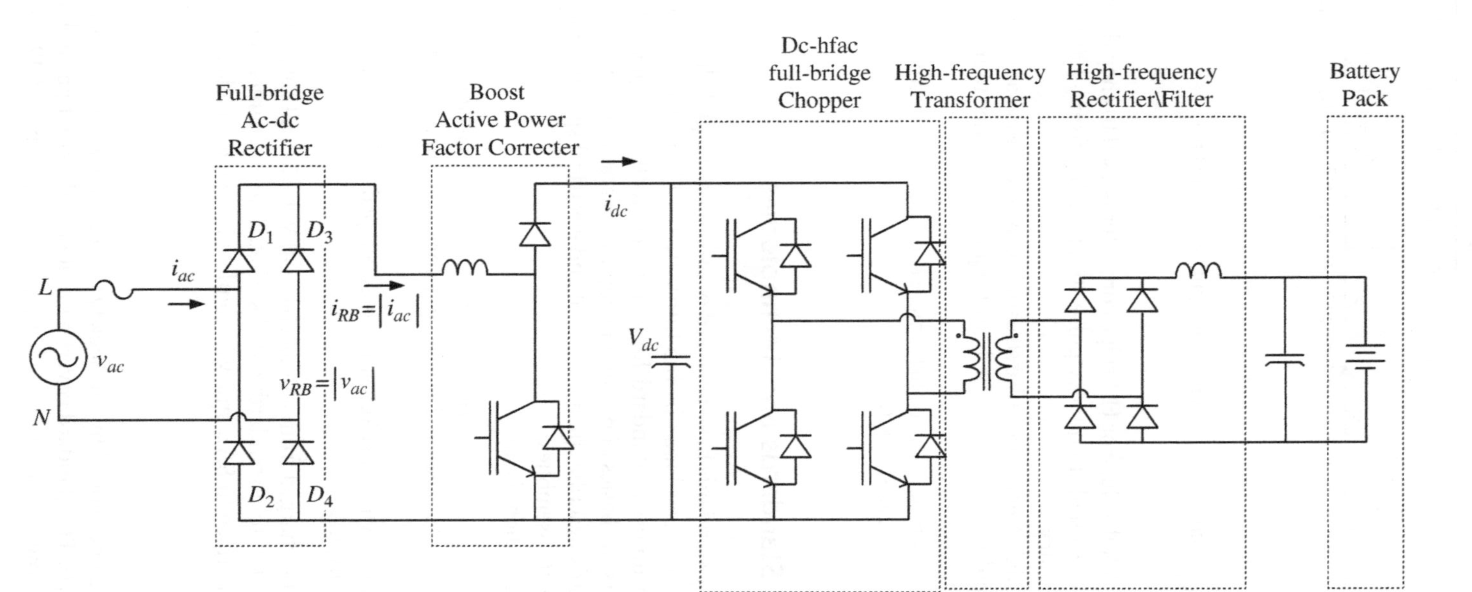

Figure 14.11 Automotive standard charger.

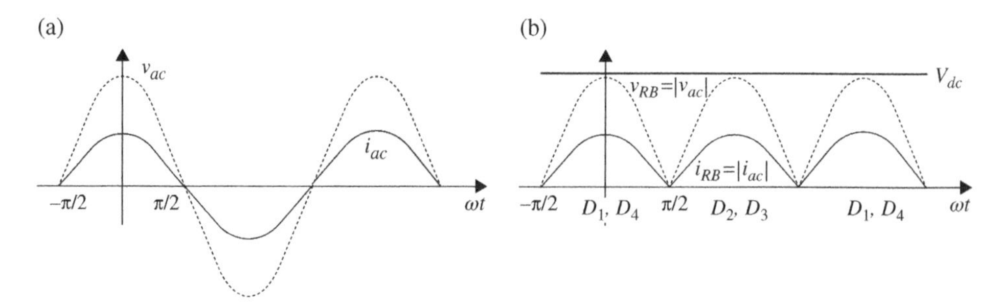

Figure 14.12 Power-factor-corrected waveforms: (a) input and (b) rectifier.

- **Dc-hfac chopper**: A full-bridge or H-bridge converter is used to chop the nominal 400 V dc link voltage into a high-frequency pulse stream going from -400 V to +400 V at the switching frequency.
- **Transformer**: the high-frequency pulse stream is galvanically isolated for safety by the transformer. The pulse stream must be high frequency in order to minimize the size and weight of the transformer.
- **Rectifier-filter**: The output of the transformer secondaries are rectified and filtered to create dc current to charge the battery.

14.5 Charging Standards and Technologies

A number of charging standards have emerged or are emerging globally. A global standard, IEC 62196, has been developed by the International Electrotechnical Commission (IEC) and acts as an umbrella standard for a number of the charging standards. The global standard covers the basics of power and communication interfaces, while the various charging standards describe the mechanical and electrical specifications of the particular plug and socket assemblies.

The main charging standards are

1) SAE J1772 for use in North America and for ac and dc charging.
2) VDE-AR-E 2623-2-2 for use in Europe and for single-phase and three-phase ac charging.
3) JEVS G105-1993, known as CHAdeMo and developed in Japan, for use globally for high-power dc charging.

Tesla vehicles can be charged from a dedicated 240 V Tesla wall charger or by using a standard plug connected to a standard 240 V socket. The ac charger is on-board. The Tesla vehicles can interface to SAE and VDE outlets by using an adapter.

14.5.1 SAE J1772

This standard has been developed by the Society of Automotive Engineers (SAE) for use in North America [1,2]. The standard covers a number of different power levels. **Level 1** charging is for low-power convenience charging using a standard 120 V outlet and

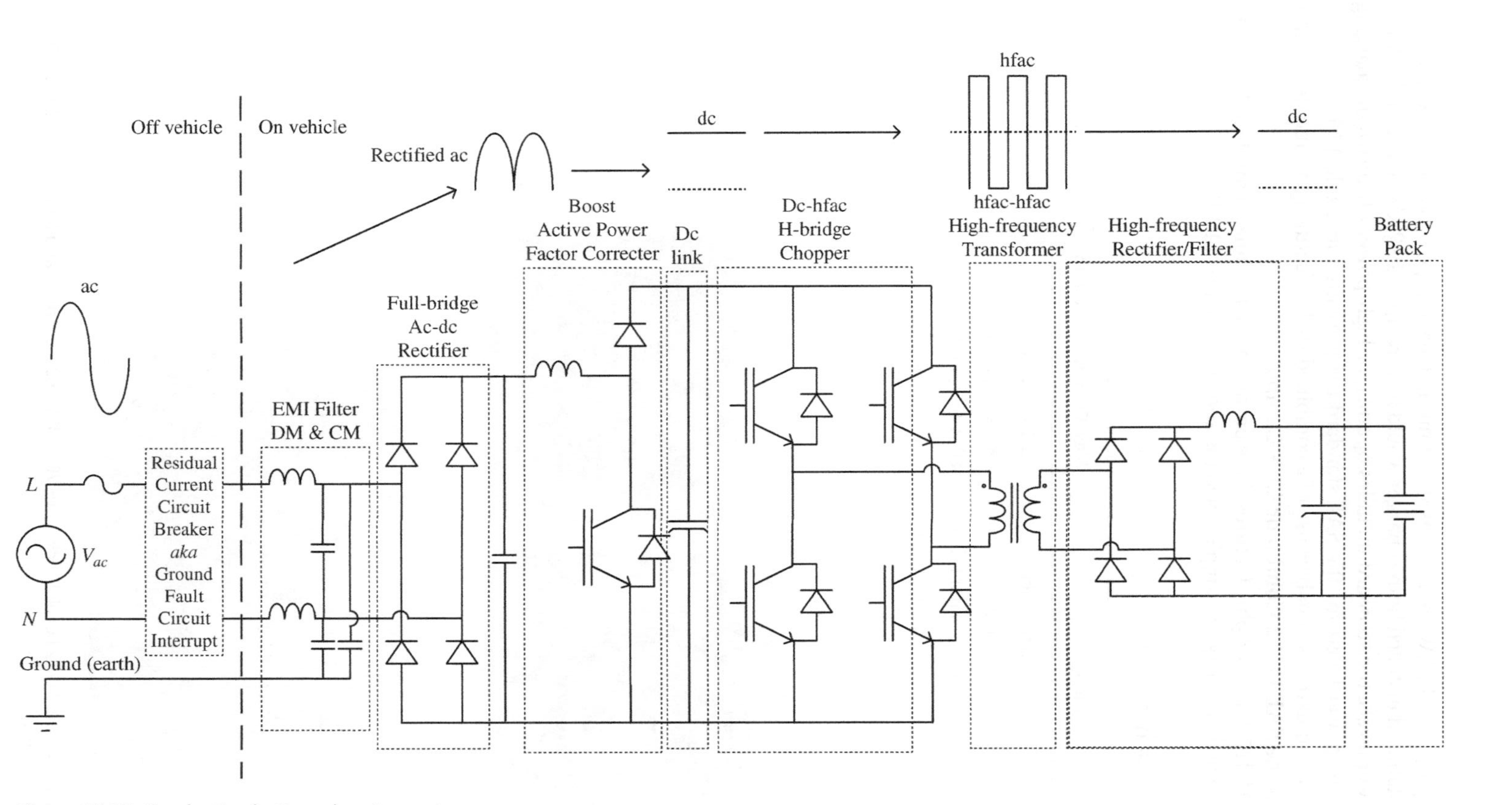

Figure 14.13 Conductive battery charging system.

supplying up to 1.44 kW or 1.96 kW maximum, depending on the outlet, whereas the **Level 2** standard installed home charger would feature up to 19.2 kW, if available. Above these power levels, the standard includes options for high-power off-board dc charging. The various power levels for the SAE standard are presented in Table 14.1.

The SAE standard also enables use of a combined socket featuring ac and dc which is known as the **SAE J1772 Combo** or **CCS Combo**.

The SAE J1772 Level 2 plug features three power contacts – line, neutral, and ground – and two signal contacts. A sample socket is shown in Figure 14.14(a).

Table 14.1 SAE J1772 levels

	Voltage	Max. Continuous Current
Ac Level 1	120 V (input ac)	12/16 A (input ac)
Ac Level 2	240 V (input ac)	<80 A (input ac)
Dc Level 1	50–500V (output dc)	80 A (output dc)
Dc Level 2	50–500V (output dc)	200 A(output dc)

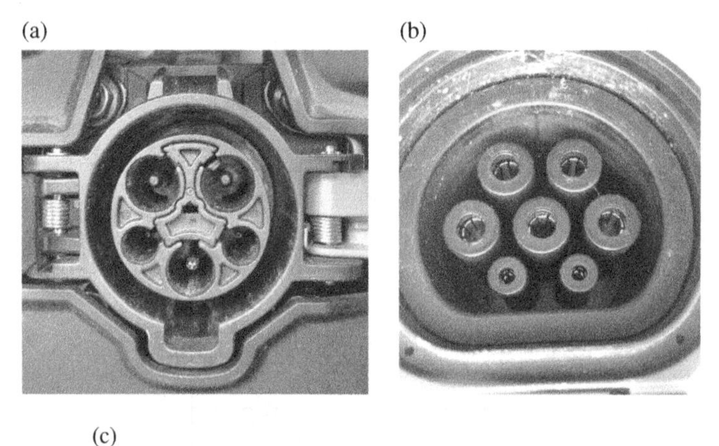

Figure 14.14 (a) SAE J1772 Level 2 socket, (b) VDE-AR-E 2623-2-2 plug, and (c) VDE-AR-E 2623-2-2 plus SAE Combo socket.

14.5.2 VDE-AR-E 2623-2-2

The VDE-AR-E 2623-2-2 standard has been developed for use in Europe. It facilitates the use of three phase, in addition to single phase, as three phase is widely available in parts of Europe. The standard includes five power wires – three lines, a neutral, and a ground – and two signal wires, as shown in Figure 14.14(b), and is often known as the **Mennekes** connector. This standard can be matched with the SAE high-power dc charging standard socket, as shown in Figure 14.14(c), to create a version of the CCS Combo.

14.5.3 CHAdeMo

Several thousand high-power chargers have been installed globally using the CHAdeMO standard. The basic high-power charger is 40 kW. The 40 kW charger, plug, and socket are all shown in Figure 14.15.

14.5.4 Tesla

The Tesla residential charger is designed to operate using commonly provided outlets. The vehicle comes with an adapter set which allows the charger to interface to the available power or off-board charging outlet as shown in Figure 14.16(a). The charger cable plug to the vehicle is shown in Figure 14.16(b).

14.5.5 Wireless Charging

14.5.5.1 Inductive

Wireless charging or inductive charging is a method of transferring electrical power from the source to the load magnetically rather than by direct ohmic contact. The technology offers the advantages of galvanic isolation, safety, connector robustness, and durability in

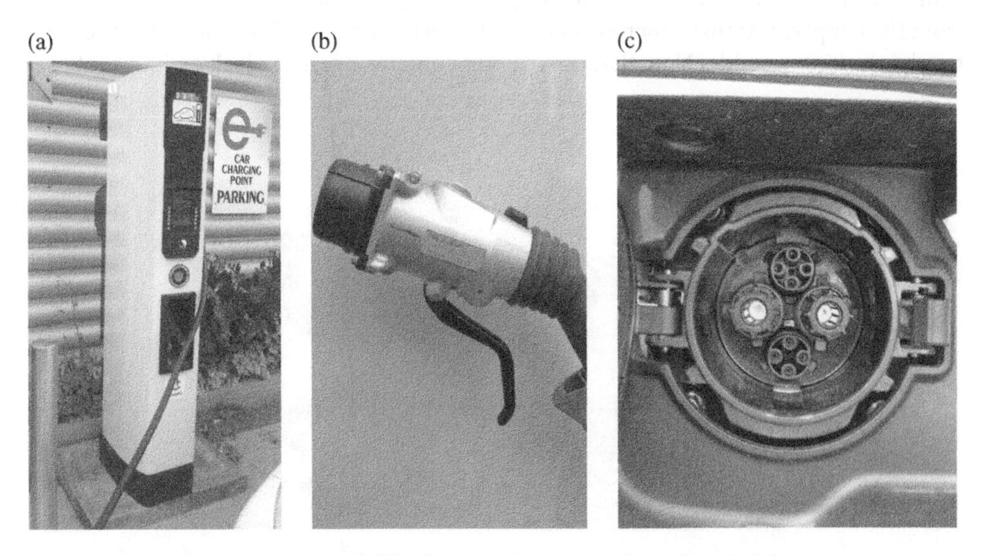

Figure 14.15 CHAdeMO: (a) off-board dc charger, (b) plug, and (c) on-board socket.

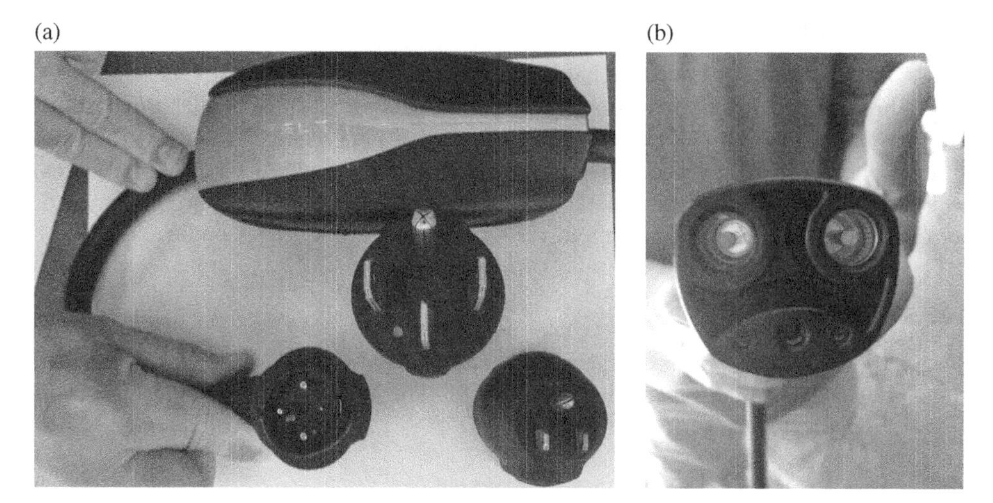

Figure 14.16 Tesla plug adapters and charging plug.

power delivery applications where harsh or hazardous environmental conditions may exist. Examples of these applications are mining and sub-sea power delivery and EV battery charging. The General Motors EV1 featured many new technologies including a radically new design for inductively coupled battery charging.

The basic principle underlying inductive coupling is that the two halves of the inductive coupling interface are the primary and secondary of a separable two-part transformer. When the charge coupler (i.e., the primary) is juxtaposed with the vehicle inlet (i.e., the secondary), power can be transferred magnetically with complete electrical isolation, as with a standard transformer. The coupler and vehicle inlet featured in the EV1 are shown in Figure 14.17(a). The coupler is attached via the cable to the off-vehicle charging module. When the coupler is inserted into the vehicle inlet, power from the coupler is transformer-coupled to the secondary, rectified, and fed to the battery by the battery cable. Note that the coupler contains a ferrite block or "puck" at the

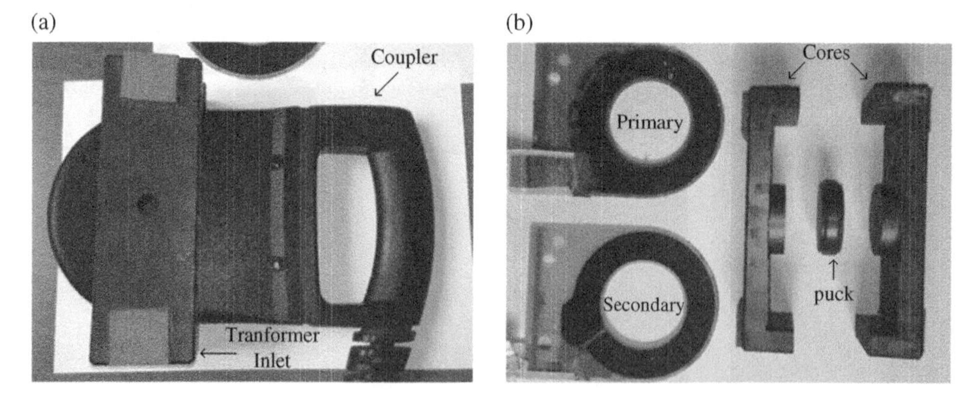

Figure 14.17 EV1 coupler and vehicle inlet.

center of the primary winding to complete the magnetic path when the coupler is inserted into the vehicle inlet. The disassembled transformer components are shown in Figure 14.17(b).

An off-vehicle high-frequency power converter feeds the cable, coupler, vehicle-charging inlet, and battery load. The EV user physically inserts the coupler into the vehicle inlet where the high-frequency power is transformer-coupled, rectified, and fed to the battery. The technology was researched and productized at levels ranging from a few kilowatts to tens of kilowatts, with a high-power demonstration at 120 kW [3–6]. A recommended practice for inductive charging of EVs, SAE J1773, was published by the SAE. The specifications, as outlined in SAE J1773, for the coupler and vehicle inlet characteristics were to be considered when selecting a driving topology. Among the most critical parameters are the frequency range, the low magnetizing inductance, the high leakage inductance, and the significant discrete parallel capacitance. The off-vehicle EV1 charge module features a frequency-controlled series-resonant converter. Driving the SAE J1773 vehicle interface with the series-resonant converter results in a four-element topology with many desirable features. This resonant topology is discussed in detail in Chapter 12, Section 12.4.

14.5.5.2 Wireless

The inductive charging system developed by GM is no longer used as the market shifted to conductive standards. However, new wireless charging standards are being developed for EVs as wireless charging is once again being viewed positively for developing the EV market. An interesting application of wireless charging is driverless vehicles, as charging can be facilitated without human actions. Recent interest in wireless charging has been in loosely coupled transformer systems. The principles are similar to the inductive coupling just described, with the difference that the transformer primary and secondary assemblies are spaced many centimeters apart and have relatively greater leakage inductances.

Worldwide standards are being developed. SAE J2954 is the SAE standard. The technology is also dependant on the types of resonant circuits discussed in Chapter 12 for application to inductive coupling. Additional safety issues must be addressed, such as the effects of radiation on humans and animals and the presence of metal objects in the magnetic fields.

14.6 The Boost Converter for Power Factor Correction

The front-end of the charger is a power-factor-correction stage utilizing a boost converter. An example of an automotive charger is shown in Figure 14.18. The charger features an interleaved boost and so has two boost inductors, shown with an L. The input (IP) power first flows through the EMI filter (EMI) and is then rectified (R) and boosted (Q+D, L). The charger requires a significant electrolytic bulk capacitor stage (C) in order to filter the 50/60 Hz ripple. The full-bridge converter stage features the power transformers (Xo), the full-bridge switches and diodes (Qo+Do), the output rectifier (Ro), the output inductor (Lo), and output filtering (EMIo). The full-bridge converter is discussed in depth in Chapter 12. The focus of this section is on the boost PFC, as shown in Figure 14.19(a).

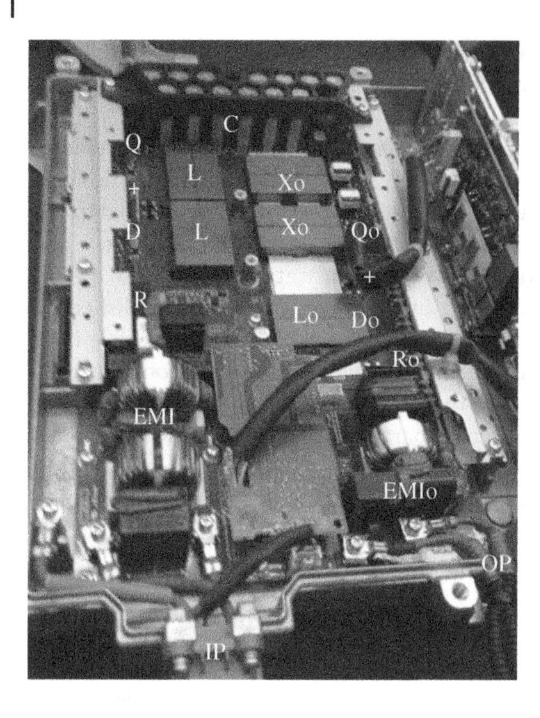

Figure 14.18 Automotive EV charger.

14.6.1 The Boost PFC Power Stage

The input current is controlled to be in phase with the supply voltage. This is achieved by using two control loops: an inner current loop and an outer voltage loop. The converter typically achieves very high power factors with values greater than 0.99 being reasonable at full load [7–9].

The ac voltages and currents are rectified by the input bridge. Let the ac voltage be

$$v_{ac}(\theta) = \sqrt{2}V_{ph}\cos\theta \tag{14.3}$$

Using the boost PFC, the ac current is controlled to be in phase with the voltage:

$$i_{ac}(\theta) = \sqrt{2}I_{ph}\cos\theta \tag{14.4}$$

The voltage v_{RB} and current i_{RB} from the rectifier bridge are

$$v_{RB}(\theta) = \sqrt{2}V_{ph}|\cos\theta| \tag{14.5}$$

$$i_{RB}(\theta) = \sqrt{2}I_{ph}|\cos\theta| \tag{14.6}$$

The boost converter controls the inductor current i_L to be in phase with the rectified input voltage v_{RB}, as shown in Figure 14.19(b) The inductor current carries the PWM ripple current as shown in Figure 14.19(c).

The capacitor at the output of the rectifier C_R is there to filter the PWM ripple current of the inductor.

The size of the dc link capacitor in a PFC boost converter is based on three factors: (1) the desired hold-up time of the capacitors, (2) the current rating and aging of the electrolytic capacitors, and (3) the low-frequency voltage ripple. All three factors must

(a)

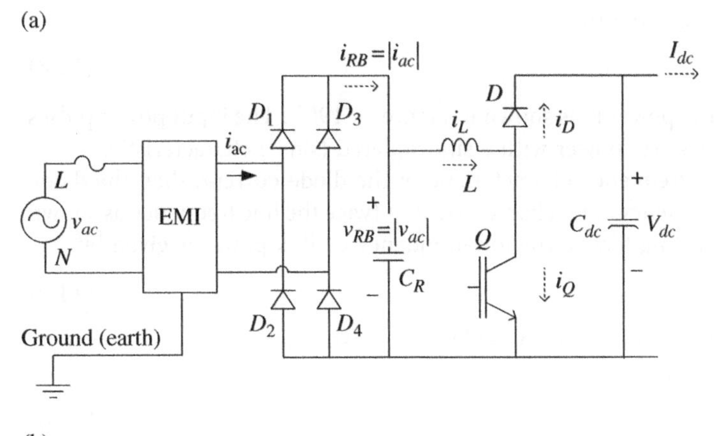

(b)

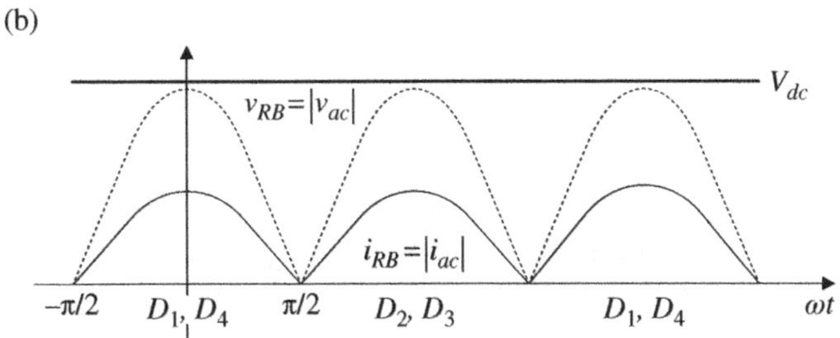

(c)

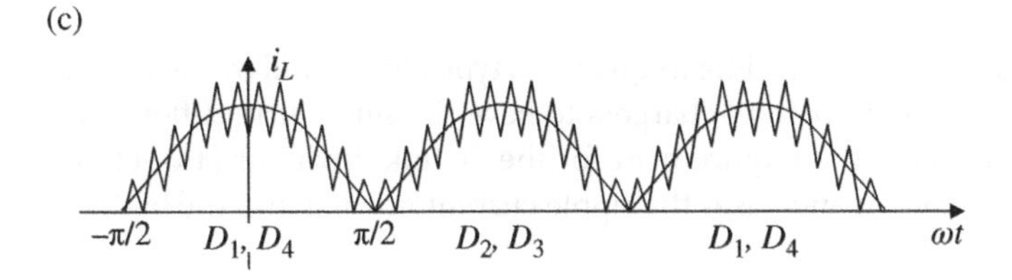

(d)

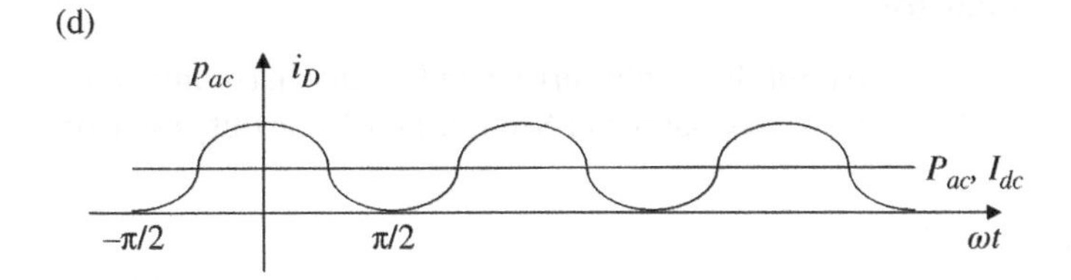

Figure 14.19 Boost PFC and waveforms.

be considered when designing the power and control stages. The dc link capacitor is usually a bank of bulky electrolytic capacitors, which are sized to minimize the low-frequency voltage ripple due to the high second harmonic of the dc link current.

The EMI stage provides common-mode and differential-mode EMI filtering in order to meet the applicable EMC standards [10–13]. It is common for EMI stages to comprise 10% to 20 % of the volume of the charger or of the generic power converter.

The power pulled from the ac source p_{ac} has a pulsing sine^2 characteristic given by

$$p_{ac}(\theta) = v_{ac}(\theta) \times i_{ac}(\theta) = \sqrt{2}V_{ph}\cos\theta \times \sqrt{2}I_{ph}\cos\theta = 2V_{ph}I_{ph}\cos^2\theta \tag{14.7}$$

with an average power P_{ac} given by

$$P_{ac} = V_{ph} I_{ph} \tag{14.8}$$

The characteristic of the power is as shown in Figure 14.19(d). The input power pulses from zero to twice the average power with a sine-squared (sine^2) characteristic.

If we consider the low-frequency characteristic of the diode current, then the diode current also has a low-frequency sine^2 characteristic at twice the line frequency as shown in Figure 14.19(d). Neglecting PWM, the low-frequency dc link power is given by

$$p_D(\theta) = V_{dc} \times i_D(\theta) \tag{14.9}$$

If we ignore the rectifier and boost power loss, the instantaneous dc link power p_D equals the input power p_{ac}:

$$p_D(\theta) = p_{ac}(\theta) \tag{14.10}$$

and so

$$i_D(\theta) = \frac{p_{ac}(\theta)}{V_{dc}} = \frac{2V_{ph}I_{ph}}{V_{dc}}\cos^2\theta \tag{14.11}$$

Thus, the boost diode has a similar low-frequency sine^2 current characteristic, with power pulses from zero to twice the average power as shown in Figure 14.19(d). The average of the diode current equals the dc link current I_{dc} and is given by

$$I_{dc} = \frac{P_{ac}}{V_{dc}} = \frac{V_{ph}I_{ph}}{V_{dc}} \tag{14.12}$$

This low-frequency current at twice the line frequency is typically filtered by the dc link capacitor C_{dc}. It is also an option in battery chargers to have the sine^2 current flow into the battery, significantly reducing the requirement for the dc link capacitor [14]. However, **sine^2 charging**, as it is known, increases the ripple current flowing into the battery.

14.6.2 Sizing the Boost Inductor

The boost inductor is sized in order to limit the ripple current and reduce harmonics and EMI. From the analysis of the CCM boost in Chapter 11 Section 11.4.1, the duty cycle of the boost switch is given by

$$d(\theta) = 1 - \frac{v_{RB}(\theta)}{V_{dc}} = 1 - \frac{\sqrt{2}V_{ph}|\cos\theta|}{V_{dc}} \tag{14.13}$$

The peak-to-peak ripple current is

$$\Delta I_{L(p-p)}(\theta) = \frac{\sqrt{2}V_{ph}|\cos\theta|}{f_s L} d(\theta) = \frac{\sqrt{2}V_{ph}}{f_s L}\left(|\cos\theta| - \frac{\sqrt{2}V_{ph}\cos^2\theta}{V_{dc}}\right) \tag{14.14}$$

and so the ripple current magnitude clearly varies with duty cycle.

14.6.2.1 Example: Sizing the Inductor

A 3.3 kW PFC boost is designed for a nominal input voltage of 230 V, 50/60 Hz with an input voltage ranging from a low line of 180 V to a high line of 265 V. Determine the value of the boost inductor and the peak inductor current if the switching frequency is 100 kHz and the peak-to-peak ripple ratio is 20% of the peak low-frequency current at the peak of the nominal input voltage. The dc link voltage is 380 V. Ignore the power loss.

Solution:
The rms input current is

$$I_{ph} = \frac{P}{V_{ph}} = \frac{3300}{230}\text{A} = 14.35\,\text{A}$$

The peak of the input current is

$$I_{ph(peak)} = \sqrt{2}I_{ph} = \sqrt{2} \times 14.35\,\text{A} = 20.29\,\text{A}$$

The peak-to-peak ripple current at the peak of the input voltage is

$$\Delta I_{L(p-p)} = r_i I_{ph(peak)} = 0.2 \times 20.29\,\text{A} = 4.06\,\text{A}$$

The duty cycle at the peak of the input voltage is

$$d(0) = 1 - \frac{\sqrt{2}V_{ph}|\cos 0|}{V_{dc}} = 1 - \frac{\sqrt{2}V_{ph}}{V_{dc}} = 1 - \frac{\sqrt{2} \times 230}{380} = 0.1440$$

The required inductance is determined by rearranging Equation (14.14):

$$L = \frac{\sqrt{2}V_{ph}|\cos 0|}{f\,\Delta I_{L(p-p)}(0)}d(0) = \frac{\sqrt{2} \times 230}{100 \times 10^3 \times 4.06} \times 0.144\,\text{H} = 115\,\mu\text{H}$$

The peak current in the inductor is the sum of the peak of the line current and half of the peak-to-peak ripple:

$$I_{L(peak)} = I_{ph(peak)} + \frac{\Delta I_{L(p-p)}}{2} \tag{14.15}$$

In this case:

$$I_{L(peak)} = I_{ph(peak)} + \frac{\Delta I_{L(p-p)}}{2} = 20.29\text{A} + \frac{4.06}{2}\text{A} = 22.32\,\text{A}$$

Once the peak and rms currents are known, the inductor can be sized as covered in Chapter 16, Section 16.3.7, using the area-product method. The rms input current can be used as a reasonable approximation for the rms inductor current.

14.6.3 Average Currents in the Rectifier

The input rectifier diodes alternately conduct. Diodes D_1 and D_4 conduct during the positive half cycle, while diodes D_2 and D_3 conduct during the negative half cycle.

The average current in the rectifier diodes $I_{R(dc)}$ for a half cycle is

$$I_{R(dc)} = I_{D1(dc)} = I_{D4(dc)} = \frac{1}{2\pi}\int_{-\frac{\pi}{2}}^{\frac{\pi}{2}} i_{RB}(\theta)\,d\theta = \frac{1}{2\pi}\int_{-\frac{\pi}{2}}^{\frac{\pi}{2}} \sqrt{2}I_{ph}\cos\theta\,d\theta = \frac{I_{ph}}{\sqrt{2}\pi}\int_{-\frac{\pi}{2}}^{\frac{\pi}{2}} \cos\theta\,d\theta = \frac{\sqrt{2}}{\pi}I_{ph} \tag{14.16}$$

The rms current in the rectifier diodes $I_{R(rms)}$ for a half cycle is

$$I_{R(rms)} = I_{D1(rms)} = I_{D4(rms)} = \sqrt{\frac{1}{2\pi}\int_{-\frac{\pi}{2}}^{\frac{\pi}{2}} (i_{RB}(\theta))^2 d\theta} = \sqrt{\frac{1}{2\pi}\int_{-\frac{\pi}{2}}^{\frac{\pi}{2}} \left(\sqrt{2}I_{ph}\cos\theta\right)^2 d\theta} = \frac{1}{\sqrt{2}}I_{ph} \tag{14.17}$$

The values of the dc and rms currents in diodes D_2 and D_3 are the same as in diodes D_1 and D_4.

14.6.3.1 Example: Input Rectifier Power Loss

Determine the power loss in the input rectifier stage of the earlier example if each rectifier has the following parameters: V_{f0} =0.8 V and r_f = 10 mΩ.

Solution:

The average current in the rectifier diodes for a half cycle is

$$I_{R(dc)} = \frac{\sqrt{2}}{\pi}I_{ph} = \frac{\sqrt{2}}{\pi} \times 14.35\,\text{A} = 6.46\,\text{A}$$

The rms current in the rectifier diodes for a half cycle is

$$I_{R(rms)} = \frac{1}{\sqrt{2}}I_{ph} = \frac{1}{\sqrt{2}} \times 14.35\,\text{A} = 10.15\,\text{A}$$

Per the section on semiconductor losses in Chapter 11, Section 11.5, the conduction loss per rectifier diode is

$$P_{R(cond)} = V_{f0}I_{R(dc)} + r_f I_{R(rms)}{}^2 \tag{14.18}$$

In this case:

$$P_{R(cond)} = V_{f0}I_{R(dc)} + r_f I_{R(rms)}{}^2 = 0.8 \times 6.46\,\text{W} + 0.01 \times 10.15^2\,\text{W} = 6.2\,\text{W}$$

The loss in the rectifier bridge P_{RB} is four times the loss in a single rectifier diode:

$$P_{RB} = 4 \times P_{R(\text{cond})} \tag{14.19}$$

For this example:

$$P_{RB} = 4 \times P_{R(\text{cond})} = 4 \times 6.2\,\text{W} = 24.8\,\text{W}$$

14.6.4 Switch and Diode Average Currents

The duty cycle of the boost switch is given by Equation (14.13).

The low-frequency time-averaged current in the switch can be approximated by

$$i_Q(\theta) = d(\theta) \times |i_{ac}(\theta)| \tag{14.20}$$

Substituting in Equation (14.4) and Equation (14.13) yields:

$$i_Q(\theta) = \left(1 - \frac{\sqrt{2}V_{ph}|\cos\theta|}{V_{dc}}\right) \times \sqrt{2}I_{ph}\cos\theta \tag{14.21}$$

or

$$i_Q(\theta) = \sqrt{2}I_{ph}\cos\theta - \frac{2V_{ph}I_{ph}\cos^2\theta}{V_{dc}} \tag{14.22}$$

The average current in the switch is given by

$$I_{Q(dc)} = \frac{1}{\pi}\int_{-\frac{\pi}{2}}^{\frac{\pi}{2}} i_Q(\theta)d\theta \tag{14.23}$$

Substituting Equation (14.22) into Equation (14.23) yields

$$I_{Q(dc)} = \frac{I_{ph}}{\pi}\int_{-\frac{\pi}{2}}^{\frac{\pi}{2}} \left(\sqrt{2}\cos\theta - \frac{2V_{ph}\cos^2\theta}{V_{dc}}\right)d\theta \tag{14.24}$$

which simplifies to

$$I_{Q(dc)} = I_{ph}\left(\frac{2\sqrt{2}}{\pi} - \frac{V_{ph}}{V_{dc}}\right) \tag{14.25}$$

since

$$\int_{-\frac{\pi}{2}}^{\frac{\pi}{2}} \cos\theta d\theta = 2 \text{ and } \int_{-\frac{\pi}{2}}^{\frac{\pi}{2}} \cos^2\theta d\theta = \frac{\pi}{2} \tag{14.26}$$

Similarly, the low-frequency time-averaged current in the boost diode is given by

$$i_D(\theta) = (1 - d(\theta)) \times |i_{ac}(\theta)| \tag{14.27}$$

which expands to

$$i_D(\theta) = \left(\frac{\sqrt{2}V_{ph}|\cos\theta|}{V_{dc}}\right) \times \sqrt{2}I_{ph}\cos\theta \tag{14.28}$$

The diode conducts the current for half the period, and the average current is

$$I_{D(dc)} = \frac{1}{\pi}\int_{-\frac{\pi}{2}}^{\frac{\pi}{2}} i_D(\theta)d\theta = \frac{1}{\pi}\int_{-\frac{\pi}{2}}^{\frac{\pi}{2}} \frac{2V_{ph}I_{ph}\cos^2\theta}{V_{dc}}d\theta = \frac{2V_{ph}I_{ph}}{\pi V_{dc}}\int_{-\frac{\pi}{2}}^{\frac{\pi}{2}} \cos^2\theta d\theta \tag{14.29}$$

which simplifies to

$$I_{D(dc)} = \frac{V_{ph}I_{ph}}{V_{dc}} \tag{14.30}$$

14.6.5 Switch, Diode, and Capacitor RMS Currents

The rms current in the boost switch can be approximated by

$$I_{Q(rms)} = \sqrt{\frac{1}{\pi}\int_{-\frac{\pi}{2}}^{\frac{\pi}{2}} d(\theta)(i_{ac}(\theta))^2 d\theta} \tag{14.31}$$

Substituting in Equation (14.4) and Equation (14.13) gives

$$I_{Q(rms)} = \sqrt{\frac{1}{\pi}\int_{-\frac{\pi}{2}}^{\frac{\pi}{2}} \left(1 - \frac{\sqrt{2}V_{ph}|\cos\theta|}{V_{dc}}\right) \times \left(\sqrt{2}I_{ph}\cos\theta\right)^2 d\theta} \tag{14.32}$$

which simplifies to

$$I_{Q(rms)} = I_{ph}\sqrt{\frac{1}{\pi}\int_{-\frac{\pi}{2}}^{\frac{\pi}{2}} \left(2\cos^2\theta - \frac{2\sqrt{2}V_{ph}|\cos\theta|\cos^2\theta}{V_{dc}}\right) d\theta} \tag{14.33}$$

Since

$$\int_{-\frac{\pi}{2}}^{\frac{\pi}{2}} |\cos\theta|\cos^2\theta d\theta = \frac{4}{3} \tag{14.34}$$

the rms current in the switch is given by

$$I_{Q(rms)} = I_{ph}\sqrt{1 - \frac{8\sqrt{2}V_{ph}}{3\pi V_{dc}}} \tag{14.35}$$

Similarly the rms current in the boost diode is given by

$$I_{D(rms)} = I_{ph}\sqrt{\frac{8\sqrt{2}V_{ph}}{3\pi V_{dc}}} \tag{14.36}$$

The rms current in the high-voltage dc link capacitor $I_{Cdc(rms)}$ is simply given by

$$\begin{aligned} I_{Cdc(rms)} &= \sqrt{I_{D(rms)}^2 - I_{D(dc)}^2} \\ &= I_{ph}\sqrt{\frac{8\sqrt{2}V_{ph}}{3\pi V_{dc}} - \frac{V_{ph}^2}{V_{dc}^2}} \end{aligned} \tag{14.37}$$

14.6.6 Power Semiconductors for Charging

While the silicon IGBT and the silicon diode are dominant for low-to-medium switching frequencies, the silicon MOSFET is typically preferred at higher switching frequencies for low- and medium-range voltages (<600 V). The conduction loss in the MOSFET can be simply modeled by the **drain-source on-resistance**, typically designated $R_{DS(on)}$. The maximum $R_{DS(on)}$ is usually specified for the component. A characteristic of the power MOSFET is that the on-resistance typically doubles between 25°C and 120°C.

The MOSFET conduction loss is given by

$$P_{Q(\mathrm{cond})} = R_{DS(on)} I_{Q(rms)}{}^2 \tag{14.38}$$

As $R_{DS(on)}$ increases with an index somewhere between the square and cube of the device voltage rating, the silicon MOSFET cannot compete with the silicon IGBT for higher voltages. However, the MOSFET is a competitive device for a high-frequency PFC boost.

A significant source of power loss for the switching pole is the reverse recovery loss of the boost diode. This loss can be eliminated by employing the more expensive wide-band-gap **silicon-carbide (SiC) Schottky diode**, which has no reverse recovery loss.

The average current in the diode and switch when switched is

$$I_{Q(sw,avg)} = I_{D(sw,avg)} = \frac{2\sqrt{2}}{\pi} I_{ph} \tag{14.39}$$

Similar to the IGBT in Chapter 11, Section 11.5.1.3, the MOSFET switching power losses remain

$$P_{Q(sw)} = f_s \left(E_{on} + E_{off}\right) \frac{V_{dc}}{V_{test}} \tag{14.40}$$

It is assumed for simplicity that the switching loss for the SiC diode is zero.

In the final part of this section, the losses are estimated for the semiconductor switches in the boost PFC.

14.6.6.1 Example: Silicon MOSFET and SiC Diode Power Losses

The PFC boost converter of the examples used in this chapter features a representative silicon MOSFET and a SiC diode, nominally the part MKE 11R600DCGFC from manufacturer IXYS [15].

The nominal maximum $R_{DS(on)}$ of the device is specified as 0.165 Ω at 25°C, but increases to about 0.375 Ω at 125°C as shown in Figure 14.20(a).

The diode conduction drops are shown in Figure 14.20(b) and can be modeled by V_{f0} = 0.8 V and r_f = 8.8 mΩ at 125°C.

The MOSFET switching losses are shown in Figure 14.21. The diode switching losses are ignored.

Determine the MOSFET and diode power losses, assuming junction temperatures of 125°C.

Solution:

The switch rms current is

$$I_{Q(rms)} = I_{ph}\sqrt{1-\frac{8\sqrt{2}V_{ph}}{3\pi V_{dc}}} = 14.35 \times \sqrt{1-\frac{8\sqrt{2}\times 230}{3\pi \times 380}}\,\mathrm{A} = 7.504\,\mathrm{A}$$

The MOSFET conduction loss is

$$P_{Q(\mathrm{cond})} = R_{DS(on)} I_{Q(rms)}{}^2 = 0.375 \times 7.504^2\,\mathrm{W} = 21.1\,\mathrm{W}$$

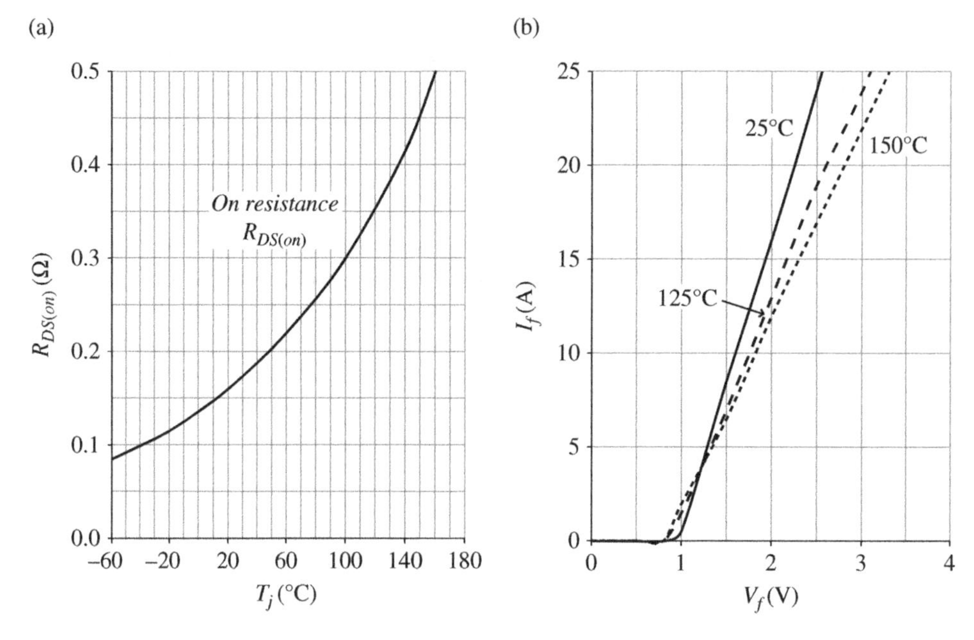

Figure 14.20 Representative 600 V, 15 A MOSFET and SiC diode conduction characteristics.

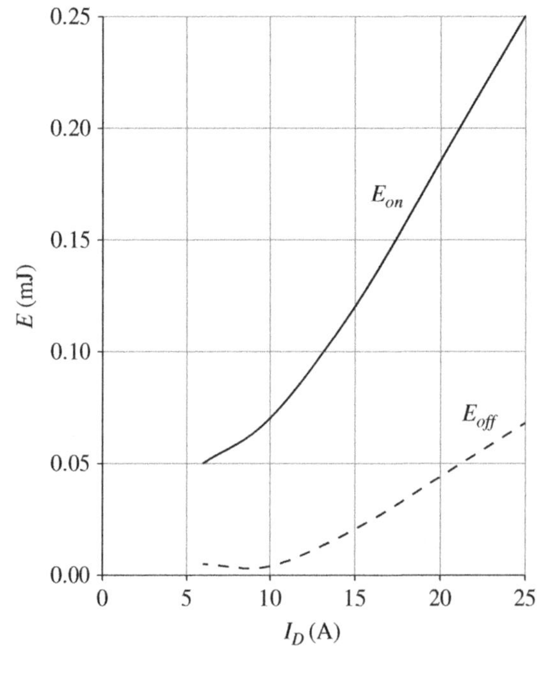

Figure 14.21 Representative 600 V, 15 A MOSFET (and SiC diode) turn-on and turn-off switching losses at 125°C and a test voltage of 380 V_{dc}.

The diode average and rms currents are

$$I_{D(dc)} = \frac{V_{ph}I_{ph}}{V_{dc}} = \frac{230 \times 14.35}{380}\,\text{A} = 8.69\,\text{A}$$

$$I_{D(rms)} = I_{ph}\sqrt{\frac{8\sqrt{2}V_{ph}}{3\pi V_{dc}}} = 14.35 \times \sqrt{\frac{8\sqrt{2} \times 230}{3\pi \times 380}}\,\text{A} = 12.23\,\text{A}$$

The diode conduction loss is

$$P_{D(\text{cond})} = V_{f0}I_{D(dc)} + r_f I_{D(rms)}{}^2 = 0.8 \times 8.69\,\text{W} + 0.0088 \times 12.23^2\,\text{W} = 8.3\,\text{W}$$

For a sinusoidal excitation, the average switching current in the switch and diode is

$$I_{Q(sw,avg)} = I_{D(sw,avg)} = \frac{2\sqrt{2}}{\pi}I_{ph} = \frac{2\sqrt{2}}{\pi}14.35\,\text{A} = 12.92\,\text{A}$$

The turn-on and turn-off power loss in the switch is

$$P_{Q(sw)} = f_s\left(E_{on} + E_{off}\right)\frac{V_{dc}}{V_{test}} = 100 \times 10^3\left(0.1 \times 10^{-3} + 0.013 \times 10^{-3}\right)\frac{380}{380}\,\text{W} = 11.3\,\text{W}$$

where E_{on} = 0.1 mJ and E_{off} = 0.013 mJ at 12.92 A from Figure 14.21.

The power losses in the switch and diode are

$$P_Q = P_{Q(cond)} + P_{Q(sw)} = 21.1\,\text{W} + 11.3\,\text{W} = 32.4\,\text{W}$$

and

$$P_D = P_{D(cond)} = 8.3\,\text{W}$$

14.6.6.2 Example: PFC Stage Losses

If the equivalent series resistance of the inductor is R_{cu} = 50 mΩ, and the combined auxiliary power and stray power loss P_{aux} is 15 W, determine the overall PFC converter power loss and efficiency.

Solution:

The inductor loss is

$$P_L = R_{cu}I_{ph}^2 = 0.05 \times 14.35^2\,\text{W} = 10.3\,\text{W}$$

The auxiliary and stray loss is

$$P_{aux} = 15\,\text{W}$$

The total loss is the sum of the losses in the inductor, the auxiliary circuits, the bridge rectifier, and the boost switch and diode:

$$P_{loss} = P_L + P_{aux} + P_{RB} + P_Q + P_D = 10.3\text{ W} + 15\text{ W} + 24.8\text{ W} + 32.4\text{ W} + 8.3\text{ W} = 90.8\text{ W}$$

The approximate efficiency is

$$\eta = \frac{P}{P + P_{loss}} \times 100\% = \frac{3300}{3300 + 90.8} \times 100\% = 97.3\%$$

References

1 *SAE Electric Vehicle and Plug in Hybrid Electric Vehicle Conductive Charge Coupler, SAE J-1772*, Society of Automotive Engineers.
2 *SAE Power Quality Requirements for Plug-In Electric Vehicle Chargers, SAE J-2894*, Society of Automotive Engineers.
3 J. G. Hayes, *Resonant Power Conversion Topologies for Inductive Charging of Electric Vehicle Batteries*, PhD Thesis, University College Cork, 1998.
4 *SAE Electric Vehicle Inductive Coupling Recommended Practice, SAE J-1773*, Society of Automotive Engineers, Draft, Feb. 1, 1995.
5 R. Severns, E. Yeow, G. Woody, J. Hall, and J. Hayes, "An ultra-compact transformer for a 100 W to 120 kW inductive coupler for electric vehicle battery charging," *IEEE Applied Power Electronics Conference*, pp. 32–38, 1996.
6 J. G. Hayes, N. O'Donovan, and M. G. Egan, "Inductance characterization of high-leakage transformers," *IEEE Applied Power Electronics Conference*, pp. 1150–1156, 2003.
7 N. Mohan, *Power Electronics A First Course*, Chapter 6, John Wiley & Sons, 2012.
8 S. Abdel-Rahman, F. Stuckler, and K. Siu, *PFC Boost Converter Design Guide 1200 W Design Example*, Infineon Application Note, 2016.
9 Texas Instruments, *UCC2817, UCC2818, UCC3817 and UCC3818 BiCMOS Power Factor Preregulator*, Unitrode Products from Texas Instruments, revised 2015.
10 M. Nave, *Power Line Filter Design for Switched-Mode Power Supplies*, Van Nostrand Reinhold, 1991.
11 H. W. Ott, *Electromagnetic Compatibility Engineering*, John Wiley & Sons, 2009.
12 P. Bardos, "Predicting the EMC performance of high-frequency inverters," *IEEE Applied Power Electronics Conference*, pp. 213–219, 2001.
13 M. Kacki, M. Rylko, J. G. Hayes, and C. R. Sullivan, "Magnetic material selection for EMI filters," *IEEE Energy Conversion Congress and Exposition*, 2017.
14 M. G. Egan, D. O'Sullivan, J. G. Hayes, M. Willers, and C. P. Henze, "Power-factor-corrected single-stage inductive charger for electric-vehicle batteries," *IEEE Transactions on Industrial Electronics*, 54 (2), pp. 1217–1226, April 2007.
15 Website of IXYS Corp., www.ixys.com.

Further Reading

1 N. Mohan, T. M. Undeland, and W. P. Robbins, *Power Electronics Converters, Applications and Design*, 3rd edition, John Wiley & Sons, 2003.
2 R. W. Erickson, *Fundamentals of Power Electronics*, Chapter 17, Kluwer Academic Publishers, 2000.
3 M. Yilmaz and P. T. Krein, "Review of battery charger topologies, charging power levels, and infrastructure for plug-in electric and hybrid vehicles," *IEEE Transactions on Power Electronics*, 28 (5), pp. 2151–2169, May 2013.
4 R. Ryan, J. G. Hayes, R. Morrison, and D. Hogan, "Digital control of an interleaved BCM boost PFC converter with fast transient response at low input voltage," *IEEE Energy Conversion Congress and Exposition*, 2017.

Problems

14.1 Determine the following for the converter example used in this chapter when the line voltage is (i) 180 V and (ii) 265 V, at an output power of 3.3 kW: input current, rectifier bridge and inductor power losses, switch and diode average and rms currents and losses, overall converter loss, and efficiency.

[Ans. (i) I_{ph} = 18.33 A, P_{RB} = 33.1 W, P_L = 16.8 W, $I_{Q(rms)}$ = 12.04 A, $I_{D(dc)}$ = 8.68 A, $I_{D(rms)}$ = 13.82 A, P_Q = 69.9 W ($E_{on} \approx 0.025$ mJ, $E_{off} \approx 0.13$ mJ), P_D = 8.6 W, P_{loss} = 143.4 W, Eff = 95.8%; (ii) I_{ph} = 12.45 A, P_{RB} = 21.0 W, P_L = 7.8 W, $I_{Q(rms)}$ = 5.02 A, $I_{D(dc)}$ = 8.68 A, $I_{D(rms)}$ = 11.39 A, P_Q = 18 W, P_D = 8.1 W, P_{loss} = 69.9 W, Eff = 97.9%]

14.2 A 2 kW PFC boost is designed for a nominal input voltage of 230 V with an input voltage ranging from a low line of 180 V to a high line of 265 V. The dc link voltage is 380 V. Assume an equivalent series resistance of the inductor of 50 mΩ and auxiliary and a stray power loss of 15 W. Use the semiconductor characteristics of Section 14.6.6.1.

i) Determine the value of boost inductor if the switching frequency is 60 kHz and the peak-to-peak ripple ratio is 15% of the peak low-frequency current at the peak of the nominal input voltage.
ii) Determine the following at the nominal voltage: input current, rectifier bridge and inductor power losses, switch and diode average and rms currents, and switch and diode losses, overall converter loss, and efficiency.

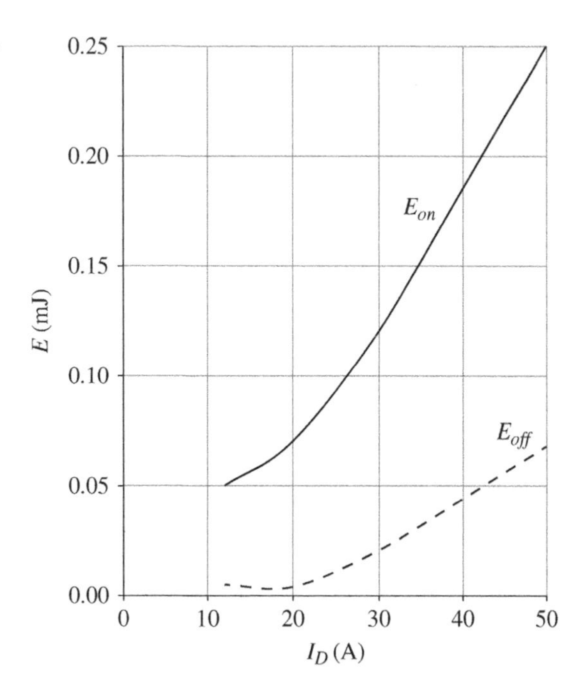

Figure 14.22 Representative 600 V, 30 A MOSFET (and SiC diode) Turn-on and turn-off energy curves at 380 V_{dc}.

[Ans. 422 μH, $I_{ph} = 8.7$ A, $P_{RB} = 14$ W, $P_L = 3.8$ W, $I_{Q(rms)} = 4.55$ A, $I_{D(rms)} = 5.26$ A, $I_{D(rms)} = 7.41$ A, $P_Q = 11.6$ W, $P_D = 4.7$ W, $P_{aux} = 15$ W, $P_{loss} = 49.1$ W, $Eff = 97.6\%$]

14.3 A 6.6 kW PFC boost is designed for a nominal input voltage of 220 V, 50/60 Hz. The switching frequency is 40 kHz and the peak-to-peak ripple ratio is 10% of the peak low-frequency current at the peak of the nominal input voltage. The dc link voltage is 380 V.

i) Determine the values of the boost inductor and the peak inductor current.
ii) Determine the power losses in the boost switch if the $R_{DS(on)}$ of the MOSFET is 0.19 Ω, and the turn-on and turn-off energy losses are given by the curves in Figure 14.22. Note that the curves are for a dc link voltage of 380 V.

[Ans. 332 μH, $I_{L(peak)} =$ 44.5 A, $P_Q = 57$ W]

Assignments

What are the voltages, currents, frequency, and wiring configurations in your region? Which charging standards are used in your region?

15

Control of the Electric Drive

"It has recently come to my attention that my good friend James Clerk Maxwell has had some difficulty with a rather trivial problem ..."

Edward John Routh (1831–1907) [1].

In this chapter, basic control theory is applied to the electric vehicle powertrain propelled by the dc machine. The basics of control apply whether the machine is dc or ac, and so the integration of the permanent-magnet (PM) dc machine is the starting point. The wound-field (WF) dc machine is then used to enable high-speed field-weakened operation of the electric vehicle. MATLAB/Simulink is used to simulate the drive.

15.1 Introduction to Control

Control is one of the key technical areas of electrical power conversion and electric drives, as we are always interested in controlling the voltage, current, frequency, torque, speed, and so on. The application of feedback control has contributed greatly to the industrial age. The first industrial controller was developed by James Watt (1736–1819) to regulate his steam engine. James Clerk Maxwell (1831–1879) is acknowledged as one of the great minds of science and mathematics, and is credited with initially developing control theory. Maxwell and his electromagnetic equations feature prominently in Chapter 16. Edward John Routh (1831–1907), a classmate of Maxwell at the University of Cambridge in England, also made some great contributions to control theory and stability analysis. In a distinguished class at Cambridge, Routh actually finished just above Maxwell in the graduating class of 1854.

Hendrik Bode (1905–1982) and Harry Nyquist (1889–1976) further developed and applied control theory as we learn it today.

From an electric drive perspective, there are two main feedback control loops on a vehicle. The outer loop is the vehicle speed loop. This loop is a relatively slow loop and is generally closed in the brain of the human driver. Obviously, the role of the driver in controlling the vehicle speed is changing with the advent of autonomous self-driving vehicles. The second main loop is the torque command, which essentially is fed into the

Electric Powertrain: Energy Systems, Power Electronics and Drives for Hybrid, Electric and Fuel Cell Vehicles,
First Edition. John G. Hayes and G. Abas Goodarzi.

Companion website: www.wiley.com/go/hayes/electricpowertrain

vehicle powertrain by stepping on the accelerator or brake. This loop is significantly faster than the speed loop, and the vehicle responds rapidly to regulate the machine torque.

This inner torque loop is essentially a current loop in the electric vehicle. Current is easily regulated and is proportional to the machine torque. The outer loop is the speed loop. As speed is proportional to the voltage, this loop can also be the voltage loop in an industrial controller.

15.1.1 Feedback Controller Design Approach

Controlling and stabilizing a feedback controller can be quite technical and mathematical, but also practical. See references [1–7] for further study.

In this investigation, we will make some simplifying assumptions in order to develop basic controllers.

There are four main steps to closing the feedback loop.

First, large-signal models are developed to characterize the system.

Second, the large-signal model is linearized around a dc operating point and investigated using small-signal analysis.

Third, the system is stabilized by applying feedback control theory to the small-signal model.

Finally, the system is tested for a dynamic response using the large-signal models.

In this system, we use two loops as shown in the control block diagram of Figure 15.1. The inner torque/current loop is very fast – typically, this loop has a crossover frequency, also known as the bandwidth, of about 1/10th of the switching frequency, for example, 500 Hz for a 5 kHz switching frequency. The outer loop is the speed loop and is much slower. In a typical industrial application, the bandwidth of the outer loop would be significantly lower than the current loop, perhaps 1/10th or less of the current-loop bandwidth. When the human driver is the regulator, the speed bandwidth is about 5 Hz, a reasonable estimate for the response time of the driver.

The outer speed loop is closed in the brain of the driver. The driver can change the speed of the vehicle by requesting more or less torque from the drive system using the accelerator or brake pedal. The speed error is amplified by the speed error amplifier $G_{c\omega}$ within the human brain. The output of the speed error amplifier is the input command to the torque/current loop, the input mechanism being the accelerator and brake pedals in the case of the vehicle.

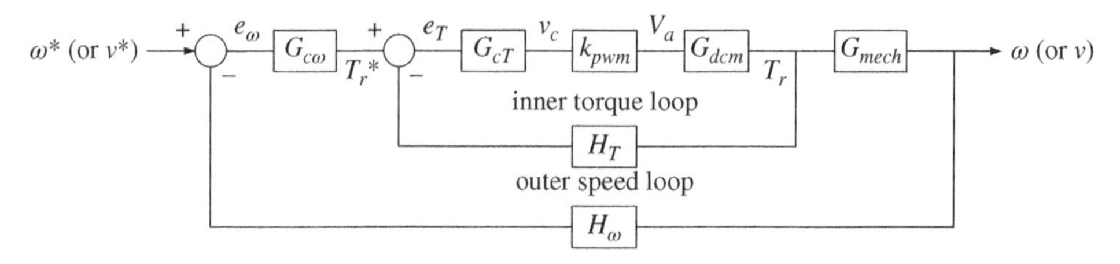

Figure 15.1 Control block diagram.

15.2 Modeling the Electromechanical System

In this section, large- and small-signal models are developed for the mechanical system, the permanent-magnet (PM) dc machine, the dc-dc converter, and the proportional-integral (PI) controller. The overall system is shown in Figure 15.2.

15.2.1 The Mechanical System

The mechanical system can be modeled using the earlier torque equation, Equation (2.26) from Chapter 2 and letting rotor torque T_r equal traction torque T_t:

$$T_r = \frac{1}{n_g \eta_g} T_{axle} = \frac{1}{n_g \eta_g}\left[r\left\{m\frac{dv}{dt} + mg\sin\theta + A + Bv + Cv^2\right\} + J_{axle}\alpha_{axle}\right]$$

The vehicle road-load force and torque are

$$F_v = A + Bv + Cv^2 + mg\sin\theta \text{ and } T_v = F_v r \tag{15.1}$$

The axle and accelerating torques, T_{axle} and T_a, respectively, are

$$T_{axle} = n_g \eta_g T_r \tag{15.2}$$

and

$$T_a = T_{axle} - T_v = rm\frac{dv}{dt} + J_{axle}\alpha_{axle} \tag{15.3}$$

We know that

$$v = r\omega_{axle} \text{ and } \alpha_{axle} = \frac{d\omega_{axle}}{dt} \tag{15.4}$$

Thus, the accelerating torque is

$$T_a = \left(r^2 m + J_{axle}\right)\frac{d\omega_{axle}}{dt} \quad \text{or} \quad d\omega_{axle} = \frac{T_a}{\left(r^2 m + J_{axle}\right)}dt \tag{15.5}$$

where J_{axle} is the equivalent moment of inertia (MOI) referenced to the drive axle.

The final part of the equation can be formatted as an integral to give the axle speed ω_{axle} in rad/s:

$$\omega_{axle} = \int \frac{T_a}{r^2 m + J_{axle}} dt \tag{15.6}$$

This expression can be written as a Laplace transform:

$$\omega_{axle}(s) = \frac{1}{\left(r^2 m + J_{axle}\right)}\frac{1}{s}T_a(s) \tag{15.7}$$

where s is the complex operator.

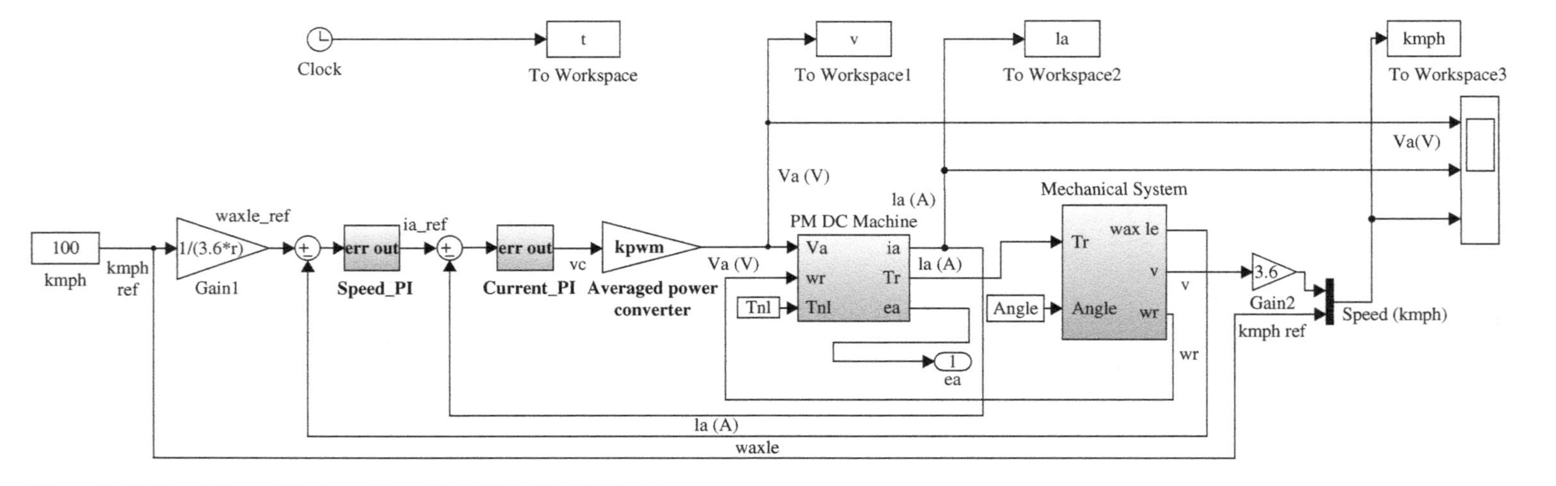

Figure 15.2 Simulink model of electric vehicle (EV) with PM dc drive.

We can substitute in the relationships between rotor speed and torque to axle speed and accelerating torque, respectively, as follows:

$$\frac{\omega_{axle}(s)}{T_a(s)} = \frac{\omega_r(s)/n_g}{n_g \eta_g T_r(s) - T_v(s)} = \frac{1}{(r^2 m + J_{axle})} \frac{1}{s} \tag{15.8}$$

The large-signal model just presented relates speed as an output to torque as an input. The large-signal mechanical system based on the earlier equations is presented in Figure 15.3.

For a small-signal perturbation, the effects of the vehicle load and slope forces can be ignored due to the large inertia of the vehicle. In such a case:

$$\frac{\omega_{axle}(s)}{T_a(s)} \approx \frac{\omega_r(s)}{n_g^2 \eta_g T_r(s)} \approx \frac{1}{(r^2 m + J_{axle})} \frac{1}{s} \tag{15.9}$$

This equation can be modified to determine the small-signal relationship between rotor torque and rotor speed. The equation when modified as follows yields the small-signal gain of the mechanical system:

$$G_{mech}(s) \approx \frac{\omega_r(s)}{T_r(s)} = \frac{n_g^2 \eta_g}{r^2 m + J_{axle}} \frac{1}{s} = \frac{1}{s J_{eq}} \tag{15.10}$$

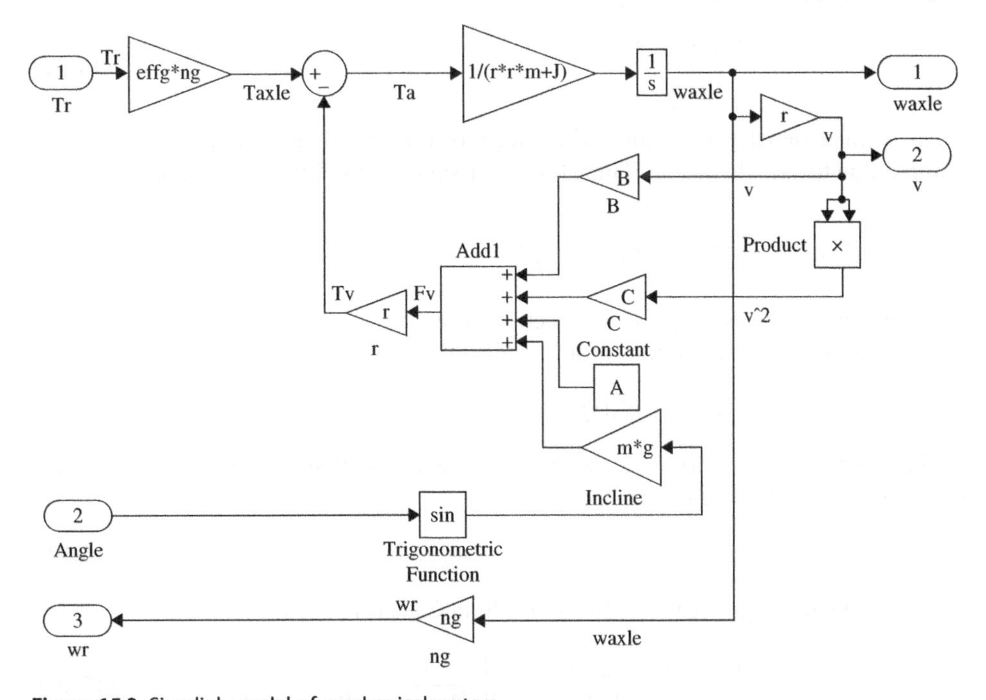

Figure 15.3 Simulink model of mechanical system.

where J_{eq} is the equivalent small-signal inertia of the system as seen from the rotor:

$$J_{eq} = \frac{r^2 m + J_{axle}}{n_g{}^2 \eta_g} \tag{15.11}$$

We now have the large- and small-signal models for the mechanical system.

15.2.2 The PM DC Machine

The PM dc machine model is now developed. The basic machine equations from Chapter 7 are as follows:

$$V_a = E_a + R_a i_a + L_a \frac{di_a}{dt} \tag{15.12}$$

$$T_{em} = kI_a = T_r + T_{nl} \tag{15.13}$$

and

$$E_a = k\omega_r \tag{15.14}$$

The circuit equation is reformatted to integral form:

$$i_a = \int \frac{V_a - E_a - R_a i_a}{L_a} dt \tag{15.15}$$

and as a Laplace transform:

$$I_a(s) = \frac{V_a(s) - E_a(s) - R_a I_a(s)}{sL_a} \tag{15.16}$$

For a small-signal perturbation, the effects of the back emf E_a (and speed) can be ignored due to vehicle inertia. Thus, the small-signal gain of the machine is given by

$$\frac{\Delta i_a}{\Delta v_a} \approx \frac{I_a(s)}{V_a(s)} = \frac{1}{R_a + sL_a} = \frac{1}{R_a} \frac{1}{(1 + s\tau_e)} \tag{15.17}$$

where τ_e, which equals L_a/R_a, is the electrical time constant of the machine.

The relationship between the rotor torque and current is

$$T_{em}(s) = T_r(s) + T_{nl}(s) = kI_a(s) \tag{15.18}$$

The effect of no-load torque can be ignored, and the small-signal relationship between rotor torque and armature current is

$$T_r(s) = kI_a(s) \tag{15.19}$$

Thus, the small-signal gain of the dc machine G_{dcm} can be modeled as

$$G_{dcm}(s) = \frac{T_r(s)}{V_a(s)} = \frac{T_r(s)}{I_a(s)} \times \frac{I_a(s)}{V_a(s)} = k \times \frac{1}{R_a} \frac{1}{(1 + s\tau_e)} = \frac{k}{R_a(1 + s\tau_e)} \tag{15.20}$$

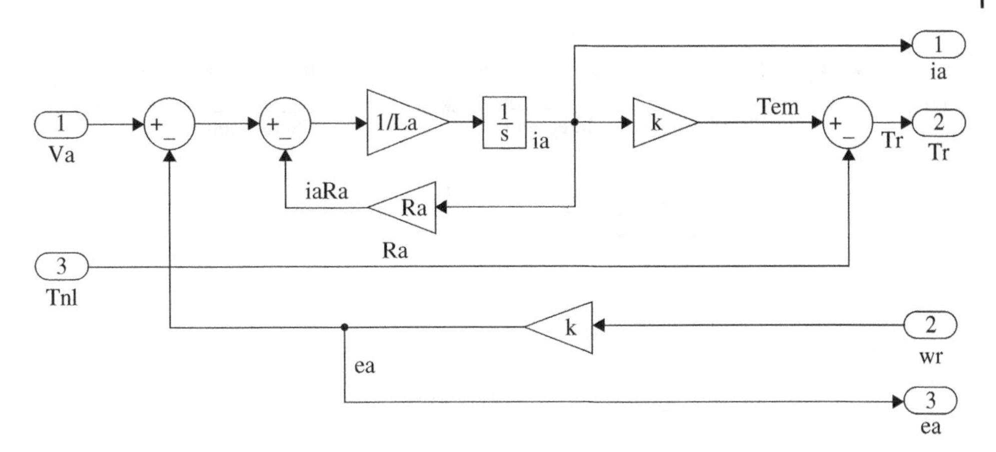

Figure 15.4 Simulink model of dc machine.

Combining Equation (15.10) and Equation (15.13), the relationship between rotor speed and armature current can be easily derived as

$$\frac{\Delta\omega_r}{\Delta I_a} \approx \frac{\omega_r(s)}{I_a(s)} = \frac{k}{sJ_{eq}} \tag{15.21}$$

The large-signal machine equations are built into a Simulink model as shown in Figure 15.4. Note that the machine moment of inertia is ignored as it is negligible compared to the vehicle inertia.

15.2.3 The DC-DC Power Converter

The full-bridge dc-dc converter is operated using current-mode control. The machine torque can be controlled by regulating the armature current.

From a control perspective, the dc-dc converter can be seen as an amplifier with a gain k_{pwm} given by

$$k_{pwm} = \frac{V_a}{V_c} = \frac{V_s}{V_{tri(pk)}} \tag{15.22}$$

where voltage V_c is the amplifier control voltage, which is amplified by gain k_{pwm} to provide the amplifier output voltage V_a, and then supplied to the armature terminals of the machine. The gain can be shown to be the ratio of the dc voltage feeding the amplifier V_s to the peak of the triangular voltage $V_{tri(pk)}$ used for pulse-width-modulation (PWM) signal generation.

15.2.4 The PI Controller

A PI controller can be used for controlling both the torque and speed loops.

The PI error amplifier is shown in Figure 15.5, and is simply modeled as

$$G_c(s) = k_p + \frac{k_i}{s} = \frac{k_i + k_p s}{s} = k_i \frac{1 + \frac{k_p}{k_i} s}{s} \tag{15.23}$$

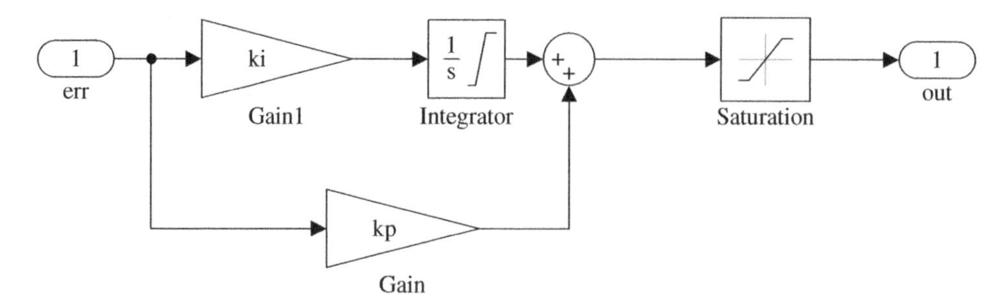

Figure 15.5 PI controller.

The error amplifier has a gain k_i with a pole at 0 Hz or dc, and a zero determined by k_p/k_i.

15.3 Designing Torque Loop Compensation

The loop gain of the torque loop $G_{LT}(s)$ is as follows:

$$G_{LT}(s) = G_{cT}(s)k_{pwm}G_{dcm}(s)H_T(s) \tag{15.24}$$

where $G_{cT}(s)$ and $H_T(s)$ are the gains of the PI error amplifier in the torque loop and of the feedback stage, respectively.

The gain of the torque-loop PI error amplifier is

$$G_{cT}(s) = k_{iT}\frac{1 + \frac{k_{pT}}{k_{iT}}s}{s} \tag{15.25}$$

For simplicity, the feedback gain $H_T(s)$ is set to unity. Thus, the torque-loop gain simplifies to

$$G_{LT}(s) = G_{cT}(s)k_{pwm}G_{dcm}(s) \tag{15.26}$$

In greater detail:

$$G_{LT}(s) = k_{iT}\frac{\left(1 + \frac{k_{pT}}{k_{iT}}s\right)}{s} \times k_{pwm} \times \frac{k}{R_a(1 + s\tau_e)} \tag{15.27}$$

For simplicity, the loop can be easily stabilized by placing the compensator zero such that it cancels the pole due to the motor. Thus, let

$$\tau_e = \frac{k_{pT}}{k_{iT}} \tag{15.28}$$

This simplifies Equation (15.27) to

$$G_{LT}(s) = \frac{k_{iT}k_{pwm}k}{R_a}\frac{1}{s} \tag{15.29}$$

which represents a single pole with a phase of −90°.

Let the bandwidth or crossover frequency of the current-control loop be ω_{cT}. The loop gain is unity at the crossover frequency. Thus, once the bandwidth is specified, the compensator parameters are easily determined:

$$|G_{LT}(j\omega_{cT})| = \frac{k_{iT} k_{pwm} k}{R_a \omega_{cT}} = 1 \tag{15.30}$$

resulting in an expression for the integral gain:

$$k_{iT} = \frac{R_a \omega_{cT}}{k_{pwm} k} \tag{15.31}$$

Now Equation (15.28) can be used to determine the proportional gain.

15.3.1 Example: Determining Compensator Gain Coefficients for Torque Loop

The system parameters of a PM dc motor supplied by a current-controlled switch-mode PWM dc-dc converter are as follows: armature resistance $R_a = 50\ \text{m}\Omega$, armature inductance $L_a = 0.5\ \text{mH}$, converter dc bus voltage $V_s = 300\ \text{V}$, amplitude of triangular waveform control voltage $V_{tri(pk)} = 3\ \text{V}$, switching frequency $f_s = 10\ \text{kHz}$, and machine constant $k = 0.77\ \text{Nm/A}$. The motor is controlled by an inner loop controlling torque (current). The feedback effects of the motor-induced back emf and the load torque on the control loop can be neglected.

Calculate the error amplifier proportional and integral gains k_{pT} and k_{iT} in order for the torque loop to have a single-pole roll-off characteristic with the crossover frequency at one tenth of the switching frequency, that is, 1000 Hz. Assume unity feedback.

Solution:

The electrical time constant and converter gain are

$$\tau_e = \frac{L_a}{R_a} = \frac{0.5 \times 10^{-3}}{50 \times 10^{-3}} \text{s} = 10\,\text{ms}$$

and

$$k_{pwm} = \frac{V_s}{V_{tri(pk)}} = \frac{300}{3} = 100$$

Assuming a single-pole roll-off, we know from Equation (15.31) that

$$k_{iT} = \frac{R_a \omega_{cT}}{k_{pwm} k} = \frac{50 \times 10^{-3} \times 2\pi \times 1000}{100 \times 0.77} = 4.08$$

and from Equation (15.28) that

$$k_{pT} = k_{iT} \tau_e = 4.08 \times 10 \times 10^{-3} = 0.0408$$

15.4 Designing Speed Control Loop Compensation

The loop gain of the speed loop $G_{L\omega}$ is as follows:

$$G_{L\omega}(s) = G_{c\omega}(s)\, G_{CLT}(s)\, G_{mech}(s) H_\omega(s) \tag{15.32}$$

where G_{CLT} is the closed-loop gain of the inner torque loop, and $H_{\omega}(s)$ is the speed feedback gain. The gain of the speed-loop PI error amplifier is

$$G_{c\omega}(s) = k_{i\omega}\frac{1+\frac{k_{p\omega}}{k_{i\omega}}s}{s} \tag{15.33}$$

As the speed loop is a much slower loop, we can assume that the torque loop is ideal and that its gain is unity, that is, $G_{CLT}(s)=1$. For simplicity, let us also set the feedback gain $H_{\omega}(s)$ to unity.

Thus, the speed-loop gain can be simplified to

$$G_{L\omega}(s) = G_{c\omega}(s)\,G_{mech}(s) \tag{15.34}$$

or

$$G_{L\omega}(s) = k_{i\omega}\frac{\left(1+\frac{k_{p\omega}}{k_{i\omega}}s\right)}{s}\frac{1}{sJ_{eq}} \tag{15.35}$$

which represents a function with two poles at dc plus a zero at a frequency determined by the controller gains. The magnitude of the loop gain is given by

$$\left|G_{L\omega}(j\omega)\right| = \frac{k_{i\omega}}{J_{eq}}\frac{\sqrt{1+\left(\frac{k_{p\omega}}{k_{i\omega}}\omega\right)^2}}{\omega^2} \tag{15.36}$$

The phase of the loop is

$$\angle G_{L\omega}(j\omega) = -180^0 + \tan^{-1}\left(\frac{k_{p\omega}}{k_{i\omega}}\omega\right) \tag{15.37}$$

as the two poles at dc provide a phase of −180°. The phase of the zero at the crossover frequency provides the speed-loop phase margin.

Thus, the phase margin PM_{ω} of the speed loop is

$$PM_{\omega} = \tan^{-1}\left(\omega_{c\omega}\frac{k_{p\omega}}{k_{i\omega}}\right) \tag{15.38}$$

If the desired phase margin is known, then the proportional gain can be determined from

$$k_{p\omega} = \frac{k_{i\omega}}{\omega_{c\omega}}\tan(PM_{\omega}) \tag{15.39}$$

The gain at the crossover frequency is

$$\left|G_{L\omega}(j\omega)\right| = \frac{k_{i\omega}}{J_{eq}}\frac{\sqrt{1+(\tan(PM_{\omega}))^2}}{\omega_{c\omega}{}^2} = 1 \tag{15.40}$$

Thus, the integral gain is given by the following equation:

$$k_{i\omega} = \frac{\omega_{c\omega}{}^2 J_{eq}}{\sqrt{1 + (\tan(PM_\omega))^2}} \tag{15.41}$$

15.4.1 Example: Determining Compensator Gain Coefficients for Speed Loop

The vehicle of the earlier example is controlled by an outer speed loop with a bandwidth of 5 Hz and a phase margin of 60°. Let the mass m = 1645 kg, wheel radius r = 0.315 m, gear ratio n_g = 8.19, gear efficiency η_g= 95%, and the axle-referenced MOI J_{axle} = 3 kgm^2. Assume a feedback of unity for the speed loop.

Calculate the error amplifier proportional and integral gains $k_{p\omega}$ and $k_{i\omega}$.

Solution:

The equivalent MOI is

$$J_{eq} = \frac{r^2 m + J_{axle}}{n_g{}^2 \eta_g} = \frac{0.315^2 \times 1645 + 3}{8.19^2 \times 0.95} = 2.61\,\text{kgm}^2$$

The PI coefficients are

$$k_{i\omega} = \frac{\omega_{c\omega}{}^2 J_{eq}}{\sqrt{1 + (\tan(PM_\omega))^2}} = \frac{(2\pi \times 5)^2 2.61}{\sqrt{1 + (\tan 60^o)^2}} = 1288$$

and

$$k_{p\omega} = \frac{k_{i\omega}}{\omega_{c\omega}} \tan(PM_\omega) = \frac{1288}{2\pi \times 5} \tan(60^o) = 71$$

The PM dc drive simulation in the next section is controlled using the parameters derived in the two previous examples.

15.5 Acceleration of Battery Electric Vehicle (BEV) using PM DC Machine

A controlled drive acceleration of the vehicle can now be simulated as the control coefficients are known, and can be used in the Simulink model of Figure 15.2. A simple acceleration of the vehicle from 0 to 100 km/h is simulated and plotted in Figure 15.6. The machine armature voltage V_a, armature current I_a and vehicle speed in km/h are shown in Figure 15.6.

As discussed in Chapter 7, Section 7.1.7, the conventional PM dc machine cannot be field weakened, and so it is not possible to operate it significantly beyond the rated speed of about 39.56 km/h for the example. As the vehicle accelerates, the machine back emf increases from 0 V to 300 V at 50 km/h. As the maximum voltage of 300 V has been reached, the vehicle cannot speed up anymore, and the speed is clamped at about 50 km/h. The armature current drops significantly from the initial high value to enable acceleration to the much lower value required to cruise at about 50 km/h.

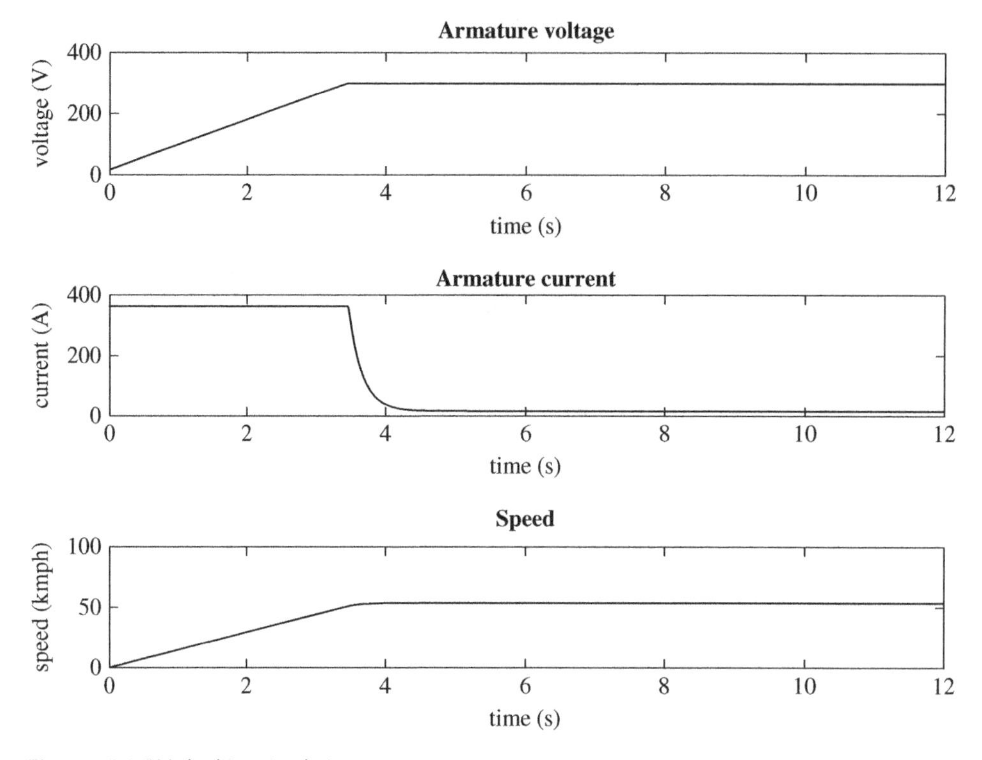

Figure 15.6 PM dc drive simulation outputs.

15.6 Acceleration of BEV using WF DC Machine

The PM dc machine model of Figure 15.2 is now replaced with a WF dc machine. The WF dc machine model is as shown in Figure 15.7 and Figure 15.8. Note that the control coefficients are not changed in this simple simulation.

As discussed in Chapter 7, Section 7.1.8, the field current I_f is reduced proportionately with speed above the rated speed to enable high-speed operation by field weakening. The machine constant is given by

$$k = \frac{p}{2} L_f I_f \tag{15.42}$$

where the field current is speed dependent as follows:

$$I_f = I_{f(rated)} \text{ below the rated speed,} \quad \text{and } I_f = I_{f(rated)} \frac{\omega_{r(rated)}}{\omega_r} \text{ above the rated speed.} \tag{15.43}$$

As the WF dc machine can be field weakened to the maximum speed of 150 km/h, the vehicle can provide full acceleration to 100 km/h, as shown in Figure 15.9. This simple example illustrates the benefits of field-weakening an electrical machine to enable high-speed operation.

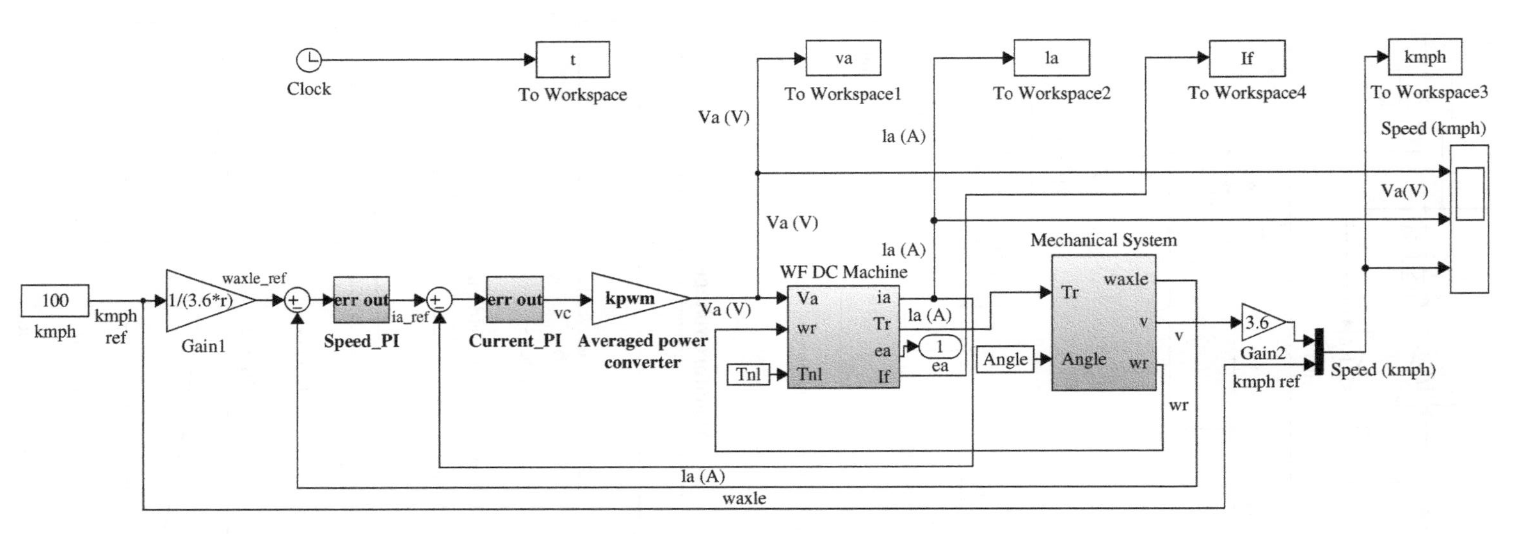

Figure 15.7 Simulink model of WF dc drive.

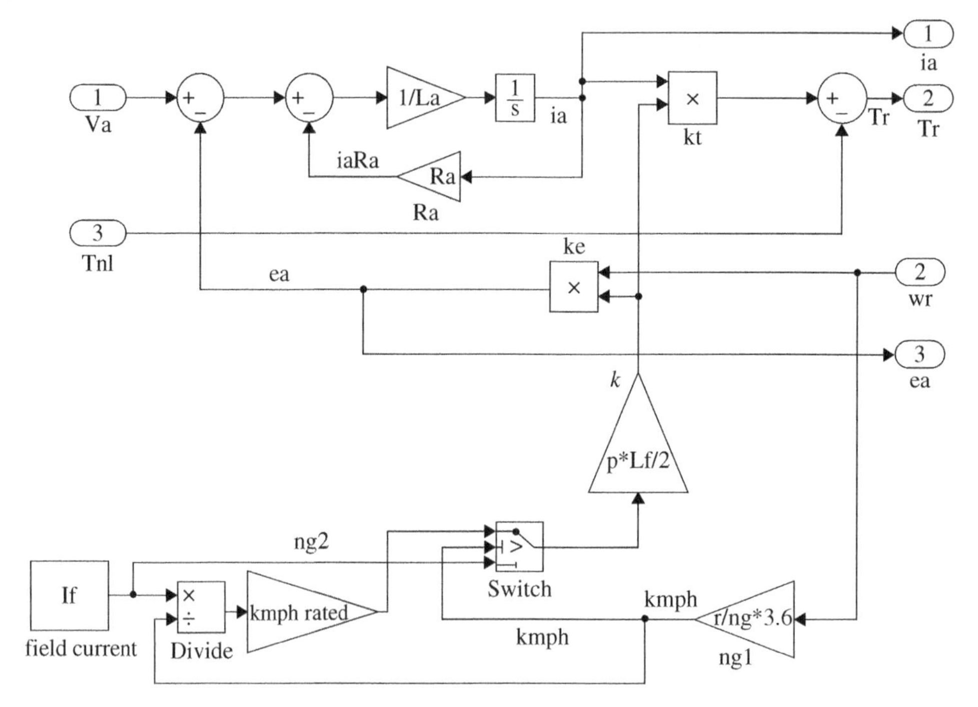

Figure 15.8 Simulink model of WF dc machine.

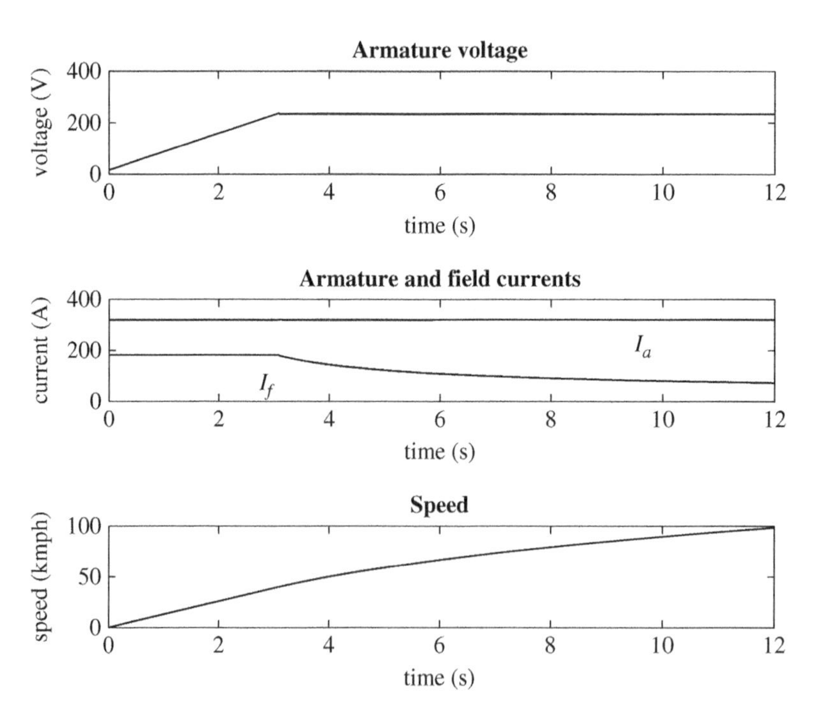

Figure 15.9 WF dc drive simulation scope outputs.

As shown in Figure 15.9, the vehicle can speed up above the rated speed and can maintain maximum current flow into the machine as the machine speed is not limited by the back emf due to the field weakening.

References

1 J. G. Truxal, *Introductory System Engineering*, McGraw-Hill, 1972, p. 527.
2 Staff of University of Michigan, Carnegie Mellon University, University of Detroit Mercy, *Control Tutorials for Matlab & Simulink.*
3 K. Dutton, S. Thompson, and B. Barraclough, *The Art of Control Engineering*, Pearson Education, 1997.
4 K. Ogata, *Modern Control Engineering*, Pearson Education, 2010.
5 N. Mohan, *Power Electronics: A First Course*, John Wiley & Sons, 2012.
6 N. Mohan, T. M. Undeland, and W. P. Robbins, *Power Electronics Converters, Applications and Design*, 3rd edition, John Wiley & Sons, 2003.
7 R. W. Erickson, *Fundamentals of Power Electronics*, Chapter 17, Kluwer Academic Publishers, 2000.

Problems

15.1 Recalculate the error amplifier gains of the example in Section 15.3.1 if the feedback is not unity but has a transducer outputting 5 V for 400 Nm.

In this problem, the simplified loop gain becomes

$$G_{LT}(s) = k_{iT}\frac{\left(1+\dfrac{k_{pT}}{k_{iT}}s\right)}{s} \times k_{pwm} \times \frac{k}{R_a(1+s\tau_e)} \times H_T \tag{15.44}$$

[Ans. $k_{iT} = 326.4$, $k_{pT} = 3.264$]

15.2 Recalculate the error amplifier gains of Problem 15.1 assuming a fuel cell voltage source outputting 150 V.

[Ans. $k_{iT} = 652.8$, $k_{pT} = 6.53$]

15.3 Recalculate the error amplifier gains of Example 15.4.1 if the vehicle is autonomous and features a bandwidth of 20 Hz with a phase gain of 45°.

[Ans. $k_{i\omega} = 29140$, $k_{p\omega} = 232$]

15.4 The system parameters of a PM dc motor supplied by a current-controlled switch-mode PWM dc-dc converter are as follows: armature resistance R_a = 20 mΩ, armature inductance L_a = 0.2 mH, converter dc bus voltage V_s = 360 V, amplitude of triangular waveform control voltage $V_{tri(pk)}$ = 3 V, switching frequency f_s = 10 kHz, machine constant k = 0.6 Nm/A, gear ratio n_g = 9.73, gear efficiency η_g= 96%, axle-referenced MOI J_{axle} = 3 kgm^2, mass m = 2155 kg, and wheel radius $r = 0.3$ m. The drive has a torque transducer outputting 5 V for 1200 Nm. The motor is controlled by an inner loop controlling torque (current). The feedback effects of the motor-induced back emf and the load torque on the control loop can be neglected. The speed loop has a feedback of unity.

i) Calculate the error amplifier proportional and integral gains k_{pT} and k_{iT} in order for the torque loop to have a single-pole roll-off characteristic with the crossover frequency at one tenth of the switching frequency.
ii) Calculate the error amplifier proportional and integral gains $k_{p\omega}$ and $k_{i\omega}$ in order for the speed loop to have a bandwidth of 15 Hz and a phase margin of 60°.

[Ans. $k_{iT} = 418.9$, $k_{pT} = 4.19$, $k_{i\omega} = 9624$, $k_{p\omega} = 177$]

Assignment and Sample MATLAB Codes

Model the acceleration of your vehicle from the Chapter 2 assignment using a WF dc machine. The following code is the setup for the Simulink model. The code is easily modified for your specific vehicle.

```
% Parameters for Cascaded (Torque & Speed) Control of 2015 Nissan
  Leaf EV with
% PM or WF dc Motor (John Hayes). The program can be modified by
  changing the vehicle or
% control parameters. The algorithm uses a simple method based on
  assumed efficiency to estimate Ra and Pcfw. The rated field
  current is estimated at 50 % of rated armature current.

%Initialisation section
clear variables;
close all;
clc;
format short g;
format compact;

%Vehicle Parameters needed
Prrated    = 80000;    % Maximum output power at rotor of motor (W)
Trrated    = 254;      % Maximum motor positive torque range [Nm]
p          = 8;        % Number of poles
ng         = 8.19;     % Gear ratio from EPA spreadsheet
r          = 0.315     % Wheel radius (m)
kmphmax    = 150;      % Maximum speed (km/h)
Aepa       = 29.97;    % Coastdown A from EPA test data
Bepa       = 0.0713;   % Coastdown B from EPA test data
Cepa       = 0.02206;  % Coastdown C from EPA test data
m          = 1645;     % Test weight of vehicle from EPA test
                         data (kg)
```

```
Angle        = 0;           % Incline angle in radians
g            = 9.81;        % Acceleration due to gravity (ms-2)

% Control parameters
fcT          = 1000;        % Bandwidth of torque loop (Hz)
fcw          = 5;           % Bandwidth of speed loop(Hz)
PMw          = pi/3;        % Phase margin of speed loop

% Assumptions
Eamin        = 220;         % Minimum battery voltage (V)
Effrated     = 0.92;        % Machine efficiency at rated\max condition
Loss_split   = 0.9;         % Armature copper loss as percentage of
                              total machine loss
Vs           = 300;         % Battery voltage (V)
Vtri         = 3;           % Peak of triangular voltage (V)
J            = 3;           % Axle-referenced MOI
effg         = 0.97;        % Gearing efficiency

%Vehicle Calculations
A            = Aepa*4.448            % Coastdown A in metric units
B            = Bepa*9.95             % Coastdown B in metric units
C            = Cepa*22.26            % Coastdown C in metric units

wrrated      = Prrated/Trrated       % Base angular speed (rad/s)
Nrrated      = wrrated*60/(2*pi)     % Base speed in [rpm] Used for field
                                       weakening
frrated      = wrrated/(2*pi)        % Base rotor frequency (Hz)
kmphrated    = 3.6*r*wrrated/ng      % Base speed (km/h)
wrmax        = kmphmax/3.6/r*ng      % Maximum motor speed [rad/s]
Nrmax        = wrmax*30/pi;          % Maximum motor speed [rpm]
k            = Eamin/wrrated         % Machine constant k (Nm/A)
Lambdaf      = k*2/p                 % Lambda
Ia           = Trrated/k             % Armature current at rated
                                       torque (A)
Iamax        = Trrated/k             % Armature current at max
                                       torque (A)
If           = Ia*0.5                % Let rated If equal 50 % of rated
                                       Ia (A)
Lf           = Lambdaf/If            % armature inductance (H)
Pmloss       = Prrated*(1/Effrated-1);      % machine power loss at
                                              rated (W)
Pcu          = Loss_split*Pmloss            % copper loss (W)
Pcfw         = (1-Loss_split)*Pmloss;       % core, friction and
                                              windage loss (W)
Ra           = Pcu/(Ia^2)            % armature resistance (ohm)
Tnl          = Pcfw/wrrated          % no-load torque (Nm)
```

```
La       = Lf/2                        % Let La equal 50 % of
                                         Lf (H)
Jeq      = (r^2*m+r*J)/(ng^2*effg);    % equivalent MOI (kg m2)
kpwm     = Vs/Vtri;                    % gain of pwm stage
te       = La/Ra;                      % armature time constant
                                         (H/ohm)

wcT      = 2*pi*fcT;          % torque crossover frequency
                                (rad/s)
kiT      = wcT*Ra/kpwm/k      % integral gain of torque loop
kpT      = kiT*te             % proportional gain of torque loop

wcw      = 2*pi*fcw;          % speed crossover frequency in
                                rad/s
kiw      = Jeq*wcw^2/(sqrt(1+(tan(PMw))^2))     % integral gain
                                                  of speed loop
kpw      = kiw * tan(PMw)/wcw % proportional gain of speedloop

disp('Vehicle parameters loaded...')
```

Part 4

Electromagnetism

16

Introduction to Electromagnetism, Ferromagnetism, and Electromechanical Energy Conversion

> "Why, sir, there is every possibility that you will soon be able to tax it." Michael Faraday (1791–1867) responding in the 1850s to the British Prime Minister William Gladstone on the possibilities for his new electrical inventions.
>
> "Tandem felix" (Latin for "Happy at last") – on the gravestone of André-Marie Ampère (1775–1836).

In this chapter, the student is introduced to the theory of electromagnetism, ferromagnetism, and electromechanical energy conversion. An understanding of these topics is important for the engineering student to understand the theory of operation of electromechanical machines and electromagnetic devices such as inductors, transformers, motors, and capacitors. Although this chapter is not essential reading for the casual reader, it is worthwhile to review the material as it provides an overview on the important laws and some engineering background on important enabling materials such as permanent magnets.

The chapter begins with a review of Maxwell's equations and the basics of electromagnetism in Section 16.1. The magnetic properties of ferromagnetic materials are presented in Section 16.2. Inductor theory is discussed in Section 16.3. An example based on the power inductor of a hybrid electric vehicle is presented. An electromagnet is also presented as an example of an inductor with an air gap, which aids in the understanding of the dc machine in Chapter 7. Hard ferromagnetic materials and permanent magnets are discussed in Section 16.4. Permanent magnets are used for the high-power-density machines featured in Chapters 7, 9, and 10. An introduction to transformers is presented in Section 16.5, and is supported by some basic examples which are useful for the earlier Chapter 8 on the induction machine and Chapter 12 on the isolated power converters. Section 16.6 discusses general capacitor technology and presents an example on the sizing of power film capacitors. Electromechanical energy conversion is presented in Section 16.7 and provides the basic theory required for the machines described in Chapters 6 to 10.

Electric Powertrain: Energy Systems, Power Electronics and Drives for Hybrid, Electric and Fuel Cell Vehicles, First Edition. John G. Hayes and G. Abas Goodarzi.

Companion website: www.wiley.com/go/hayes/electricpowertrain

16.1 Electromagnetism

Electromagnetic and electromechanical devices are made up of coupled electric and magnetic circuits. By a magnetic circuit, we mean a path for magnetic flux, just as an electric circuit provides a path for electric current. Sources of magnetic flux are electromagnets and permanent magnets. Examples of electromagnetic devices are transformers, inductors, and capacitors.

Electromechanical machines combine mechanical action and electromagnetism. Examples of these machines are electric motors, generators, and contactors. In these electrical machines, current-carrying conductors interact with magnetic fields, themselves arising from electric currents in conductors or from permanent magnets, resulting in electromechanical energy conversion.

The operation of these devices is based on electromagnetism, whose key equations were summarized elegantly by **James Clerk Maxwell** (1831–1879). Maxwell's equations are the basis of the operation of all electromagnetic and optical devices such as motors, wireless transmitters and receivers, radar, telescopes, and more. Reference [1] discusses the history of the development of the equations and the important roles played by **Oliver Heaviside** (1850–1925) in putting the equations in their present form, and **Heinrich Hertz** (1857–1894) in publishing the earliest work in 1888 validating the equations.

Although the chapter will cover Maxwell's equations and related laws in a bit of depth, we will initially review what these laws mean in practical terms.

Gauss' law for electricity deals with the capacity of a capacitor to store energy in the form of electric charge.

Gauss' law for magnetism describes a key characteristic of magnetic fields. Basically, this law rules that a magnetic field must have North *and* South poles. This is not the case for electric fields, as positive and negative charges can exist independently.

Faraday's law of electromagnetic induction relates to the generation of electricity by the relative motion of a magnetic field.

The **Ampere–Maxwell law** relates to the production of a magnetic field by a changing electric field or current. In simple terms, this law states that a magnetic field can be created by electric current or by changing electric flux in the case of a capacitor.

16.1.1 Maxwell's Equations

This section presents the various laws of electromagnetism for application to electric machines and devices. The laws are initially presented in integral form, but simpler forms are used later. The first two laws are named for Johann Carl Friedrich Gauss (1777–1855), a brilliant German mathematician.

Gauss' law for electricity describes the relationship between charge and an electric field. In equation form,

$$\oint \mathbf{E} d\mathbf{A} = {}^{q}/_{\varepsilon} \tag{16.1}$$

where **E** is the electric field strength vector, q is the electric charge, ε is the permittivity, and **A** is a vector representing the area.

This law describes the ability of a capacitor of capacitance C to store energy in the form of an electric charge of magnitude Q when subjected to a voltage V with the simple relationship

$$Q = CV \tag{16.2}$$

Gauss' law for magnetism describes a key characteristic of magnetic fields. In equation form:

$$\oint \mathbf{B} d\mathbf{A} = 0 \tag{16.3}$$

where **B** is the magnetic flux density vector, and **A** is a vector representing the area.

In simple terms, this law states that a magnetic field must have North *and* South poles. This is not the case for electric fields, as positive and negative charge can exist independently.

Faraday's law of electromagnetic induction relates the production of an electric field to a changing magnetic field:

$$\oint \mathbf{E} d\mathbf{l} = -\frac{d\Phi}{dt} \tag{16.4}$$

where Φ is the magnetic flux, t is time, and **l** is a vector representing length.

Michael Faraday (1791–1867) was a brilliant English scientist who discovered many of the principles of electromagnetism. Heinrich Lenz (1804–1865) corrected the original law of Faraday to include the negative sign, in effect integrating Newton's third law into Faraday's law.

The **Ampere–Maxwell law** relates the production of a magnetic field to a changing electric field or current:

$$\oint \mathbf{H} d\mathbf{l} = \varepsilon \frac{d\Phi_E}{dt} + I \tag{16.5}$$

where **H** is the magnetic field strength, I is the electric current, Φ_E is the electric flux, and ε is the permittivity.

In simple terms, this law states that a magnetic field can be created by electric current or by changing electric flux in the case of a capacitor. The law is named for **André-Marie Ampère** (1775–1836), a French physicist and mathematician who lived through the French Revolution and the Napoleonic era and contributed greatly to electromagnetism. The unit of current is the ampere (symbol A), and is named in his honor.

These equations are presented above in integral form. In our study of power devices in the following sections, we will generally use these equations in their magnetostatic form, where we neglect the effects of changing electric flux.

16.1.1.1 Ampere's Circuital Law (Based on Ampere–Maxwell Law)

The mutual relationship between an electric current I and a magnetic field is given by **Ampere's circuital law**:

$$\oint \mathbf{H} d\mathbf{l} = I \tag{16.6}$$

where **H** is the **magnetic field strength** vector or the magnetic field intensity vector, and **l** is a vector of magnitude l in the direction of the path. The closed-integration symbol denotes the line integral over a closed path. Ampere's law was modified by Maxwell to include the changing electric field component as shown in Equation (16.5), as Ampere originally based his law only on the effects of current.

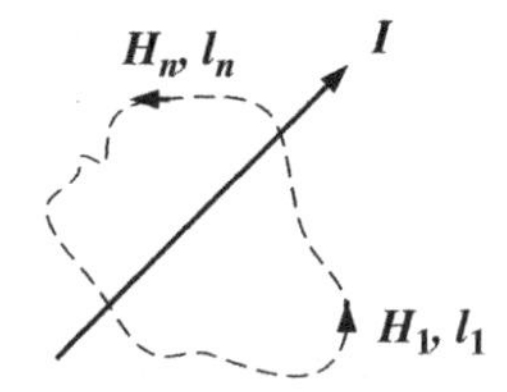

Figure 16.1 Closed contour of magnetic field strength H around conductor of current I.

A useful form of the above equation is to express it in a piecewise fashion by summing the contributions around a closed path, as shown in Figure 16.1. The line integral of **H** around any closed path of integration is equal to the current I enclosed by the path, independent of the surrounding magnetic medium. In equation form:

$$\sum H_n l_n = I \tag{16.7}$$

where H_n and l_n are the magnetic field strength and length of the differential lengths about the contour.

This simple expression for Ampere's law is commonly used for studying power devices, such as inductors and motors.

If the path of integration encloses N conductors, each carrying a current I in the same direction, then

$$\oint \mathbf{H} d\mathbf{l} = NI \tag{16.8}$$

where NI is the number of ampere-turns enclosed. If the path of integration does not enclose any current, or if it encloses equal currents flowing in opposite directions, then

$$\oint \mathbf{H} d\mathbf{l} = 0 \tag{16.9}$$

The units of magnetic field strength are A/m or A-turns/m.

16.1.1.2 Right Hand Screw Rule: Direction of Magnetic Flux

Consider a conductor carrying current out of the page toward you in the direction of the head of an arrow coming at you, as illustrated in Figure 16.2(a), and returning into the

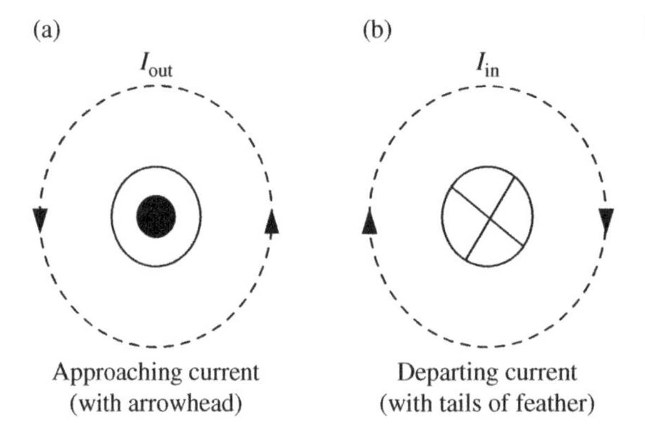

Figure 16.2 Right hand rule.

page from you such that you are seeing the feathered tail of the arrow going away from you, as shown in Figure 16.2(b). Grip the conductor with the right hand, with the thumb outstretched parallel to the conductor and in the direction of the current; the fingers then point in the direction of the magnetic flux around the conductor.

16.1.1.3 Magnetic Flux Density Vector (B)

The magnetic field is principally represented by the magnetic flux density vector (symbol **B**).

The **B** and **H** vectors are related by

$$\mathbf{B} = \mu \mathbf{H} \tag{16.10}$$

where μ is the **absolute permeability** of the medium. Permeability is a measure of the ability of a material to conduct magnetic flux.

The unit of magnetic flux density is the **tesla** (symbol T). Nikola Tesla (1856–1943), a Serb, was the inventor of the ac induction motor and the ac transmission system, making him one of the greatest contributors to the modern world. Tesla Motors company was named in his honor.

Permeability (symbol μ) has the dimensions of H/m, where H is the henry, the unit of inductance.

In air or free space:

$$\mathbf{B} = \mu_0 \mathbf{H} \tag{16.11}$$

where μ_0 has the value of $4\pi \times 10^{-7}$ H/m.

The **relative permeability** μ_r is defined as the ratio of the absolute permeability of the material μ to the permeability of free space μ_0 (note that the subscript is zero). Therefore:

$$\mu_r = \frac{\mu}{\mu_0} \tag{16.12}$$

and is dimensionless.

Hence:

$$\mathbf{B} = \mu_r \mu_0 \mathbf{H} \tag{16.13}$$

For non-ferromagnetic materials such as air, wood, copper, aluminum, and so on, μ_r is nearly unity. For ferromagnetic materials, μ_r can be very high and nonlinear, with values ranging from hundreds to tens of thousands.

Since the **B** and **H** vectors are parallel, the flux lines giving the direction of the **B** vector also give the direction of **H**.

16.1.1.4 Magnetic Flux

The **magnetic flux** Φ through a given open or closed surface **A** is the flux of **B** through that surface:

$$\Phi = \int_A \mathbf{B} d\mathbf{A} \tag{16.14}$$

This equation is commonly written as

$$\Phi = \int_A B \cos\theta dA \tag{16.15}$$

where θ is the angle between $\mathbf{B}$ and the normal to the surface $\mathbf{n_A}$. Note that the surface vector $\mathbf{A}$ has a magnitude A and a direction given by $\mathbf{n_A}$, as shown in Figure 16.3.

Magnetic flux is a scalar quantity and is measured in webers (unit symbol Wb).

If the flux density B is uniform over a surface of area A and is perpendicular to the surface, then

$$\Phi = BA \tag{16.16}$$

16.1.1.5 Gauss' Law for Magnetism

This law simply states that the net magnetic flux entering or leaving a closed surface is zero:

$$\oint \mathbf{B} d\mathbf{A} = 0 \tag{16.17}$$

All the flux which enters the surface enclosing a volume, as shown in Figure 16.4, must leave that volume on some other portion of the surface because magnetic flux lines form closed loops.

This is the mathematical expression of the fact that no magnetic monopoles can exist, unlike electric fields.

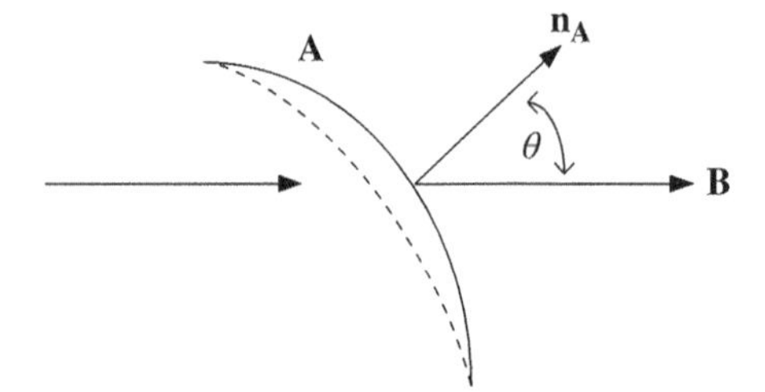

Figure 16.3 Flux and surface vectors.

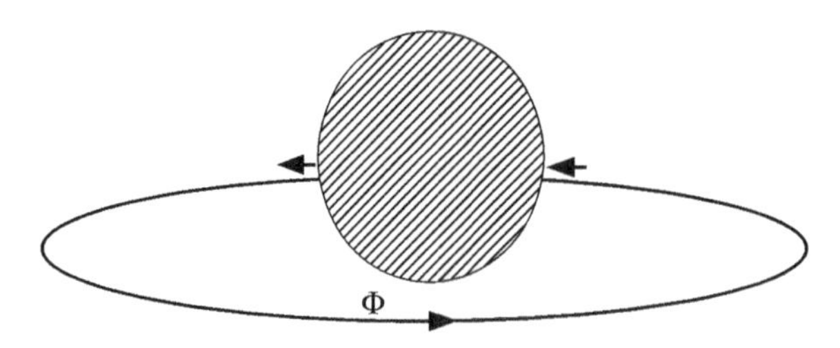

Figure 16.4 Surface and flux.

16.2 Ferromagnetism

Ferromagnetic materials guide magnetic fields and, due to their high permeability, require a small magnetic field strength to produce the desired flux density.

Thus, magnetic components such as inductors, transformers, and motors are built using ferromagnetic materials such as laminated silicon steel, which is used for low-frequency 50/60 Hz applications or motors, or ferrite, which is used for high-frequency (>10 kHz) applications. Samples of the Toyota Prius motor laminations, various cores, and a completed inductor are shown in Figure 16.5.

An example of a disassembled electromagnetic device is shown in Figure 16.6(a), where the copper conductors on the left are combined with the CCTT shaped ferromagnetic materials in the middle and placed into the metal housing on the right to make the assembled inductive device shown in Figure 16.6(b).

The stator of the 2007 Toyota Camry motor is shown in Figure 16.7. The three copper conductors for the three-phase power are shown to the right of the picture. The conductors are wound into the slots in the ferromagnetic material.

16.2.1 Magnetism and Hysteresis

These ferromagnetic materials exhibit multivalued nonlinear behavior as shown by their *B-H* characteristics in Figure 16.8. These closed loops are known as **hysteresis loops**.

Figure 16.5 Various ferromagnetic core shapes.

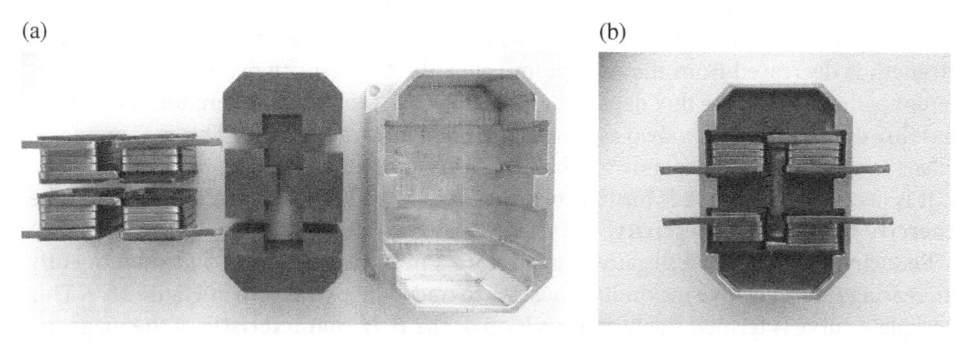

Figure 16.6 (a) Disassembled and (b) assembled magnetic device [2].

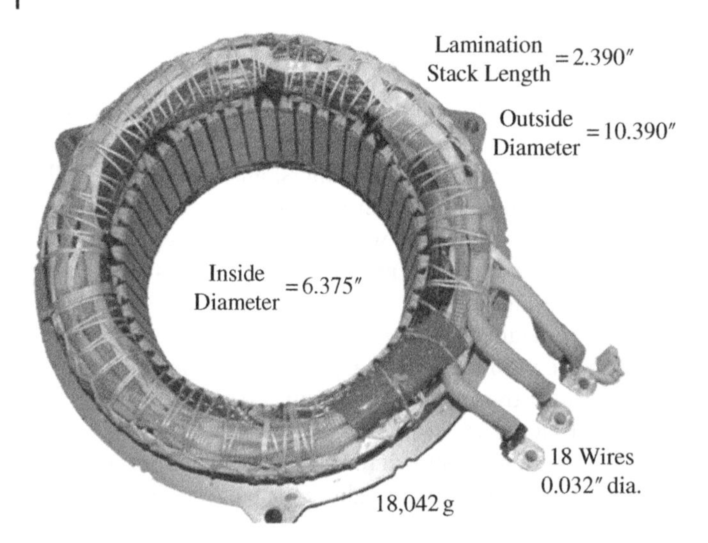

Figure 16.7 Stator of 2007 Toyota Camry hybrid electric vehicle (HEV) traction motor [3]. (Courtesy of Oak Ridge National Laboratory, US Dept. of Energy.)

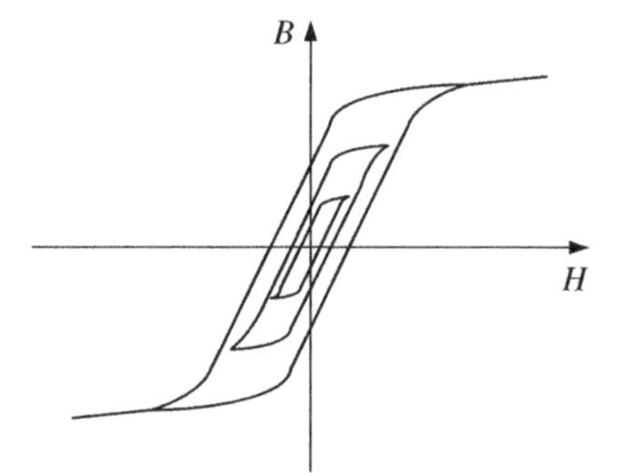

Figure 16.8 Nested *B-H* curves or hysteresis loops.

From Ampere's circuital law, current flow into a magnetic component results in a magnetic field strength H and an associated magnetic flux density B.

As the magnetic field strength H is slowly increased from zero, the flux density B follows the path shown in Figure 16.9 to its maximum value. However, as the magnetic field strength is decreased from the maximum value, the flux density decreases along a different path such that the flux density has a finite value known as the **remanent** or **residual flux density** B_R, also known as **remanance**, when H reaches zero. This is a "memory" effect as the final state depends on the earlier excitation.

It is necessary to reduce H further to reduce B to zero. This value of H is known as the **coercive force** H_C or **coercivity**.

Reducing H further in a negative direction to a negative maximum and subsequently increasing H to a positive maximum causes B to trace a loop as shown in Figure 16.9. This complete curve is termed the hysteresis loop or the ***B-H* characteristic** of the magnetic material. The word **hysteresis** is a Greek word describing a lagging or delay of an

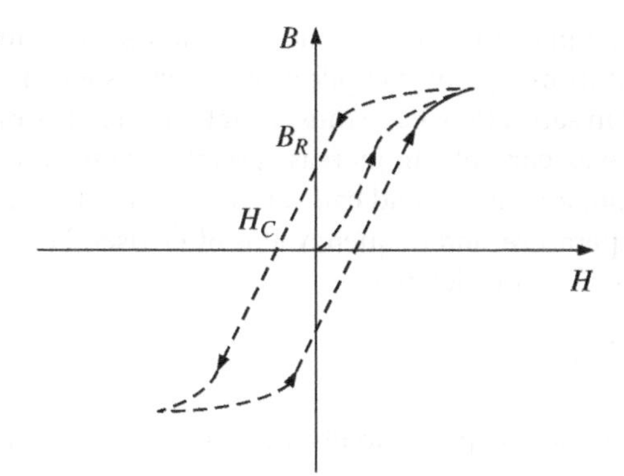

Figure 16.9 Hysteresis loop.

effect – in this case, the flux density B lags with respect to its stimulus, the magnetic field intensity H.

Completing the loop once results in a net dissipation of energy within the material, causing power loss within the magnetic material known as **hysteresis loss**, which will be discussed in Section 16.3.5.

Increasing the peak value of the magnetic field H results in a bigger hysteresis loop. If hysteresis loops are determined for different maximum values of H, they are found to lie nested within one another, as shown in Figure 16.8. Joining the peaks of the hysteresis loops, we can approximate the B-H characteristic by a single curve, as shown in Figure 16.10. This is known as the **magnetization curve** of the material.

In Figure 16.10, the quasi-linear relationship with a constant μ_m is approximately valid until the knee of the curve is reached, beyond which the material begins to "**saturate.**" When the material is saturated with magnetic flux, further increases in H result in only relatively minor increases in B. In the saturated regions, the incremental permeability approaches μ_0, the permeability of air, as shown by the slope of the curve. Thus, the material incrementally loses its magnetic property of high permeability when it becomes saturated. Caution should be exercised in selecting values of permeability as the permeability can vary greatly over the working range of the material.

Ferromagnetic materials can be operated up to the **saturation flux density B_{Sat}**. Many applications are designed to operate below the knee in the quasi-linear region. Above the

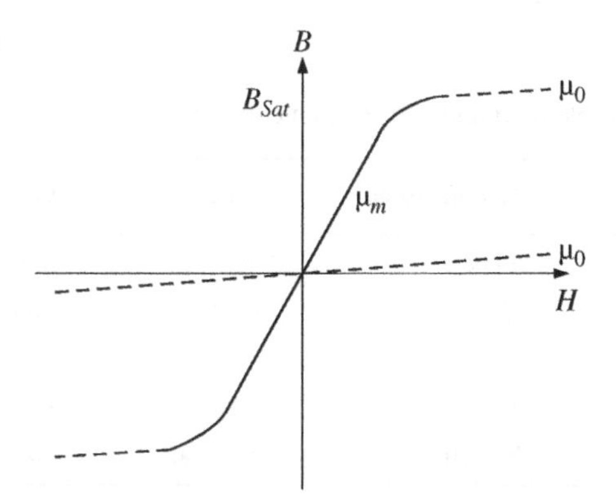

Figure 16.10 Magnetization curve.

knee, many more ampere-turns are required to increase the flux density slightly. However, it is common in many other applications, such as electric motors, to operate the magnetic material in saturation. Care must be taken when the magnetic device is excited by a voltage source as current can increase greatly due to saturation.

Permeability is not a constant and can vary with many factors, including flux density, field strength, temperature, and frequency. It is often useful to consider the **incremental permeability** (μ_Δ), which is defined as

$$\mu_\Delta = \frac{\Delta B}{\Delta H} = \mu_{\Delta r}\mu_0 \tag{16.18}$$

The relative incremental permeability ($\mu_{\Delta r}$) is again with respect to that of air or free space.

16.2.2 Hard and Soft Ferromagnetic Materials

Ferromagnetic materials can be divided into magnetically "**soft**" materials like annealed iron, which can be magnetized but does not remain magnetized when the source of magnetization is removed, and magnetically "**hard**" materials, which remain magnetized. Permanent magnets are made from hard ferromagnetic materials, such as alnico and ferrite, that are subjected to special processing in a powerful magnetic field during manufacture to align their internal microcrystalline structures, making them very hard to demagnetize. Hard materials have high coercivities, whereas soft materials have low coercivities. Table 16.1 provides some values for the remanence and coercivity of common hard and soft materials. While the remanence values are comparable, the coercivities of the hard materials are tens of thousands of times greater than those of the soft materials.

16.2.2.1 Soft Ferromagnetic Materials

There are many different types of ferromagnetic materials. In this section, we briefly review a sample of commonly used ferromagnetic materials and note some of the key differences between them.

These materials are principally based on **iron. Cobalt** and **nickel**, which appear next to iron on the periodic table, as shown in Table 16.2, are the other commonly used ferromagnetic materials. Iron has the highest saturation flux density at approximately 2.2 T, followed by cobalt at approximately 1.8 T, and then nickel at approximately 0.6 T.

Table 16.1 Sample hard and soft magnetic materials.

Material	Manufacturer	Type	Class	B_R (T)	H_C (A/m)
Ferrite	TDK	3C92	Soft	0.52	15
Silicon steel	JFE	JNHF	Soft	1.88	20
Ferrite	TDK	FB13B	Hard	0.475	380,000
Nd-Fe-B	Hitachi	Neomax	Hard	1.45	1,042,000

Table 16.2 Ferromagnetic metals in the periodic table.

26	27	28
Fe	Co	Ni
Iron	Cobalt	Nickel
56	59	59

A basic description of the physics of these magnet materials follows. The magnetic material consists of regions of magnetic alignment, known as domains – in effect, micro magnets within the material. These domains are based on the magnetic properties of the elements within the material. In an unmagnetized material, these domains exist but are not aligned and leave the bulk material unmagnetized. If an external magnetic field is applied, the magnetic domains align, and the key characteristic of a permanent magnet is that these domains remain aligned after the removal of the external magnetic field.

When we repeatedly realign the domains in a magnetic component, heat is generated due to the hysteresis loss which was mentioned earlier. Heat is additionally generated because the material is electrically conductive. The challenge for ferromagnetic materials is to be a great conductor of magnetic flux while also being as poor a conductor of electric current as possible, that is, by having a high resistivity. The resistivity can be maximized by adding air gaps, glass, or silicon, as all these options increase the resistivity. The combined loss due to the hysteresis loss and the resistivity loss, termed the **eddy current loss,** is known as **core** loss.

16.2.2.2 A Review of Commonly Used Soft Ferromagnetic Materials

Let us consider some commonly used ferromagnetic materials from various global manufacturers [4–10]. These types of materials feature as options for high-frequency magnetic components such as inductors and transformers.

Table 16.3 features ferrite of type 3C93 from Ferroxcube, iron-based amorphous metal of type 2605SA1 from Hitachi Metals, 6.5% silicon steel of type 10JNHF600 from JFE,

Table 16.3 Properties of sample soft magnetic materials.

Type	Material	Composition	B_{Sat} @ 25 °C (T)	P_{core} (0.1 T, 20 kHz) (kW/m^3)	P_{core} (1 T, 400 Hz) (kW/m^3)
Silicon steel	10JNHF600	6.5% Fe-Si	1.88	150	
Silicon steel	JNEH1500	6.5% Fe-Si	1.79		96
Powder core	Mega Flux	Fe-Si	1.6	186	
Iron-based amorphous metal	2605SA1	Fe-B-Si	1.56	70	
Nanocrystalline	Vitroperm500F	Fe-based	1.2	5	
Ferrite	3C93	MnZn ferrite	0.52	5	

powdered core of type Mega Flux from Chang Sung, and nanocrystalline of type Vitroperm500F from Vacuumschmelze.

Ferrite is composed of a high-resistivity iron-based ceramic which is pressed and sintered into the final core shape. **Sintering** is the process of compaction of a material using temperature and pressure. Ferrite is brittle as its manufacturing process is similar to that of pottery ceramics.

The **iron-based amorphous metal** and **nanocrystalline** materials are both blends of iron with glass. They are manufactured as rolls of tape and wound into shape. The tape is very thin with nominal material thicknesses of 25 μm and 20 μm for the amorphous metal and nanocrystalline materials shown, respectively. The tape laminations are wound over each other, and the overall thickness builds up in order to get the required cross-sectional area to conduct the magnetic flux.

Silicon is actually added to steel to create the **silicon steel** material. The addition of the silicon reduces the electrical conductivity of the material – electrical conductivity being an undesirable attribute for a magnetic material. Silicon steel is also manufactured as a tape roll and has a relatively thick lamination, specified as 100 μm for the given material. The silicon steel laminations are stamped into the desired shape, glued together, and stacked up in order to get the required cross-sectional area.

The **powder core** is composed of ferrous alloy powders with small distributed ceramic insulation layer air gaps, which are pressed and heated into the final shape.

Ferrite and powder cores do not require laminations.

There are two general types of silicon steel: grain-oriented and non-grain-oriented. A grain-oriented material is manufactured to have the magnetic characteristics optimized for a certain physical direction, while non-grain-oriented materials have similar magnetic characteristics in all directions. A non-grain-oriented silicon-steel-type material is also included, JNEH1500 from JFE, as an example of a rotating-machine lamination. The lamination for an electric motor is non-grain-oriented due to the varying flux patterns, unlike the grain-oriented 10JNHF600. Rotating machines typically operate with a high flux density. A grain-oriented material is likely to be used in an inductor or transformer.

Silicon steel has the highest nominal saturation flux at 1.88 T, followed by the powder core at 1.6 T, amorphous metal at 1.56 T, nanocrystalline at 1.2 T, and ferrite at 0.52 T. All these saturation flux densities are for a temperature of 25 °C, but the saturation flux densities can drop significantly with increasing temperature – and we often operate these materials at high temperatures.

Nanocrystalline tends to have the highest relative permeability, followed by ferrite. The powder-core material has a low relative permeability due to the air gaps designed into the material. The other materials likely require discrete separate air gaps, especially if used in inductors, resulting in a lower overall effective permeability for the magnetic device.

The generation of heat due to magnetic core loss can be a defining factor of a material. In general, the thinner the lamination, the lower is the core loss. Specific core loss is cited for the various materials at 20 kHz and 0.1 T in Table 16.3. The highest losses are in the powder core, followed by the silicon steel and amorphous metal. The lowest losses are in the ferrite and nanocrystalline materials. The value for JNEH1500 is for conditions of 1 T and 400 Hz.

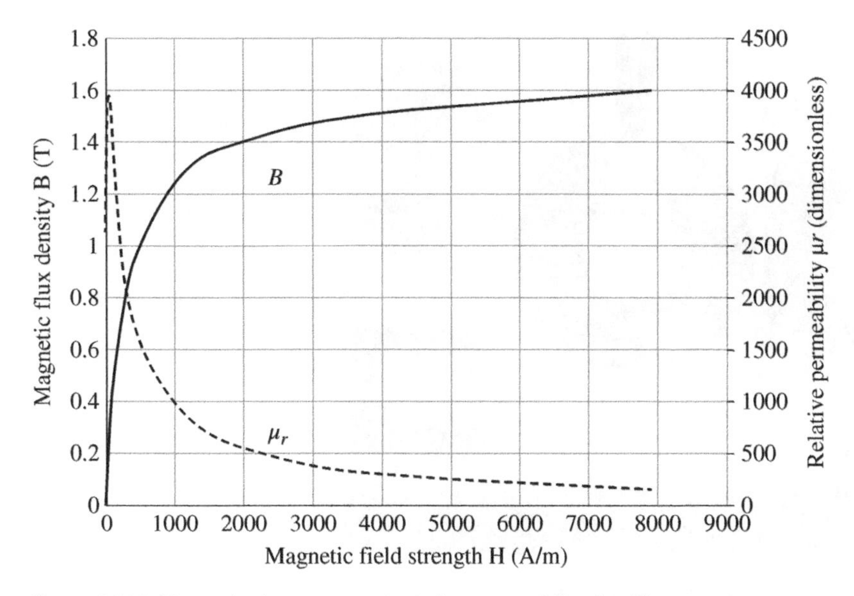

Figure 16.11 Magnetization curve and relative permeability for silicon steel core.

Table 16.4 Magnetization curve and relative permeability data for silicon steel core.

B (T)	0	0.1	0.3	0.7	1.0	1.3	1.6
H (A/m)	0	30	60	220	500	1200	7900
μ_r	—	2653	3979	2532	1592	862	161

A representative magnetization curve for 10JNHF600 silicon steel is presented in Figure 16.11. A related table of values is shown in Table 16.4. It can clearly be seen from both the curves and the data set in the table that the magnetization curve can be quite nonlinear with significant changes in the relative permeability – dropping from a peak of 3979 at 0.3 T to 161 at 1.6 T.

16.3 Self-Inductance

An inductor is an electromagnetic device which uses electric current to generate magnetic flux. The laws of Faraday and Lenz explain the inductor. The operation of the inductor is closely related to that of the transformer, which is discussed in the next section. Inductors have many uses in engineering, usually as energy storage devices or as filter components to reduce ripple in current. Thus, power inductors have many uses on electric vehicles in applications such as the power-factor-corrected boost converter in the charger, the full-bridge converter for auxiliary power, and the dc-dc converters in the hybrid electric vehicle (HEV) or fuel cell electric vehicle (FCEV). A selection of power inductors is shown in Figure 16.12.

Figure 16.12 A selection of power inductors.

16.3.1 Basic Inductor Operation

The inductor basically works as follows – current flowing into the inductor induces a magnetic flux in the core of the inductor. If the current and resulting flux are changing with time, then according to the laws of Faraday and Lenz, a back emf e is induced across the coil which acts to oppose the changing current. The symbol of the inductor is shown in Figure 16.13. Top and side views of an elementary gapped toroidal inductor are shown in Figure 16.14. A current i flows into the coil of N turns, distributed evenly around the magnetic core. A resulting flux Φ flows in the magnetic core.

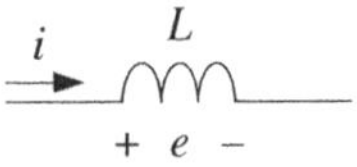

i L

$+\ e\ -$

Figure 16.13 Inductor symbol.

The core is gapped to have two identical air gaps, each of length l_g. The mean circumference l_c of the toroid is simply given by

$$l_c = 2\pi r \tag{16.19}$$

where r is the mean radius of the toroid.

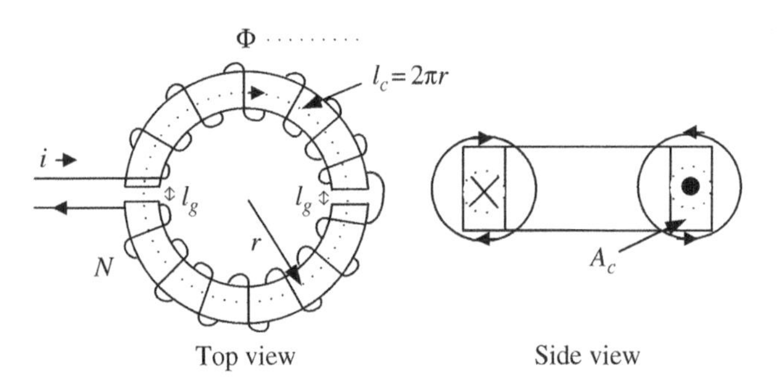

Figure 16.14 Elementary gapped toroidal inductor.

The law of Biot and Savart relates the generation of a magnetic field to the magnitude, direction, length, and proximity to an electric conductor. (Frenchmen Jean-Baptiste Biot and Felix Savart discovered this important relationship in the early 1800s.) Per the law of Biot and Savart, the magnetic field strength and flux density are constant within the core for a given radius. However, the flux density along the mean radius is typically a reasonable estimation for the flux density across the core. Hence, we use the mean magnetic path length for the flux in our calculations. See Problems 16.1 and 16.2 at the end of chapter in order to investigate the relationships further.

The magnetic core has a cross-sectional area A_c. Due to a magnetic phenomenon known as **fringing**, the effective area of the air gap through which the flux travels may not be the same as the cross-sectional area of the air gap. Thus, the cross-sectional area of the air gap is designated A_g. In the examples in this chapter, fringing is ignored, in which case $A_g = A_c$.

If the inductor has a **flux linkage** $\lambda = N\Phi$ when the current is I, then the coefficient of self-inductance L is defined as

$$L = \frac{\lambda}{I} = \frac{N\Phi}{I} = \frac{NB_cA_c}{I} \tag{16.20}$$

where B_c is the flux density of the core.

The **inductance** L of a conducting system may be defined as the ratio of the magnetic flux linkage to the current causing the flux.

If no ferromagnetic material is present, then λ is directly proportional to I, and hence L is a constant, as is the case for an air core.

The inductance is a measure of the flux linkage per ampere and has the unit of henry (symbol H), named for American engineer Joseph Henry (1797–1868).

If the current in the circuit is changing, there is an induced emf e in the circuit due to the changing flux linkage. The inductance can be defined in terms of this induced emf by Faraday's law:

$$e = -\frac{d\lambda}{dt} = -N\frac{d\Phi}{dt} = -L\frac{di}{dt} \tag{16.21}$$

In electric circuit theory, L is called **the self-inductance** of the element, and e is called the voltage of self-inductance or the **back emf** of the inductor.

Note that if the back emf has the polarity shown in Figure 16.13, where the back emf opposes the current, then Lenz's law is implicit to the circuit element, and the back emf equation is written as

$$e = L\frac{di}{dt} \tag{16.22}$$

16.3.2 Inductor Equations

Applying Ampere's circuital law around the magnetic path of a core with two air gaps, as shown in Figure 16.14, results in the following equation:

$$\oint \mathbf{H}.d\mathbf{l} = NI = H_cl_c + 2H_gl_g \tag{16.23}$$

where H_c and H_g are the magnetic field strengths of the core and air gaps, respectively.

The effectiveness of the coil as a producer of flux depends on the amperes-turns of the coil, just as the effectiveness of an electric circuit as a producer of current depends on the magnitude of the voltage source or electromotive force. Consequently, the ampere-turns NI of the coil is termed the **magnetomotive force** (**mmf**) of the circuit.

The relationships between the flux densities of the core and air gap and the magnetic field strengths are given by

$$B_c = \mu_r \mu_0 H_c \text{ and } B_g = \mu_0 H_g \tag{16.24}$$

where B_c and B_g are the magnetic flux densities of the core and air gaps, respectively.

Expressing Equation (16.23) in terms of the flux density, we get

$$NI = \frac{B_c}{\mu_r \mu_0} l_c + 2\frac{B_g}{\mu_0} l_g \tag{16.25}$$

From Gauss' law, the flux in the core Φ is given by

$$\Phi = B_c A_c = B_g A_g \tag{16.26}$$

Rearranging Equation (16.25) to express the mmf in terms of flux yields

$$NI = \Phi\left(\frac{l_c}{\mu_r \mu_0 A_c} + \frac{2l_g}{\mu_0 A_g}\right) \tag{16.27}$$

or the flux in terms of the mmf

$$\Phi = \frac{NI}{\dfrac{l_c}{\mu_r \mu_0 A_c} + \dfrac{2l_g}{\mu_0 A_g}} \tag{16.28}$$

By substituting Equation (16.28) into Equation (16.20), we get the following expression for the inductance of a device with two identical air gaps:

$$L = \frac{N\Phi}{I} = \frac{N^2}{\dfrac{l_c}{\mu_r \mu_0 A_c} + \dfrac{2l_g}{\mu_0 A_g}} \tag{16.29}$$

If the core is of high permeability, the inductance is dominated by the two air gaps, and the preceding equation can be simplified to

$$L \approx \frac{\mu_0 N^2 A_g}{2l_g} \quad \text{– two air gaps} \tag{16.30}$$

It is also common to have inductors with a single air gap, in which case

$$L \approx \frac{\mu_0 N^2 A_g}{l_g} \quad \text{– single air gap.} \tag{16.31}$$

For an inductor with no air gap, the inductance is given by

$$L = \frac{\mu_r \mu_0 N^2 A_c}{l_c} \quad \text{– no air gap.} \tag{16.32}$$

16.3.2.1 Example: A Gapped Inductor

The high-current inductor for a hybrid electric vehicle is specified at 400 μH. The inductor has 28 turns, a core cross-sectional area of 12 cm^2, and a mean magnetic path length of 0.22 m. Assume a silicon steel core with a maximum core flux density of 1.3 T, and the magnetization curve of Figure 16.11 and Table 16.4. Ignore fringing ($A_c = A_g$).

i) What is the air gap length if there are two air gaps?
ii) What are the maximum current, mmf, flux, and magnetic field strengths when the flux density in the core reaches a peak value $B_{max} = 1.3$ T?

Solution:

i) Rearranging Equation (16.29) gives an expression for the air gap:

$$l_g = \frac{\mu_0 A_c N^2}{2L} - \frac{l_c}{2\mu_r} = \frac{4\pi \times 10^{-7} \times 12 \times 10^{-4} \times 28^2}{2 \times 400 \times 10^{-6}}\text{m} - \frac{0.22}{2 \times 862}\text{m} = 1.35\text{mm}$$

ii) Rearranging Equation (16.20) gives an expression for current:

$$L = \frac{N\Phi}{I} = \frac{NB_{max}A_c}{I} \Rightarrow I = \frac{NB_{max}A_c}{L} = \frac{28 \times 1.3 \times 12 \times 10^{-4}}{400 \times 10^{-6}}\text{A} = 109.2\text{A}$$

The mmf is

$$NI = 28 \times 109.2\text{A-turns} = 3057.6\text{A-turns}$$

Flux is simply given by

$$\Phi = B_{max}A_c = 1.3 \times 12 \times 10^{-4}\text{Wb} = 1.56\text{mWb}$$

The magnetic field strengths are

$$H_c = 1.2\,\text{kAm}^{-1} \qquad (\text{from Table 16.4})$$

and

$$H_g = \frac{B_g}{\mu_0} = \frac{1.3}{4\pi \times 10^{-7}}\text{Am}^{-1} = 1034.5\text{kAm}^{-1}$$

Note that the air gap magnetic field strength is significantly higher than that of the core – as is to be expected due to the high relative permeability of the core.

16.3.2.2 Inductance Variation with Magnetization Curve

As the magnetization curve and relative permeability are nonlinear for ferromagnetic materials, it follows that the inductance for a given design varies with the exciting current. A new data set is created in Table 16.5 which includes estimates for the inductance and exciting current as the relative permeability varies. The inductance is calculated on the basis of Equation (16.29). The exciting current is calculated by rearranging Equation (16.20) to get

$$I = \frac{NB_cA_c}{L} \qquad (16.33)$$

Table 16.5 Magnetization curve and relative permeability for silicon steel core.

B (T)	0	0.1	0.3	0.7	1.0	1.3	1.6
H (A/m)	0	30	60	220	500	1200	7900
μ_r	—	2653	3979	2532	1592	862	161
L (μH)	—	425	429	424	416	400	291
I (A)	—	7.9	23.5	55.5	80.7	109.2	184.9

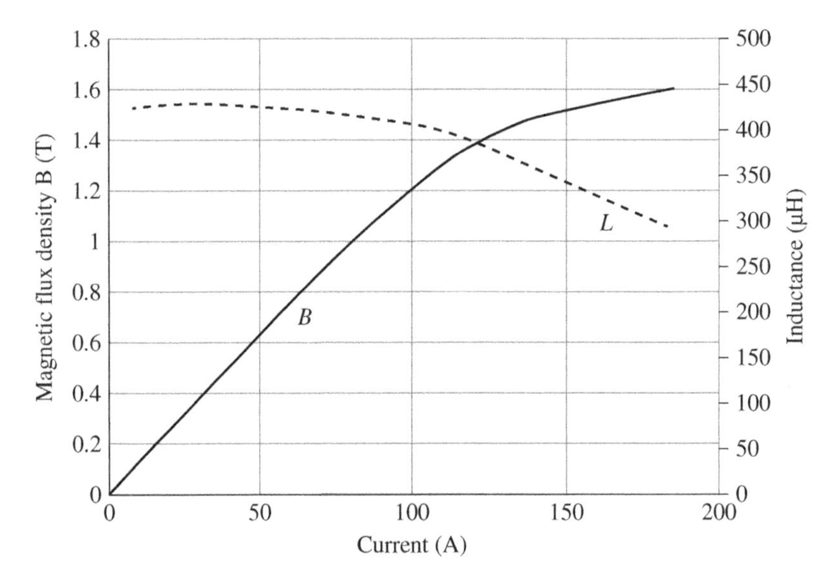

Figure 16.15 Variation of flux density and inductance with exciting current.

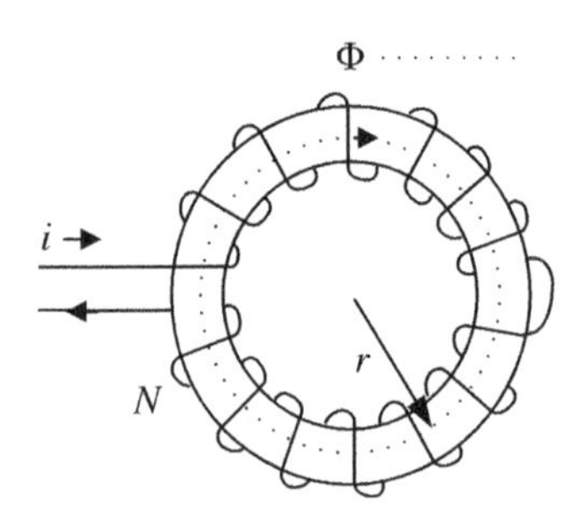

Figure 16.16 Gapless toroidal inductor.

The flux density and inductance are plotted as a function of the exciting current in Figure 16.15. It can be seen that as the current increases, the flux density also increases but the inductance drops.

16.3.3 Reluctance

Per Gauss' law of magnetism, magnetic flux lines form closed paths, as shown in Figure 16.16, for a gapless toroid wound with a current-carrying coil.

The flux in the toroid can be calculated by selecting a circular area A_c in a plane perpendicular to the direction of the flux lines. As previously discussed, it is reasonable to assume a uniform *H*, and hence a uniform flux density *B*, throughout the core's cross section. Thus, the total flux within the core is

$$\Phi = B_c A_c = \frac{\mu_r \mu_0 N I A_c}{l_c} \tag{16.34}$$

Rearranging this equation yields

$$\Phi = \frac{NI}{\left(\dfrac{l_c}{\mu_r \mu_0 A_c}\right)} = \frac{mmf}{\Re} \tag{16.35}$$

where NI equals the ampere-turns or **magnetomotive force (mmf)** applied to the core, and the term in the denominator is called the **reluctance** $\Re$ of the core. Thus, the reluctance is given by

$$\Re = \frac{mmf}{\Phi} = \frac{l_c}{\mu_r \mu_0 A_c} \tag{16.36}$$

The units of reluctance are ampere-turn/webers or the per henry, written as A/Wb or H^{-1}. The equation shows that the reluctance of a magnetic structure, for example, the toroid, is proportional to its magnetic path and inversely proportional to both its cross-sectional area and the permeability of its material.

The relationship of reluctance $\Re$, mmf NI, and flux Φ in a magnetic circuit is analogous to the relationship between the resistance R, emf E or V, and current I in an electrical circuit. Thus, we get the following relationships:

$$\textbf{Magnetic circuit: } \Re = \frac{mmf}{\Phi} = \frac{l}{\mu_r \mu_0 A} \tag{16.37}$$

where l and A are the length and area of the magnetic path.

$$\textbf{Electrical circuit: } R = \frac{V}{I} = \frac{l_{cu}}{\sigma A_{cu}} = \frac{\rho l_{cu}}{A_{cu}} \tag{16.38}$$

where l_{cu} and A_{cu} are the length and cross-sectional area of the electrical conductor, respectively, and σ and ρ are the material conductivity and resistivity, respectively.

Once the reluctance of a magnetic circuit is calculated, the inductance can easily be determined using the following relationships:

$$L = \frac{\lambda}{I} = \frac{N\Phi}{I} = \frac{N \times N\Phi}{N \times I} = N^2 \frac{\Phi}{NI} = \frac{N^2}{\Re} \tag{16.39}$$

or

$$L = \frac{N^2}{\Re} \tag{16.40}$$

Thus, the inductance of a magnetic circuit is directly proportional to the square of the turns and inversely proportional to the reluctance.

As is done with electrical circuits, we can draw a magnetic circuit as shown in Figure 16.17.

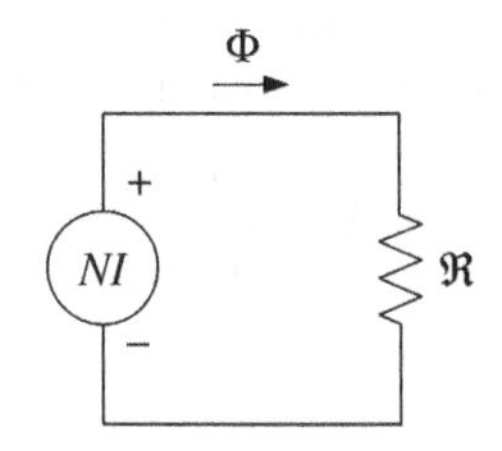

Figure 16.17 Magnetic circuit.

16.3.3.1 Example: A Gapless Inductor

The inductor in the previous example is assembled without the gap.

i) Determine the reluctance and inductance of the gapless structure.
ii) What is the maximum current when the flux density in the core reaches a peak value of 1.3 T?

Solution:

Per Equation (16.36):

$$\Re_c = \frac{l_c}{\mu_r \mu_0 A_c} = \frac{0.22}{862 \times 4\pi \times 10^{-7} \times 12 \times 10^{-4}} \mathrm{H}^{-1} = 169.25\,\mathrm{kH}^{-1}$$

Thus:

$$L = \frac{N^2}{\Re_c} = \frac{28^2}{169.25 \times 10^3} \mathrm{H} = 4630\,\mu\mathrm{H}$$

and

$$I = \frac{N B_c A_c}{L} = \frac{28 \times 1.3 \times 12 \times 10^{-4}}{4630 \times 10^{-6}} \mathrm{A} = 9.43\,\mathrm{A}$$

We can now easily see the benefit of the air gap. By introducing two air gaps of 1.35 mm into a structure with a magnetic path length of 0.22 m, the inductor does not reach the peak flux density until the current reaches 109.2 A, whereas the maximum is reached at 9.43 A without the air gap. Of course, we also see that the inductance drops significantly by introducing the air gap. The gapped inductor is 400 μH, whereas the ungapped inductor is 4630 μH.

16.3.3.2 Reluctance of Gapped Magnetic Structures

To study the effects of air gaps, let us consider the earlier magnetic structure of Figure 16.14 consisting of an N-turn coil on a ferromagnetic core. Rewriting Equation (16.28):

$$\Phi = \frac{NI}{\dfrac{l_c}{\mu_r \mu_0 A_c} + \dfrac{2 l_g}{\mu_0 A_g}}$$

We can recognize from this equation that the two terms in the denominator are the reluctances of the core and two air gaps. Therefore, the effective reluctance $\Re$ of the magnetic structure in the path is the sum of the two reluctances. Thus:

$$\Re = \frac{NI}{\Phi} \tag{16.41}$$

where

$$\Re = \Re_c + 2\Re_g = \frac{l_c}{\mu_r \mu_0 A_c} + \frac{2 l_g}{\mu_0 A_g} \tag{16.42}$$

and

$$\Re_g = \frac{l_g}{\mu_0 A_g} \tag{16.43}$$

16.3.3.3 Example: Reluctances of Gapped Inductor

Determine the core, air gap, and total reluctances for the earlier 400 μH inductor. Ignore fringing.

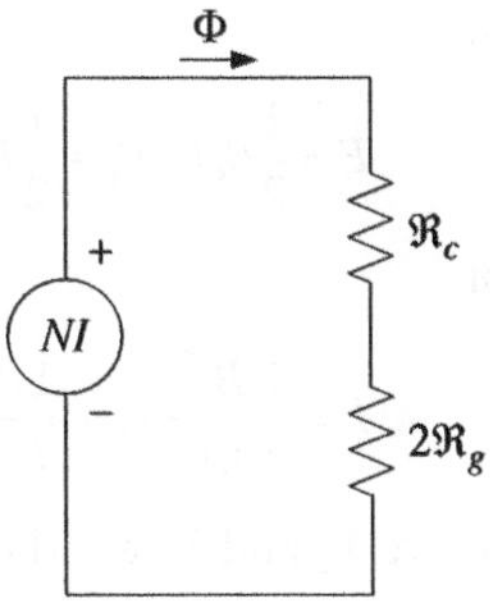

Solution:

The core reluctance is the same as in the previous example and is 169.25 kH^{-1}.

The reluctance for each air gap is given by

$$\Re_g = \frac{l_g}{\mu_0 A_c} = \frac{1.35 \times 10^{-3}}{4\pi \times 10^{-7} \times 12 \times 10^{-4}} H^{-1} = 895.25\,kH^{-1}$$

The total reluctance is the sum of the core and air gap reluctances:

$$\Re = \Re_c + 2\Re_g = 169.25 kH^{-1} + 2 \times 895.25 kH^{-1} = 1959.75 kH^{-1}$$

The magnetic circuit for this example is shown in Figure 16.18.

Figure 16.18 Magnetic circuit with core reluctance and two air gaps in series.

16.3.4 Energy Stored in Magnetic Field

If the inductor carries a current I, then the energy E stored in the magnetic field is given by

$$\begin{aligned} E &= \frac{1}{2}LI^2 = \frac{1}{2}\frac{\lambda}{I}I^2 = \frac{1}{2}\lambda I = \frac{1}{2}N\Phi I \\ &= \frac{1}{2}\Phi(NI) \end{aligned} \tag{16.44}$$

Since

$$H_c l_c + 2H_g l_g = NI \tag{16.45}$$

the energy is given by

$$E = \frac{1}{2}\Phi\left(H_c l_c + 2H_g l_g\right) \tag{16.46}$$

Given that

$$\Phi = B_c A_c = B_g A_g \tag{16.47}$$

we get

$$\begin{aligned} E &= \frac{1}{2}B_c H_c A_c l_c + \frac{1}{2}B_g H_g A_g \times 2l_g \\ &= \text{Energy stored in core} + \text{Energy stored in air gaps} \end{aligned} \tag{16.48}$$

or

$$E = \frac{1}{2}B_c H_c V_c + \frac{1}{2}B_g H_g V_g \tag{16.49}$$

or

$$E = \frac{1}{2}\frac{B_c^{\ 2}}{\mu_r \mu_0}V_c + \frac{1}{2}\frac{B_g^{\ 2}}{\mu_0}V_g \tag{16.50}$$

where V_c and V_g are the volumes of the core and gaps, respectively, and are given by

$$V_c = A_c l_c \text{ and } V_g = 2A_g l_g \tag{16.51}$$

Typically, the bulk of the energy storage in an inductor is in the air gap.

16.3.4.1 Example: Inductor Energy Storage

For the earlier 400 μH inductance, calculate the energy storage in the inductor at 1.3 T. Determine the distribution of energies in the core and air gaps.

Solution:

First, the inductor energy storage at the maximum flux density occurs at 109.2 A and is given by

$$E = \frac{1}{2}LI^2 = \frac{1}{2}400 \times 10^{-6} \times 109.2^2\,\text{J} = 2.385\,\text{J}$$

The core and air gap energies are given by

$$E_c = \frac{1}{2}B_c H_c A_c l_c = \frac{1}{2} \times 1.3 \times 1200 \times 0.0012 \times 0.22\,\text{J} = 0.206\,\text{J}$$

$$E_g = \frac{1}{2}B_g H_g A_g \times 2l_g = \frac{1}{2} \times 1.3 \times 1034500 \times 0.0012 \times 2 \times 0.00135\,\text{J} = 2.179\,\text{J}$$

Thus, there is a significant increase in energy storage due to the gap for the same flux density. This is one of the big benefits of gapping a magnetic core. In this case, the energy stored in the gap of the gapped inductor is 2.179 J and is more than ten times greater than the energy of 0.206 J stored in the core.

16.3.5 Core Loss

Core loss is the energy loss in a ferromagnetic material due to hysteresis and eddy current losses.

16.3.5.1 Hysteresis Loss

Completing the *B-H* or hysteresis loop once results in a net dissipation of energy within the material, causing power loss within the magnetic material, known as **hysteresis loss.** The hysteresis loss per unit volume can be shown to be equal to the area of the hysteresis loop. In general, it can be shown that the area of the hysteresis loop is proportional to the peak of the alternating flux density $B_{ac(pk)}$ of the loop:

$$\text{hysteresis loss} \propto \text{area of hysteresis loop} \propto B_{ac(pk)}^{\ \ h} \tag{16.52}$$

where h is a material parameter usually in the range 1.5 to 2.5. The power dissipated within the core additionally increases with (1) the frequency of magnetization f and (2) the actual volume of the magnetic material V_c. Note that $B_{ac(pk)}$ relates to the peak of the ac component of flux density only. It is common in magnetic devices, such as inductors, to have a large dc flux density component.

Thus, the hysteresis power loss P_h is described by the equation

$$P_h = k_h V_c f B_{ac(pk)}^{\ h} \tag{16.53}$$

where k_h is a constant for a given specimen of material and a given range of flux density and frequency.

16.3.5.2 Eddy Current Loss

Another source of magnetic core power loss is due to eddy currents. According to Faraday's law, a changing current in the winding results in a changing flux in the core. However, the changing flux induces circulating voltages within the magnetic core, which result in circulating or ***eddy currents*** within the core to oppose the flux changes, as shown in Figure 16.19. All electrical materials have a finite electrical resistivity (ideally, it would be infinite for magnetic materials). The core resistance, together with eddy currents within the material, result in **eddy current loss**.

The eddy current loss within a material is reduced by two methods. First, in many applications, the material, such as ferrite, can be designed to have a very high resistivity, thus reducing the eddy current magnitude and the associated power loss.

Second, the eddy current loss can be reduced by fabricating the core from thin ferromagnetic laminations, which are insulated from each other by means of thin layers of varnish. Because of the insulation between the laminations, the loop area is reduced, resulting in smaller induced voltages and eddy currents in the core and lower eddy current losses.

In general, for a core of volume V_c, the eddy current power loss P_e is given by

$$P_e = k_e V_c f^2 B_{ac(pk)}^{\ 2} \tag{16.54}$$

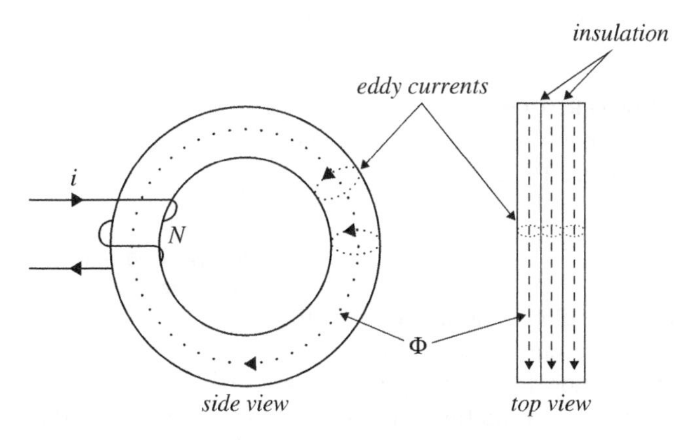

Figure 16.19 Eddy currents in a laminated toroid.

where k_e is a parameter of the magnetic material. Ferromagnetic materials for 50/60 Hz inductors and transformers and for many electromechanical machines typically use laminated materials.

16.3.5.3 Core Loss

The hysteresis loss and the eddy current loss are usually lumped together to give the overall core loss:

$$P_c = P_h + P_e \tag{16.55}$$

The effects of the hysteresis and eddy current losses are usually represented in a single core loss equation, known as the **Steinmetz** equation:

$$P_c = k V_c f^m B_{ac(pk)}{}^n \tag{16.56}$$

where k, m, and n are the Steinmetz coefficients.

The equation and coefficients are named for Charles Steinmetz (1865–1923), a Prussian mathematician and engineer who contributed greatly to the understanding of ac power systems and core loss.

16.3.5.4 Example: Core Loss

Calculate the core loss of the material in the previous example assuming that the input current has an ac component of 9.2 A peak with a frequency of 10 kHz. The core material has the following Steinmetz coefficients: $k = 3.81$, $m = 1.486$, and $n = 1.630$.

Solution:

First, let us calculate the core volume:

$$V_c = A_c l_c = 0.0012 \times 0.22\,\text{m}^3 = 260 \times 10^{-6}\text{m}^3$$

Next, let us determine the peak value of the ac flux density from Equation (16.33):

$$B_{ac(pk)} = L\frac{I_{ac(pk)}}{NA_c} = 400 \times 10^{-6} \times \frac{9.2}{28 \times 0.0012}\,\text{T} = 0.1095\,\text{T}$$

The core loss is then given by

$$P_c = k V_c f^m B_{ac(pk)}{}^n = 3.81 \times 260 \times 10^{-6} \times 10000^{1.486} \times 0.1095^{1.63}\,\text{W} = 23.7\,\text{W}$$

16.3.5.5 Core Loss Equivalent Parallel Resistance

The core loss can be modeled as an equivalent parallel resistance across the inductor as shown in Figure 16.20.

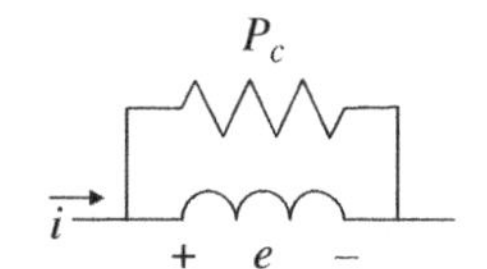

Figure 16.20 Inductor with core loss.

16.3.6 Copper Loss

The coils of wire which form the winding of an inductor, or the primary and secondary windings of a transformer, have an associated resistance. This resistance is modeled as an equivalent series resistance (*ESR* or R_s or R_{cu}). When an rms current $I_{L(rms)}$ flows through the wire, the power loss is R_{cu} $I_{L(rms)}{}^2$. Copper is typically used as the conductor, and so this conduction power loss

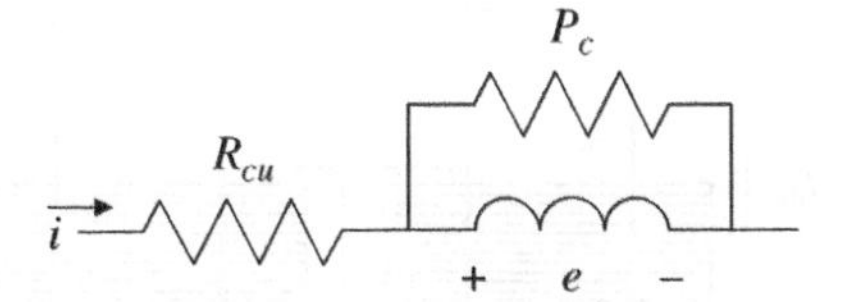

Figure 16.21 Inductor with core and copper loss.

is often termed the **copper loss**. Aluminum is a competitive option for some applications. The copper loss is typically illustrated in circuit form as shown in Figure 16.21.

16.3.6.1 Copper Loss of Wire

The resistance of the wire is given by

$$R_{cu} = \rho \frac{l_{cu}}{A_{cu}} \tag{16.57}$$

where l_{cu} is the total length of the wire, A_{cu} is the cross-sectional area of the wire, and ρ is the resistivity of copper. The length l_{cu} can be approximated by multiplying the turns N by the mean length per turn (MLT) of wire wound around the core.

Thus, the resistance can be calculated from

$$R_{cu} = \rho \frac{N \times MLT}{A_{cu}} \tag{16.58}$$

When conducting rms current $I_{L(rms)}$, the copper loss P_{cu} is given by

$$P_{cu} = R_{cu} I_{L(rms)}^2 \tag{16.59}$$

16.3.6.2 Example: Copper Loss

Calculate the copper loss of the inductor in the previous example, assuming the MLT = 0.188 m, resistivity = 1.725 × 10^{-8}Ωm, and copper cross-sectional area = 15.6 mm^2. Let the rms current $I_{L(rms)}$ be 100 A.

Solution:

$$R_{cu} = \rho \frac{N \times MLT}{A_{cu}}$$

$$= 1.725 \times 10^{-8} \times \frac{28 \times 0.188}{15.6 \times 10^{-6}} \text{m}\Omega = 5.82 \text{m}\Omega$$

$$P_{cu} = R_{cu} I_{L(rms)}^2 = 5.82 \times 10^{-3} \times 100^2 \text{W} = 58.2 \text{W}$$

16.3.6.3 Copper Loss of CC Core with Helical Winding

The earlier 400 μH inductor features a CC core rather than a toroid, as shown in Figure 16.22. Toroids and CC cores are common core types for high-power inductors. Solid helical windings are commonly used for high-power medium-frequency inductors. Foil, stranded wires, or litz wires may be required for high-frequency inductors in order to reduce the copper loss.

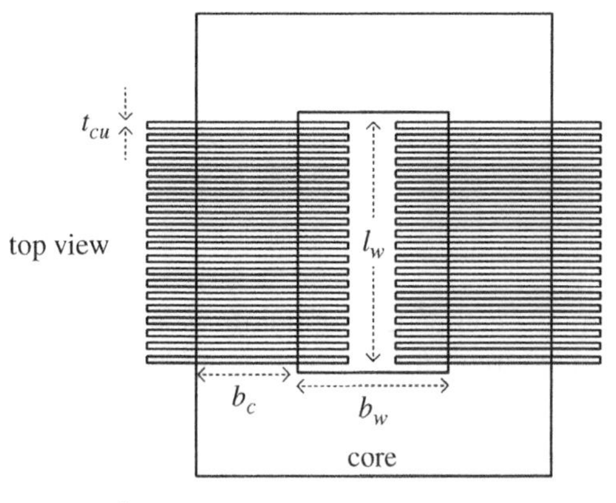

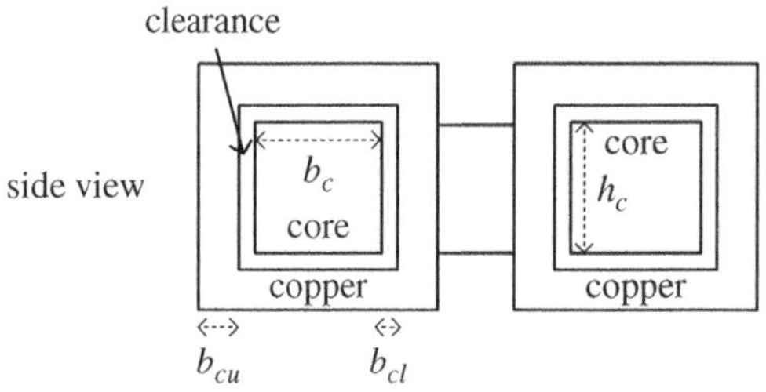

Figure 16.22 Top and side views of CC-core inductor.

If we assume that the copper winding is arranged as a rectangular helix, we can calculate the resistance of the winding and the associated copper losses.

The mean length per turn is given by

$$MLT = 2b_c + 2h_c + 4b_{cu} + 8b_{cl} \tag{16.60}$$

where b_c and h_c are the width and height of the core, respectively; b_{cu} is the width of the copper; and b_{cl} is the clearance between the copper winding and the core.

The area of the rectangular cross section of the wire is

$$A_{cu} = b_{cu} \times t_{cu} \tag{16.61}$$

where t_{cu} is the thickness of the wire.

16.3.6.4 Example: MLT of Winding

Calculate the MLT of the 400 μH inductor, assuming the dimensions given in Table 16.6. The inductor features two helical windings in parallel, and so the cross-sectional area of the winding is twice that of an individual helix.

Table 16.6 Inductor parameters.

N	b_c	b_w	h_c	l_w	b_{cu}	t_{cu}	b_{cl}
Parallel	(mm)	(mm)	(mm)	(mm)	(mm)	(mm)	(mm)
28 × 2	20	20	50	50	6	1.3	3

The MLT is

$$MLT = 2b_c + 2h_c + 4b_{cu} + 8b_{cl} = 2 \times 20\,\text{mm} + 2 \times 50\,\text{mm} + 4 \times 6\,\text{mm} + 8 \times 3\,\text{mm}$$
$$= 0.188\,\text{m}$$

The winding cross-sectional area is

$$A_{cu} = 2 \times b_{cu} \times t_{cu} = 2 \times 6 \times 10^{-3} \times 1.3 \times 10^{-3}\,\text{m}^2 = 15.6\,\text{mm}^2$$

The factor of 2 is due to the two parallel wires.

16.3.7 Inductor Sizing using Area Product

The design of an inductor typically uses an iterative approach in order to arrive at the optimum design. The simplest starting place when designing an inductor is to use the **area product** (AP) method.

Rearranging the basic inductor equation in terms of the cross-sectional area of the magnetic core A_c gives

$$L = \frac{N\Phi}{I} = \frac{NB_{max}A_c}{I_{L(max)}} \Rightarrow A_c = \frac{LI_{L(max)}}{NB_{max}} \tag{16.62}$$

The area required within the magnetic structure for the winding is the **window area**, and is given by

$$A_w = \frac{NI_{L(rms)}}{k_{cu}J_{cu}} \tag{16.63}$$

The window area is directly proportional to the number of turns and the rms current. It is inversely proportional to the **copper fill factor** k_{cu} and the winding **current density** J_{cu}.

To put it simply, the size of a magnetic device is related to the product of these two areas. The core area is dependent on flux density, while the window area is dependent on the current density.

The area product is defined as

$$AP = A_c A_w = \frac{LI_{L(rms)}I_{L(max)}}{k_{cu}J_{cu}B_{max}} \tag{16.64}$$

The details of the inductor design depend on the relative size of the two areas. It is often an iterative exercise to determine the optimum design.

For simplicity, let the two areas be equal; then

$$A_c = A_w = \sqrt{AP} \tag{16.65}$$

The number of turns is then given by

$$N = \frac{LI_{L(max)}}{B_{\max}A_c} \tag{16.66}$$

The rest of the inductor parameters can then be determined. See Example 11.6.1 in Chapter 11 for a related example.

16.3.8 High-Frequency Operation and Skin Depth

The phenomenon of skin depth is one of the great challenges of magnetics, and conduction in general. When conducting dc current, the current is uniform across the cross section of the conductor, and it is relatively easy to determine the resistance of the conductor, as was done in the earlier example. An example of a circular conductor carrying dc current is shown in Figure 16.23(a). If the current is varying with time, then the current distribution is not uniform but is forced to the outside surface, or skin, of the conductor, as shown in Figure 16.23(b). This high-frequency effect reduces the distribution of current at the center of the conductor, and is explained by the laws of Faraday and Lenz. The practical result for a conductor is that the current is concentrated along the surface and reduces exponentially away from the surface. A skin depth, with symbol δ, is the distance below the surface at which the current density has reduced to $1/e$, where $e = 2.718$, or to 36.8% of its value at the surface.

Skin depth is an active design consideration when sizing cables or conductors. For example, it is common for high-power ac conductors at 50 or 60 Hz to have a hollow core. High-frequency conductors at tens or hundreds of kilohertz may be constructed using litz (short for *litzendraht*, German for "braid" or "woven") wire. Litz wire was invented by Sebastian de Ferranti in 1888 and is composed of bundles of multiple strands of smaller insulated wires which are braided together, as shown in Figure 16.23(c).

The skin depth for a given material and frequency f can be calculated using the following formula:

$$\delta = \sqrt{\frac{\rho}{\pi f \mu}} \qquad (16.67)$$

where ρ is the resistivity, and μ is the permeability.

The skin depth δ of copper is calculated for various frequencies of interest in Table 16.7. The nominal values of 1.725×10^{-8} Ωm and $4\pi \times 10^{-7}$ Hm^{-1} are used for the resistivity and permeability of copper at 25 °C.

The skin depth is nominally 9.35 mm at 50 Hz, and decreases to 93.5 μm at 500 kHz.

Practical inductor and transformer design is complex, and significant care must be paid to many factors.

As we have just seen, high frequencies can induce significant additional copper loss due to skin depth. There are significant related effects known as layering and proximity, which can also significantly increase the power loss and reduce the efficiency.

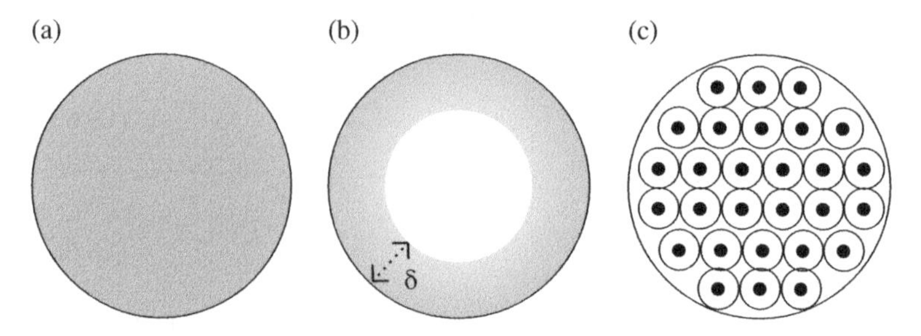

Figure 16.23 Conductors: (a) solid with dc, (b) solid with ac, and (c) litz.

Table 16.7 Copper skin depth at 25 °C.

Frequency (Hz)	Skin depth (mm)
50	9.35
5,000	0.935
500,000	0.0935

Core loss can increase significantly due to dc bias and high-frequency operation. Thermal management is crucial. Audible noise can be an issue for some materials. Safety standards with required levels of insulation, clearance, and creepage are required for many components.

See the Further Reading section at the end of this chapter for references.

16.4 Hard Ferromagnetic Materials and Permanent Magnets

Hard or permanent-magnet materials are also generally iron based, but the iron is combined with other elements, such as rare earth elements, to achieve a very high coercivity. The automotive-grade permanent magnets commonly used for high-power-density traction motors are usually neodymium-iron-boron (Nd-Fe-B). Rare earth materials such as samarium and neodymium cluster in the period table, as shown in Table 16.8. The supply and cost of these rare earth elements can be volatile, and this volatility has been a motivating factor in the development of machines without permanent magnets, such as the widely used induction motor, or permanent-magnet machines using magnets without rare earth elements, such as high-performance ceramic or ferrite magnets.

Sample permanent-magnet materials from Hitachi Metals are shown in Table 16.9 [9].

The earlier basic relationship between **B** and **H** was presented as follows:

$$\mathbf{B} = \mu_r \mu_0 \mathbf{H}$$

Table 16.8 Rare earth materials in the periodic table.

60 Nd Neodymium 141	61 Pm Promethium 145	62 Sm Samarium 150

Table 16.9 Sample permanent magnetic materials from Hitachi Metals.

Material	Manufacturer	Type	B_R (T)	H_{cB} (kA/m)	H_{cJ} (kA/m)	*(BH)max*	μ_{rec}
Nd-Fe-B	Hitachi	NMX-S41EH	1.24–1.31	923–1018	> = 1990	294–334	1.05
Ferrite	Hitachi	NMF-12 J	0.43–0.46	300–350	> = 430	35–40	1.09

When discussing permanent magnets, this equation is more appropriately expressed as

$$\mathbf{B} = \mu_0(\mathbf{H} + \mathbf{M}) \tag{16.68}$$

where **M** is the **magnetization** of the material. The magnetization has the same units as the magnetic field strength and is related to the intrinsic flux density of the material.

If there is no external applied field to the magnet, that is, $\mathbf{H} = 0$, then the magnetized material is defined by

$$\mathbf{B_i} = \mathbf{J} = \mu_0\mathbf{M} \tag{16.69}$$

where $\mathbf{B_i}$ is called the **intrinsic magnetization, intrinsic induction,** or **magnetic polarization**. The symbol **J** is also used.

In the absence of an external magnetic field, the intrinsic flux density is plotted against the magnetic field strength in Figure 16.24(a). The normal *B-H* loop for the permanent-magnet material when excited by an external magnetic field *H* has the form shown in Figure 16.24(b).

Sample normal and intrinsic hysteresis loops for the 2010 Toyota Camry traction motor permanent magnets are shown in Figure 16.25 for a high temperature of 114 ° C [3]. Temperature is a significant design consideration when using permanent magnets, as excessive temperatures can permanently degrade the magnet.

We also note that the units shown in Figure 16.25 are not metric. It is common in the electrical machine industry to use other systems of units. Table 16.10 converts some of the commonly used non-metric units to their metric equivalents.

16.4.1 Example: Remanent Flux Density

Determine the remanent flux density of the permanent magnet shown in Figure 16.25.

Solution:

The remanent flux density B_R occurs at $H = 0$.

$$B_R = \frac{12{,}200\,\text{G}}{10{,}000\,^{\text{G}}/_{\text{T}}} = 1.22\,\text{T}$$

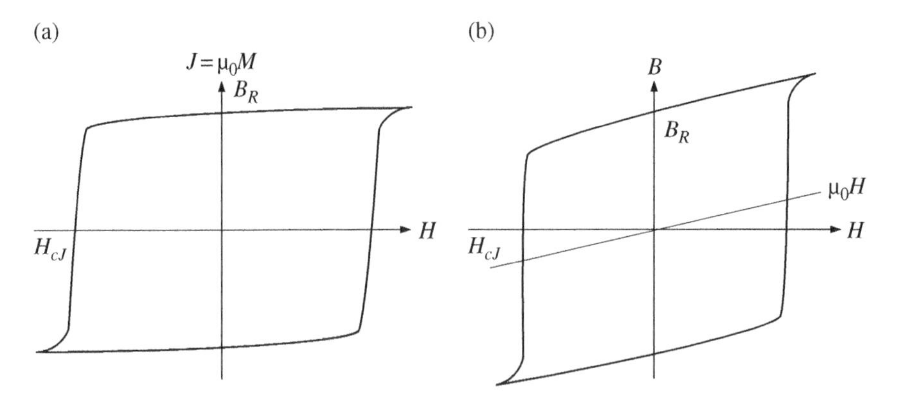

Figure 16.24 (a) Intrinsic and (b) normal hysteresis loops for hard material.

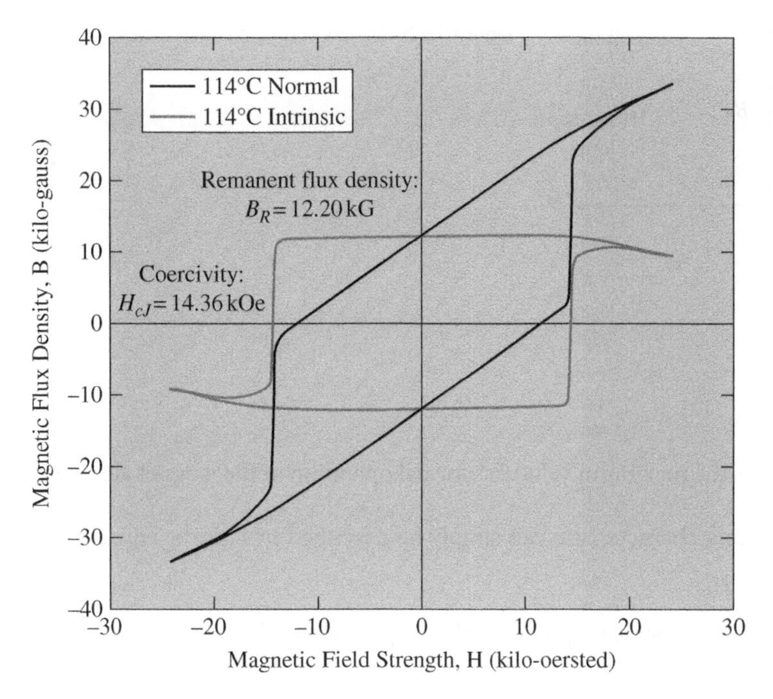

Figure 16.25 Hysteresis loops for 2010 Toyota Camry traction motor permanent magnets at 114 °C [3]. (Courtesy of Oak Ridge National Laboratory, US Dept. of Energy.)

Table 16.10 Unit conversions for magnetic quantities.

Parameter	Conversion to metric
B	1 gauss (G) = 10^{-4} T
H	1 oersted (Oe) = 79.58 A/m

Determining the operating point for a permanent magnet can be quite complex as the magnet is subject to **demagnetization**. The magnet can actually lose its magnetization due to excessive external magnetic fields, excessive temperatures, or mechanical stress. The operating range for the magnet must be constrained in order to ensure the long life and stability that is necessary for automotive applications.

Fortunately, many hard magnet materials have extremely high coercivities and can operate with great stability along the **recoil line**. The recoil line is based on the *BH* loop and essentially runs tangentially to the *BH* characteristic. The recoil line is linear within the quadrant of operation. A typical plot for the magnet operating characteristic is shown in Figure 16.26.

The slope of the recoil line is very low and is slightly greater than the permeability of air. The slope is termed the **recoil relative permeability** μ_{rec}, and is about 1.05 for the Hitachi material. The coercivity used for the intrinsic *JH* loop H_{cJ} is not the coercivity specified along the recoil line. A lower value of coercivity H_{cB} based on the normal

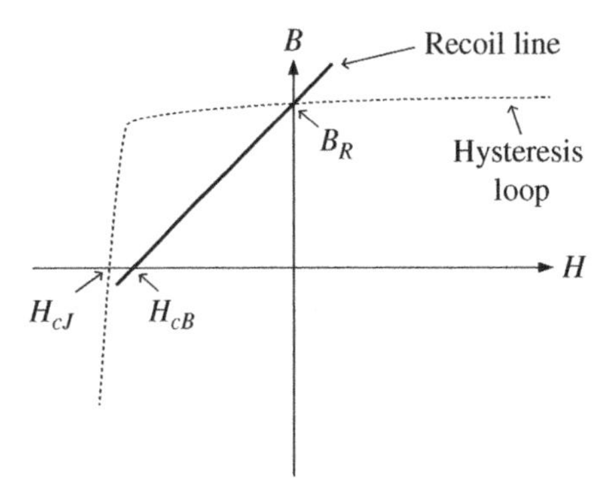

Figure 16.26 Magnet recoil characteristics.

BH loop is specified as the maximum value for normal operation of the magnet along the recoil line.

The magnet operating characteristic can simply be described by a linear equation:

$$B_m = B_R + \mu_{rec}\mu_0 H_m \tag{16.70}$$

where B_m and H_m are the operating points of the magnet along the recoil line.

16.4.2 Example: The Recoil Line

Determine the remanent flux density, coercivity, and slope of the recoil line for the Hitachi Nd-Fe-B material in Table 16.9.

Solution:

From Table 16.9, these values are in the range 1.24 to 1.31 T for B_R, 923 to 1018 kA/m for H_{cB}, and 1.05 for μ_{rec}.

The operating point of the magnet depends on the characteristic of the load. A permanent magnet is typically loaded with an air gap and ferromagnetic material to conduct the flux from the magnet. The interception of the load line and the recoil line is the operating point of the magnet, as shown in Figure 16.27. We next work through a common example – that of the dc motor.

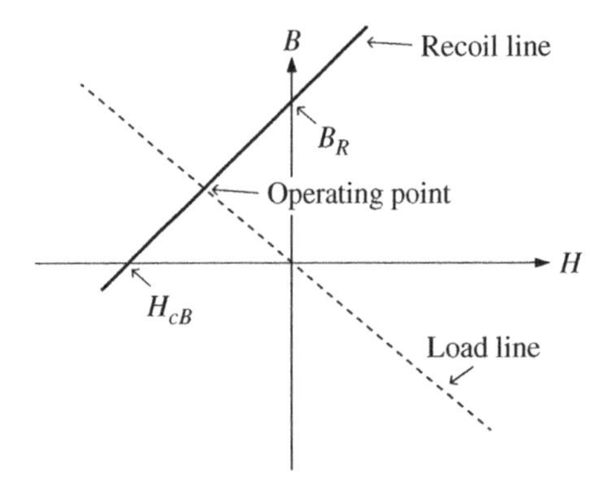

Figure 16.27 Magnet operating points.

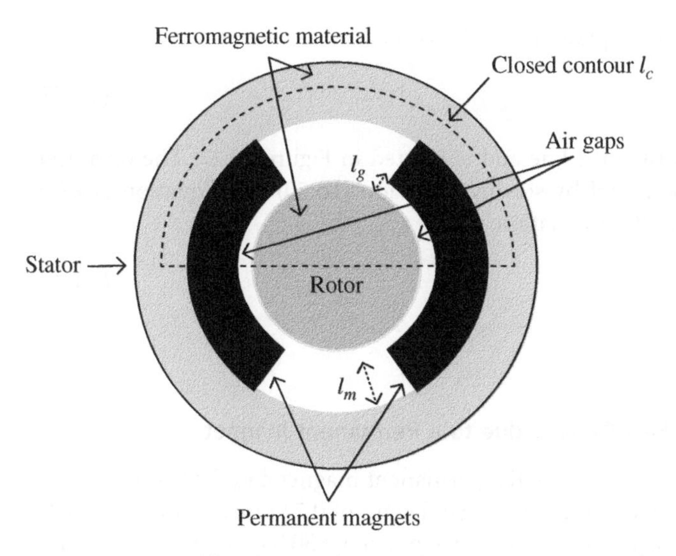

Figure 16.28 Simplified diagram showing the magnetic materials of a permanent-magnet dc machine.

The basic structure of the permanent-magnet dc machine is as shown in Figure 16.28. The current-carrying components are not shown. The lightly shaded areas are made of ferromagnetic material, and the black areas are the permanent magnets of thickness l_m. The center part is the spinning part of the machine shown as the **rotor**. An air gap l_g exists between the magnets and the rotor. The **stator** is the stationary part of the machine.

As the ferromagnetic material has a far higher permeability than either the magnets or the air, its effect on the magnetic path can be neglected for a simple analysis.

Let B_g and H_g be the magnetic flux density and the magnetic field strength of the air gap, and let H_c be the magnetic field strength of the ferromagnetic core.

Applying Ampere's law to the structure around the closed contour yields the following equation:

$$\oint \mathbf{H}.d\mathbf{l} = 2H_m l_m + 2H_g l_g + H_c l_c = 0 \tag{16.71}$$

If we ignore the core due to the high permeability, we can rearrange the preceding equation to get

$$H_g = -\frac{l_m}{l_g} H_m = \frac{B_g}{\mu_0} \tag{16.72}$$

The flux generated by the magnet Φ is

$$\Phi = B_m A_m = B_g A_g \tag{16.73}$$

Rearranging yields

$$B_g = \frac{A_m}{A_g} B_m \tag{16.74}$$

Solving Equation (16.72) and Equation (16.74) gives a load line in terms of B_m and H_m:

$$B_m = -\mu_0 \frac{l_m}{l_g} \frac{A_g}{A_m} H_m \tag{16.75}$$

This equation represents the load line and is plotted in Figure 16.27. The operating point of the magnet can be found by solving Equation (16.70) and Equation (16.75), resulting in the following solution equation:

$$B_m = \frac{B_R}{1 + \mu_{rec} \dfrac{A_m}{A_g} \dfrac{l_g}{l_m}} \tag{16.76}$$

16.4.3 Example: Air Gap Flux Density due to a Permanent Magnet

Determine the magnetic flux densities of the permanent magnet and the rotor air gap if the machine has the following parameters: air gap length of 0.73 mm and magnet thickness of 6.5 mm. Let the cross-sectional area of the air gap be 50% larger than the area of the permanent magnet.

Solution:

Using the Nd-Fe-B magnet parameters from Table 16.9, the magnet flux density is

$$B_m = \frac{B_R}{1 + \mu_{rec} \dfrac{A_m}{A_g} \dfrac{l_g}{l_m}} = \frac{1.24}{1 + 1.05 \times \dfrac{1}{1.5} \times \dfrac{0.73}{6.5}} \mathrm{T} = 1.15\,\mathrm{T} \tag{16.77}$$

The air gap flux density B_g is given by

$$B_g = \frac{A_m}{A_g} B_m = \frac{1}{1.5} \times 1.15\,\mathrm{T} = 0.766\,\mathrm{T} \tag{16.78}$$

16.4.4 Maximum Energy Product

The **maximum energy product** is a term used to describe the capacity of a magnet. It is often denoted by $\mathbf{(BH)_{max}}$. As the product of B and H represents the energy stored within the magnet, the stored energy is maximized, and the magnet volume minimized, when the product of B and H is a maximum. To put it simply, if we consider the BH characteristic to be the recoil line, then the product is a maximum at the midpoint of the recoil line at $(H_{cB}/2, B_R/2)$ for $(\mathrm{BH})_{\max} = (H_{cB}\, B_R)/4$, as shown in Figure 16.29. The magnets do not necessarily operate at the maximum energy point for the basic rotating machine as the objective is usually to maximize the air gap flux density rather than the magnet's energy density.

16.4.5 Force due to Permanent Magnet

We now explore the lifting force of a permanent magnet or an electromagnet which is related to the ability of the magnet to store energy in an air gap.

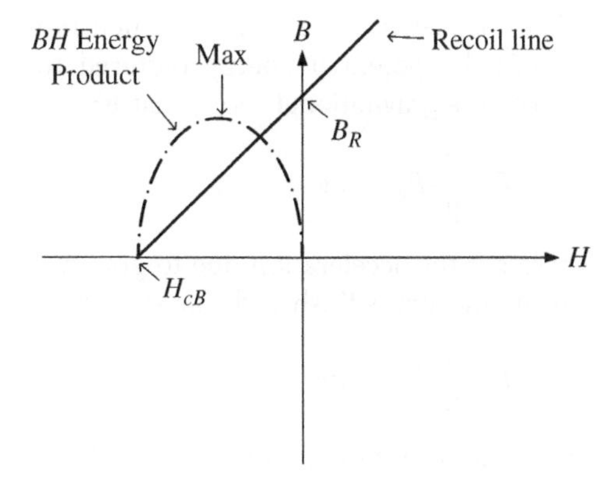

Figure 16.29 BH energy product and the recoil line.

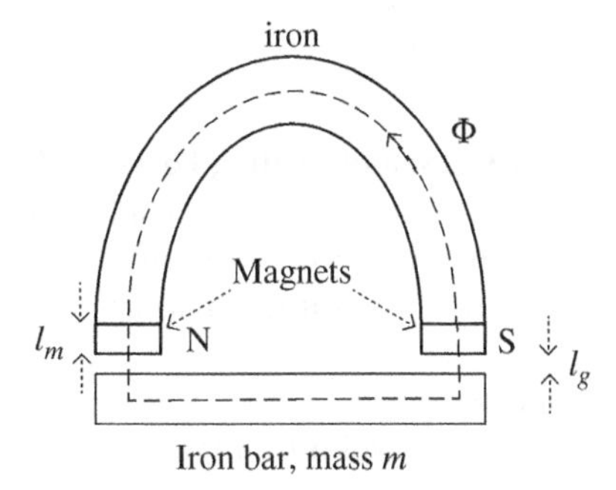

Figure 16.30 A horseshoe-shaped magnet lifting an iron bar.

Consider a horseshoe-shaped magnet in close proximity to an iron bar of mass m, as shown in Figure 16.30. The magnet poles are each of area A_m.

As discussed earlier, energy is stored throughout the magnetic path length, shown by the contour, in the air gaps, in the iron, and in the magnet. The energy is distributed as follows:

$$E = \frac{1}{2}\frac{B_m^2}{\mu_m}V_m + \frac{1}{2}\frac{B_c^2}{\mu_c}V_c + \frac{1}{2}\frac{B_g^2}{\mu_0}V_g \tag{16.79}$$

If we assume that the iron core has an infinite permeability, we can ignore the energy storage in the iron. The energy stored in the air gap is given by

$$E_g = \frac{1}{2}\frac{B_g^2}{\mu_0}V_g \tag{16.80}$$

If we consider the incremental energy ΔE_g stored in the two air gaps with an incremental air gap length Δl_g,

$$\Delta E_g = \frac{1}{2}\frac{B_g^2}{\mu_0}A_g\Delta l_g \times 2 = \frac{B_g^2 A_g \Delta l_g}{\mu_0} \tag{16.81}$$

The magnetic force F required to store the incremental energy is given by

$$F = \frac{\Delta E_g}{\Delta l_g} = \frac{A_g}{\mu_0}B_g^2 \tag{16.82}$$

The magnet lifts the iron bar when the energy stored magnetically in the air gaps exceeds the potential energy required to lift the mass – or if the magnetic force exceeds the gravitational force; that is:

$$F = \frac{A_g}{\mu_0} B_g^2 > mg \tag{16.83}$$

where g is the acceleration due to gravity.

If the magnet is flush with the surface of the iron such that there is no air gap, then

$$F = \frac{A_g}{\mu_0} B_g^2 = mg \tag{16.84}$$

and the maximum mass that can be lifted by the magnet is given by

$$m = \frac{A_g}{\mu_0 g} B_g^2 \tag{16.85}$$

16.4.5.1 Example: Lifting Force of Magnet with no Gap

How heavy an iron mass can a Nd-Fe-B horseshoe-shaped magnet lift if each pole has an area of 6 mm by 3 mm = 18 mm^2, a magnet thickness l_m of 3 mm, and a residual flux density of 1.3 T? Assume no air gap and no fringing.

Solution:

The magnet will lift a mass of

$$m = \frac{A_g}{\mu_0 g} B_g^2 = \frac{18 \times 10^{-6}}{4\pi \times 10^{-7} \times 9.81} \times 1.3^2 = 2.47\,\text{kg}$$

Note that the combined area of the two magnetic poles is about one third of the area of your little finger's nail – and yet can lift over 2 kg.

From Equation (16.76), we can easily determine the air gap flux density for small air gaps. The effective flux density can be more difficult to determine for large air gaps, due to fringing.

16.4.5.2 Example: Lifting Force of Magnet with Gap

Determine the air gap when the horseshoe magnet can lift the iron bar if the iron bar weighs 1 kg. Ignore fringing ($A_m = A_g$).

Solution:

The required magnetic flux density in the gap can be determined by expressing Equation (16.85) as

$$B_g = \sqrt{\frac{\mu_0 g m}{A_g}} = \sqrt{\frac{4\pi \times 10^{-7} \times 9.81 \times 1}{18 \times 10^{-6}}}\,\text{T} = 0.827\,\text{T} \tag{16.86}$$

The air gap length can be determined by expressing Equation (16.76) as

$$l_g = \frac{A_g l_m}{A_m \mu_{rec}}\left(\frac{B_R}{B_m} - 1\right) = \frac{1}{1} \times \frac{3 \times 10^{-3}}{1.05}\left(\frac{1.3}{0.827} - 1\right)\text{m} = 1.63\,\text{mm}$$

16.4.6 Electromagnet

The magnetic field in an electromechanical machine can be provided by a permanent magnet or an **electromagnet**. The **wound-field dc machine** uses an electromagnet to generate the field.

The windings of the electromagnet are typically split in half and wound across each of the poles as shown in Figure 16.31. In this case, the field winding is supplied with a field current I_f which generates a magnetic field across the air gap.

Per the earlier analysis of Section 16.3.2, the inductance of the field winding is

$$L_f = \frac{{N_f}^2}{\dfrac{l_c}{\mu_r \mu_0 A_c} + \dfrac{2l_g}{\mu_0 A_{pole}}} \tag{16.87}$$

which can be simplified to the following equation if the core material is of high permeability:

$$L_f \approx \frac{\mu_0 {N_f}^2 A_{pole}}{2l_g} \tag{16.88}$$

where A_{pole} is the area of the pole.

16.4.6.1 Example: Air Gap Flux Density due to Field Winding

Determine the field winding inductance and the field current that must be supplied if the field winding is to generate an air gap flux density $B_g = 0.766$ T, similar to the earlier example of Section 16.4.3, with the permanent magnet. The machine has the following parameters: air gap length $l_g = 0.73$ mm, $N_f = 14$ turns, rotor radius $r = 8.24$ cm, and rotor length $l = 6.2$ cm. Assume that the pole covers 2/3 of the rotor surface. Assume a high-permeability core.

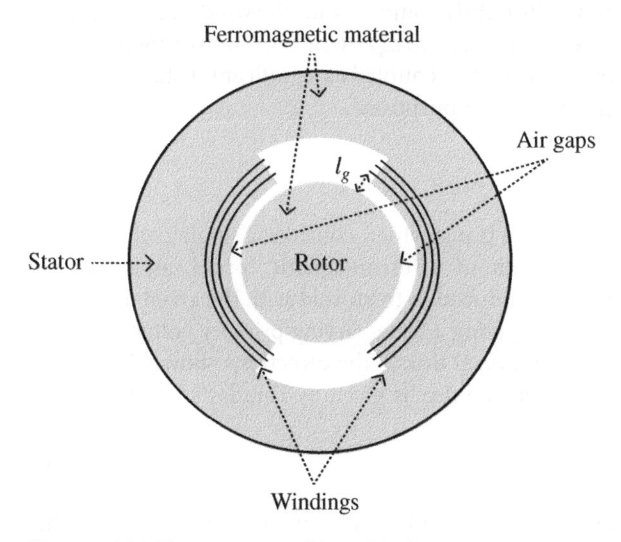

Figure 16.31 Elementary machine with electromagnet.

***Solution*:**
First, the pole area is calculated as follows:

$$A_{pole} \approx {}^{2}/_{3}\pi r l = {}^{2}/_{3} \times \pi \times 8.24 \times 10^{-2} \times 6.2 \times 10^{-2}\,\text{m}^2 = 107\,\text{cm}^2$$

The field inductance is then

$$L_f \approx \frac{\mu_0 N_f{}^2 A_{pole}}{2l_g} = \frac{4\pi \times 10^{-7} \times 14^2 \times 107 \times 10^{-4}}{2 \times 0.73 \times 10^{-3}}\,\text{H} = 1.8\,\text{mH}$$

Since

$$\lambda = N_f \Phi = N_f B_g A_{pole} = L_f I_f$$

we get the following expression for the field current:

$$I_f = \frac{N_f B_g A_{pole}}{L_f} = \frac{14 \times 0.766 \times 107 \times 10^{-4}}{0.0018}\,\text{A} = 63.7\,\text{A} \tag{16.89}$$

16.5 The Transformer

Transformers play a critical role in electrical and electronic engineering. In the modern ac power grid, transformers enable the safe and efficient transmission and distribution of electrical energy. Coupled inductors are inversely coupled transformers which play a far smaller but still important role in the world of power conversion and control.

A transformer is an electromagnetic device having two or more coupled stationary coils. All or a significant portion of the flux produced by one coil (winding) links one or more other windings. Transformers are essential for the transmission and distribution of electric power because of the ease with which alternating voltage can be increased (stepped up) or decreased (stepped down) safely.

Conventional transformers are generally tightly coupled, with the windings being close to each other while meeting various electrical safety requirements. Transformers for use in inductive or wireless power transfer are loosely coupled as significant distance may be required between the windings for operational purposes.

16.5.1 Theory of Operation

To understand the operating principles of a transformer, consider a single coil or winding of N_p turns, wound around a portion of a ferromagnetic toroid, as shown in Figure 16.32(a). A magnetizing flux Φ_m is induced in the toroid with the given direction. Current i_m is the magnetizing current. Applying a time-varying primary voltage $v_p(t)$ to the winding results in current flow and induced flux in the directions shown. The relationship of the primary voltage to the induced flux is given by Faraday's law as

$$e_p(t) = N_p \frac{d\Phi_m(t)}{dt} \tag{16.90}$$

If a second winding is now added to the toroid as shown in Figure 16.32(b), the construction of the windings and toroid results in transformer operation. According

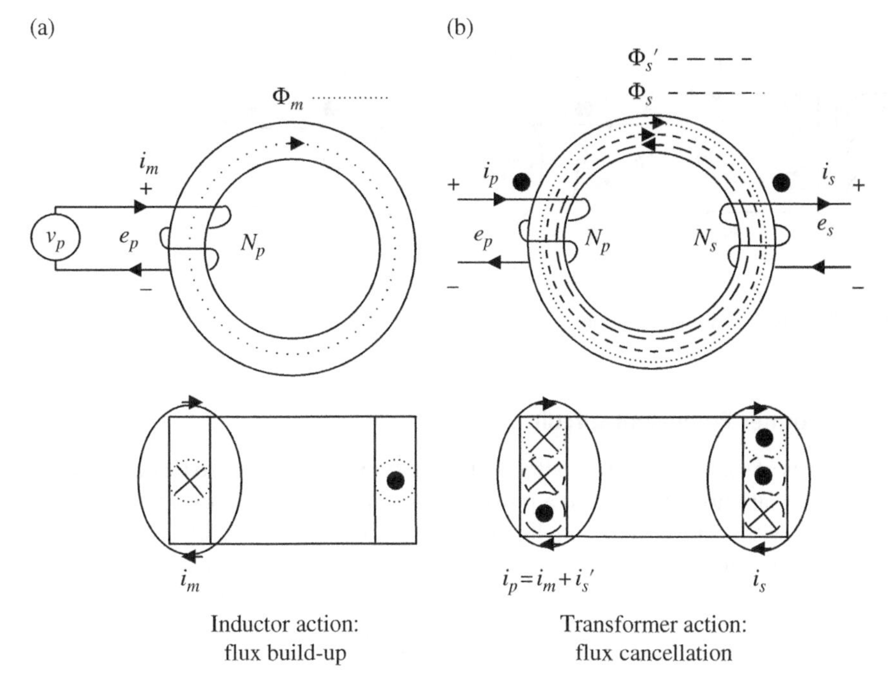

Figure 16.32 (a) Inductor and (b) transformer operation.

to Faraday's law, the primary-induced time-varying flux Φ_m linking the toroid is mutual to the secondary and induces an emf in the secondary winding given by

$$e_s(t) = N_s \frac{\mathrm{d}\Phi_m(t)}{\mathrm{d}t} \tag{16.91}$$

The direction of the second winding emf e_s is given by Lenz's law. The induced emf is always in such a direction that it generates a current i_s to create a flux Φ_s to oppose the flux Φ_m responsible for inducing the emf in the first place.

A counter flux Φ_s' is induced in the core by transformer action to cancel out Φ_s, and so the net flux in the core is Φ_m. A primary current i_s' is induced in the primary winding by flux Φ_s'.

The basic components of the ideal transformer described above are (1) the **primary winding** N_p, so called because it connects to the voltage source; (2) the **secondary winding** N_s which supplies the load; and (3) the ferromagnetic core.

The ideal transformer is drawn as shown in Figure 16.33(a), with a toroidal representation in Figure 16.33(b). The dots in the figure convey the information that the winding voltages are of the same polarity at the dotted terminals with respect to their un-dotted terminals. For example, if Φ_m is increasing with time, the voltages at both dotted terminals are positive with respect to the corresponding un-dotted terminals. The advantage of using this dot convention is that the winding orientations on the core need not be shown in detail.

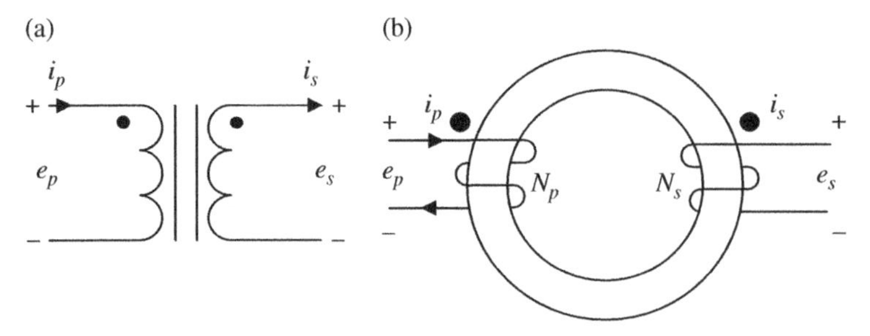

Figure 16.33 Polarity and dot convention.

16.5.2 Transformer Equivalent Circuit

Additional details of the second-order effects of a real transformer have to be understood in order to appreciate how it works in a power circuit. In a real transformer, not all the flux is contained within the core. A small portion of the flux generated by each winding "leaks" from the winding as shown in Figure 16.34. Flux Φ_{lp} leaks from the primary winding, and flux Φ_{ls} leaks from the secondary. The flux mutual or common to the two windings is Φ_m. The effect of the mutual inductance between the windings is illustrated in the electrical model by adding an inductance in parallel to the primary or the secondary winding. This inductance is usually termed the magnetizing inductance L_m.

The operation of the real transformer can now be more fully understood with reference to the circuit model in Figure 16.35. Primary current i_p flows through the primary leakage inductance L_{lp} and has two components. The first component is i_m, the magnetizing current flowing into the primary-referenced magnetizing inductance L_m. The second current is i_s', and this current is coupled perfectly to the secondary by the ideal transformer resulting in current i_s (= $N_p\, i_s' \,/\, N_s$) flowing in the secondary winding. The superscript ' implies the value is reflected through the transformer from the secondary to the primary.

Current i_m and the magnetizing inductance L_m can now be seen from a new perspective. In order for the ideal transformer of a real transformer to perfectly couple current from primary to secondary, the transformer core must be *magnetized*. This magnetization requires the magnetizing current i_m to flow into the magnetizing inductance L_m. The back emf across the magnetizing inductance is denoted by e_m.

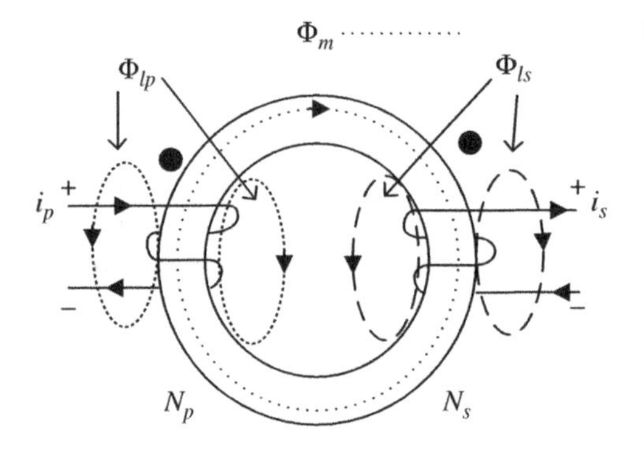

Figure 16.34 Leakage flux.

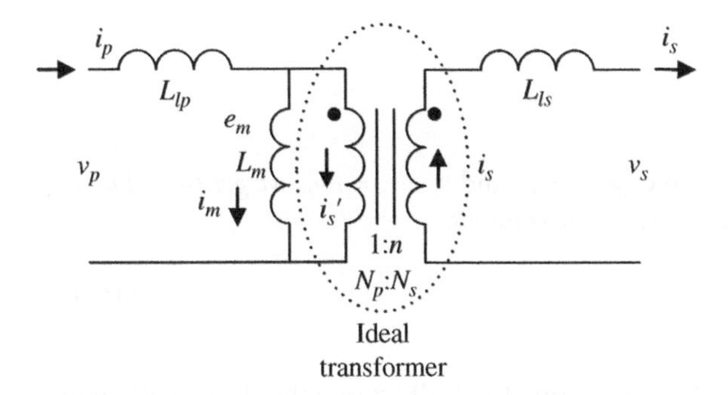

Figure 16.35 Transformer model.

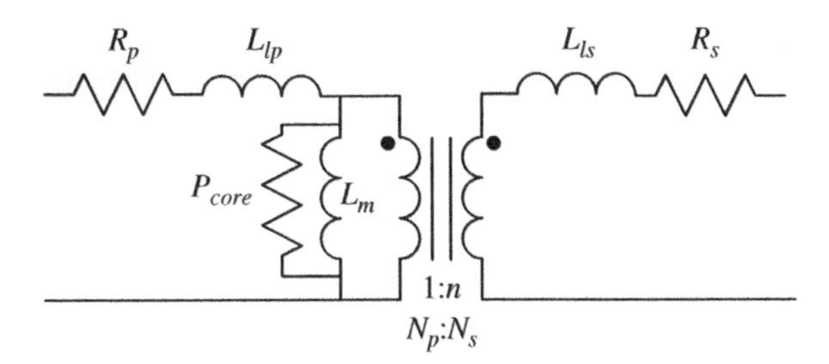

Figure 16.36 Full equivalent circuit.

The full transformer equivalent circuit is shown in Figure 16.36. This model includes the primary and secondary winding resistances, R_p and R_s, plus the core loss P_{core}.

16.5.3 Transformer Voltages and Currents

Rearranging Equation (16.90) and Equation (16.91) yields

$$\frac{d\Phi_m(t)}{dt} = \frac{e_p(t)}{N_p} = \frac{e_s(t)}{N_s} = \frac{\text{volts}}{\text{turn}} \tag{16.92}$$

which shows that in each winding, the volts per turn are the same, because of the common coupling.

We represent this relationship by the ideal transformer, which relates the voltages in the two windings by the **turns ratio** n:

$$\frac{e_s}{e_p} = \frac{N_s}{N_p} = n \tag{16.93}$$

or

$$\frac{e_p}{N_p} = \frac{e_s}{N_s} \tag{16.94}$$

or

$$e_s = n\, e_p \tag{16.95}$$

or

$$\frac{\text{volts}}{\text{turn}} = \text{constant} \tag{16.96}$$

If the leakages are ignored, in which case $v_p = e_p$ and $v_s = e_s = nv_p$, *we get the following relationship between the transformer terminal voltages:*

$$\frac{v_s}{v_p} = \frac{N_s}{N_p} = n \tag{16.97}$$

Because the transformer is ideal, the evaluation of $\oint \mathbf{H} \bullet \mathbf{dl}$ around the closed path of the toroid must be zero. If i_s' and i_s are the reflected secondary and secondary currents, respectively, then

$$\oint \mathbf{H} \bullet d\mathbf{l} = N_p i_s' - N_s i_s = 0 \tag{16.98}$$

resulting in

$$\frac{i_s}{i_s'} = \frac{N_p}{N_s} = \frac{1}{n} \tag{16.99}$$

or

$$N_p i_s' = N_s i_s \tag{16.100}$$

or

$$i_s' = n\, i_s \tag{16.101}$$

or

$$\text{ampere} \times \text{turns} = \text{constant} \tag{16.102}$$

From Faraday's law, the magnetizing current is related to the back emf by

$$e_m = L_m \frac{di_m}{dt} \tag{16.103}$$

and the magnetizing current is given by integrating the voltage over time:

$$i_m = \frac{1}{L_m} \int e_m dt \tag{16.104}$$

The primary current is the sum of the magnetizing current and the reflected secondary current:

$$i_p = i_m + i_s' = i_m + \frac{N_s}{N_p} i_s \tag{16.105}$$

We next consider two simple but common examples of voltage sources feeding the transformer primary: a square wave and a sine wave. In these examples, the leakage inductances are ignored, and it is assumed that the supply voltage equals the primary back emf. Refer to Figure 16.37.

Figure 16.37 Transformer equivalent circuit without leakage inductances or resistive losses.

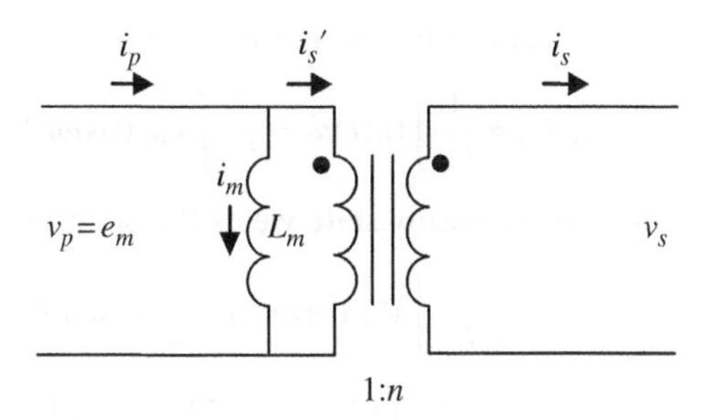

16.5.3.1 Exciting the Transformer with Sinusoidal Wave

The power transformer is commonly supplied with a sine or cosine voltage waveform. The waveforms are shown in Figure 16.38(a). The transformer primary is fed by a 50 Hz or 60 Hz ac cosine wave. Thus, let the supply voltage be

$$v_p(t) = V_{pk} \cos\omega t \tag{16.106}$$

where V_{pk} is the peak of the input voltage, and ω is the angular frequency given by

$$\omega = 2\pi f \tag{16.107}$$

where f is the electrical frequency.

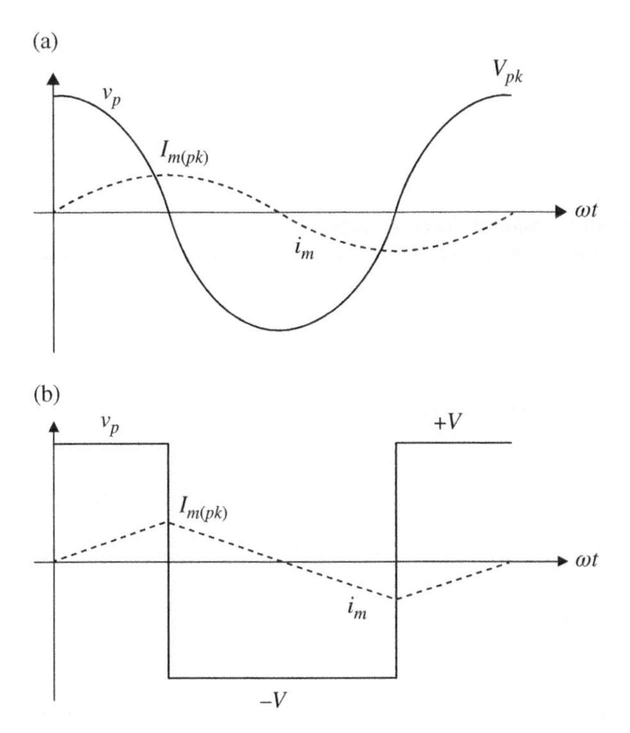

Figure 16.38 (a) Sinusoidal and (b) square wave voltage sources and resulting magnetizing currents.

The magnetizing current is given by

$$i_m(t) = \frac{1}{L_m}\int v_p(t)dt = \frac{1}{L_m}\int V_{pk}\cos\omega t dt \tag{16.108}$$

which in the steady state yields the solution

$$i_m(t) = \frac{1}{L_m}\int V_{pk}\cos\omega t\, dt = \frac{V_{pk}}{\omega L_m}\sin\omega t = I_{m(pk)}\sin\omega t \tag{16.109}$$

where $I_{m(pk)}$ is the peak of the magnetizing current and is given by

$$I_{m(pk)} = \frac{V_{pk}}{\omega L_m} \tag{16.110}$$

16.5.3.2 Example: Induction Machine Magnetizing Current

The induction motor is also an example of a transformer and is discussed in detail in Chapter 8. If the supply voltage is 120 V_{rms} at 200 Hz and the magnetizing inductance is 2 mH, determine the peak and rms magnetizing currents.

Solution:

Since the peak value of a sine wave is $\sqrt{2}$ times the rms value, the peak value of the magnetizing current is given by

$$I_{m(pk)} = \frac{\sqrt{2}V}{\omega L_m} = \frac{\sqrt{2}\times 120}{2\pi\times 200\times 2\times 10^{-3}}\text{A} = 67.52\ \text{A}$$

and the rms value is given by

$$I_m = \frac{I_{m(pk)}}{\sqrt{2}} = \frac{67.52}{\sqrt{2}}\text{A} = 47.75\ \text{A}$$

16.5.3.3 Exciting the Transformer with a Square Wave Voltage

The square wave is a very common excitation waveform in high-frequency power conversion, as shown in Figure 16.38(b).

Let the supply voltage be an oscillating square wave

$$v_p(t) = \pm V \tag{16.111}$$

where V is the value of the positive and negative levels.

The current is given by

$$i_m(t) = \frac{1}{L_m}\int v_p(t)dt \tag{16.112}$$

In the steady state, the current changes from zero to the peak in one quarter cycle as the square wave is applied.

The maximum magnetizing current is

$$I_{m(pk)} = \frac{1}{L_m}\int_0^{T/4} Vdt = \frac{1}{L_m}[Vt]_0^{T/4} = \frac{VT}{4L_m} = \frac{V}{4fL_m} \tag{16.113}$$

16.5.3.4 Example: High-Frequency Transformer

A high-frequency power converter (for an on-vehicle auxiliary power converter) features a transformer which is fed by a +/− 380 V square wave at 100 kHz. The transformer magnetizing inductance L_m is 1 mH, and the turns ratio n is 0.04.

i) Determine the primary current if the secondary current is 168 A when the magnetizing current is at a positive peak.
ii) What are the values of the square wave voltage on the secondary?

Ignore transformer leakage and parasitics.

Solution:

i) The peak value of the magnetizing current is given by

$$I_{m(pk)} = \frac{V}{4fL_m} = \frac{380}{4 \times 100 \times 10^3 \times 1 \times 10^{-3}} \text{A} = 0.95\ \text{A}$$

Knowing the magnetizing current, the primary current can be determined as the secondary current is known.
The reflected secondary current is

$$i_s' = n\, i_s = 0.04 \times 168\ \text{A} = 6.72\ \text{A}$$

The primary current is

$$i_p = i_m + i_s' = 0.95\ \text{A} + 6.72\ \text{A} = 7.67\ \text{A}$$

The transformer circuit is then as shown in Figure 16.39. The instantaneous currents are noted.

ii) The secondary voltage is given by

$$v_s = n\, v_p = 0.04 \times \pm 380\ \text{V} = \pm 15.2\ \text{V}$$

16.5.4 Sizing the Transformer using the Area-Product (AP) Method

First, let us rearrange the basic transformer equation in terms of the cross-sectional area of the magnetic core A_c. For a square wave, Equation (16.113) can be rearranged to show that

$$L_m I_{m(pk)} = \frac{V}{4f} = N_p \Phi_{max} = N_p B_{max} A_c \tag{16.114}$$

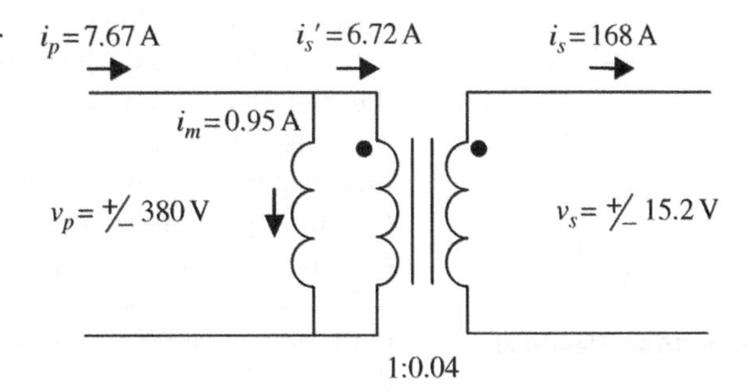

Figure 16.39 Transformer currents.

where Φ_{max} and B_{max} are the peak flux and flux density in the core.

Thus, the core area is given by

$$A_c = \frac{V}{4f\,N_p B_{\max}} \tag{16.115}$$

The area required within the magnetic structure for the winding is the **window area** A_w:

$$A_w = \frac{N_p I_p + N_s I_s}{k_{cu} J_{cu}} \tag{16.116}$$

where I_p and I_s are the primary and secondary rms currents, respectively.

The window area is directly proportional to the number of turns and the rms current. It is inversely proportional to the **copper fill factor** k_{cu} and the winding **current density** J_{cu}.

The size of a magnetic device is related to the product of these two areas. The core area is dependent on flux density, while the window area is dependent on the current density.

The AP for a transformer fed by a square wave is given by

$$AP = A_c A_w = \frac{V}{4fN_p B_{\max}} \frac{N_p I_p + N_s I_s}{k_{cu} J_{cu}} = \frac{V\left(I_p + nI_s\right)}{4fB_{\max} k_{cu} J_{cu}} \tag{16.117}$$

A more generic expression for the area product of a duty-cycle-controlled voltage source supplying a transformer is

$$AP = \frac{DV\left(I_p + nI_s\right)}{2fB_{\max} k_{cu} J_{cu}} \tag{16.118}$$

where D is the duty cycle.

See the examples in Sections 12.2.3.1 and 12.3.2.1 in Chapter 12.

16.6 The Capacitor

Gauss' law states that the surface integral over any closed surface is proportional to the enclosed charge. This law can be simply interpreted for the standard power capacitor as illustrated in Figure 16.40.

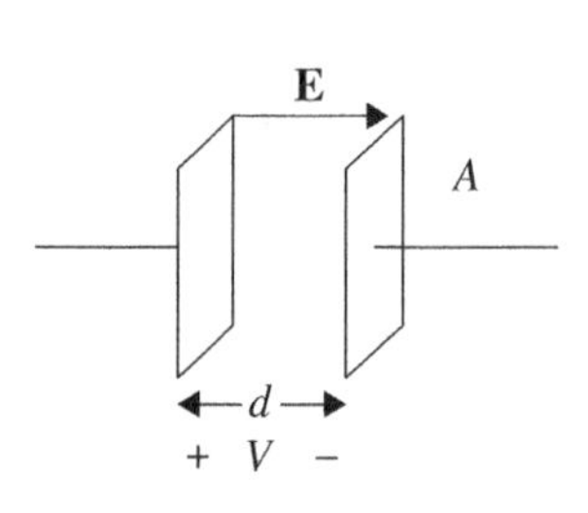

Figure 16.40 Elementary capacitor.

The capacitor is shown as having two parallel plates, each of area A. An insulating material known as the dielectric material is of thickness d and maintains a fixed distance between the plates. Many different materials are used as the dielectric material between the capacitor plates, and so, in designing electric circuits, we use a variety of capacitors depending on the application. Electrolytic, paper, tantalum, polyester, polypropylene, ceramic, and tantalum are among the many varieties used.

For the parallel-plate capacitor, the electric field E is equal to the voltage difference V between the plates divided by the distance d between the plates. The plates are each of area A.

$$E = \frac{V}{d} \tag{16.119}$$

Assuming that a charge Q is built up on each plate:

$$\oint \mathbf{E} d\mathbf{A} = \frac{V}{d} A = \frac{Q}{\varepsilon} \tag{16.120}$$

or

$$Q = \varepsilon \frac{A}{d} V = CV \tag{16.121}$$

where C is the capacitance of the device. The unit of capacitance is the farad (with symbol F), named in honor of Michael Faraday.

Thus:

$$C = \varepsilon \frac{A}{d} \tag{16.122}$$

for the parallel-plate capacitor. The constant ε is known as the permittivity of the dielectric material. Similar to our earlier discussion of μ, we also define a permittivity ε_r relative to that of free space:

$$\varepsilon_r = \frac{\varepsilon}{\varepsilon_0} \tag{16.123}$$

where ε_0 is 8.85×10^{-12} F/m.

Permittivity is a measure of the ability of a dielectric to store charge. A variety of capacitors is shown in Figure 16.41.

It is notable that relative permittivity for commonly used dielectrics is low – values in the single digits are usual. This is unlike the situation for the relative permeability of

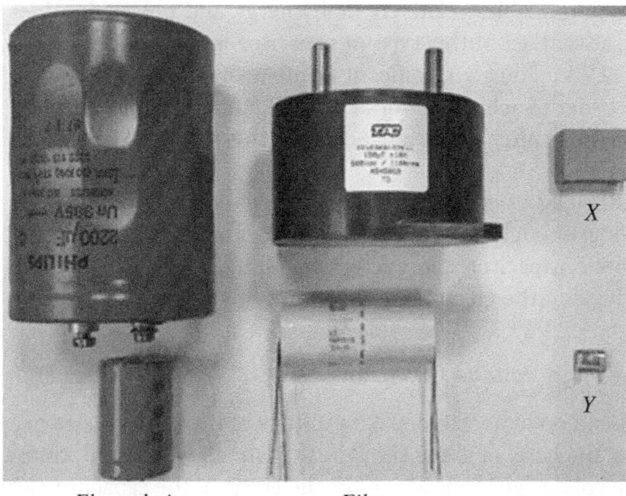

Figure 16.41 Images of electrolytics, film, and X and Y caps.

magnetic materials, which tend to be very high – with values in the hundreds and thousands. The low relative values together with the extremely low absolute value for ε_0 at 8.85×10^{-12} F/m compared to μ_0 at $4\pi \times 10^{-7}$H/m helps explain why magnetic fields are much more effective than electric fields at producing mechanical force.

Capacitors are ubiquitous in signal and power circuits, where they have two main roles. First, just like inductors, capacitors are used to filter ac waveforms. Second, capacitors, unlike inductors, store dc energy.

The typical capacitor used in an electronic circuit can have a value ranging from pico (10^{-12}) to milli (10^{-3}). The choice of a capacitor type is influenced by electrical factors including the value, voltage, current, frequency, temperature, and charge/discharge rates, while being influenced by other factors such as size, cost, manufacturability, failure modes, flammability, lifetime, termination/connection, structure, leakage current, corona and arcing, and the other second-order electric parasitics of resistance and inductance. Specific technologies, often known as *X* capacitors, which are across the high-voltage input, and *Y* capacitors, which are from high-voltage to ground, are required for electromagnetic interference (EMI) noise suppression capacitors in safety critical roles.

Electrolytic capacitors have been a mainstay of the power industry for many years. Electrolytics are typically used in power applications where relatively large storage, for example, millifarads at hundreds of volts, is required. While electrolytics were more commonly used in earlier EVs, they typically now only feature prominently as the dc bus capacitor on the output of the power-factor-corrected boost front-end of the charging system. While the electrolytic can feature a relatively high energy storage, the device is often limited to particular applications due to the ripple current and lifetime requirements [11]. See Chapter 14, Section 14.6.

In the past few decades, a type of capacitor known as a **supercapacitor** has been developed and productized [12]. Supercapacitors are now widely applied and can have values ranging from a few farads to thousands of farads. These types of capacitor are also known as ultracapacitors and double-layer capacitors. Supercapacitors are playing a role in energy storage as the construction enables relatively high energy and power densities and a long lifetime [13]. A disadvantage of this type of capacitor is the very low rated voltage – it is typically about 2.7 V. Thus, a significant number of capacitors must be arranged in series in a pack in order to achieve a workable high voltage. Typically, these packs require significant electric circuitry in order to balance the voltages across each capacitor.

Polypropylene film capacitors play a significant role as the dc bus capacitor for dc-dc converters and dc-ac inverters. Although this type of film capacitor does not have the energy density of the electrolytic capacitor, the current-handling ability and lifetime of the film capacitor exceed those of the electrolytic capacitor.

16.6.1 Sizing Polypropylene High-Voltage Capacitor

The size of a capacitor can depend on many factors. An estimate of the physical size of a film capacitor can be made on the basis of some simple calculations and assumptions.

The dielectric strength is the voltage gradient which the dielectric of the capacitor can withstand without breaking down. It is a critical parameter for a capacitor: the higher the dielectric strength, the smaller the device.

Thus, if V is the maximum voltage of the capacitor and DS is the dielectric strength, the thickness of the dielectric is given by

$$d = \frac{V}{DS} \tag{16.124}$$

The area of the capacitor can then be determined

$$A = \frac{Cd}{\varepsilon_r \varepsilon_0} = lw \tag{16.125}$$

where l and w are the length and width of the capacitor, respectively.

See Section 11.6.2.1 for a related example.

16.7 Electromechanical Energy Conversion

This section reviews the basic theory of electromechanical energy conversion. The reader is directed to the various machine chapters for applications and worked examples.

16.7.1 Ampere's Force Law

Consider a conductor carrying a current I through an external magnetic field of magnetic flux density $\mathbf{B}$, as shown in Figure 16.42. The force exerted on the conductor by the magnetic field is given by

$$\mathbf{F} = I\mathbf{l} \times \mathbf{B} \tag{16.126}$$

where $\mathbf{l}$ is a vector of magnitude l in the direction of the current.

For the basic machine where the conductor of length l is perpendicular to a uniform magnetic field B, the this formula simplifies to the following equation:

$$F = BIl \tag{16.127}$$

16.7.1.1 Fleming's Left Hand Rule

The vector cross product emphasizes that the force is perpendicular to both the conductor and the magnetic field, and reverses in direction if either the field or the current is reversed.

A simple mnemonic to help remember Fleming's left hand rule is to rely on the **F**ederal **B**ureau of **I**nvestigations (FBI) to solve the problem (see Figure 16.42 for an explanation)! Hold your left hand such that the flux is in the direction of the first finger and the current is in the direction of the index finger. The resulting force on the conductor due to the external magnetic field is in the direction of the thumb.

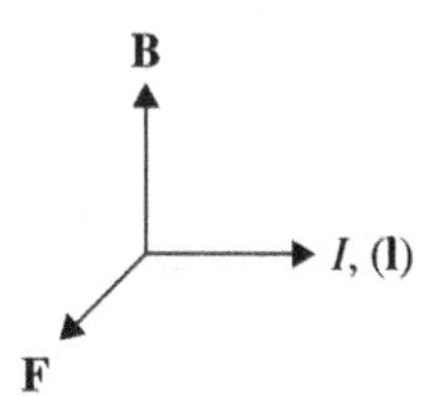

Figure 16.42 Fleming's left hand rule.

The relationship can be easily constructed by considering the lines of flux as shown in Figure 16.43:

In this figure, the conductor is running orthogonal to the magnetic field running from the north pole to the south pole. The conductor is carrying current out of the page and generates a counterclockwise magnetic field.

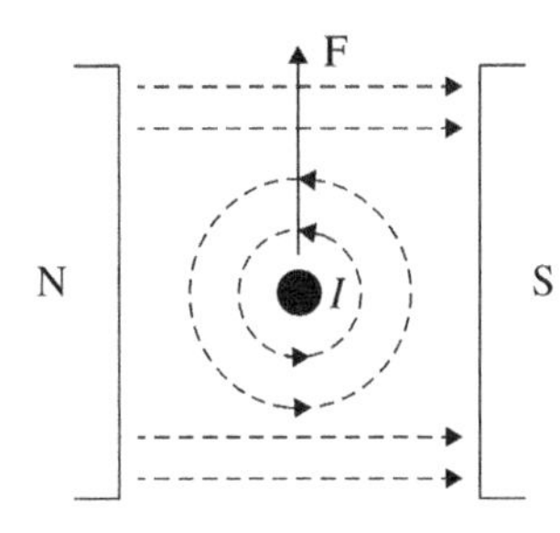

Figure 16.43 Interaction of field resulting in motion.

The buildup of flux on one side of the wire can be seen as analogous to a buildup of pressure on an aircraft wing. The higher flux on one side and the lower flux on the other side forces the wire in the direction of lower flux concentration.

16.7.2 General Expression for Torque on Current-Carrying Coil

Consider a coil set of radius r and carrying a dc current of I amperes, which is placed in a steady external magnetic field (not due to the coil itself). The external magnetic field can be generated by a permanent magnet or an electromagnet, as shown in Figure 16.44.

In general, the coil experiences a magnetic torque about the axis of rotation. The field flux direction through the coil from the north to south poles is regarded as the positive direction for flux, and is related to the positive current direction around the coil by the right hand screw rule. Consequently, both the current I and flux are positive. From Ampere's force law, the force on the conductor, F, is given by Equation (16.127).

The force $+F$ on the approaching-current conductor can be considered to have a tangential force vector only, since the faces of the magnetic poles have been curved and shaped around the circumference of the coil's rotation to eliminate the radial force which would otherwise exist. The tangential vector acts to torque or "twist" the coil of wire around its axis of rotation. Thus, the electromagnetic torque T_{em} about the axis of rotation is given by

$$T_{em} = 2Fr \tag{16.128}$$

or

$$T_{em} = 2BIlr \tag{16.129}$$

For the simple machine, the flux density B is uniform over a surface of area A and is perpendicular to the surface. Assuming a two-pole machine, with one north pole and

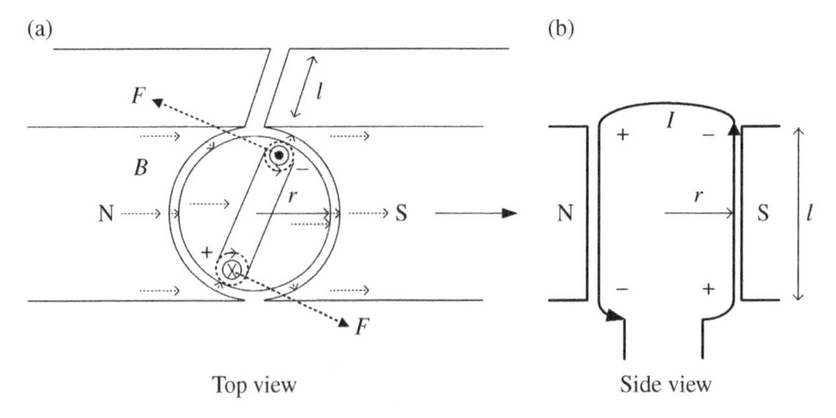

Figure 16.44 Elementary dc machine.

one south pole, and the poles spanning 180°, then the flux per pole is the pole flux density multiplied by the area of the pole:

$$\Phi = BA = B\pi rl \tag{16.130}$$

Thus, Equation (16.130) can be substituted into Equation (16.129) to give an expression for torque in terms of flux per pole and current:

$$T_{em} = \frac{2\Phi}{\pi} I \tag{16.131}$$

16.7.3 Torque, Flux Linkage, and Current

The relationship between torque and current is more commonly described in machine theory by using flux linkage with symbol λ rather than flux with the symbol Φ. The **flux linkage** is defined as the product of the number of turns of the coil and the flux which the coil links. The units of flux linkage are webers or weber-turns.

For simple magnetic structures such as inductors and transformers, the flux linkage is simply the product of the turns and the linking flux:

$$\lambda = N\Phi \tag{16.132}$$

Although flux linkage is easily determined for stationary magnetic components such as inductors and transformers, it is often more complex for machines, as different coils are likely linking different flux patterns because the magnet itself is spinning. The flux linkage is also easily determined for the elementary machine and is now investigated.

The instantaneous variations of flux density and flux linkage with the coil are as shown in Figure 16.45. The back emf developed across the coil has the same wave shape as the flux density seen by the coil. The flux density and related back emf are represented by the solid line. The flux linkage is represented by the dotted line and is maximum at $+\Phi$ when the coil is oriented at $\theta = 0°$, is minimum at $-\Phi$ when the coil is at $\theta = +/-\pi$, and is zero when the coil is aligned with the direction of the magnetic field at $\theta = -\frac{\pi}{2}$ and $+\frac{\pi}{2}$.

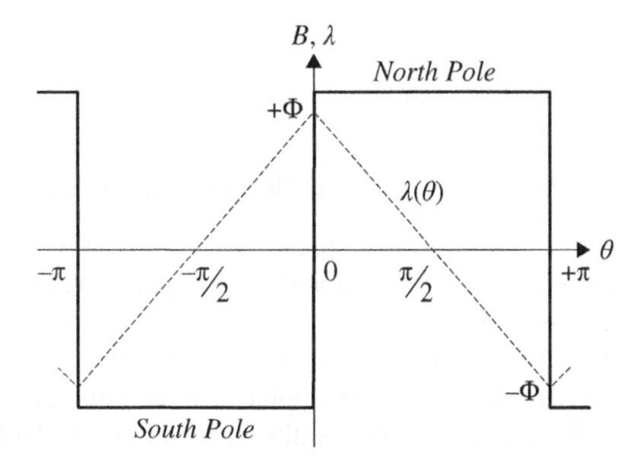

Figure 16.45 Flux density and flux linkage distributions of the elementary dc machine.

The flux linkage λ is the flux linked by the coil as it rotates 360° or 2π. As the coil rotates, the coil links the flux leaving the north pole and entering the south pole for a variation of 2Φ over the half cycle π. Thus, for a single turn:

$$\lambda = \frac{2\Phi}{\pi} \tag{16.133}$$

Thus, the expression for torque in Equation (16.131) can be expressed in terms of the flux linkage λ:

$$T_{em} = \lambda I \tag{16.134}$$

These expressions for torque can be generalized for p poles and N conductors per pole:

$$T_{em} = pBNIlr \tag{16.135}$$

or

$$T_{em} = \frac{p}{2} N \frac{2\Phi}{\pi} I \tag{16.136}$$

resulting in the following commonly used expressions for torque:

$$T_{em} = \frac{p}{2} \lambda I \tag{16.137}$$

and

$$T_{em} = kI \tag{16.138}$$

where k is the **machine constant**.

16.7.4 Faraday's Law of Electromagnetic Induction

When the magnetic flux linkage within a circuit is altered in any manner, an emf is induced as long as the flux linkage is changing. A common expression of Faraday's law is

$$e = -N\frac{d\Phi(t)}{dt} = -\frac{d\lambda(t)}{dt} \tag{16.139}$$

for a rotating machine.

16.7.5 Lenz's Law and Fleming's Right Hand Rule

Per Lenz's law, the induced emf is always in such a direction that it tends to set up a current which will oppose the change of flux linkage or current responsible for inducing the emf. This is a similar statement to Newton's third law of motion – for every action there is an equal and opposite reaction.

This rule relates to the polarity of the induced or back emf and is complementary to the earlier left hand rule, as illustrated in Figure 16.46 by Fleming's right hand rule.

In the elementary machine:

$$\omega = \frac{d\theta}{dt} \tag{16.140}$$

Thus:

$$e = -\frac{d\lambda(\theta)}{d\theta}\omega \tag{16.141}$$

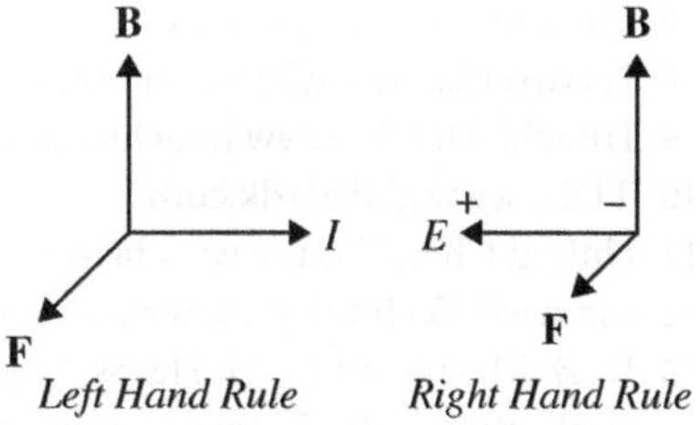

resulting in

$$e = -\frac{2\Phi}{\pi}\omega \text{ for } -\pi \le \theta \le 0$$

and

$$e = +\frac{2\Phi}{\pi}\omega \text{ for } 0 \le \theta \le \pi \tag{16.142}$$

B

B

I

E

F

F

Left Hand Rule *Right Hand Rule*

Figure 16.46 Fleming left and right hand rules.

The preceding expression can be generalized for p poles and N conductors per pole:

$$E = pBNlr\omega \tag{16.143}$$

or

$$E = \frac{p}{2}N\frac{2\Phi}{\pi}\omega \tag{16.144}$$

or

$$E = \frac{p}{2}\lambda\omega \tag{16.145}$$

or

$$E = k\omega \tag{16.146}$$

where k is again the machine constant, and E is the winding back emf.

We have now established the important equations underlying machine operation. The reader is referred to the machine chapters for further reading.

References

1 J. C. Rautio, "The Long Road to Maxwell's Equations," *IEEE Spectrum Magazine*, pp. 22–37, December 2014.
2 K. J. Hartnett, J. G. Hayes, M. G. Egan, and M. Rylko, "CCTT-core split-winding integrated magnetic for high-power dc-dc converters," *IEEE Transactions on Power Electronics*, 28, pp. 4970–4974, November 2013.
3 T. A. Burress, S. L. Campbell, C. L. Coomer, C. W. Ayers, A. A. Wereszczak, J. P. Cunningham, L. D. Marlino, L. E. Seiber, and H. T. Lin, *Evaluation of the 2010 Toyota Prius Hybrid Electric Drive System*, Oak Ridge National Laboratory report, 2011.
4 JFE, www.jfe-steel.co.jp
5 Metglas, www.metglas.com
6 Magnetics Inc., www.mag-inc.com

7 Vacuumschmelze, www.vacuumschmelze.de
8 Ferroxcube, www.ferroxcube.com
9 Hitachi Metals, www.hitachi-metals.co.jp
10 TDK, www.global.tdk.com
11 United ChemiCon, www.chemi-con.com
12 Maxwell Technologies, www.maxwell.com
13 D. B. Murray and J. G. Hayes, "Cycle testing of supercapacitors for long-life robust applications," *IEEE Transactions on Power Electronics, Special Edition on Robust Design and Reliability in Power Electronics*, 30, pp. 2505–2516, May 2015.
14 Finite Element Method Magnetics, a free software by David Meeker: www.femm.info.
15 K. Davis and J. G. Hayes, *FEMM Modelling of an Inductor*, University College Cork, 2017 (files available on Wiley book web site).

Further Reading

1 N. Mohan, T. M. Undeland, and W. P. Robbins, *Power Electronics Converters, Applications and Design*, 3rd edition, John Wiley & Sons, 2003.
2 W. G. Hurley and W. H. Wolfle, *Transformers and Inductors for Power Electronics Theory, Design and Applications*, John Wiley & Sons, 2013.
3 Colonel W. T. McLyman, *Transformer and Inductor Design Handbook*, 2nd edition, Marcel Dekker, Inc., 1988.
4 A. Van den Bossche and V. C. Valchev, *Inductors and Transformers for Power Electronics*, CRC Press, Taylor & Francis Group, 2005.
5 R. W. Erickson, *Fundamentals of Power Electronics*, Kluwer Academic Publishers, 2000.
6 B. Lyons, *High-Current Inductors for High-Power Automotive DC-DC Converters*, PhD Thesis. University College Cork, 2008.
7 M. Rylko, B. Lyons, J. G. Hayes, and M. G. Egan, "Revised magnetics performance factors and experimental comparison of high-flux magnetic materials for high-current DC-DC inductors," *IEEE Transactions on Power Electronics*, 26, pp. 2112–2126, August 2011.
8 K. J. Hartnett, J. G. Hayes, M. S. Rylko, B. C. Barry, and J. W. Maslon, "Comparison of 8 kW CCTT IM and discrete inductor interleaved boost converter for renewable energy applications," *IEEE Transactions on Industry Applications*, 51, pp. 2455–2469, May/June 2015,.
9 B. G. You, J. S. Kim, B. K. Lee, G. B. Choi, and D. W. Yoo, "Optimization of powder core inductors of buck-boost converters for hybrid electric vehicles," in *Proc. IEEE Vehicle Power and Propulsion Conf.*, 2009, pp. 730–735.
10 J. Hu and C. R. Sullivan, "AC resistance of planar power inductors and the quasi-distributed gap technique," *IEEE Transactions on Power Electronics*, 16, pp. 558–567, 2001.
11 M. Gerber, J. A. Ferreira, I. W. Hofsajer, and N. Seliger, "A high-density heat-sink-mounted inductor for automotive applications," *IEEE Transactions on Industry Applications*, 40, pp. 1031–1038, July/August 2004.
12 W. Shen, F. Wang, D. Boroyevich, and C. W. Tipton, "Loss characterization and calculation of nanocrystalline cores for high-frequency magnetics applications," *IEEE Transactions on Power Electronics*, 23 (1), pp. 475–484, January 2008.

13 J. G. Hayes, D, Cashman, M. G. Egan, N. Wang, and T. O'Donnell, "Comparison of test methods for characterization of high-leakage two-winding transformers," *IEEE Transactions on Industry Applications*, pp. 1729–1741, September–October 2009.
14 C. R. Sullivan and R. Y. Zhang, "Simplified design method for Litz wire," *IEEE Applied Power Electronics Conference*, pp. 2667–2674, 2014.
15 M. Kacki, M. Rylko, J. G. Hayes, and C. R. Sullivan, "Magnetic Material Selection for EMI Filters," *IEEE Energy Conversion Congress and Exposition*, 2017.

Further Viewing

1 *Shock and Awe: The Story of Electricity*, British Broadcasting Corp., 2011.
2 *The Elegant Universe*, Nova 2003.
3 *Tesla: Master of Lightning*, PBS Home Video, 2004.
4 *The Current War*, 101 Studios, 2019.

Problems

16.1 A toroidal coil has a square cross section of sides 1.5 cm and a mean radius of 3 cm. The coil has 50 turns, closely wound, and carries 2 A.

Determine the mmf, magnetic field strength, flux density, flux, and inductance.

Assume an air core.

[Ans. 100 A-turns, 530.5 A/m, 666.6 μT, 0.15 μWb, 3.75 μH]

16.2 A toroidal coil has a square cross section of sides h =1.5 cm and a mean radius of 3 cm. The coil has 50 turns, closely wound, and carries 2 A. The internal radius of the toroid r_i = 2.25 cm, and the external radius r_o = 3.75 cm.

The flux across the core is given by

$$\Phi = \frac{\mu_r \mu_0}{2\pi} NIh \log_e(r_o/r_i)$$

Determine the mmf, flux, and inductance.

Calculate the percentage error incurred by the assumption of a uniform magnetic field in the previous problem.

Again assume an air core.

[Ans. 100 A-turns, 0.1532 μWb, 3.83 μH, −2.1%]

16.3 A toroidal coil is wound on a gapless ferrite core, of relative permeability μ_r = 2000, with a square cross section of sides h =1 cm. The inner radius r_i of the toroid is 1.5 cm, and the outer radius r_o is 2.5 cm. A 10-turn coil is wound uniformly and tightly around the toroid and carries a current of 0.25 A.

Assuming that the magnetic field strength within the toroid is uniform and equal to the value at the mean radius, determine (i) the magnetic field strength, (ii) the flux density in the core, (iii) the total flux, (iv) the reluctance, and (v) the self-inductance of the coil.

[Ans. 19.89 Am^{-1}, 50 mT, 5 μWb, 500 kH^{-1}, 200 μH]

16.4 An inductor with an iron core of relative permeability $\mu_r = 150$ has a cross-sectional of 6.25 cm^2 and a mean magnetic path of 0.4 m. The coil has 1000 turns, closely wound, and carries 1.5 A.

i) Calculate the magnetic field strength and flux density along the mean magnetic path of the inductor.
ii) Calculate the approximate flux through the core and the inductance.

Note that $\mu_0 = 4\pi \times 10^{-7}$ H/m.

[Ans. 3750 A/m, 0.71 T, 0.444 mWb, 0.296 H]

16.5 Determine the inductance value for the inductor of Section 16.3.2.1 if the air gap area is 20% greater than the core area due to fringing while the discrete air gap remains at 1.35 mm.

[Ans. 472 μH]

16.6 Determine the inductance and exciting current for the inductor of Section 16.3.2.1 at a flux density of 0.5 T and a related magnetic field strength of 100 A/m. Ignore fringing.

[Ans. 429 μH, 39.2 A]

16.7 A high-current inductor is specified at 45 μH. The inductor has 12 turns (two 6-turn windings in series), a cross-sectional area of 9 cm^2, and a mean magnetic-path length of 0.22 m. Assume an amorphous metal core with a relative permeability of 2000 at a flux density of 1 T. Ignore fringing.

i) What is the air gap length if there are two air gaps?
ii) At what current will the flux density in the core reach 1.0 T?
iii) Determine the core and air gap reluctances.
iv) Determine the energy storage in the air gaps, in the core material, and in the total inductor.
v) What are the inductance, maximum current, and energy storage if there are no air gaps in the structure?
vi) Calculate the core loss of the material assuming that the input current is a sinusoidal wave of 60 A peak with a frequency of 16 kHz. The core material has the following Steinmetz coefficients: $k = 2.71$, $m = 1.489$, and $n = 1.936$.
vii) Calculate the MLT and copper loss of the 45 μH inductor, assuming the dimensions shown in Table 16.11, if the inductor is carrying 180 A.

Table 16.11 Inductor parameters.

L (μH)	N (series)	b_c (mm)	b_w (mm)	h_c (mm)	l_w (mm)	b_{cu}(mm)	t_{cu} (mm)	b_{cl} (mm)
45	6 × 2	26	15	35	38	4.5	2.5	4.5

[Ans. 1.755 mm, 240 A, 97.3 kH^{-1}, 2 × 1551.4 kH^{-1}, 1.296 J, 0.039 J, 1.257 J, 1.48 mH, 7.3 A, 0.039 J, 66.7 W, 0.176 m, 105 W]

16.8 What is the rotor air gap flux density if a similar sized ferrite magnet is used rather than the Nd-Fe-B magnet in Section 16.4.3?

Use the ferrite magnet parameters from Table 16.9.

[Ans. 0.265 T]

16.9 Determine the magnetic flux densities of a Nd-Fe-B permanent magnet if the magnet is ten times thicker than the 1 mm rotor air gap. Let the cross-sectional area of the air gap be equal to the area of the permanent magnet.

[Ans. 1.12 T]

16.10 The traction motor of the 2010 Toyota Camry features 16 magnets, each with an area of 60.6 × 19.1 mm^2 and 6.5 mm thickness.

i) Estimate the lifting force if the combined magnets in the 2010 Toyota Camry are arranged in a horseshoe magnet. Assume eight magnets per pole, and assume a remanent flux density of 1.2 T and a recoil relative permeability of 1.05. Ignore fringing.
ii) How much weight can the magnet lift with a 2 mm air gap?

[Ans. 10610 N, 617.9 kg]

16.11 A horseshoe magnet has two poles, each with an area of 100 cm^2, a remanent flux density B_R of 1.2 T, a magnet thickness l_m of 6.2 mm, and a recoil relative permeability μ_{rec} of 1.06.

i) Estimate the mass which can be lifted if there is no air gap.
ii) Determine the air gap flux density and the maximum air gap at which the above magnet can lift 500 kg. Ignore fringing.

[Ans. 1168 kg, 0.785 T, 3.1 mm]

16.12 Determine the wound-field inductance of the example in Section 16.4.6.1 if the effective path length and area of the core are 40 cm and 20 cm^2, respectively. Assume a relative permeability of 500.

[Ans. 0.459 mH]

16.13 A line transformer sources a magnetizing current by 0.5 Arms when fed by 230 Vrms at 50 Hz. What is the value of the magnetizing inductance?

[Ans. 1.464 H]

16.14 A high-frequency power converter (for an on-board EV battery charger) features a transformer which is fed by a +/− 380 V square wave at 100 kHz. The transformer magnetizing inductance L_m is 0.5 mH, and the turns ratio $n = 1.053$.

Determine the primary current if the secondary current is 15.75 A when the magnetizing current is at a positive peak.

What are the values of the square wave voltage on the secondary?

[Ans. 18.48 A, +/− 400 V]

16.15 A high-frequency power converter (for an inductive battery charging power converter) features a transformer which is fed by a +/− 380 V square wave at 180 kHz. The transformer magnetizing inductance L_m is 45 μH, and the turns ratio n is 1.

Determine the primary current if the secondary current is +30 A when the magnetizing current is at a negative peak.

What are the values of the square wave voltage on the secondary?

[Ans. 18.27 A, +/− 380 V]

Assignments

The student is encouraged to experiment with finite-element analysis software packages, such as FEMM, Maxwell, Quickfield, and others. FEMM is a free 2D package and is widely used [14]. The commercial packages typically offer 3D and enhanced features. See reference [15] for an example of an inductor.

Reference Conversion Table

The following Table A.1 can be of use when performing common conversions between US and Imperial units and the metric system.

Table A.1 Conversions from commonly used US and Imperial units to metric units.

Parameter	Conversion to metric
Distance	1 mile = 1.609 km
Volume	1 gallon (US) = 3.785 L
	1 gallon (Imp) = 4.546 L
Mass	1 pound (lb) = 0.4536 kg
Power	1 hp = 0.7457 kW
Energy	1 kWh = 3,600,000 J
	1 calorie = 4.184 J
	1 British Thermal Unit (BTU) = 1055 J
Force	1 lbf = 4.448 N
Torque	1 lbf ft = 1.3558 Nm

Index

Electric Powertrain: Energy Systems, Power Electronics and Drives for Hybrid, Electric and Fuel Cell Vehicles, First Edition. John G. Hayes and G. Abas Goodarzi.

Companion website: www.wiley.com/go/hayes/electricpowertrain

C

d

Printed and bound by CPI Group (UK) Ltd, Croydon, CR0 4YY

Printed and bound by CPI Group (UK) Ltd, Croydon, CR0 4YY

16/04/2025

14658381-0004